Quark–Gluon Plasma 5

Quark–Gluon Plasma 5

Editor

Xin-Nian Wang

Central China Normal University
Lawrence Berkeley National Laboratory

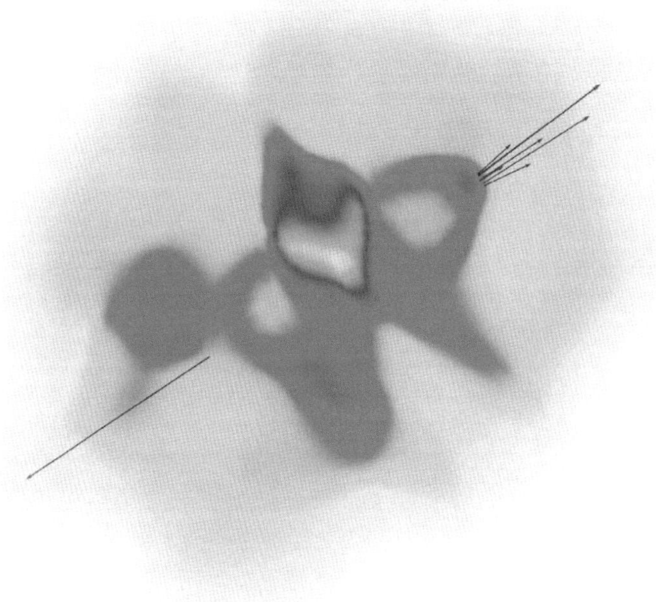

World Scientific

NEW JERSEY • LONDON • SINGAPORE • BEIJING • SHANGHAI • HONG KONG • TAIPEI • CHENNAI • TOKYO

Published by

World Scientific Publishing Co. Pte. Ltd.
5 Toh Tuck Link, Singapore 596224
USA office: 27 Warren Street, Suite 401-402, Hackensack, NJ 07601
UK office: 57 Shelton Street, Covent Garden, London WC2H 9HE

Library of Congress Cataloging-in-Publication Data
Names: Wang, Xin-Nian, editor.
Title: Quark-gluon plasma 5 / edited by Xin-Nian Wang (Central China Normal University,
China & Lawrence Berkeley National Laboratory, USA).
Other titles: Quark-gluon plasma five
Description: New Jersey : World Scientific, [2016] | Includes bibliographical references and index.
Identifiers: LCCN 2015039607| ISBN 9789814663700 (hardcover ; alk. paper) |
ISBN 9814663700 (hardcover ; alk. paper)
Subjects: LCSH: Quark-gluon plasma. | Heavy ion collisions. | Quantum chromodynamics.
Classification: LCC QC793.5.Q252 Q36 2016 | DDC 539.7/548--dc23
LC record available at http://lccn.loc.gov/2015039607

British Library Cataloguing-in-Publication Data
A catalogue record for this book is available from the British Library.

Cover image: Courtesy of CCNU and LBNL.
Description: Simulation of dijet propagation from the center of Au+Au collisions as given by the CoLBThydro model.

Typeset by Stallion Press
Email: enquiries@stallionpress.com

Printed in Singapore

Preface

In 1990, Rudy Hwa edited the first volume of this book series on the advances in theoretical and phenomenological studies of quark–gluon plasma (QGP). During the past 25 years, three more volumes of the series aimed at providing comprehensive reviews for both beginners and experts in the field were published. In the meantime, one witnessed the tremendous progresses in the experimental and theoretical studies of the new form of matter in high-energy heavy-ion collisions at the Relativistic Heavy-Ion Collider (RHIC) and the Large Hadron Collider (LHC). The discovery of the strongly coupled quark–gluon plasma (sQGP) at RHIC and the heavy-ion collisions at unprecedented high energies at LHC have redefined the focus of current and future studies of the properties of sQGP.

I am very honored to take over the baton from Rudy and edit the fifth volume of the series. The collection of articles in this volume provides an up-to-date, detailed overview of the advances driven by the most recent experimental data at RHIC and LHC. The articles are both pedagogical and comprehensive which can be helpful for the new researchers as well as the experienced ones.

Xin-Nian Wang
Berkeley, California
Wuhan, China
June 2015

Contents

Thermodynamics of Strong-Interaction Matter from Lattice QCD

Heng-Tong Ding

Key Laboratory of Quark & Lepton Physics (MOE),
Institute of Particle Physics,
Central China Normal University,
Wuhan, 430079, P. R. China

Frithjof Karsch

Physics Department,
Brookhaven National Laboratory,
Upton, NY 11973, USA

Fakultät für Physik, Universität Bielefeld,
D-33615 Bielefeld, Germany

Swagato Mukherjee

Physics Department,
Brookhaven National Laboratory,
Upton, NY 11973, USA

We review results from lattice QCD calculations on the thermodynamics of strong-interaction matter with emphasis on input these calculations can provide to the exploration of the phase diagram and properties of hot and dense matter created in heavy ion experiments. This review is organized in sections as follows: (1) Introduction, (2) QCD thermodynamics on the lattice, (3) QCD phase diagram at high temperature, (4) Bulk thermodynamics, (5) Fluctuations of conserved charges, (6) Transport properties, (7) Open heavy flavors and heavy quarkonia, (8) QCD in external magnetic fields, (9) Summary.

1. Introduction

It has long been recognized that under extreme conditions of high temperature or densities, matter interacting through the strong force — strong-interaction matter — cannot exist as nuclear matter formed out of hadrons. The copious production of new resonances[1] and the intrinsic size of the nucleons put a natural limit on the range of validity of hadron physics.[2,3] In the context of thermodynamics of strong-interaction matter this found its expression in the formulation of the hadron resonance gas model[1,4] which led to the postulate of a limiting temperature — Hagedorn temperature — for the thermodynamics of ordinary nuclear matter.

With the formulation of Quantum Chromodynamics (QCD)[5,6] as the fundamental theory of strong interactions it has soon been realized that matter formed out of strongly interacting particles may be converted into a new form of matter[7,8] — the quark–gluon plasma[9] — in which the dominant degrees of freedom are the constituents of hadrons, i.e., quarks and gluons. At low temperatures and densities they are confined in hadrons. However, they can move "freely" over macroscopic distances in bulk strong-interaction matter at high temperature and densities.

It has been speculated that the transition from confined to deconfined matter, like in condensed matter systems, goes along[10] with the restoration of global symmetries of the underlying microscopic theory describing the interaction among constituents of strong-interaction matter, i.e., QCD. The analysis of these symmetries led to the phenomenology of a complex phase structure of strong-interaction matter, leading to homogeneous and inhomogeneous as well as color-superconducting phases. This bears many analogies to phase structures known in condensed matter systems.[11,12] Only little of this complex structure is verified through first principle lattice QCD calculations. In Fig. 1 we show a sketch of the QCD phase diagram highlighting only the aspects that we believe can be addressed at present through lattice QCD calculations. We will discuss the phase diagram in detail in Sec. 3.

Phase transitions as well as properties of matter close to such transitions arise from complex long range interactions in multi-particle systems on multiple length scales and at the same time often develop simple universal behavior. Their analysis

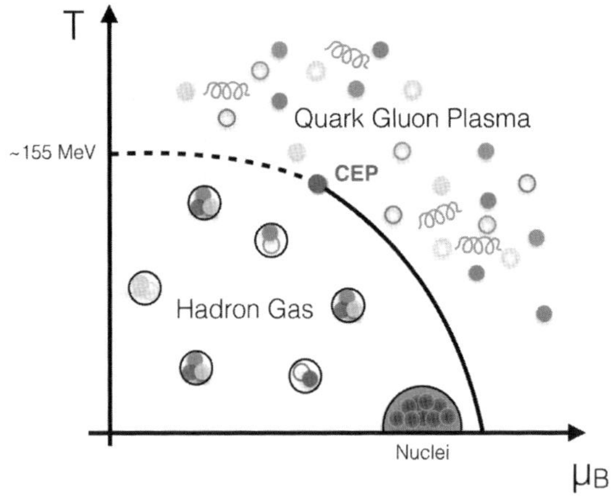

Fig. 1. (Color online) QCD phase diagram in the temperature (T) and baryon chemical potential (μ_B) plane. At vanishing μ_B lattice QCD calculations show that the transition is not a phase transition but a continuous crossover, reflecting pseudo-critical behavior in the vicinity of the true chiral transition. At $\mu_B > 0$ a second order critical end point (CEP) may exist, which would be followed by a line of first order transitions at larger values of μ_B. Constraints on the location of a CEP will come from lattice QCD calculations.

requires non-perturbative techniques. With the formulation of lattice QCD, i.e., the introduction of a discrete space–time lattice as regulator for quantum field theories like QCD,[13] a framework for new non-perturbative approaches to the study of strong-interaction matter had been established. It became possible to perform first-principle numerical calculations — lattice QCD simulations[14] — by which a systematic study of the phase structure and basic bulk properties of strong-interaction matter[15–17] could be performed for the first time. We will give a brief introduction to lattice QCD in Sec. 2. Today these calculations are highly advanced and continuum extrapolated results for the QCD transition temperature and the equation of state exist at vanishing net baryon number density. We will review these results and report on the status of the lattice QCD program that aims at an extension of these calculations to nonzero net baryon number density or equivalently nonzero baryon chemical potential in Sec. 4.

Properties of strong-interaction matter are studied intensively in large experimental programs at the Relativistic Heavy Ion Collider (RHIC) of Brookhaven National Laboratory (BNL), USA, and the Large Hadron Collider (LHC) as well as the Super Proton Synchrotron (SPS) of the European Research Center (CERN). Experiments at these accelerators are devoted to the exploration of the properties of hot and dense matter created in collisions of ultra-relativistic heavy ions. Through the variation of beam energies at RHIC, i.e., the beam energy scan (BES), it also became possible to explore systematically the phase structure of strong-interaction matter at nonzero net baryon number density. This allows one to search systematically for the chiral critical point or critical end point (CEP) in the QCD phase diagram. The CEP is a postulated second order phase transition point which is expected to mark the endpoint of a line of first order phase transitions that separates the low temperature, low density hadronic phase from a low temperature, large baryon number quark-gluon plasma phase (also the existence of this first order line at present is not confirmed through lattice QCD calculations). The existence of a CEP in the QCD phase diagram is imprinted in the properties of fluctuations of conserved charges, e.g., net baryon number, electric charge or strangeness fluctuations. These conserved charge fluctuations can be calculated in lattice QCD at vanishing baryon chemical potential and the calculations can be extended to nonzero values of the chemical potential using Taylor expansions of thermodynamic observables. They can directly be compared to experimental measurements of conserved charge fluctuations performed at the LHC and the BES at RHIC and may provide insight into the existence and location of a CEP in the QCD phase diagram. We will report in Sec. 5 on the status of lattice QCD calculations of conserved charge fluctuations and their usage to analyze freeze-out conditions in heavy ion experiments.

Heavy ion experiments at RHIC provided striking evidence for the highly correlated, non-perturbative structure of strong-interaction matter in the high temperature phase. In particular, the observation of strong elliptic flow and very efficient energy loss of high momentum quarks traversing through the medium (jet quenching) paved the way for a picture of a strongly interacting medium that has been called an "almost perfect liquid". These measurements suggest that strong-interaction matter at temperatures close to but above the transition temperature has quite unique transport properties, (i) a small shear viscosity to entropy ratio that may be close to the conformal limit value $\eta/s = 1/4\pi$ calculated in conformal field theories (AdS/CFT limit), (ii) a large bulk viscosity that may diverge in the massless QCD limit at the chiral phase transition temperature, and (iii) small electrical conductivity and diffusion coefficients. These transport properties can systematically be analyzed in lattice QCD, although this requires the reconstruction of continuous spectral functions from a finite set of observables, which is in general an ill-posed problem. The calculation of transport properties in lattice QCD will provide crucial input to the modeling of the hydrodynamic expansion of hot and dense matter formed in heavy ion collisions. Spectral functions also carry information on in-medium modifications of hadrons. The dissociation of heavy quark bound states has long been advocated as one of the striking signatures for the formation of a deconfined color screened medium[18] while the melting of light quark bound states is intimately related with the restoration of chiral symmetry.[19] We will report on the status of calculations of transport coefficients and in-medium hadron properties in Secs. 6 and 7.

Finally, in Sec. 8 we will briefly review a relatively new but very exciting field of QCD thermodynamics that addresses the behavior of strong-interaction matter in strong external magnetic fields. Strong, static magnetic fields will certainly influence the thermodynamics and lead to modifications of the transition temperature as well as the equation of state. Strong fields are generated in the early phase of heavy ion collisions[20,21] but weaken rapidly during the subsequent expansion phase. While lattice QCD calculations will allow one to understand basic effects that occur in matter exposed to strong fields, it is at present not clear to what extent equilibrium lattice QCD calculations can contribute to a quantitative analysis of experimental data.

2. QCD Thermodynamics on the Lattice

2.1. *Path integral formulation of QCD thermodynamics*

The equilibrium thermodynamics of elementary particles interacting only through the strong force is controlled by the QCD partition function which can be expressed in terms of a Euclidean path integral. The grand canonical partition function $Z(T, V, \vec{\mu})$ is given as an integral over the fundamental quark ($\bar{\psi}$, ψ) and gluon fields (A_ν). In addition to its dependence on volume (V), temperature (T) and a set of N_f chemical potentials $\vec{\mu} \equiv (\mu_u, \mu_d, \mu_s, \ldots)$, the partition function implicitly depends on the masses of the N_f different quark flavors $\vec{m} \equiv (m_u, m_d, m_s, \ldots)$. In this review we will most of the time discuss the thermodynamics of QCD with 2 light quarks (u, d) which are assumed to have degenerate masses $m_u = m_d$ and a heavier strange quark (s), with mass m_s. This often is referred to as (2+1)-flavor QCD, or $N_f = 2 + 1$.

In Euclidean space–time, which is obtained from the Minkowski formulation by substituting $t \to -i\tau$ with $\tau \in \mathbb{R}$, the QCD Lagrangian is given by

$$\begin{aligned}\mathcal{L}^E_{\rm QCD} &= \mathcal{L}^E_{\rm gluon} + \mathcal{L}^E_{\rm fermion} \\ &= -\frac{1}{4}F^{\mu\nu}_a(x)F^a_{\mu\nu} - \sum_{f=u,d,s,\ldots} \bar{\psi}^\alpha_f(x)\left(\not{D}^E_{\alpha\beta} + m_f\delta_{\alpha\beta}\right)\psi^\beta_f(x),\end{aligned} \tag{1}$$

where Greek letters are spinor indices, $a = 1, \ldots, N_c^2 - 1$ is the color index, N_c is the number of colors ($N_c = 3$ for QCD) and m_f is the mass of quarks with flavor f. The covariant derivative $\not{D}^E$ and the field strength tensor $F^a_{\mu\nu}$ are given by

$$\not{D}^E = \gamma^E_\mu D^E_\mu = \left(\partial_\mu + ig\frac{\lambda_a}{2}A^a_\mu\right)\gamma^E_\mu, \tag{2}$$

$$F^a_{\mu\nu} = \partial_\mu A^a_\nu - \partial_\nu A^a_\mu - gf^{abc}A^b_\mu A^c_\nu. \tag{3}$$

Here A^a_μ are the gauge fields, ψ^α_f ($\bar{\psi}^\alpha_f$) are the quark (anti-quark) fields, λ^a are the generators of SU(N_c), f_{abc} are the corresponding structure constants, γ^E_μ are the Euclidean Dirac matrices obeying $\{\gamma^E_\mu, \gamma^E_\nu\} = 2\delta_{\mu\nu}$, and g is the bare coupling constant.

In the Euclidean path-integral formalism, the partition function of QCD is then given by

$$\mathcal{Z}(T, V, \vec{\mu}) = \int \prod_\mu \mathcal{D}A_\mu \prod_{f=u,d,s,\ldots} \mathcal{D}\psi_f \mathcal{D}\bar{\psi}_f \; e^{-S_E(T,V,\vec{\mu})}, \tag{4}$$

with the Euclidean action

$$S_E(T, V, \vec{\mu}) \equiv -\int_0^{1/T} dx_0 \int_V d^3\mathbf{x}\, \mathcal{L}^E(\vec{\mu}). \tag{5}$$

Here we have suppressed the dependence of the Euclidean Lagrangian and action on the fields $(A_\mu, \bar{\psi}_f, \psi_f)$ but have stressed explicitly their dependence on the various quark chemical potentials that couple to the conserved quark number currents

$$\mathcal{L}^E(\vec{\mu}) = \mathcal{L}^E_{\rm QCD} + \sum_{f=u,d,s,\ldots} \mu_f \bar{\psi}_f \gamma_0 \psi_f. \tag{6}$$

The thermal expectation value of physical observables $\mathcal{O}$ can be obtained through

$$\langle \mathcal{O} \rangle = \frac{1}{Z(T, V, \vec{\mu})} \int \prod_\mu \mathcal{D}A_\mu \prod_f \mathcal{D}\psi_f \mathcal{D}\bar{\psi}_f \, \mathcal{O}\, e^{-S_E(T,V,\vec{\mu})}. \tag{7}$$

Basic thermodynamic quantities like the pressure (P), energy density (ϵ) or net-quark number density n_f can be obtained from the logarithm of the partition function using standard thermodynamic relations,

$$\frac{P}{T^4} = \frac{1}{VT^3} \ln Z(T, V, \vec{\mu}), \tag{8}$$

$$\frac{\epsilon}{T^4} = -\frac{1}{VT^4} \left. \frac{\partial \ln Z(T, V, \vec{\mu})}{\partial 1/T} \right|_{\vec{\mu}/T\text{fixed}}, \tag{9}$$

$$\frac{n_f}{T^3} = \frac{1}{VT^3} \frac{\partial \ln Z(T, V, \vec{\mu})}{\partial \hat{\mu}_f}, \tag{10}$$

where we introduced the chemical potentials in units of temperature, $\hat{\mu}_f = \mu_f/T$. In fact, the QCD partition function depends on chemical potentials only through these dimensionless combinations, which are the logarithms of the fugacities, $z_f \equiv \exp(\mu_f/T)$.

As will become clear in the next subsection, the numerical analysis of thermodynamic observables at non-vanishing chemical potential, $\vec{\mu} \neq 0$, is difficult in lattice regularized QCD. A viable approach that circumvents the so-called sign-problem in lattice QCD at $\vec{\mu} \neq 0$, suitable for moderate values of the chemical potentials, is to consider Taylor expansions for thermodynamic observables. The starting point for such an analysis is the Taylor expansion of the pressure. In (2+1)-flavor QCD it reads

$$\frac{P}{T^4} = \sum_{i,j,k} \frac{1}{i!j!k!} \chi_{ijk}^{uds}(T) \left(\frac{\mu_u}{T}\right)^i \left(\frac{\mu_d}{T}\right)^j \left(\frac{\mu_s}{T}\right)^k, \tag{11}$$

where the expansion coefficients are dimensionless, generalized susceptibilities that can be evaluated at $\vec{\mu} = 0$,

$$\chi_{ijk}^{uds}(T) = \left. \frac{\partial^{i+j+k} P/T^4}{\partial \hat{\mu}_u^i \partial \hat{\mu}_d^j \partial \hat{\mu}_s^k} \right|_{\vec{\mu}=0}. \tag{12}$$

We will present results from Taylor expansions of QCD thermodynamics in several sections of this review. Other thermodynamic observables and expressions more suitable for the calculation of P/T^4 and ϵ/T^4 in lattice regularized QCD will be introduced in later sections where they appear.

2.1.1. *The high temperature, ideal gas limit*

At very high temperatures, all quark masses are small on the scale of the temperature and massless QCD becomes a good approximation. Furthermore, because of asymptotic freedom interactions among quarks and gluons become small at high temperature. QCD thermodynamics thus approaches that of a non-interacting, massless quark–gluon gas. In the temperature range accessible to heavy ion collisions the massless limit of 3-flavor QCD is most relevant ($N_f = 3$). In this limit the pressure is given by

$$\left(\frac{P}{T^4}\right)_{\rm ideal} = \frac{8\pi^2}{45} + \sum_{f=u,d,s} \left[\frac{7\pi^2}{60} + \frac{1}{2}\left(\frac{\mu_f}{T}\right)^2 + \frac{1}{4\pi^2}\left(\frac{\mu_f}{T}\right)^4\right], \tag{13}$$

where the first term gives the contribution of gluons and the sum yields the contribution of quarks with different flavor degrees of freedom.

The thermodynamics of QCD at high temperatures can be systematically analyzed in perturbation theory. This, however, becomes complicated beyond $\mathcal{O}(g^2)$ due to the appearance of non-perturbative length scales of $\mathcal{O}(gT)$ and $\mathcal{O}(g^2T)$ reflecting electric and magnetic screening lengths,[22] which require the resummation of diagrams leading to the so-called hard thermal loop perturbation theory,[23,24] or the explicit integration over hard scales leading to the dimensional reduction scheme.[25,26]

2.1.2. *The low temperature hadron resonance gas approximation*

At low temperature quarks and gluons are confined in colorless hadrons, i.e., baryons and mesons are the relevant degrees of freedom. In fact, it turns out that even at temperatures close to the transition region from hadronic matter to the high temperature quark–gluon plasma phase, a non-interacting gas constructed from all experimentally known resonances does provide quite a good description of QCD thermodynamics. This had been anticipated by Hagedorn when he formulated the hadron resonance gas (HRG)[1,4] model prior to QCD.

In an HRG model the pressure and other thermodynamic observables are easily obtained from the logarithm of the partition function,

$$\ln \mathcal{Z}_{\rm HRG}(T, V, \vec{\mu}) = \sum_{i\in \text{ mesons}} \ln \mathcal{Z}^{M}_{m_i}(T, V, \vec{\mu}) + \sum_{i\in \text{ baryons}} \ln \mathcal{Z}^{B}_{m_i}(T, V, \vec{\mu}) \ , \tag{14}$$

where $\vec{\mu} = (\mu_B,\ \mu_Q,\ \mu_S)$ is the set of baryon number, electric charge and strangeness chemical potentials, respectively. The partition functions for mesons (M) or baryons (B) are given by

$$\ln \mathcal{Z}^{M/B}_{m_i}(T, V, \vec{\mu}) = \mp \frac{V}{2\pi^2} \int_0^\infty dk\, k^2 \ln(1 \mp z_i e^{-\varepsilon_i/T}) \quad , \tag{15}$$

with energies $\varepsilon_i^2 = k^2 + m_i^2$ and fugacities

$$z_i = \exp((B_i\mu_B + Q_i\mu_Q + S_i\mu_S)/T). \tag{16}$$

Of course, $B_i = 0$ for all mesons and $B_i = \pm 1$ for baryons. The set of hadron chemical potentials, related to conserved quantum numbers, and the set of quark flavor chemical potentials are easily related to each other,

$$\mu_u = \frac{1}{3}\mu_B + \frac{2}{3}\mu_Q,$$
$$\mu_d = \frac{1}{3}\mu_B - \frac{1}{3}\mu_Q,$$
$$\mu_s = \frac{1}{3}\mu_B - \frac{1}{3}\mu_Q - \mu_S. \tag{17}$$

With this it is straightforward to rewrite the Taylor series given in Eq. (11) in terms of quark chemical potentials, also in terms of baryon number, electric charge and strangeness chemical potentials,

$$\frac{P}{T^4} = \sum_{i,j,k} \frac{1}{i!j!k!}\chi_{ijk}^{BQS}(T)\left(\frac{\mu_B}{T}\right)^i\left(\frac{\mu_Q}{T}\right)^j\left(\frac{\mu_S}{T}\right)^k, \tag{18}$$

where the expansion coefficients[a] χ_{ijk}^{BQS} can again be evaluated at $\vec{\mu}=0$,

$$\chi_{ijk}^{BQS}(T) = \left.\frac{\partial^{i+j+k}P/T^4}{\partial\hat{\mu}_B^i\partial\hat{\mu}_Q^j\partial\hat{\mu}_S^k}\right|_{\vec{\mu}=0}. \tag{19}$$

2.2. *Lattice regularization and continuum limit*

Lattice gauge theory suggested by Wilson in 1974[13] is based on the Euclidean path integral formalism and provides a particular regularization scheme for QCD by introducing a finite lattice spacing as cut-off. In this way space–time is discretized and the path integral becomes a finite, yet high dimensional integral over the gauge and fermion field variables. The discretization of space and time introduces a systematic cut-off dependence in all observables, which vanishes when the lattice spacing is taken to zero, i.e., in the continuum limit. Many theoretical and technical details can be found in excellent textbooks such as Refs. 27–30. Here we will only give a brief introduction that may help to make this review self-contained.

By introducing a hyper-cubic lattice of size $N_\sigma^3 \times N_\tau$ with a small but finite lattice spacing a, the calculation of observables given by Eq. (7) reduces to the evaluation of finite, but high dimensional integrals. Rather than calculating the partition function directly one usually is concerned with the calculation of expectation values, Eq. (7). This can be done by exploiting well-known Monte Carlo simulation techniques.[14] In order to preserve gauge invariance of the discretized action, one introduces the gauge degrees of freedom as variables living on links between neighboring sites of the lattice, $U_{x,\hat{\mu}} \equiv \exp(iagA_\mu(x))$, while the fermionic degrees of freedom, $\bar{\psi}_x$ and ψ_x, are defined on the sites of the lattice. The latter are anti-commuting Grassmann variables and the former are $N_c \times N_c$ matrices that are elements of the $SU(N_c)$ color group.

[a] We will in general suppress super- and sub-scripts of the generalized susceptibilities, if a sub-script is zero, i.e., $\chi_{102}^{BQS} \equiv \chi_{12}^{BS}$ or $\chi_{200}^{BQS} \equiv \chi_2^B$.

The volume V and the temperature T are related to the spatial and temporal extents of the lattice, respectively,

$$V = (aN_\sigma)^3, \quad T = \frac{1}{aN_\tau}. \tag{20}$$

Here N_σ and N_τ are the number of sites in spatial and temporal directions. A consequence of introducing a discrete space–time lattice is that aside from Lorentz symmetry also some of the symmetries of QCD get explicitly broken by the finite lattice cut-off. They will be recovered in the continuum limit, $a \to 0$ at fixed V and T. Asymptotically, for small values of the bare gauge coupling g this is controlled by the QCD β-function

$$a\Lambda_L = \left(\frac{1}{b_0 g^2}\right)^{b_1/2b_0^2} e^{-1/2b_0 g^2}, \tag{21}$$

with $b_0 = (\frac{11}{3}N_c - \frac{2}{3}N_f)/16\pi^2$ and $b_1 = (\frac{34}{3}N_c^2 - (\frac{10}{3}N_c + \frac{N_c^2-1}{N_c})N_f)/(16\pi^2)^2$.

Keeping physical observables constant when approaching the continuum limit also requires a proper tuning of the bare quark masses. In the discretized version of QCD (lattice QCD), all quark masses are naturally expressed in units of the lattice spacing, $\hat{m}_f \equiv m_f a$. Taking the continuum limit thus requires one to take the limit $g^2 \to 0$ and $\hat{m}_f \to 0$ while keeping some physical observables, i.e., one per flavor degree of freedom, constant. This defines "lines of constant physics" (LCP) in the space of quark masses and the gauge coupling. In the case of (2+1)-flavor QCD with degenerate light quark masses, $m_u = m_d$, one thus can determine the bare quark mass parameters $\hat{m}_l \equiv \hat{m}_u$ and the strange quark mass $\hat{m}_s$ using two physical observables. An often used approach is to fix the strange quark mass using the (fictitious) $\eta_{s\bar{s}}$ meson mass, which only contains strange quarks and is quite insensitive to light quark mass values. This mass is matched to the lowest order chiral perturbation theory estimate $m_{\eta_{s\bar{s}}} = \sqrt{2m_K^2 - m_\pi^2}$. The light quark mass is then fixed using a constant ratio $\hat{m}_s/\hat{m}_l$, the value corresponding to the physical pion mass value is $\hat{m}_s/\hat{m}_l = 27.5$.[31]

The lattice regularized partition function may be written as

$$\begin{aligned}\mathcal{Z}(T, V, \vec{\mu}) &= \int \prod_{x,\hat{\mu}} dU_{x,\hat{\mu}} \prod_{x,f} d\psi_{x,f} d\bar{\psi}_{x,f}\, e^{-S_f - S_g} \\ &= \int \prod_{x,\hat{\mu}} dU_{x,\hat{\mu}} \prod_{f} \det M_f(\mu_f)\, e^{-S_g},\end{aligned} \tag{22}$$

where S_g and S_f are the discretized versions of the gluonic and fermonic part of the action S_E. As the action is bilinear in the quark fields these can be integrated out, which gives rise to the determinant of fermion matrix M_f. Quite often exploratory lattice QCD studies are performed in the so-called "quenched approximation". This refers to the calculation of observables that contain fermionic degrees of freedom as valence quarks. However, back-reaction of fermions on the gauge fields (virtual quark loops) are not included in the observables. This is achieved by fixing

$\det M_f = const.$ in Eq. (22), i.e., the determinant does not contribute in the generation of gauge field configurations nor does it contribute in the calculation of expectation values.

The fermion matrix depends on the quark chemical potential μ_f which is introduced on the time-like links of the lattice.[32] Although other formulations are possible[33] a common approach is to replace the link variables $U_{x,\hat{0}} \rightarrow \exp(\mu_f a) U_{x,\hat{0}}$ and $U^\dagger_{x,\hat{0}} \rightarrow \exp(-\mu_f a) U^\dagger_{x,\hat{0}}$. The fermion determinant is positive definite only for $\mu_f \equiv 0$. For nonzero values of the chemical potential $\det M_f(\mu_f)$ is a complex function. Still the partition function is real, i.e., integration over the gauge fields eliminates the imaginary contributions. However, the real part of the fermion determinant still changes sign, which prohibits the application of conventional Monte Carlo simulation techniques. This is known as the "sign-problem" in finite density QCD. There are various attempts to circumvent this sign-problem in numerical simulations performed directly at non-vanishing $\vec{\mu}$. At present complex Langevin simulation techniques[34–36] and the integration over Lefschetz thimbles[37–39] are most actively being explored. In this review we will focus on the Taylor expansion approach[40,41] for the evaluation of thermodynamic observables at nonzero chemical potential. This only requires numerical simulations at vanishing baryon chemical potential. However, it is naturally limited in applicability through the radius of convergence of the Taylor series which remains to be determined for QCD.

2.3. *Fermion discretization schemes*

When discretizing the Euclidean action one has quite some freedom. Discretization schemes that differ by higher order corrections in the cut-off are equivalent to the extent that they yield the same physical answers in the continuum limit. The basic construction principle of a lattice discretized version of QCD is that one wants to start with a theory that is manifestly gauge invariant. In addition, one tries to choose discretization schemes that preserve as many of the symmetries of QCD as possible already at nonzero values of the lattice spacing. The discretization of the gluonic part of the action, S_g, is rather straightforward and well-established improvement schemes (Symanzik improvement[42,43]) are known that allow to eliminate systematically higher order cut-off errors.

Some difficulties already arise in the naive discretization of the fermionic part of the Euclidean action, S_f. When one simply approximates the covariant derivative in the fermion action by nearest-neighbor differences of the quark fields additional, unphysical degrees of freedom arise even in the continuum limit owing to the periodicity of the fermion dispersion relation on the lattice. This is known as the fermion doubling problem which is closely related to the explicit breaking of chiral symmetry in most commonly used fermion discretization schemes.[44]

There are several discretization schemes for the fermionic part of the action such as Wilson fermions,[13] staggered fermions,[45] Domain Wall fermions[46–49] as well as overlap fermions.[50,51] They differ to the extent they preserve chiral symmetry and/or eliminate the problem of fermion doublers. Wilson fermions avoid the doubling problem by adding a dimension five operator to the naively discretized fermion action.[13] A variant of this is the so-called clover-improved Wilson fermion action where lattice cut-off effects are reduced from $\mathcal{O}(a)$ to $\mathcal{O}(a^2)$.[52] Most commonly used in QCD thermodynamics calculations is the staggered fermion discretization scheme in which the Dirac 4-spinor is spread over several lattice sites.[45] This reduces the doubler problem but does not eliminate it completely. However, it preserves a $U(1)_{\text{even}} \times U(1)_{\text{odd}}$ remnant of chiral symmetry, which allows independent rotations of the quark fields on even and odd sites of the lattice, respectively. This is of great advantage in the study of chiral symmetry breaking at nonzero temperature as it allows one to introduce an order parameter, the chiral condensate, that is sensitive to the spontaneous breaking of this continuous symmetry. The reduction of the doubler problem in the staggered fermion approach leads to the introduction of additional heavier states for each fermion flavor. In the continuum limit, these so-called taste symmetry partners will become degenerate with the Goldstone mode that corresponds to the broken $U(1)_{\text{even}} \times U(1)_{\text{odd}}$ symmetry. The Goldstone boson thus has another 15 heavier partners (tastes) which become degenerate with the Goldstone boson only in the continuum limit. To reduce the taste symmetry breaking, i.e., reduce the mass difference between physical states and their heavier taste partners two improved staggered actions, the Highly Improved Staggered Quarks (HISQ)[53] and the stout smeared action,[54] are commonly used in lattice studies of QCD thermodynamics. In Fig. 2 (left) we show some results for the cut-off dependence of the root-mean-square mass values M_π^{RMS} of the Goldstone particle and its 15 taste partners calculated in different staggered fermion discretization schemes. As seen from the left panel of Fig. 2 the 4stout and HISQ actions have the smallest M_π^{RMS}. The right panel of this figure shows the influence of cut-off effects on

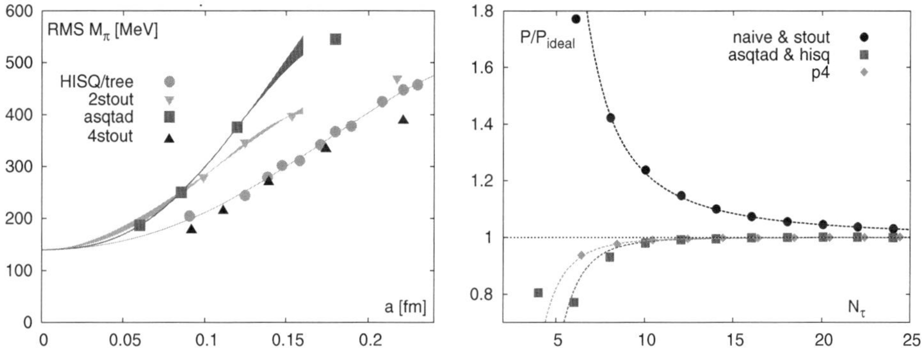

Fig. 2. (Color online) (Left) Root-mean-squared pion mass (M_π^{RMS}) as a function of lattice spacing a. RMS M_π is defined as the rooted sum of the mass squared of 16 pseudo scalar states divided by 4. The lattice cut-off effects are smaller with smaller M_π^{RMS}. (Right) Ratio of quark contribution to the pressure, obtained from lattice QCD calculations in the infinite temperature, ideal gas limit, to the pressure of an ideal quark gas in the continuum (P_{ideal}) as a function of N_τ. P/P_{ideal} approaches unity in the continuum limit.

the calculation of the fermion contribution to the pressure in the high temperature ideal gas limit. These cut-off effects are independent of the taste improvement schemes and solely arise from the strategy used to discretize the covariant derivatives appearing in the fermion action. The stout-smeared actions utilize the naive 1-link discretization scheme that leads to $\mathcal{O}(a^2)$ errors in the ideal gas limit. In the HISQ, asqtad and p4[55] actions 3-link terms are added to the action that eliminate the leading order cut-off effects in the ideal gas limit. In these cases discretization errors only start at $\mathcal{O}(a^4)$.

Domain Wall fermions (DWF) and overlap fermions allow for an exact representation of chiral symmetry even at nonzero lattice spacing. In the Domain Wall discretization scheme the physical chiral Dirac fermions are constructed on two 4-dimensional hyper-surfaces at the edges of a 5-dimensional lattice with extent L_s in that fifth direction. These hyper-surfaces represent the usual 4-dimensional space–time and on each of the surfaces fermions with one given chirality exist. There persists a small explicit breaking of chiral symmetry which vanishes exponentially in the limit $L_s \to \infty$. In DWF calculations this gives rise to a small additive renormalization of the quark masses, the so-called residual mass[56] $m_{\rm res}$. Controlling and reducing these residual mass effects is one of the important improvement steps in calculations with DWF.[57] Overlap fermions on the other hand preserve exact chiral symmetry on a 4-dimensional lattice by obeying the Ginsparg–Wilson relation at nonzero lattice spacing.[58]

Numerical calculations with Domain Wall as well as overlap fermions are quite time consuming. However, with increasing speed of super-computers calculations with physical light quark masses become feasible and chiral fermions have been used recently also for QCD thermodynamics studies.[59–62] In particular, in the analysis of subtle aspects of the QCD transition related to the temperature dependence of the axial anomaly, calculations with fermions obeying exact chiral symmetry already at nonzero values of the cut-off are mandatory.

3. QCD Phase Diagram at High Temperature

Our thinking about the phase structure of strong-interaction matter centers around two very basic concepts in strong interaction physics — confinement and chiral symmetry breaking. The former expresses the fact that only colorless states, baryons and mesons, can exist in the vacuum and are observed experimentally. This gave rise to the concept of a linearly rising, confining potential exhibited between quarks and anti-quarks,

$$V_{q\bar{q}}(r) = -\frac{\alpha(r)}{r} + \sigma r, \tag{23}$$

with σ being the string tension and $\alpha(r)$ the running coupling of QCD.

Chiral symmetry breaking, on the other hand is a mandatory feature of strong interactions needed to explain the appearance of a light, almost massless, particle in the hadron spectrum — the pions. In the limit of vanishing quark masses the

QCD Lagrangian has a build-in chiral symmetry. It is invariant under independent global rotations of quarks with left- and right-handed chirality in flavor space as well as chiral rotations of single flavor quark spinors. This gives rise to the $U(N_f)_L \times U(N_f)_R$ chiral symmetry which is equivalent to $SU(N_f)_L \times SU(N_f)_R \times U(1)_A \times U_V(1)$. The latter $U_V(1)$ reflects the conservation of baryon number, and the axial $U(1)_A$ symmetry, although an exact symmetry of the classical QCD Lagrangian, is explicitly broken by quantum fluctuations. The global $SU(N_f)_L \times SU(N_f)_R$ flavor symmetry, on the other hand, is broken spontaneously in massless QCD, giving rise to massless Goldstone modes and a non-vanishing chiral condensate,

$$\langle\bar{\psi}\psi\rangle_f = \frac{T}{V}\frac{\partial \ln Z}{\partial m_f}, \quad f = 1, \ldots, N_f. \tag{24}$$

The masses of the light pseudo-scalar (Goldstone) pions are then understood as arising from the small nonzero values of the light up and down quark masses. This is reflected in the Gell-Mann–Oakes–Renner relation between the pion mass (m_π), the quark masses (m_u, m_d), the pion decay constant (f_π) and the non-vanishing chiral condensate $\langle\bar{\psi}\psi\rangle = \langle\bar{\psi}\psi\rangle_u + \langle\bar{\psi}\psi\rangle_d$ that arises from the spontaneous breaking of chiral symmetry,

$$f_\pi^2 m_\pi^2 = \frac{1}{2}(m_u + m_d)\langle\bar{\psi}\psi\rangle \,. \tag{25}$$

The fact that QCD describes confinement as well as spontaneous breaking of chiral symmetry is not evident from the QCD Lagrangian or the perturbative treatment of strong interactions described by it. It requires a non-perturbative analysis – lattice QCD calculations — to firmly establish the confining and symmetry breaking features of QCD.

Asymptotic freedom of QCD suggests that non-perturbative effects are suppressed at high temperatures and hot strong-interaction matter approaches ideal gas behavior asymptotically. It thus is expected that non-perturbative condensates also disappear at high temperatures. For exact global symmetries of QCD this is expected to happen through a true phase transition. Basic features of this transition can be understood by invoking universality arguments.[10]

The nature of the QCD transition depends crucially on the values of the quark masses and the number of flavors (N_f). Figure 3 shows a sketch of the nature of the QCD transition as functions of the quark masses for a theory with two degenerate light (up and down) quarks with masses, $m_{u,d} \equiv m_l = m_u = m_d$, and a heavier strange quark with mass, m_s, at zero baryon chemical potential. In the limit $m_l \to \infty$ and $m_s \to \infty$ fermions decouple and the thermodynamics of a pure $SU(3)$ gauge theory ($N_f = 0$) is recovered. The $SU(3)$ gauge theory has an exact $Z(3)$ symmetry and the deconfimement transition is first order. This first order region extends to lower quark masses and ends at a second order critical line that belongs to the universality class of the 3-d, $Z(2)$ symmetric Ising model.[63,64] For three degenerate flavors $N_f = 3$ with small quark masses $m_l = m_s \to 0$ the chiral transition is known to be first order.[65,66] Recent lattice QCD studies[67,68]

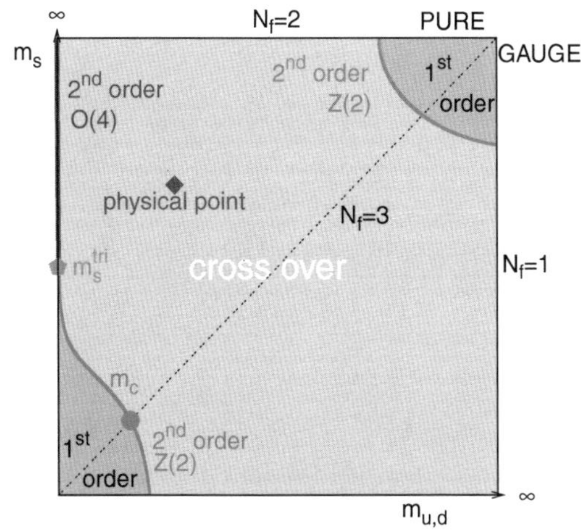

Fig. 3. (Color online) A sketch of the nature of the QCD transition as functions of the two degenerate light (up and down) quarks with masses, $m_{u,d} \equiv m_l$, and a heavier strange quark with mass, m_s, at zero baryon chemical potential.

with improved actions suggest that the extent of this first order region is quite small, i.e., limited to $m_l = m_s \lesssim m_s^{\rm phys}/270$ where $m_s^{\rm phys}$ is the physical value of the strange quark mass.

An additional ingredient in the discussion of the order of the transition in the chiral limit arises from the role of the axial anomaly. The nature of the chiral transition for the massless $N_f = 2$ theory, i.e., for $m_l \to 0$ and $m_s \to \infty$, depends on the magnitude of the axial $U_A(1)$ symmetry breaking. If this remains significant close to the transition temperature then the relevant symmetry becomes isomorphic to that of the 3-d $O(4)$ spin model and the transition is expected to be second order belonging to that universality class.[10,69] However, if $U_A(1)$ symmetry breaking becomes negligible near the chiral transition temperature, the relevant symmetry becomes isomorphic to $O(2) \times O(4)$ and the transition be either first order[10] or second order.[70,71] In the intermediate quark mass region there is no true phase transition, rather a crossover takes place from the hadronic to the quark–gluon plasma phase.

All the first order regions are separated from the crossover region by lines of second order phase transitions belonging to the 3-d $Z(2)$ universality class. The first order region for the $N_f = 2+1$ case, the second order $Z(2)$ line separating the $N_f = 2+1$ first order and the crossover regions and the second order $O(4)$ line for the $N_f = 2$ case are supposed to meet at a tri-critical point characterized by a certain value of the strange quark mass, $m_s^{\rm tric}$. Although, it is well established that in the real world, i.e., for the physical values of the quark masses, the transition is a crossover,[61,72] the location of the physical point with respect to $m_s^{\rm tric}$ has not been established and even $m_s^{\rm tric} \to \infty$ cannot be ruled out. More specifically, it is yet unclear whether $m_s^{\rm phys} > m_s^{\rm tric}$ or $m_s^{\rm phys} = m_s^{\rm tric}$ or $m_s^{\rm phys} < m_s^{\rm tric}$. If $m_s^{\rm phys} > m_s^{\rm tric}$

then in the limit of $m_l \to 0$ one should find a second order transition belonging to the 3-d $O(4)$ universality class, if $m_s^{\rm phys} = m_s^{\rm tric}$ the tri-critical point is a Gaussian fixed point of the 3-d ϕ^6 model and its critical exponents take the mean field values[73] and if $m_s^{\rm phys} < m_s^{\rm tric}$ then in the $m_l \to 0$ limit one may cross through a second order transition belonging to the 3-d $Z(2)$ universality class and then may end up in the first order transition region.

In the following discussion we assume $m_s^{\rm phys} > m_s^{\rm tric}$.

3.1. *Chiral transition in (2 + 1)-flavor QCD*

In the vicinity of the chiral phase transition, the free energy density may be expressed as a sum of a singular and a regular part

$$f = -\frac{T}{V} \ln Z \equiv f_{\rm sing}(t,h) + f_{\rm reg}(T, m_l, m_s, \vec{\mu}). \tag{26}$$

The parameter h represents the dimensionless explicit chiral symmetry breaking (magnetic) field and t incorporates all the 'thermal' variables that do not explicitly break the chiral symmetry, e.g., at leading order,

$$t = \frac{1}{t_0}\left[\frac{T - T_c^0}{T_c^0} + \kappa_q \left(\frac{\mu_q}{T}\right)^2\right], \quad \text{and} \quad h = \frac{1}{h_0}\frac{m_l}{m_s}, \tag{27}$$

where $\mu_q \equiv \mu_u = \mu_d$. T_c^0 denotes the (unknown) phase transition temperature in the chiral limit $m_l \to 0$ and for $\mu_q = 0$. The scaling variables t, h are normalized by two unknown parameters t_0 and h_0. The other unknown parameter κ_q controls how the chiral transition temperature T_c^0 changes as a function of μ_q. All these four unknown quantities $(T_c, \kappa_q, t_0, h_0)$ are unique to QCD, similar to the low energy constants in the chiral Lagrangian. These parameters can be determined by analyzing the scaling behaviors, arising from the singular part of the free energy, of the chiral order parameter and susceptibilities using lattice QCD.

3.1.1. *Pseudo-critical behavior at vanishing chemical potential*

Sufficiently close to the QCD chiral transition the renormalization group invariant dimensionless order parameter M_b, constructed out of the light quark chiral condensate, defined in Eq. (24), and the strange quark mass obeys the scaling relation

$$M_b \equiv \frac{m_s \langle \bar{\psi}\psi \rangle_l}{T^4} = h^{1/\delta} f_G(z) + \text{regular terms}, \tag{28}$$

in terms of a single scaling variable $z = t/h^{1/\beta\delta}$. The critical exponents β and δ and the scaling function $f_G(z)$ uniquely characterize the universality class of the chiral phase transition. As discussed before, for continuum QCD the relevant universality class is expected to be the same as that of 3-d $O(4)$ spin models. However, the situation is more subtle for QCD on the lattice. For example, since for staggered fermions away from the continuum limit there is only one Goldstone boson in the chiral limit the relevant universality class is that of the 3-d $O(2)$ spin model. Table 1

Table 1. Critical exponents α, β, γ, δ for the 3-dimensional $O(N)$ universality classes.[74,75] Only two of the four critical exponents are independent. They are related to each other through the scaling relations $\gamma = \beta(\delta - 1)$ and $\alpha + 2\beta + \gamma = 2$. The last two columns give the location of the maxima of the scaling functions $f'_G(z)$ and $f_\chi(z)$, respectively.[74,75]

N	α	β	γ	δ	z_t	z_p
2	−0.017	0.349	1.319	4.779	0.46 (20)	1.56 (10)
4	−0.213	0.380	1.453	4.824	0.73 (10)	1.35 (3)

summarizes the critical exponents and other relevant quantities for the 3-d $O(2)$ and $O(4)$ universality classes.

Figure 4 shows results for the chiral order parameter M_b calculated in (2+1)-flavor QCD[76] for several values of the light to strange quark mass ratio m_l/m_s while keeping m_s fixed to its physical value $m_s^{\rm phys}$. In the continuum limit the physical value of this ratio is $m_l^{\rm phys}/m_s^{\rm phys} = 1/27.5$.[31] The left panel in Fig. 4 shows the variation of M_b with temperature calculated on rather coarse lattices ($N_\tau = 4$) for several values of m_l/m_s. In the continuum limit the smallest of these ratios would correspond to pion masses of about 80 MeV. Clearly, the transition sharpens as the light quark masses become smaller. The right panel of Fig. 4 shows a comparison of the scaled order parameter $M_b/h^{1/\delta}$ (see Eq. (28)) with the $O(2)$ scaling function. As can be seen, for small enough light quark masses, $m_l \lesssim m_s^{\rm phys}/20$, the scaled order parameter for different values of the light quark mass collapses on a unique curve and compares well with the $O(2)$ scaling function. While these lattice results suggest that the chiral transition for $N_f = 2$ theory belongs to the 3-d $O(N)$ universality class one should bear in mind that these results are not continuum extrapolated and discretization effects may change the conclusion. In fact, lattice studies performed with the unimproved staggered action even further away from the continuum limit suggest that in this case the transition may even be first order.[77]

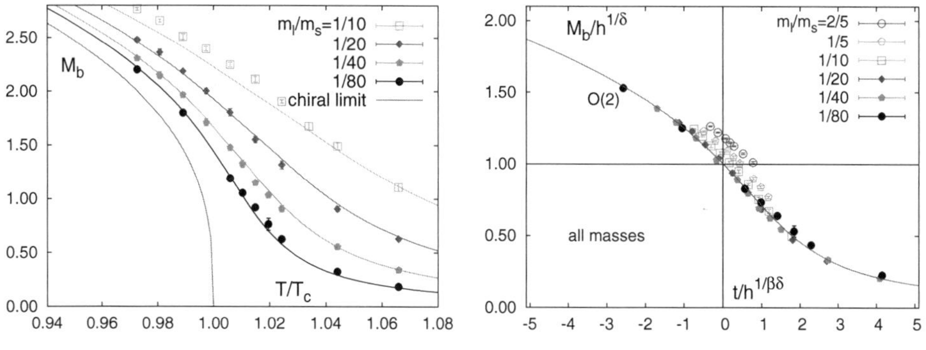

Fig. 4. (Color online) $O(N)$ scaling of the chiral order parameter M_b introduced in Eq. (28) for (2+1)-flavor QCD.[76]

The chiral order parameter M_b can be used to define two susceptibilities obtained by either taking a derivative with respect to 'thermal' variable t, which give the mixed susceptibility $\chi_t \sim \partial M_b/\partial t \sim \partial^2 f_s/\partial t \partial h$, or with respect to the symmetry breaking variable, $\chi_h \sim \partial M_b/\partial h \sim \partial^2 f_s/\partial h^2$. These two susceptibilities are the only two second derivatives of the free energy of an $O(N)$ symmetric theory like QCD that diverge at the critical point,[b] i.e., at $T = T_c^0$ for $m_l = 0$. Their singular behavior is controlled by two scaling functions, $f'_G(z) = df_G(z)/dz$, and $f_\chi(z) = (f_G(z) - (z/\beta) f'_G(z))/\delta$, respectively. Away from the chiral limit, these two scaling functions have maxima at universal values of z, i.e., $z = z_t$ and $z = z_p$ (see Table 1), which can be used to define pseudo-critical temperatures.

For example, the scaling behavior of the chiral susceptibility for two light, degenerate flavors,

$$\chi_{m,l}(T) = 2\frac{\partial \langle \bar{\psi}\psi \rangle_l}{\partial m_l}, \tag{29}$$

is related to that of χ_h. The renormalization group invariant product of the chiral susceptibility $\chi_{m,l}$ and the square of the strange quark mass is related to the scaling functions by,

$$\frac{m_s^2 \chi_{m,l}}{T^4} = \frac{1}{h_0} h^{1/\delta - 1} f_\chi(z) + \text{regular terms}. \tag{30}$$

At vanishing quark chemical potential the chiral susceptibility thus diverges at the chiral critical temperature T_c^0 in the chiral limit $h \to 0$. The peak location of the chiral susceptibility will then be associated with the peak of the scaling function $f_\chi(z)$, located at $z = z_p$, and defines the chiral pseudo-critical temperature T_c for $h > 0$,

$$T_c = T_c^0 \left[1 + \frac{z_p}{z_0} \left(\frac{m_l}{m_s} \right)^{1/\beta\delta} \right] + \text{regular terms}, \tag{31}$$

with $z_0 = t_0/h_0^{1/\beta\delta}$. As illustrated in Fig. 5, the chiral crossover temperature for physical QCD with $m_s = m_s^{\rm phys}$ and $m_l^{\rm phys} = m_s^{\rm phys}/27.5$ was studied in detail[78] by using such a scaling analysis of the chiral susceptibility, including the influence of regular terms. By performing scaling fits to the chiral susceptibility, extrapolating to the physical value of light quark mass and subsequently taking the continuum limit one obtains a value of

$$T_c = (154 \pm 9)\,\text{MeV} \tag{32}$$

for the chiral crossover temperature in (2+1)-flavor physical QCD.[78] Studies with other improved staggered fermion discretization schemes,[79–81] which used other

[b]In general, there is a third susceptibility, the specific heat, C_V, which is obtained as a second derivative of the free energy with respect to the 'thermal' variable t, $C_V \sim \partial^2 f_s/\partial t^2$. In $O(N)$ symmetric theories, however, the relevant critical exponent α is negative. The specific heat, thus, does not diverge at T_c in the chiral limit.

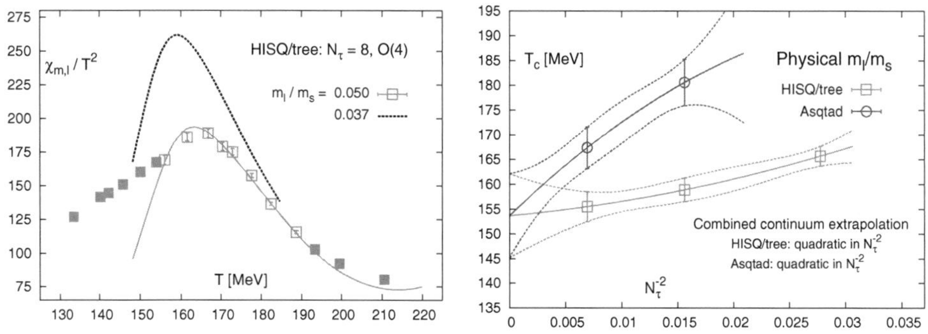

Fig. 5. (Color online) (Left) $O(4)$ scaling of the chiral susceptibility and its extrapolation to the physical light quark mass for (2+1)-flavor QCD.[78] The peak location of the chiral susceptibility defines the chiral crossover temperature. (Right) Continuum extrapolation of the chiral crossover temperature for (2+1)-flavor physical QCD, yielding $T_c = (154 \pm 9)$ MeV.[78]

criteria, not directly related to criticality, to define a crossover temperature, and with chiral Domain Wall Fermions[61] also yielded compatible results for the QCD chiral crossover temperature.

3.1.2. *Curvature of the pseudo-critical line*

The scaling behavior of the mixed susceptibility, χ_t, can be utilized to determine the chiral phase boundary in the T–μ plane.[82] As introduced in Eq. (27), to leading order the chemical potential μ_q does not contribute to the explicit chiral symmetry breaking field h and can be treated as part of the 'thermal' scaling variable t. Thus, the derivative of the order parameter with respect to the chemical potential, $\chi_{m,\mu} = \partial^2 M_b/\partial(\mu_q/T)^2$, is equivalent to the mixed susceptibility, χ_t. The scaling behavior of the renormalization group invariant combination of $\chi_{m,\mu}$ multiplied with the strange quark mass, which combines a 'thermal' and 'field-like' derivative of the free energy, is given by

$$\frac{m_s \chi_{m,\mu}}{T^2} = \frac{2\kappa_q}{t_0} h^{-(1-\beta)/\beta\delta} f'_G(z) + \text{regular terms}. \tag{33}$$

In the chiral limit, $h \to 0$, this susceptibility diverges at T_c^0 for $\mu_q = 0$. For $\mu_q > 0$, the chemical potential dependence of the chiral transition temperature $T_c^0(\mu_q)$ is defined by the peak location, $z = z_t$, of the scaling function $f'_G(z)$ and in the leading order in $(\mu_q/T)^2$ it is given by

$$T_c^0(\mu_q) = T_c^0 \left[1 - \kappa_q \left(\frac{\mu_q}{T_c^0}\right)^2 + \mathcal{O}\left[\left(\frac{\mu_q}{T_c^0}\right)^4\right]\right] + \text{regular terms}. \tag{34}$$

As illustrated in Fig. 6 (left), the curvature of the chiral phase transition line, κ_q, can be determined by fitting the scaled susceptibility $m_s t_0 h^{(1-\beta)/\beta\delta}\chi_{m,\mu}/T^2$ to the universal scaling function $f'_G(z)$.[82] Such a scaling analysis provides a value of the

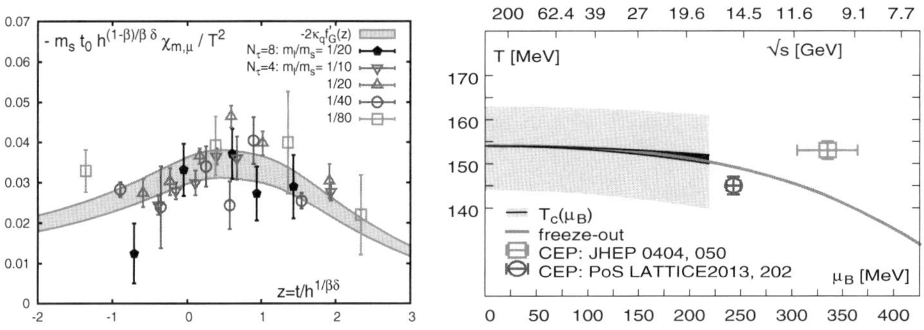

Fig. 6. (Color online) (Left) $O(N)$ scaling of the susceptibility $\chi_{m,\mu}$ and determination of the curvature, κ_q, of the chiral phase boundary in the T–μ_q plane.[82] (Right) Chiral crossover temperature as function of baryon chemical potential, $T_c(\mu_B)$, compared with the freeze-out temperature in heavy ion collisions. The parametrization of the freeze-out line is taken from Ref. 86, with the freeze-out temperature in the limit of infinite collision energy adjusted to the crossover temperature at vanishing chemical potential, $T_{fo}(\sqrt{s} \to \infty) = 154\,\mathrm{MeV}$. The black band for the crossover line, $T_c(\mu_B)$, reflects the systematic uncertainty arising from a factor of 2 difference in current estimates from calculations with real[82,83] and imaginary[84,85] μ_B (see text). Data points are from a reweighting analysis[87] and Taylor expansion[88] (see text). They also are normalized to $T_c = 154\,\mathrm{MeV}$.

curvature $\kappa_q = 0.059(6)$.[82] In terms of the baryon chemical potential $\mu_B = 3\mu_q$, the chiral transition temperature is then given by

$$T_c^0(\mu_B) = T_c^0 \left[1 - 0.0066(7)\left(\frac{\mu_B}{T_c^0}\right)^2 + \mathcal{O}\left[\left(\frac{\mu_B}{T_c^0}\right)^4\right]\right]. \tag{35}$$

This result is in very good agreement with the independent determination of the continuum extrapolated results of the curvature of the chiral phase boundary for (2+1)-flavor of physical QCD,[83] using different criteria. However, analytic continuation from purely imaginary chemical potential yields a factor two larger values for the curvature.[84,85] Figure 6 (right) shows the proximity of the QCD phase boundary to the freeze-out line in heavy ion collisions[86] in the T–μ_B plane.

3.1.3. *The critical end point*

Information on the location of a critical end point (CEP) in the QCD phase diagram (Fig. 1) from lattice QCD is ambiguous. Although considerable progress in developing algorithms that would allow direct calculations in lattice QCD at $\mu_B > 0$ have been made recently (see Ref. 36 for a recent review) these techniques are not yet applicable to realistic QCD parameter values.

The first calculations that provided hints for the existence of a critical point[87] used a reweighting technique that is applicable only on rather small lattices. It had been performed on coarse lattices with temporal extent $N_\tau = 4$ using only the naive 1-link staggered fermion action. Taste violation effects thus are large and the reweighting is known to fail when used too far away from the parameters actually used in the calculations. It has been argued that this 'overlap problem' may have led to the spurious identification of a signal for a critical point.[89] Calculations performed with an imaginary chemical potential do not find any evidence for the existence of a critical point.[90] However, these calculations have also been performed on coarse lattices and utilized the naive staggered 1-link action.

The conclusions drawn in Ref. 90 are based on an analysis of the μ_B-dependence of the location of the boundary line that separates the first order transition region shown in Fig. 3 at small quark masses from the crossover region. These calculations, performed with staggered fermions using an imaginary chemical potential,[90] suggest that this boundary moves to smaller quark masses with increasing chemical potential. This disfavors the existence of a critical point connected to the $Z(2)$ boundary line at vanishing chemical potential. A similar analysis, performed with clover-improved Wilson fermions, comes to the opposite conclusion[91] and thus does favor the existence of a critical point. The apparent differences between calculations performed with staggered fermions and Wilson fermions on coarse lattices underscores that better control over systematic errors is needed before a definite conclusion on the existence of a critical point can be drawn.

Analyzing the convergence properties of the Taylor series for the pressure at nonzero chemical potential[40,92] (Eq. (18)) does, in principle, allow one to relate the location of the CEP to the radius of convergence of this series. Although this approach has been used closer to the continuum limit $(N_\tau = 8)$[88,93] than the studies discussed above, the current estimates still are based on calculations with the naive staggered 1-link action. They still suffer from large taste symmetry violations. These calculations yield for the location of the critical point[88] $(T^E/T_c, \mu_B^E/T^E) = (0.94(1), 1.68(5))$[93] which is significantly lower than the reweighting result $(T^E/T_c, \mu_B^E/T^E) = (0.99(1), 2.2(2))$.[87] We show these two estimates for the location of the CEP in Fig. 6 (right). The systematic errors of these estimates are at present difficult to estimate.

Further information on the location of a critical point comes from the calculations of (i) the equation of state at $\mu_B > 0$ performed with improved staggered fermions, which we will discuss in Sec. 4, as well as (ii) the analysis of freeze-out parameters based on Taylor expansions of cumulants of charge fluctuations, which are also performed with improved staggered actions and which we will discuss in Sec. 5, both suggest that a CEP located at $\mu_B/T < 2$ is unlikely.

3.2. *Deconfining aspect of QCD*

Pure $SU(N_c)$ Yang–Mills theory without quarks possess an exact global $Z(N_c)$ center symmetry, which gets spontaneously broken in the high temperature deconfining phase.[15,16] The thermal expectation value of the renormalized Polyakov loop

$$L_{\rm ren}(T) = e^{-c(g^2)N_\tau} \cdot \frac{1}{VN_c} \sum_{\vec{x}} \left\langle {\rm Tr} \prod_{x_0=1}^{N_\tau} U_{(x_0,\vec{x}),\hat{0}} \right\rangle , \qquad (36)$$

is not invariant under the $Z(N_c)$ center symmetry. The Polyakov loop can be interpreted as the free energy difference, $F_\infty(T)$, of a thermal system with and without an infinitely heavy static quark–anti-quark pair separated by infinite distance, $L_{\rm ren}(T) = \exp(-F_\infty(T)/2T)$. The renormalization constant $c(g^2)$ can be fixed[94] by demanding that in the short distance limit the heavy quark free energy coincides, up to a trivial additive constant, with the Coulombic short distance behavior of the zero temperature heavy quark potential defined in Eq. (23). In the confined phase $F_\infty(T) = \infty$, as a static quark and anti-quark pair cannot be separated by infinite distance in this phase. Thus in the $Z(N_c)$ symmetric confined phase $L_{\rm ren}(T) = 0$. On the other hand, in the spontaneously $Z(N_c)$ broken deconfined phase a static quark–anti-quark pair can be separated by infinite distance due to the presence of color screening and $L_{\rm ren}(T) \neq 0$. Thus, for a pure $SU(N_c)$ gauge theory, i.e., in the limit $m_l \to \infty$, the Polyakov loop serves as the order parameter for the deconfinement transition.[95] However, since the mere presence of quarks explicitly breaks the $Z(N_c)$ center symmetry, the Polyakov loop does not serve as an order parameter for QCD with realistic light quarks. Furthermore, since the Polyakov loop is not related to a derivative of the QCD partition function with respect to the thermal or the symmetry breaking field, its change or fluctuation across the QCD transition may not capture the true singularities of the QCD partition function in any limit. Thus, a deconfining temperature defined from the change of the Polyakov loop may not reflect the pseudo-critical properties of QCD with realistic light quark masses. As shown in Fig. 7 (left), the change of the Polyakov loop within the chiral crossover region, $T_c = 154(9)\,\text{MeV}$, is rather gradual and smooth.

Deconfinement is associated with the liberation of degrees of freedom, manifested by a rapid rise in the bulk thermodynamic observables such as the pressure, energy density, etc. Among such bulk thermodynamic observables fluctuations and correlations of different conserved quantum numbers, e.g., the baryon number

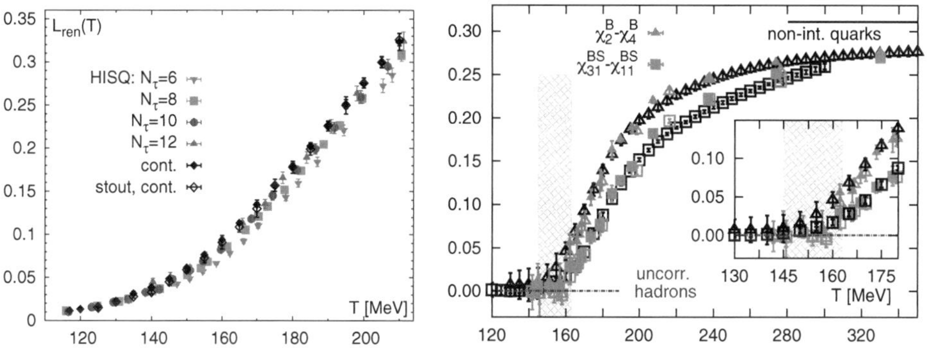

Fig. 7. (Color online) (Left) The renormalized Polyakov loop in (2+1)-flavor QCD.[96] (Right) Appearance of the fractionally charged degrees of freedom in the chiral crossover region $T_c = 154(9)\,\text{MeV}$ (shaded region) for the light as well as the strange quark. The black points show results obtained using the stout action[97] and the other points have been obtained using the HISQ action.[98]

(B), electric charge (Q) and strangeness (S), directly probe the liberation of the quark degrees of freedom and the consequent appearance of fractionally charged quantum number carriers. For example, both for an uncorrelated hadron gas and in a free quark gas the difference of second to fourth order cumulants of baryon number fluctuations probes the baryon number of degrees of freedom, specifically $\chi_2^B - \chi_4^B \propto B^2 - B^4$. For hadronic degrees of freedom carrying baryon number $B = \pm 1$, one thus has $\chi_2^B - \chi_4^B = 0$, but for quark-like degrees of freedom with $B = \pm 1/3$, one finds $\chi_2^B - \chi_4^B \neq 0$. Similarly, for the strange quark sector higher order baryon-strangeness correlation such as $\chi_{31}^{BS} - \chi_{11}^{BS} \propto (B^3 - B)S$ is non-vanishing if the strangeness carrying degrees of freedom are associated with quark-like baryon number $B = \pm 1/3$ and vanishes when the strangeness is carried by hadronic degrees of freedom with $B = \pm 1$. Thus, such combinations of higher order fluctuations and correlations of quantum numbers are sensitive probes of appearance of fractionally charged degrees of freedom in a deconfined medium.[98–100] The right panel of Fig. 7 shows lattice QCD results for $\chi_2^B - \chi_4^B$ and $\chi_{31}^{\mathrm{BS}} - \chi_{11}^{\mathrm{BS}}$ from the BNL–Bielefeld collaboration[98] as well as the Budapest–Wuppertal collaboration.[97] Clearly, both quantities start deviating strongly from zero in the chiral crossover region, $T_c = 154(9)\,\mathrm{MeV}$, indicating that fractionally charged degrees of freedom start to appear at these temperatures and onset of deconfinement takes place for the light as well as the strange quarks. Note, however, that these fluctuation observables are not order parameters in the strict sense.

3.3. *Axial symmetry of QCD at high temperature*

Although, the axial $U_A(1)$ symmetry is not an exact symmetry of QCD, as mentioned before, the magnitude of its breaking is expected to influence the order of the chiral phase transition. Hence, knowledge of the temperature dependence of the $U_A(1)$ breaking is essential for a comprehensive understanding of the QCD chiral transition. The axial $U_A(1)$ symmetry of the massless QCD Lagrangian is broken due to quantum fluctuations, resulting in non-conservation of axial current[101,102] as well as explicit breaking of the global $U_A(1)$ symmetry induced by topologically non-trivial gauge field configurations, such as the instantons.[103] Due to color screening the instanton density gets suppressed as the temperature increases[104] and in the $T \to \infty$ limit the $U_A(1)$ symmetry becomes exact.

Since $U_A(1)$ is not an exact symmetry of QCD, one cannot define an order parameter associated with its breaking. However, for two massless flavors, the pion (π) and the iso-vector scalar δ (a_1) mesons transform into each other under a $U_A(1)$ rotation. Since the presence of an exact $U_A(1)$ will render these meson states degenerate, the difference of the integrated two-point correlation functions of pion and δ meson,

$$\chi_\pi - \chi_\delta = \int d^4x \, \left(\langle \pi^+(x)\pi^-(0) \rangle - \langle \delta^+(x)\delta^-(0) \rangle \right), \tag{37}$$

will also vanish in the limit of exact $U_A(1)$ symmetry. Corrections due to small non-vanishing light quark masses will only contribute at $\mathcal{O}(m_l^2)$. Thus, this quantity can be taken as a measure of the $U_A(1)$ breaking.[105]

This measure of $U_A(1)$ breaking was studied in detail using improved p4 staggered fermions,[106] and it was found that $\chi_\pi - \chi_\delta$ remains non-vanishing for $T \lesssim 1.2T_c$. However, for staggered fermions the issue of axial anomaly is quite subtle and the correct anomaly may only emerge in the continuum limit.[107,108] On the other hand, emergence of the axial anomaly is more straightforward for the chiral DWF formulation.[49] For DWF, axial symmetry is broken by the same topologically nontrivial configurations as in the continuum. Lattice artifacts appear only at $\mathcal{O}(m_{\text{res}}^2)$, due to the explicit chiral symmetry breaking residual mass,[56] m_{res}, arising from finiteness of the fifth dimension ($L_s < \infty$). Thus, the DWF action turns out to be a natural candidate for investigation of the temperature dependence of $U_A(1)$ breaking in QCD. The temperature dependence of the $U_A(1)$ breaking measure, $\chi_\pi - \chi_\delta$, has been extensively studied using the DWF formalism for several volumes as well as quark masses.[59,61,109] As depicted in Fig. 8, calculations with DWF clearly show that $\chi_\pi - \chi_\delta$ does not vanish around the chiral crossover temperature T_c and remains non-vanishing for $165\,\text{MeV} \lesssim T \lesssim 195\,\text{MeV}$, independent of the light quark masses. These results indicate that $U_A(1)$ symmetry may remain broken at these temperatures even in the chiral limit.

Since $U_A(1)$ breaking arises due to topology of the gauge fields, it is intimately related to the infrared modes of the Dirac fermions. In the limit of infinite volume, both the chiral order parameter, $\langle\bar{\psi}\psi\rangle_l$, and the $U_A(1)$ breaking measure, $\chi_\pi - \chi_\delta$,

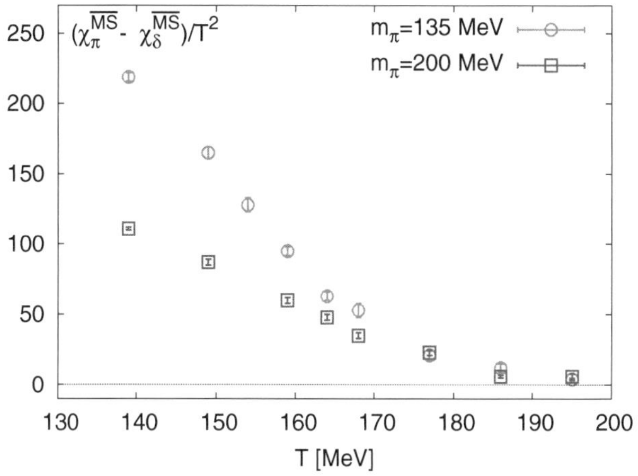

Fig. 8. (Color online) The $U_A(1)$ breaking measure $\chi_\pi - \chi_\delta$ renormalized in the $\overline{\text{MS}}$ scheme as function of temperature. Results are from a calculation with Domain Wall fermions.[61] The difference is non-vanishing for $T \geq T_c = 154(9)$ MeV. It becomes independent of quark mass for $T \gtrsim 168$ MeV and rises with decreasing quark mass at lower temperature. In fact, it will diverge in the chiral limit for $T < T_c$.

can be represented in terms of the eigenvalue density, $\rho(\lambda)$, of the Dirac fermions

$$\langle\bar{\psi}\psi\rangle_l = \int_0^\infty d\lambda \, \frac{2m_l \, \rho(\lambda)}{\lambda^2 + m_l^2}, \quad \text{and} \quad \chi_\pi - \chi_\delta = \int_0^\infty d\lambda \, \frac{4m_l^2 \, \rho(\lambda)}{(\lambda^2 + m_l^2)^2}. \tag{38}$$

In the chiral symmetric phase and in the chiral limit $\langle\bar{\psi}\psi\rangle_l$ must vanish, but $\chi_\pi - \chi_\delta$ can remain nonzero. Identification of the infrared fermionic modes, i.e., the form of $\rho(\lambda)$, that give rise to such a phenomenon naturally leads to the underlying non-perturbative mechanism of axial symmetry breaking. Studies with DWF[59,109] as well as with overlap fermions,[62] possessing even better chiral properties and an exact index theorem, suggest that an accumulation of near-zero eigenmodes of the form $\rho(\lambda) \sim m_l^2 \delta(\lambda)$ may largely account for the observed $U_A(1)$ breaking in $\chi_\pi - \chi_\delta$ at high temperature. Such accumulation of the near-zero modes are depicted in Fig. 9 (top). More detailed studies[62] of the space–time profiles, as illustrated in Fig. 9 (bottom), localization properties and distributions of these near-zero modes indicate that their behavior is consistent with underlying presence of a dilute instanton gas — a gas of widely separated, weakly interacting, small instantons and anti-instantons. This suggests that even at temperatures $T \sim 1.5T_c$ weakly interacting instanton anti-instanton pairs are largely responsible for the $U_A(1)$ breaking.

The analysis of $U_A(1)$ symmetry breaking close to the chiral limit of 2 or (2+1)-flavor QCD, however, is far from being settled. Detailed systematic studies of the

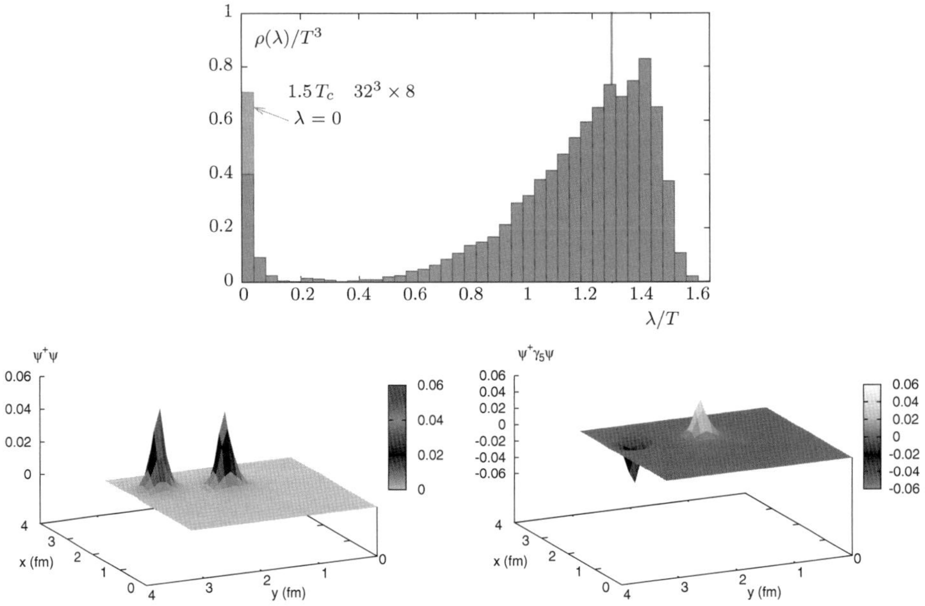

Fig. 9. (Color online) (Top) Accumulation of near-zero eigenmodes of the Overlap-Dirac fermion matrix at $T = 1.5T_c$.[62] (Bottom) Spatial profile of a typical near-zero mode (left) and the spatial profile of the chirality of the same near-zero mode (right) at $T = 1.5T_c$.[62]

quark mass and volume dependence as well as the cut-off dependence are still missing. Moreover, results obtained with chiral fermions are still controversial. Calculations performed with dynamical overlap fermions[60,110] and so-called optimal domain wall fermions[111] at present suggest that $U_A(1)$ does get restored already at T_c, which would still allow for the occurrence of a first order transition when approaching the chiral limit.

4. Bulk Thermodynamics

4.1. *The QCD equation of state at vanishing chemical potential*

Lattice QCD calculations of the equation of state, or more general, of basic bulk thermodynamic observables like the pressure (P), energy density (ϵ) or entropy density (s), have reached now a precision where continuum extrapolated results for physical light and strange quark masses can be obtained. This also allows detailed comparisons with perturbative calculations at high temperature and model calculations at low temperature.

The procedure to calculate bulk thermodynamic observables in lattice QCD is well established. The starting point for these calculations is the evaluation of the trace anomaly, $\Theta^{\mu\mu}(T) \equiv \epsilon - 3P$. The trace anomaly is directly obtained as a temperature derivative of P/T^4 which also is related to the QCD partition function, $Z(T,V)$,

$$\frac{\epsilon - 3P}{T^4} = T\frac{dP/T^4}{dT}, \tag{39}$$

with $P/T^4 = \lim_{V\to\infty}(VT^3)^{-1}\ln Z(T,V)$. As the trace anomaly arises as a derivative of the logarithm of the QCD partition function with respect to temperature, it can be expressed in terms of expectation values of rather simple observables, e.g., a combination of the gauge action and light and strange quark chiral condensates. Nonetheless, the calculations become rather time consuming because a subtraction of the zero temperature contributions is needed to eliminate divergent vacuum contributions.[c]

As can be seen from Eq. (39) up to an integration constant at a temperature T_0 the pressure can be obtained from the trace anomaly,

$$\frac{P(T)}{T^4} - \frac{P(T_0)}{T_0^4} = \int_{T_0}^{T} dT' \frac{\epsilon - 3P}{T'^5}. \tag{40}$$

This allows one to reconstruct the energy density as well as the entropy density $s/T^3 = (\epsilon + P)/T^4$.

The determination of thermodynamic quantities in QCD is a parameter free calculation. All input parameters needed in the calculation, e.g., the quark masses ($m_u = m_d$, m_s) and the relation between the lattice cut-off, a, and the bare gauge

[c]For more details we refer, for instance, to Ref. 112.

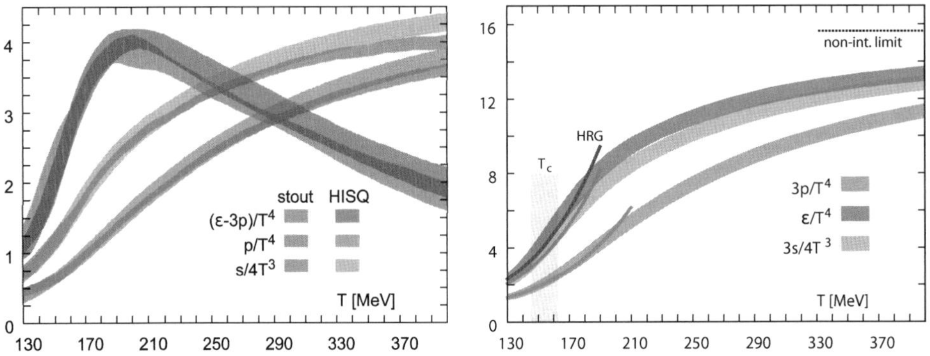

Fig. 10. (Color online) (Left) Comparison of the trace anomaly $(\epsilon - 3P)/T^4$, pressure and entropy density calculated with the HISQ (colored)[114] and stout (grey)[113] discretization schemes for staggered fermions. (Right) Continuum extrapolated results for pressure, energy density and entropy density obtained with the HISQ action.[114] Solid lines on the low temperature side correspond to results obtained from hadron resonance gas (HRG) model calculations. The dashed line at high temperatures shows the result for a non-interacting quark–gluon gas.

coupling, $\beta = 6/g^2$, are determined through calculations at zero temperature. Likewise, there is only a single independent thermodynamic observable that is calculated in a lattice QCD calculation, for instance the trace anomaly, $\Theta^{\mu\mu}(T)$. All other bulk thermodynamic observables are obtained from $\Theta^{\mu\mu}(T)$ through standard thermodynamic relations. In Fig. 10 (left) we show recent results for the trace anomaly of (2+1)-flavor QCD[113,114] obtained with two different discretization schemes by two different groups. The results are extrapolated to the continuum limit and are obtained with a strange quark mass tuned to its physical value and light quark masses that differ slightly ($m_s/m_l = 27$[113] and 20[114]). The right panel in this figure shows results for the pressure, energy density and entropy density obtained from the trace anomaly by using Eqs. (39) and (40).

Also shown in Fig. 10 are results obtained from a hadron resonance gas (HRG) model calculation of bulk thermodynamics. As can be seen this describes the QCD equation of state quite well also in the transition region, although it may be noted that the HRG calculations yield results for all observables that are at the lower error band of the current QCD results. It has been speculated that this may indicate contributions from additional, experimentally not yet observed resonances which could contribute to the thermodynamics.[115] Indeed evidence for the contribution of a large number of strange baryons has recently been found in lattice QCD calculations of conserved charge fluctuations[116] (see also the discussions in Secs. 5 and 7).

In Fig. 11 we show the energy density in the low temperature region. The box highlights the transition region characterized by the crossover temperature, $T_c = (154 \pm 9)\,\text{MeV}$. From this we deduce the energy density in the crossover region, $\epsilon_c = (0.34 \pm 0.16)\,\text{GeV/fm}^3$. This is a rather small value for the energy density needed to convert ordinary hadronic matter into a medium made up from quarks and gluons. It may be compared to the energy density of ordinary nuclear

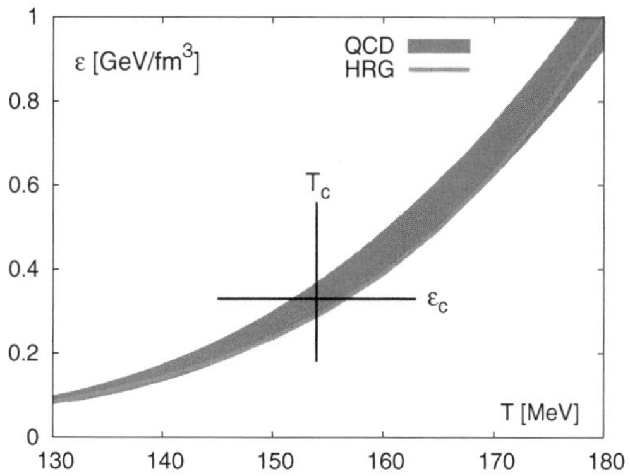

Fig. 11. (Color online) The critical energy density ϵ_c in (2+1)-flavor QCD. The band gives the continuum extrapolated result for the energy density taken from Ref. 114. The HRG curve is based on all resonance with mass less than 2.5 GeV listed by the Particle Data Group.[31]

matter, $\epsilon^{\text{nuclear matter}} \simeq 0.15\,\text{GeV/fm}^3$ or the energy density inside a nucleon, $\epsilon^{\text{nucleon}} \simeq 0.45\,\text{GeV/fm}^3$, assuming the radius of nucleon $R_N \simeq 0.8\,\text{fm}$. In fact, ϵ_c is close to the energy density reached in the dense packing limit of nucleons with radius R_N.

The simple bulk thermodynamic observables like pressure and energy density give the impression that an HRG model also provides a good description of the equation of state for temperatures above the crossover region, i.e., for $T \gtrsim 165\,\text{MeV}$. However, as discussed in the previous section the analysis of conserved charge fluctuations clearly shows that at temperatures $T \gtrsim 160\,\text{MeV}$ thermodynamics can no longer be described in terms of hadronic degrees of freedom (see for instance Fig. 7 (right) and the discussion in Sec. 7). This also becomes evident in second order derivatives of the QCD partition function with respect to temperature. The speed of sound, $c_s^2 = dp/d\epsilon$, is related to the inverse of the specific heat, $C_V = d\epsilon/dT$,

$$c_s^2 = \frac{dp}{d\epsilon} = \frac{dp/dT}{d\epsilon/dT} = \frac{s}{C_V}, \tag{41}$$

$$\frac{C_V}{T^3} = \left.\frac{\partial\epsilon}{\partial T}\right|_V \equiv \left(4\frac{\epsilon}{T^4} + T\left.\frac{\partial(\epsilon/T^4)}{\partial T}\right|_V\right). \tag{42}$$

Both quantities are shown in Fig. 12. The specific heat does not develop a pronounced peak in the transition region as one could have expected from pseudo-critical behavior of energy density fluctuations close to a critical point. This may be understood[114] from the temperature dependence of the two terms contributing to C_V/T^3. The dominant singular contribution arises from the temperature derivative of ϵ/T^4, which has a peak. This, however, is overwhelmed by the large energy density contribution at high temperature which reflects the liberation of many partonic degrees of freedom. Furthermore, even in the chiral limit, where

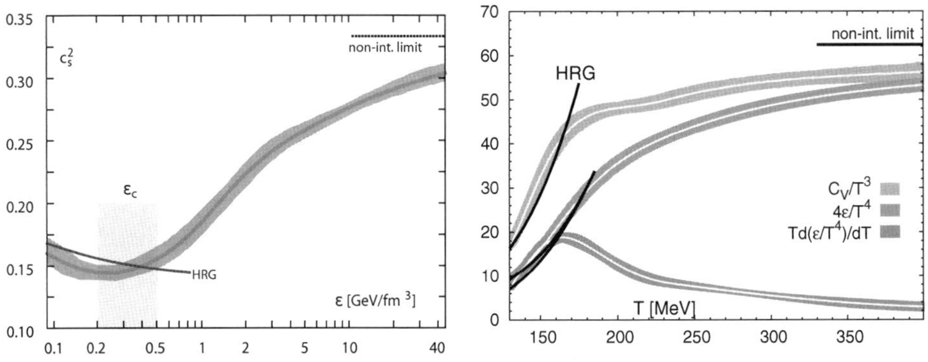

Fig. 12. (Color online) The velocity of sound in (2+1)-flavor QCD (left) and the specific heat C_V/T^4 together with the two components (see Eq. (42)) contributing to it (right). Solid black lines in the low and high temperature regions show the corresponding hadron resonance gas (HRG) and non-interacting quark–gluon gas results, respectively.

QCD is expected to have a second order phase transition belonging to the universality class of 3-d, $O(4)$ symmetric spin models, the specific heat will not diverge as the relevant critical exponent $\alpha \simeq -0.2$ that controls its singular behavior, $C_V/T^3 \sim (|T - T_c|/T_c)^{-\alpha} + \text{const.}$, is negative for this universality class (see Table 1). The speed of sound will therefore stay nonzero at T_c also in the chiral limit.

4.2. *The QCD equation of state at non-vanishing chemical potential*

In Sec. 3 we have discussed lattice QCD results on the dependence of the QCD crossover temperature on the baryon chemical potential and its relation to the freeze-out temperatures determined in heavy ion experiments. These experiments, in particular the beam energy scan program performed at RHIC, will probe properties of strong-interaction matter at non-vanishing baryon chemical potential in the temperature range $0.9 \lesssim T/T_c \lesssim 2$ and $0 \lesssim \mu_B/T \lesssim 3$, with T_c denoting the crossover temperature at $\mu_B = 0$. For the hydrodynamic modeling of matter in this (T, μ_B) regime it thus is of importance to also know the equation of state at non-vanishing μ_B/T. As direct numerical calculations at nonzero μ_B are not yet possible, a viable approach is to analyze the equation of state using a Taylor expansion in terms of chemical potentials[117,118] as given in Eq. (18). In this way some results for the equation of state at nonzero baryon chemical potential have already been obtained on coarse lattices.[40,41,118,119] Continuum extrapolated results for Taylor expansion coefficients of the pressure at $\mathcal{O}(\mu_f^2)$ have been obtained.[120,121] These have been used to construct the EoS at $\mathcal{O}(\mu_B^2)$[122] which implements the strangeness neutrality constraint[119] ($\langle n_S \rangle = 0$) and the electric charge to baryon number relation ($\langle n_Q \rangle = 0.4 \langle n_B \rangle$) suitable for conditions met in heavy ion collisions.

We will discuss here only the case $\mu_Q = \mu_S = 0$ in some detail in order to illustrate the relative importance of higher order corrections in different temperature and μ_B regions. The Taylor series for the pressure is given by,

$$\frac{P(T,\mu_B) - P(T,0)}{T^4} = \frac{1}{2}\chi_2^B(T)\left(\frac{\mu_B}{T}\right)^2\left(1 + \frac{1}{12}\frac{\chi_4^B(T)}{\chi_2^B(T)}\left(\frac{\mu_B}{T}\right)^2\right) + \mathcal{O}(\mu_B^6). \quad (43)$$

The leading order correction to the pressure thus is proportional to the quadratic fluctuations of net baryon number. The next to leading order corrections are proportional to the quartic fluctuations. When written in the form given in Eq. (43) it becomes easy to identify the relative importance of leading and next to leading order corrections. In Fig. 13 we show $\chi_2^B(T)$ (left) and $\chi_2^B(T)/\chi_4^B(T)$ (right). With

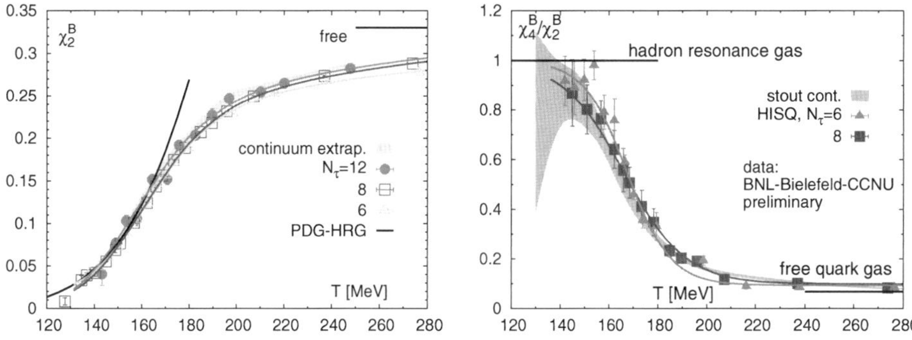

Fig. 13. (Color online) Expansion coefficients of the pressure at nonzero baryon chemical potential. The left hand figure shows the leading order correction[121] and the right hand figure shows the relative contribution of the next to leading order correction. The continuum extrapolated result obtained with the stout action is taken from Ref. 123.

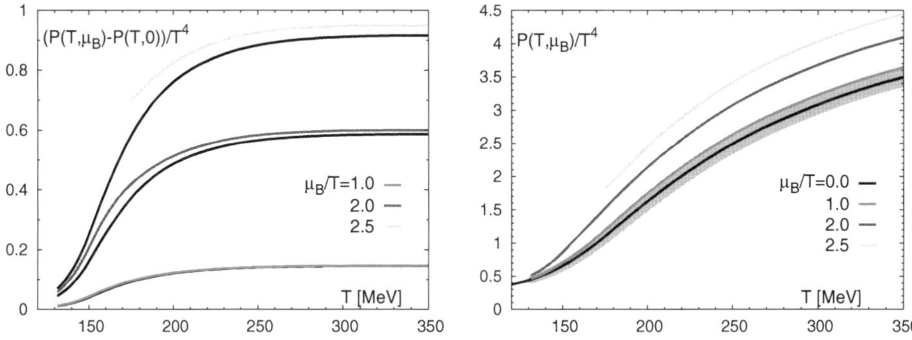

Fig. 14. (Color online) (Left) The μ_B-dependent part of the pressure at $\mathcal{O}((\mu_B/T)^2)$ (black) and $\mathcal{O}((\mu_B/T)^4)$ (colored).[124] The latter is shown only in the temperature regime where neglected corrections at $\mathcal{O}((\mu_B/T)^6)$ are estimated to contribute less than 10%. (Right) When combined with the $\mu_B = 0$ contribution for the pressure shown in Fig. 10 these neglected terms contribute less than 3% to the total pressure.[124] The grey band gives the uncertainty on $P(T,0)/T^4$ and the central line in the band is the parametrization of $P(T,0)/T^4$ given in Ref. 114.

increasing temperature the $\mathcal{O}(\mu_B^4)$ correction rapidly looses importance relative to the leading $\mathcal{O}(\mu_B^2)$ term.

In Fig. 14 we show preliminary results for the μ_B-dependent contribution to the total pressure evaluated for different values of μ_B/T and taking into account corrections up to $\mathcal{O}((\mu_B/T)^4)$.[124] These results suggests that an $\mathcal{O}((\mu_B/T)^4)$ Taylor expansion of the pressure (and energy density) is well controlled for all values of the chemical potential below $\mu_B/T = 2$. This covers a wide range of the QCD phase diagram accessible in the beam energy scan (BES) at RHIC, i.e., the region of beam energies $\sqrt{s} \geq 20\,\text{GeV}$.

4.3. *Perturbation theory, hadron resonance gas and the strongly interacting liquid*

In our discussion of bulk thermodynamic observables, we have compared with HRG model calculations. We have seen that the HRG does provide a rather good description of bulk thermodynamics below and even in the crossover region. However, the HRG does depend on the resonance spectrum used in the calculation. This is of some significance when one looks at more selective observables like, e.g., fluctuations and correlations of strange baryons (see Secs. 5 and 7). Spectrum independent HRG results are, for instance, related to fluctuations of net baryon number. The ratio of fourth to second order net-baryon number cumulants shown in Fig. 13 (right) is unity in any HRG model irrespective of details of the spectrum. This figure thus shows that an HRG model calculation can be expected to provide a good description of QCD thermodynamics on a 10% level only at temperatures less than about $T \simeq (140 - 145)\,\text{MeV}$.

At asymptotically high temperatures hard thermal loop, perturbative calculations[24] or dimensionally reduced QCD (EQCD)[125] provide a well-established framework for the analysis of bulk thermodynamics and fluctuations of conserved

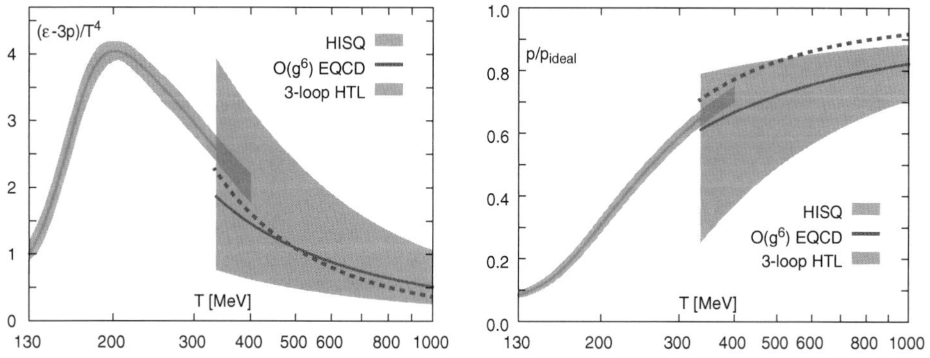

Fig. 15. (Color online) Comparison of the (2+1)-flavor calculation[114] of the trace anomaly (left) and pressure (right) with HTL and EQCD (dashed line) calculations. The black line corresponds to the HTL calculation[24] with renormalization scale $\mu = 2\pi T$. Note that this solid line would move up for the trace anomaly and move down for the pressure if the scale μ in HTL is reduced.

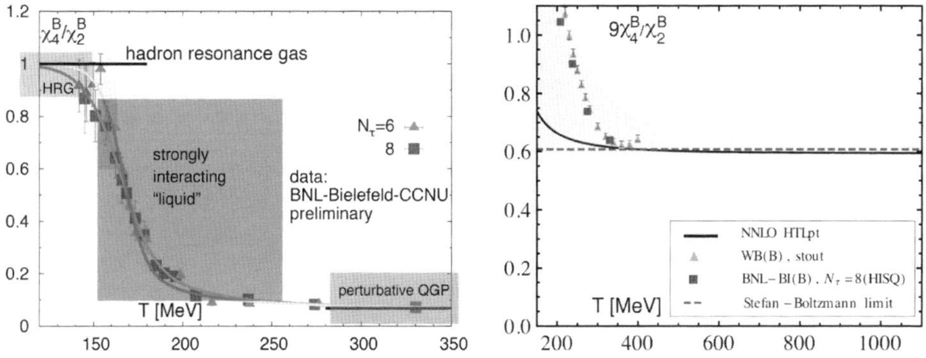

Fig. 16. (Color online) The ratio of quartic and quadratic net-baryon number fluctuations versus temperature. The left panel shows temperature ranges in which HRG and resummed perturbative calculations, respectively, provide good approximations to lattice QCD results. The right panel shows the result from a HTL-resummed calculations.[24]

charges. We show in Fig. 15 a comparison of the trace anomaly (left) and the pressure (right) calculated in lattice QCD with HTL and EQCD calculations. These calculations seem to give a good description of bulk thermodynamics at temperatures $T \gtrsim 400\,\mathrm{MeV}$. However, even in this regime there remain differences between HTL and EQCD calculations that are of the order of 10%. Moreover, within the HTL calculations there seems to be no unique choice for the renormalization scale μ, that would allow to match all observables simultaneously. Similar conclusions can be drawn from the HTL analysis of net-baryon number fluctuations shown in Fig. 16 (right). Although observables that are dominated at high temperature by quark rather than gluon contributions seem to approach perturbative behavior earlier, it still is evident that agreement with lattice QCD results on the 10% level only is possible for $T \gtrsim (250-300)\,\mathrm{MeV}$. In general the temperature range $T_c \le T \le 2\,T_c$ is highly non-perturbative and obviously not accessible to hadronic model calculations. This is highlighted in the left panel of Fig. 16. We will discuss in the following sections properties of strong-interaction matter in this temperature range.

5. Fluctuations of Conserved Charges

Proximity of a second order criticality, such as the $O(4)$ chiral phase transition or the QCD critical point, is universally manifested through long-range correlations at all length scales, resulting in increased fluctuations of the order parameter. These fluctuations can be quantified through the Gaussian (variance) as well as non-Gaussian (skewness, kurtosis, etc.) cumulants of the distribution of the order parameter. Higher order non-Gaussian cumulants become increasingly sensitive to proximity of a critical point[100] which is reflected in their grows with higher powers of the correlation length.[126] Moreover, even qualitative features, such as the sign change and the associated non-monotonicity, of these non-Gaussian cumulants can encode the presence of a nearby critical region.[127–129] Non-Gaussian cumulants can be accessed in heavy ion experiments via the event-by-event fluctuations of various conserved charges and particle multiplicities.[130–134] In this vein, a major focus of the Beam Energy Scan (BES) program[135] at the RHIC is measurements of the event-by-event fluctuations of particle multiplicities and conserved charges.[136–141]

Among several conserved charges, the net electric charge is of special interest. Cumulants of net electric charge fluctuations can be measured in experiments[138,141] and are calculable in lattice QCD.[120,121] Cumulants of net electric charge fluctuations and correlations with other conserved charges are well defined in lattice QCD at vanishing chemical potentials where standard lattice QCD techniques can be used to compute them. Furthermore, Taylor expansions in powers of μ_B around $\mu_B = 0$, can be employed to obtain generalized susceptibilities for $\mu_B > 0$,

$$\chi_n^Q(T,\mu_B) = \sum_{k=0}^{\infty} \frac{1}{k!}\, \chi_{kn}^{BQ}(T) \left(\frac{\mu_B}{T}\right)^k . \tag{44}$$

These susceptibilities directly relate to cumulants of the net electric charge fluctuations,

$$\begin{aligned}
\chi_1^Q(T,\mu_B) &= \frac{1}{VT^3}\langle N_Q\rangle, \\
\chi_2^Q(T,\mu_B) &= \frac{1}{VT^3}\langle(\delta N_Q)^2\rangle, \\
\chi_3^Q(T,\mu_B) &= \frac{1}{VT^3}\langle(\delta N_Q)^3\rangle, \\
\chi_4^Q(T,\mu_B) &= \frac{1}{VT^3}[\langle(\delta N_Q)^4\rangle - 3\langle(\delta N_Q)^2\rangle^2],
\end{aligned} \tag{45}$$

where N_Q is the net (positive minus negative) charge and $\delta N_Q = N_Q - \langle N_Q\rangle$.

On the other hand, through the measurements of the event-by-event distributions of the net electric charge, heavy ion experiments provide various cumulants, mean (M_Q), variance (σ_Q), skewness (S_Q), and kurtosis (κ_Q), of the electric charge fluctuations for given beam energy ($\sqrt{s}$)[138]

$$\begin{aligned}
M_Q(\sqrt{s}) &= \langle N_Q\rangle, \quad \sigma_Q^2(\sqrt{s}) = \langle(\delta N_Q)^2\rangle, \\
S_Q(\sqrt{s}) &= \frac{\langle(\delta N_Q)^3\rangle}{\sigma_Q^3}, \quad \kappa_Q(\sqrt{s}) = \frac{\langle(\delta N_Q)^4\rangle}{\sigma_Q^4} - 3.
\end{aligned} \tag{46}$$

Thus, the charge susceptibilities obtained from lattice QCD calculations and the cumulants measured in the heavy ion experiments are directly related to each other through the appropriate volume-independent ratios[142]

$$\frac{M_Q(\sqrt{s})}{\sigma_Q^2(\sqrt{s})} = \frac{\chi_1^Q(T,\mu_B)}{\chi_2^Q(T,\mu_B)} \equiv R_{12}^Q, \tag{47a}$$

$$\frac{S_Q(\sqrt{s})\sigma_Q^3(\sqrt{s})}{M_Q(\sqrt{s})} = \frac{\chi_3^Q(T,\mu_B)}{\chi_1^Q(T,\mu_B)} \equiv R_{31}^Q. \tag{47b}$$

As in many other cases, one still has to worry about directly confronting thermodynamic QCD calculations performed in equilibrium and in the thermodynamic limit to the rapidly expanding dense matter created in heavy ion collisions in a small volume. In the analysis of higher order cumulants additional complications may arise from more complicated efficiency corrections and the limited experimental acceptance windows.[143–145]

In heavy ion collision experiments the only tunable parameter is the beam energy, $\sqrt{s}$. However, to gain access to the information regarding the QCD phase diagram this only tunable parameter needs to be related to the thermodynamic variables, e.g., temperature and baryon chemical potential. Traditionally, this $\sqrt{s} \leftrightarrow (T, \mu_B)$ mapping has been done by relying on the statistical hadronization model based analysis.[86] Recent advances in heavy ion experiments as well as in lattice QCD calculations have placed us in a unique situation where, for the first time, this mapping can now be obtained through direct comparisons between the experimental results and rigorous (lattice) QCD calculations. Recently, it has been shown[142,146] that by directly comparing lattice QCD calculations for R_{31}^Q (Eq. (47b)) and R_{12}^Q (Eq. (47a)) with their corresponding cumulant ratios measured in heavy ion experiments it is possible to extract the thermal parameters, namely the freeze-out temperature, T^f, and the freeze-out baryon chemical potential μ_B^f. The feasibility of such a procedure has been demonstrated in Refs. 123, 147, 148. Figure 17[149] illustrates a recent example of such a comparison and subsequent determination of the freeze-out parameters.

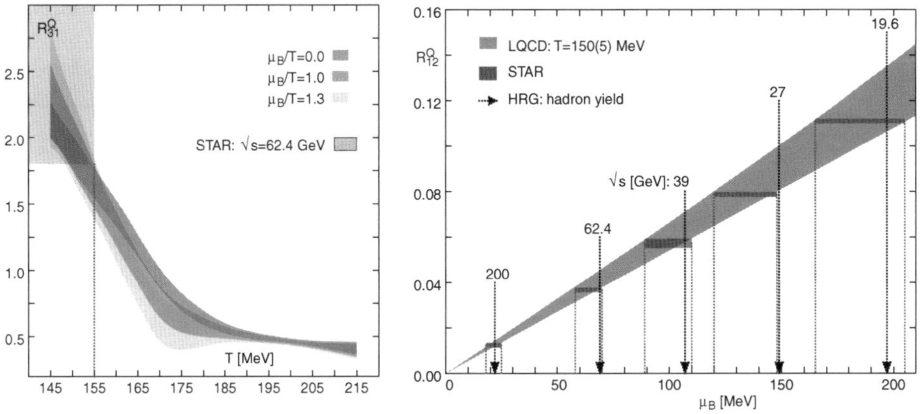

Fig. 17. (Color online) (Left) A comparison between the lattice QCD results[149] for R_{31}^Q and the STAR data[138] for $(S_Q\sigma_Q^3)/M_Q$ at $\sqrt{s} = 62.4\,\text{GeV}$. The overlap of the experimental results with the lattice QCD calculations provides an upper bound on the freeze-out temperature $T^f \leq 155\,\text{MeV}$. (Right) Lattice QCD results[149] for R_{12}^Q as a function of μ_B compared with the STAR data[138] for M_Q/σ_Q^2 in the temperature range $T^f = 150(5)\,\text{MeV}$. The overlap regions of the experimentally measured results with the lattice QCD calculations provide estimates for the freeze-out chemical potential μ_B^f for a given $\sqrt{s}$. The arrows indicate the corresponding values of μ_B^f obtained from the statistical hadronization model fits to experimentally measured hadron yields.[86]

As can be seen from Fig. 17, due to large errors on the experimental results for $(S_Q\sigma_Q^3)/M_Q$, at present, only an upper limit on the freeze-out temperature can be determined using the method described above. Thus, a complementary procedure for the determination of T^f, relying on a separate observable that can be extracted both from heavy ion experiments and lattice QCD calculations, is certainly welcome. Recently, such a complementary procedure for the determination of T^f has been proposed in Ref. 116. This procedure takes advantage of the fact that the initially colliding nuclei in heavy ion collisions are free of net strangeness. Thus, the conservation of strangeness under strong interaction ensures that the QGP medium created during the collisions of these heavy ions is also strangeness neutral.

By Taylor expanding the net strangeness density, $\langle n_S\rangle(\mu_B, \mu_S)$, in μ_B and μ_S and subsequently imposing the strangeness neutrality condition, $\langle n_S\rangle(\mu_B, \mu_S) = 0$, for a homogeneous thermal medium the strangeness chemical potential, μ_S, can be obtained as[116]

$$\frac{\mu_S}{\mu_B} = s_1(T) + s_3(T)\left(\frac{\mu_B}{T}\right)^2 + \mathcal{O}\left[\left(\frac{\mu_B}{T}\right)^4\right]. \tag{48}$$

The coefficients s_1, s_3, etc., consist of various generalized baryon, charge and strangeness susceptibilities defined at vanishing chemical potentials and can be calculated through standard lattice QCD computations at zero chemical potentials.[116] Figure 18 (left) shows the leading order contribution to μ_S/μ_B, i.e., $s_1(T)$. A comparison of the lattice result with the predictions from the hadron resonance gas model reveal that the inclusion of only experimentally observed hadrons, as listed by the Particle Data Group,[31] fails to reproduce the lattice results around the crossover region. Note that, while μ_S/μ_B is unique in QCD, for a hadron gas it depends on the relative abundances of the open strange baryons and mesons. For

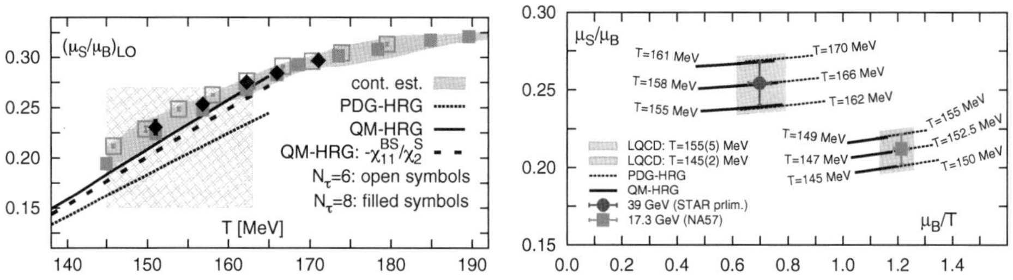

Fig. 18. (Color online) (Left) Lattice QCD results[116] for μ_S/μ_B at the leading order, i.e., $s_1(T)$ (see Eq. (48)). The dotted line (PDG-HRG) shows the results of a hadron resonance gas model containing only hadrons listed by the Particle Data Group.[31] The solid line (QM-HRG) depicts the result for a hadron gas when additional, experimentally yet unobserved, quark model predicted strange hadrons[150,151] are included. The shaded region indicates the chiral crossover region $T_c = 154(9)$ MeV. (Right) A comparison between experimentally extracted values of $(\mu_S^f/\mu_B^f, \mu_B^f/T^f)$ (filled symbols) and lattice QCD results for μ_S/μ_B (shaded bands).[116] The lattice QCD results are shown for $\mu_B/T = \mu_B^f/T^f$. The temperature range where lattice QCD results match with μ_S^f/μ_B^f provide the values of T^f, i.e., $T^f = 155(5)$ MeV and 145(2) MeV for $\sqrt{s} = 39$ GeV and 17.3 GeV, respectively.

fixed T and μ_B, a strangeness neutral hadron gas having a larger relative abundance of strange baryons over open strange mesons naturally leads to a larger value of μ_S. Astonishingly, the inclusion of additional, unobserved strange hadrons predicted within the quark model[150,151] provides a much better agreement with lattice results, hinting that these additional hadrons become thermodynamically relevant close to the crossover temperature.[116] As will be discussed in Sec. 7 , other lattice thermodynamics studies also indicate that additional, unobserved charm hadrons do become thermodynamically relevant close to the QCD crossover.[152]

On the other hand, the experimentally measured yields of the strangeness S anti-baryon to baryon ratios, R_H, at the freeze-out are determined by the thermal freeze-out parameters (T^f, μ_B^f, μ_S^f),[86]

$$R_H(\sqrt{s}) = \exp\left[-\frac{2\mu_B^f}{T^f}\left(1 - \frac{\mu_S^f}{\mu_B^f}|S|\right)\right]. \tag{49}$$

By fitting the experimentally measured values of R_Λ, R_Ξ and R_Ω, corresponding to $|S| = 1, 2$ and 3, the values of μ_S^f/μ_B^f and μ_B^f/T^f, as 'observed' in a heavy ion experiment at a given $\sqrt{s}$, can easily be extracted. Matching these experimentally extracted values of μ_S^f/μ_B^f with the lattice QCD results for μ_S/μ_B as a function of temperature, one can determine the freeze-out temperature T^f. Figure 18 (right) illustrates this procedure. Once again, the inclusion of additional unobserved strange hadrons in the hadron resonance gas model (QM-HRG) leads to very similar values of the freeze-out temperatures as obtained using the lattice data. However, including only the hadrons listed by the Particle Data Group[31] (PDG-HRG) yields freeze-out temperatures that are 5–8 MeV larger.

6. Transport Properties

As we have seen in previous sections the analysis of bulk thermodynamics and charge fluctuations provides plenty of evidence that thermodynamics of strong-interaction matter above the crossover transition temperature, T_c, and up to temperatures of about $(1.5-2)T_c$ is highly non-perturbative. In this temperature range the velocity of sound is significantly smaller than that in a non-interacting quark–gluon gas suggesting also significant modifications of its transport properties. One of the striking results of heavy ion experiments at RHIC and LHC is that strong-interaction matter in this temperature range shows features of a strongly coupled medium with a small shear viscosity to entropy density ratio and small diffusion coefficients. This led to the notion of strong-interaction matter close to the crossover transition being a 'nearly perfect liquid'.

Transport properties of strong-interaction matter can be explored in lattice QCD calculations through the analysis of current-current correlation functions, e.g., for some current $J(\tau, \vec{x})$,

$$G_J^{\alpha\beta}(\tau, \mathbf{p}) = \int d^3x \, e^{i\mathbf{p}\mathbf{x}} \langle J^\alpha(0,0) J^{\beta\dagger}(\tau, \mathbf{x})\rangle = \int_0^\infty \frac{d\omega}{2\pi} \rho_J^{\alpha\beta}(\omega, \mathbf{p}) \, K(\omega, \tau), \tag{50}$$

where the integration kernel $K(\omega,\tau)$ is

$$K(\omega,\tau) = \frac{\cosh(\omega(\tau - 1/2T))}{\sinh(\omega/2T)}, \tag{51}$$

and α, β are appropriately chosen (scalar, vector, tensor) indices specifying the current.

In Euclidean space–time these correlation functions have a spectral representation. The low frequency and low momentum structure of the spectral functions $\rho_J^{\alpha\beta}(\omega,\mathbf{p})$ then gives access to transport properties of strong-interaction matter while at higher frequencies the spectral functions provide information on in-medium properties of bound states. We will discuss the latter in the Sec. 7. Extracting the spectral function from a Euclidean correlation function is, however, difficult. This is due to the fact that Euclidean correlation functions are calculated in finite temperature lattice QCD only at a few discrete Euclidean time separations, e.g., $0 \le \tau \le N_\tau$.[d] It requires the application of inversion methods like the Maximum Entropy Method (MEM)[153] or a modeling of the structure of the spectral function at low frequencies.

According to the Kubo relations, which express the response of a medium to small perturbations, transport coefficients can be determined from spectral functions calculated in equilibrium thermodynamics. They are given by the slope of the corresponding spectral functions at vanishing energy. For instance, using in Eq. (50) the electromagnetic current $J^\mu(t,\mathbf{x}) \equiv J_V(t,\mathbf{x}) = \bar{\psi}(t,\mathbf{x})\gamma^\mu\psi(t,\mathbf{x})$ for a fermionic charge carrier (quarks) with charge e (e.g., $Q_d = -e/3$ for a down quark) one can extract the electrical conductivity σ which quantifies the response of the QGP to small perturbations induced by an electromagnetic field,

$$\sigma = \frac{C_{\rm em}}{6} \lim_{\omega\to 0} \lim_{\mathbf{p}\to 0} \sum_{i=1}^{3} \frac{\rho^{ii}(\omega,\mathbf{p},T)}{\omega}. \tag{52}$$

Here $C_{\rm em}$ is the sum of the square of the elementary charges of the quark flavor $f, C_{\rm em} = \sum_f Q_f^2$. The electrical conductivity and the vector spectral function are also related to the emission rate of soft thermal photons,

$$\lim_{\omega\to 0} \omega \frac{dR_\gamma}{d^3p} = \frac{3}{2\pi^2}\, \sigma(T)\, T\, \alpha_{\rm em}, \tag{53}$$

and the contribution of quark–anti-quark annihilation to the thermal dilepton rate,

$$\frac{dW}{d\omega\, d^3\mathbf{p}} = \frac{5\alpha_{\rm em}^2}{54\pi^3} \frac{1}{\omega^2(e^{\omega/T} - 1)} \rho_V(\omega,\mathbf{p},T), \tag{54}$$

where $\alpha_{\rm em}$ is the electromagnetic fine structure constant.

[d]Note that Eq. (51) is symmetric around $\tau = 1/2T$ and as a consequence only half the number of data points of the Euclidean temporal correlation function, i.e., $N_\tau/2$, provide independent information to extract the spectral function.

Using the vector current $J_V(\tau, \mathbf{x})$ for heavy quarks one obtains from Eq. (50) the corresponding spectral function of heavy quarkonium in the vector channel, ρ_V^{ii}, which allows to determine the heavy quark diffusion coefficient D,

$$D = \frac{1}{6\chi_2^h} \lim_{\omega \to 0} \lim_{\mathbf{p} \to 0} \sum_{i=1}^{3} \frac{\rho_V^{ii}(\omega, \mathbf{p}, T)}{\omega}. \tag{55}$$

Here χ_2^h is the heavy quark number susceptibility which is defined through the zeroth component of the temporal meson correlator in the vector channel. For example, for charm quarks this is just the net charm number susceptibility introduced in Eq. (12). Similarly, in the light quark sector, the electric charge diffusion coefficient D_Q is defined as the ratio of σ to electric charge susceptibility χ_2^Q defined in Eq. (19), i.e., $D_Q = \sigma/\chi_2^Q$.

The heavy quark diffusion coefficient D can be related to the momentum diffusion coefficient κ and drag coefficient η through the Einstein relation

$$D = \frac{2T^2}{\kappa} = \frac{T}{M\eta}. \tag{56}$$

It is used to describe the Brownian motion of a heavy quark in the hot medium and can also be related to the ratio of shear viscosity over entropy density η/s.[154]

The shear (η) and bulk (ζ) viscosities of strong-interaction matter are extracted from correlation functions of the energy–momentum tensor, $T^{\mu\nu} = F^{\mu\alpha}F^{\nu\alpha} - \frac{1}{4}\delta_{\mu\nu}F^{\rho\sigma}F^{\rho\sigma}$. That is, using the current $J^{\mu\nu}(\tau, \mathbf{x}) \equiv T^{\mu\nu}(t, \mathbf{x})$ in Eq. (50), one obtains

$$\eta = \pi \lim_{\omega \to 0} \lim_{\mathbf{p} \to 0} \frac{\rho^{12,12}(\omega, \mathbf{p})}{\omega}, \tag{57}$$

$$\zeta = \frac{\pi}{9} \lim_{\omega \to 0} \lim_{\mathbf{p} \to 0} \sum_{k,l} \frac{\rho^{kk,ll}(\omega, \mathbf{p})}{\omega}. \tag{58}$$

At present only a few calculations of transport coefficients have been performed in QCD with dynamical quark degrees of freedom. Most calculations utilized the quenched approximation at vanishing baryon number density. In the following subsections we will summarize current results for the transport coefficients introduced above. For further details on the calculation of transport coefficients on the lattice and earlier results see, for instance, Ref. 155.

6.1. *Electrical conductivity, charge diffusion and dilepton rates*

Probably the best analyzed spectral function is that of the light quark vector current which gives access to the electrical conductivity, $\sigma(T)$ (Eq. (52)), soft photon emission rates (Eq. (53)) and thermal dilepton rates (Eq. (54)) as well as the electric charge diffusion constant $D_Q = \sigma/\chi_2^Q$. The vector spectral function $\rho_V(\omega) \equiv \rho_V(\omega, 0, T)$ has been calculated first in quenched QCD[156] using the MEM approach[153,157] in which the analysis had only been constrained at large frequencies through a default model based on the free fermion spectral function and the largest lattices used only had $N_\tau = 16$ points in the temporal direction. It has, however, been noticed[158] that this is not sufficient to get access to transport coefficients, which require the spectral functions at low frequencies to be linear in ω. Furthermore, a larger time extent[159] is needed to control the low frequency part of the spectral functions.

First continuum extrapolated results for the vector spectral function were obtained in quenched QCD using clover-improved Wilson fermions[159] at $T \simeq 1.4\,T_c$. These results have been extended to three different temperatures recently.[160–162] In these calculations the thermal spectral functions are obtained by fitting the continuum-extrapolated two-point Euclidean correlation functions to an ansatz for the spectral function that is motivated by kinetic theory and perturbation theory

$$\rho^{\mathrm{model}}(\omega) = \chi_q c_{\mathrm{BW}} \frac{\omega\Gamma}{\omega^2 + \Gamma^2/4} + \frac{3}{2\pi}(1+k)\,\omega^2 \tanh\left(\frac{\omega}{4T}\right). \tag{59}$$

Here χ_q is similar to χ_2^h appearing in Eq. (55) but for light quarks and it is obtained by summing the zeroth component of the light temporal vector correlation function over space–time coordinates. The unknown parameters are thus the amplitude c_{BW} and the width Γ of the Breit–Wigner function as well as the parameter k, which parametrizes deviations from the free theory behavior at large ω. The thermal spectral functions obtained in this way and the resulting thermal dilepton rates are shown in Fig. 19. In the high frequency region, i.e., for $\omega/T \gtrsim 7$, all the thermal spectral functions at the three temperatures above T_c are well described by leading order perturbation theory (Born rate) as well as Hard Thermal Loop (HTL) calculations. At small frequencies the lattice QCD results are significantly enhanced over the Born rate, indicating the presence of a transport peak, but are smaller than the HTL rates. In fact, the latter rises too rapidly at low frequency and would give rise to a divergent electrical conductivity.

The finite intercept of $\rho(\omega)/(\omega T)$ at $\omega = 0$ seen in Fig. 19 gives the electrical conductivity (see Eq. (52)), $\sigma/(C_{\mathrm{em}}T) \in [0.21, 0.44]$.[162] In these calculations σ/T

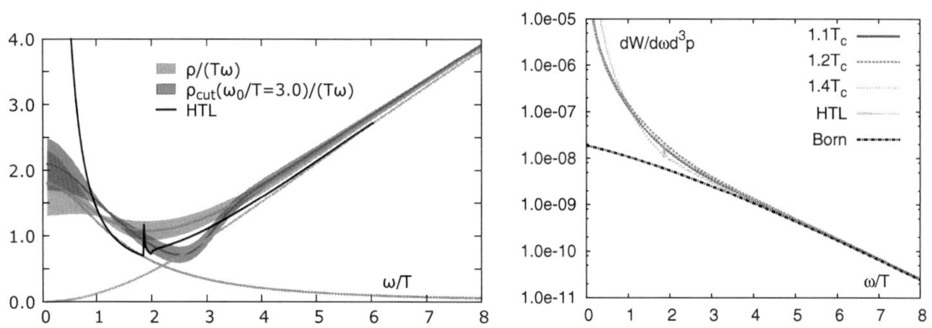

Fig. 19. (Color online) (Left) Thermal spectral function at $1.1T_c$ calculated in quenched QCD.[162] (Right) Temperature dependence of the quark–antiquark annihilation contribution to the thermal dilepton rate above T_c.[162]

does not show any significant temperature dependence[e], which may be due to the use of the quenched approximation. The results are consistent with a calculation that used staggered fermions, but no renormalized vector currents,[158] $\sigma/(C_{\rm em}T) \simeq 0.4(1)$ for $1.5 \lesssim T/T_c \lesssim 2.25$. These results are, however, substantially smaller than earlier calculations performed on coarse lattices.[163]

Recently the calculation of the electrical conductivity has also been performed in QCD with dynamical quarks. The first calculation was done in 2-flavor QCD using $\mathcal{O}(a)$ improved Wilson quarks corresponding to $m_\pi \simeq 270\,\mathrm{MeV}$. The electrical conductivity of the quark-gluon plasma at $T \simeq 250\,\mathrm{MeV}$ was found to be similar to the quenched result, $\sigma/(C_{\rm em}T) = 0.40(12)$.[164]

First results from a calculation in $(2+1)$-flavor QCD performed on anisotropic lattices with quark masses corresponding to a pion mass $m_\pi \simeq 380\,\mathrm{MeV}$[165,166] are shown in Fig. 20. The simulation is performed at one value of the lattice cut-off. Varying the temporal extent of the lattice, N_τ, from between 48 and 16 a temperature range from 0.63 T_c to 1.90 T_c is covered. The left panel of this figure shows temperature dependence of $\sigma/(C_{\rm em}T)$ (square points). It stays constant within errors in the transition region and increases with temperature for $T > T_c$. Also shown in the left panel of Fig. 20 are electrical conductivities obtained in quenched QCD[158,159,162] and 2-flavor QCD.[164] The right panel of Fig. 20 shows the temperature dependence of charge diffusion coefficient D_Q multiplied by $2\pi T$. $2\pi T D_Q$ shows a dip near T_c which arises from the rapid exponential drop of the electric

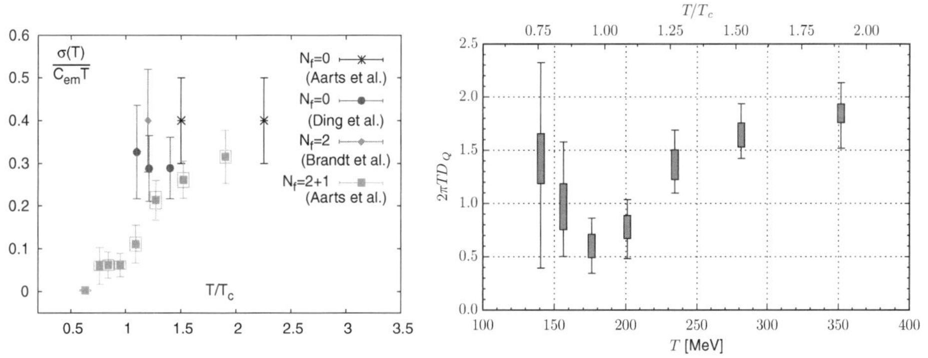

Fig. 20. (Color online) (Left) Lattice QCD results for the electrical conductivity $\sigma/(C_{\rm em}T)$ as function of T/T_c including results obtained in quenched QCD,[158,159,162] 2-flavor QCD with $m_\pi \simeq 270\,\mathrm{MeV}$[164] and (2+1)-flavor QCD with $m_\pi \simeq 380\,\mathrm{MeV}$.[165] Note that the values of T_c obtained in these calculations are different, i.e., $T_c^{N_f=0} \simeq 270\,\mathrm{MeV}$, $T_c^{N_f=2} \simeq 208\,\mathrm{MeV}$ and $T_c^{N_f=2+1} \simeq 185\,\mathrm{MeV}$. (Right) Temperature dependence of charge diffusion coefficient $2\pi T D_Q = 2\pi T\sigma/\chi_2^Q$. The results are obtained in (2+1)-flavor QCD calculations on anisotropic lattices at one value of the spatial lattice cut-off, $a \simeq 0.12\,\mathrm{fm}$ and with light quark masses corresponding to $m_\pi \simeq 380\,\mathrm{MeV}$.[165]

[e]This is also reflected in the temperature dependence of the temporal correlation functions divided by T^3 which are temperature independent for this temperature range.[162]

charge susceptibility at low temperature where the electrical conductivity of a pion gas[167] remains nonzero and vanishes only like $\sqrt{T}$.

6.2. *Heavy quark diffusion*

Contrary to earlier expectations it has been found in heavy ion collision experiments at RHIC[168,169] and LHC[170] that heavy quark mesons show a substantial elliptic flow that is comparable to that of light quark mesons. Moreover, also somewhat unexpectedly, heavy quarks are found to lose a significant amount of energy while traversing through the QGP. The latter has been called the heavy quark energy loss puzzle. Phenomenological explanations of these phenomena try to model the heavy quark diffusion in a hot and dense medium. This requires knowledge about the heavy quark diffusion coefficients D[154,171,172] which can be determined in lattice QCD calculations.

The charm quark diffusion coefficient has been calculated in quenched QCD at three temperatures in the deconfined phase[173] using lattices with temporal extent $N_\tau = 48, 32$ and 24 at a fixed value of the cut-off. Results from this calculation are shown in Fig. 21. The left panel shows the transport peak in the charmonium vector spectral function at three different temperatures above T_c. These transport peaks are obtained using a MEM analysis where an ansatz of Lorentzian form is used as the prior information for the low frequency part of the spectral function, i.e., $\omega\eta/(\omega^2+\eta^2)$. The intercept at $\omega/T = 0$ gives the charm quark diffusion coefficient

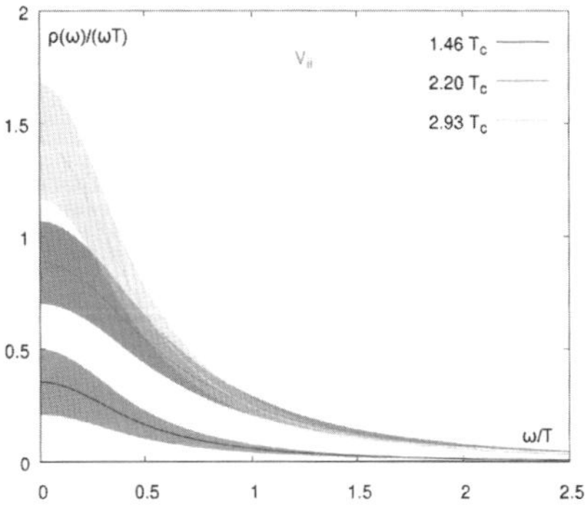

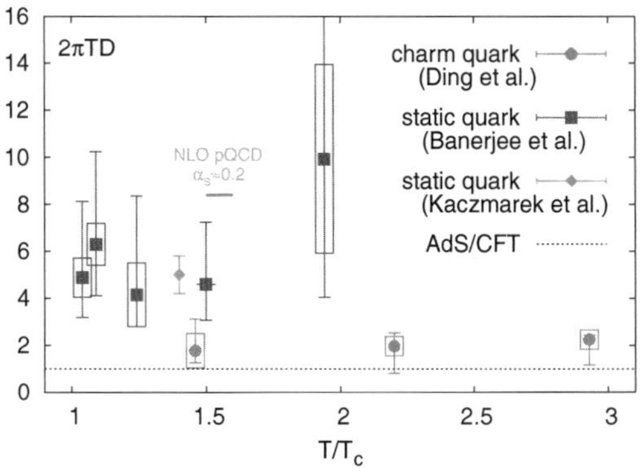

Fig. 21. (Color online) (Left) The low frequency transport part of the charmonium vector spectral function obtained from a MEM analysis of vector correlation functions in quenched lattice QCD.[173] The central lines are mean values while the bands reflect the statistical uncertainties. (Right) Temperature dependence of heavy quark diffusion coefficients multiplied by $2\pi T$ in pure gluonic matter. The charm diffusion coefficients are obtained from the spectral functions shown in the left plot.[173] Also shown are results for the static quark diffusion coefficient obtained by using the lattice discretized versions of Heavy Quark Effective theory.[176,177] The boxes show the statistical error while the error bars reflect the systematic uncertainties. The horizontal dotted line labels the value of $2\pi TD$ in the heavy quark limit from the AdS/CFT correspondence[175] and the short, horizontal solid line indicates the value of $2\pi TD$ from next-to-leading order pQCD calculations at $\alpha_s \simeq 0.2$.[174]

(c.f. Eq. (55)). It is found that $2\pi TD \simeq 2$ and, with current understanding of uncertainties, this product is independent of temperature. These values of D are much smaller than the results from perturbative calculations[174] and are very close to the value obtained from the AdS/CFT correspondence.[175] These different results are compared in the right panel of Fig. 21.

A direct determination of the diffusion coefficient for bottom quarks from bottomonium correlation functions is difficult as it would require quite small lattice spacings to accommodate the bottomonium states on the lattice. The method of choice here is to start with the infinite quark mass limit where one can use the Heavy Quark Effective Theory (HQET) to compute the diffusion coefficient of a static quark. In this approach the propagation of a heavy quark carrying color charge and its response to a colored Lorentz force can be described using linear response theory. This leads to the spectral analysis of a "color-electric correlator"[178,179]

$$G_E(\tau) = -\frac{1}{3}\sum_{i=1}^{3} \frac{\langle \mathrm{ReTr}[U(1/T;\tau) gE_i(\tau,\mathbf{0})\, U(\tau;0) gE_i(0,\mathbf{0})]\rangle}{\langle \mathrm{Re\,Tr}[U(1/T;0)]\rangle}, \tag{60}$$

where $U(\tau_2;\tau_1)$ is a Wilson line in the Euclidean time direction and $gE_i(\tau,\mathbf{x})$ denotes the color-electric field at time–space $(\tau,\mathbf{x})$. The momentum diffusion coefficient κ can then be extracted from the slope of the corresponding spectral function in the limit of vanishing frequency ω,

$$\kappa/T^3 = \lim_{\omega\to 0} \frac{2\,\rho_E(\omega)}{\omega T^2}. \tag{61}$$

Here $\rho_E(\omega)$ is the corresponding spectral function that is related to $G_E(\tau)$ via Eq. (50). In the non-relativistic limit, i.e., for a heavy quark mass $M \gg \pi T$, the momentum diffusion coefficient κ is related to the (spatial) heavy quark diffusion coefficient D via Eq. (56).

The color-electric correlator $G_E(\tau)$ is defined in terms of gluonic observables only. It thus suffers from a small signal to noise ratio, a problem similar to what one encounters in the calculation of viscosities[180] (see next subsection). In both cases efficient noise reduction techniques are needed for a calculation of the gluonic correlation functions. In a recent study[176] the correlator $G_E(\tau)$ has been calculated using multi-level updates[180,181] as well as the link integration technique.[182,183] The improved correlator $G_{\mathrm{imp}}(\tau T)$ obtained this way at $T \simeq 1.4T_c$ for four different values of the lattice cut-off in quenched QCD has been extrapolated to the continuum limit. Results are shown in the left panel of Fig. 22. In order to extract the momentum diffusion coefficient from this correlator, an ansatz for the spectral function has been used.[176] The corresponding fit is shown in the right panel of Fig. 22. It yields,

$$\kappa/T^3 = 2.5(4) \quad \text{at } T \sim 1.4T_c. \tag{62}$$

This corresponds to $2\pi TD = 5.0(8)$ which is consistent with similar lattice QCD studies using HQET.[177] However, it is a factor 2 to 3 larger than the charm quark diffusion coefficient discussed above[173] (see Fig. 21).

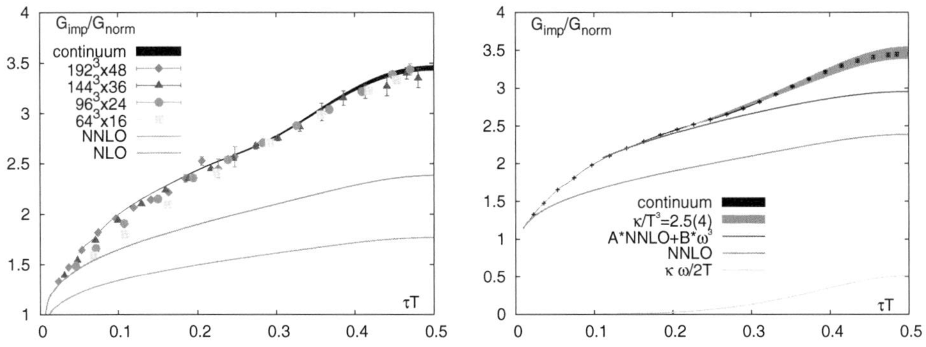

Fig. 22. (Color online) (Left) Continuum extrapolation of improved color-electric correlator $G_E(\tau) \equiv G_{\rm imp}(\tau)$ (see Eq. (60)) normalized to the high temperature, free field theory result $G_{\rm norm}(\tau)$ and the comparison with NNLO and NLO perturbative QCD calculations at $1.4T_c$ on quenched lattice QCD.[176] (Right) χ^2 fit to the continuum extrapolated $G_{\rm imp}/G_{\rm norm}$ using an ansatz $\rho_E^{\rm model}(\omega) = \max\{A\rho_{\rm NNLO}(\omega) + B\omega^3, \frac{\omega\kappa}{2T}\}$. Also shown are the separate contributions to the correlator arising from the different parts of the fit ansatz for the spectral function.[176]

The heavy quark diffusion coefficient D has also been calculated in perturbative QCD in both leading and next-to-leading order as well as from the AdS/CFT correspondence. At $\alpha_s \approx 0.2$ the leading order pQCD calculation[184] gives $2\pi TD \approx 71.2$ while $2\pi TD \approx 8.4$ is obtained in next-to-leading order calculations.[174] In the strong coupling limit $2\pi TD = 1$ is obtained from the AdS/CFT correspondence[175] (see Fig. 21).

6.3. *Shear and bulk viscosities*

The computation of shear and bulk viscosities has been attempted already in the '80s.[185] However, it turned out that the signal to noise ratio in calculating Euclidean energy–momentum correlation functions is very small and high statistics or improved algorithms are needed. The latter became available for quenched QCD calculations with the multi-level algorithms[181] which have been used for the calculation of shear and bulk viscosities.[180,186] Results from such an analysis by Meyer[180] and those with direct high statistics calculations by Nakamura and Sakai[187] are shown in Fig. 23. The ratio obtained in the former calculation are smaller than those obtained in the earlier direct simulations; $\eta/s = 0.102(56)$ at $1.24T_c$ and $0.134(33)$ at $1.65T_c$ obtained in quenched QCD on lattices with temporal extent $N_\tau = 8$.[180] This shows that noise reduction techniques like the multi-level algorithm are mandatory for a successful calculation of viscosities. Results on the bulk viscosity to entropy ratio ζ/s show that this ratio rapidly becomes small above T_c.[188] At $T \gtrsim 1.2T_c$ it is smaller than η/s and, in fact, within errors it is consistent with zero.[180,187,188]

Still the calculations of viscosities are performed on lattices with rather small temporal extent compared to those used in calculations of the electrical conductivity and diffusion constants. Systematic uncertainties in these calculations need to be

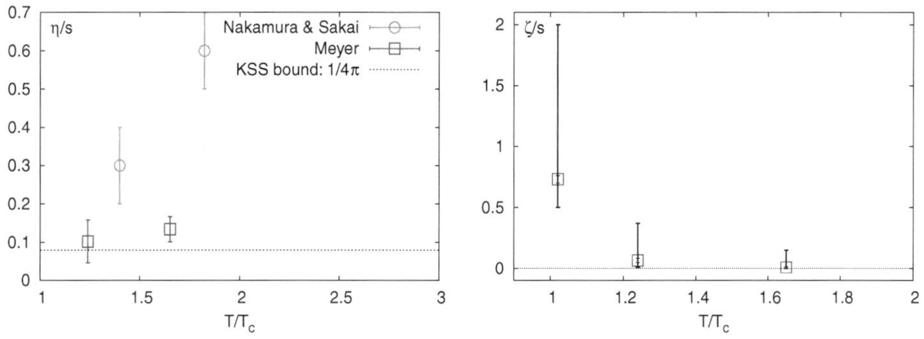

Fig. 23. (Color online) (Left) Temperature dependence of the shear viscosity, η/s, of gluonic matter obtained by using noise-reduction techniques (squares)[180] and direct high statistical calculations (circles)[187] on $N_\tau = 8$ lattices. The dotted line corresponds to the conjectured lower bound of η/s from AdS/CFT correspondence.[175] (Right) Temperature dependence of the bulk viscosity, ζ/s, of gluonic matter.[188] The square points include the statistical uncertainties while the solid black bars denote the systematic uncertainties. Results for the same quantity from Ref. 187 which are not shown are consistent with zero at $T \in [1.4T_c, 1.8T_c]$. Data are taken from Refs. 180, 187 and 188.

better controlled in future. Here it will also be helpful to combine lattice QCD calculations with information from analytic approaches that put constraints on the structure of spectral functions, e.g., QCD sum rules.[189–191]

6.4. *Transport coefficients of second order hydrodynamics*

Due to the importance of viscous effects in the evolution of hot and dense matter created in heavy ion collision there is considerable interest in extending the hydrodynamic modeling beyond leading order gradient expansions of the energy–momentum tensor. A second order gradient expansion is parameterized by additional transport coefficients,[192] which may become accessible to lattice QCD calculations.[193–195] One of these new transport coefficients[f] is κ, which controls the momentum dependence of space–like correlations of the energy–momentum tensor. It is defined as the leading order coefficient of the Taylor expansion of the retarded correlator of energy–momentum tensor in momentum $\mathbf{p}$ about zero,[191,196]

$$\begin{aligned} G^R(\omega = 0, \mathbf{p}) &= \int_{-\infty}^{\infty} dt d\mathbf{x}\, e^{i\mathbf{p}\mathbf{x}} \langle [T_{12}(\tau, \mathbf{x}), T_{12}(0,0)] \rangle \theta(t) \\ &= G^R(0) + \frac{\kappa}{2} |\mathbf{p}|^2 + \mathcal{O}(|\mathbf{p}|^4). \end{aligned} \tag{63}$$

Since the difference between the retarded correlator G^R and the Euclidean correlator at vanishing frequency $\omega = 0$ is just a contact term, which has been shown to be momentum independent,[197] κ can directly be extracted from the corresponding

[f]Although this particular second order coefficient is also called κ it should be noted that it is not related to the momentum diffusion coefficient κ introduced in Eq. (56).

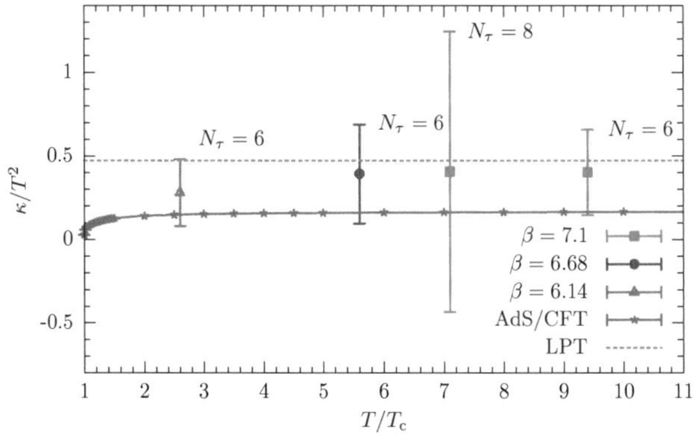

Fig. 24. (Color online) Transport coefficient of second order hydrodynamics κ/T^2 as a function of T. The lines denote the results from AdS/CFT correspondence[196] and lattice perturbation theory (LPT),[198] respectively.

Euclidean correlator,

$$\kappa = 2 \lim_{|\mathbf{p}| \to 0} \frac{dG^E(\omega = 0, \mathbf{p})}{d|\mathbf{p}|^2}. \tag{64}$$

This is considerably simpler to compute than first order coefficients such as the shear or bulk viscosity as it does not require analytic continuation from imaginary time to real time. The only complication is that large spatial lattice sizes are required to get access to small momenta in the lattice QCD calculation.

The second order hydro-coefficient κ has been computed in quenched QCD on lattices with temporal extent $N_\tau = 8$ and 6 and a large spatial size.[198] Results from this calculation are shown in Fig. 24. It yields $\kappa/T^2 = 0.36(15)$. Within the current large errors no further temperature dependence is observed in the temperature range $2T_c < T < 10T_c$ and results are compatible with both perturbative QCD[191,194] and AdS/CFT results.[199]

7. Open Heavy Flavors and Heavy Quarkonia

Owing to their large masses heavy flavors, such as charm and bottom, are produced only during the earliest stages of a heavy ion collision and get affected by the hot and dense medium during its entire evolution. They then emerge as hadrons with open heavy flavors or as quarkonia. Thus, heavy flavors provide unique identifiable probes against which the temperature and coupling of the medium created during heavy ion collisions can be calibrated. The melting of charmonium states has long been considered as an important signature for the formation of quark–gluon plasma.[18] Understanding its production rates and abundance can provide detailed information on properties of the matter formed in a heavy ion collision. We will discuss current results on the temperature dependence of quarkonium bound states in Sec. 7.2. In addition, understanding the fate of open flavor bound states also is of importance for the understanding of the strongly interacting regime in the

QGP close to the transition temperature. The question whether or not heavy flavor mesons can survive above T_c plays a crucial role in the phenomenological modeling of the heavy quark energy loss.[200–202] Lattice QCD results on the dissociation of open charm hadrons will be discussed in the next subsection.

7.1. *Melting and abundances of open charm hadrons*

Similar to the case of strangeness,[98] discussed in Sec. 3.2, the deconfinement or melting of open charm hadrons can be probed[152] by using fluctuations of the charm quantum number (C) and its correlations with baryon number (B) fluctuations. To this end one may consider appropriately chosen ratios of correlations between net-baryon and net-charm number fluctuations. One such ratio is, for instance, $\chi_{22}^{\rm BC}/\chi_{13}^{\rm BC}$.[152] Since at low temperatures, for a gas of uncorrelated charmed hadrons, the thermodynamic contributions of the much heavier $|C| = 2, 3$ baryons can be safely neglected, this particular ratio is dominated by $|C| = 1$ baryons and thus yields $\chi_{22}^{\rm BC}/\chi_{13}^{\rm BC} = |B| = 1$. On the other hand, at high temperatures, for a free gas of massive charm quarks with masses much larger than T, the above ratio is[g] $\chi_{22}^{\rm BC}/\chi_{13}^{\rm BC} = |B| = 1/3$. Thus, this quantity can be viewed as a measure for the baryon number associated with the predominant degrees of freedom that are carriers of the charm quantum number in a given temperature region. Its behavior can indicate the liberation of quark-like degrees of freedom with fractional baryon number.

Figure 25 (left)[152] shows that close to the chiral crossover temperature, $T_c = 154(9)$ MeV, the quantity $\chi_{22}^{\rm BC}/\chi_{13}^{\rm BC}$ starts deviating from unity. This indicates that beyond this temperature fractionally charged quark-like degrees of freedom start appearing and open charm hadrons start to melt. The left panel of Fig. 25 also shows a comparison with similar quantities involving both baryon-strangeness correlations as well as baryon-electric charge correlations. The first of these quantities is sensitive to the deconfinement of strange quarks and the second one is also sensitive to the deconfinement of light up and down quarks. It is clear from the figure that the onset of deconfinement for the charm, strange as well as the up/down quarks happens in the same chiral crossover region.

The right panel of Fig. 25 shows ratios of generalized susceptibilities that are constructed such that in a gas of uncorrelated hadrons the numerator would correspond to the partial pressure of baryons with quantum numbers selected by the other indices, while the denominator would correspond to the partial pressure of corresponding mesons. For example, $\chi_{13}^{\rm BC}$ will give the partial pressure of charmed baryons and $\chi_4^C - \chi_{13}^{\rm BC}$ filters out the corresponding partial pressure of charmed mesons. The ratios $\chi_{13}^{\rm BC}/(\chi_4^C - \chi_{13}^{\rm BC})$, $\chi_{112}^{\rm BQC}/(\chi_{13}^{\rm QC} - \chi_{112}^{\rm BQC})$ and $-\chi_{112}^{\rm BSC}/(\chi_{13}^{\rm SC} - \chi_{112}^{\rm BSC})$ thus correspond to ratios of partial pressures arising from charmed baryons to that from open charm mesons, charged-charmed baryons to open charm charged

[g] Here one makes use of the fact that the Boltzmann gas approximation is still a good approximation for a charm quark gas at a few times the transition temperature. This is the temperature range of interest to heavy ion collision experiments.

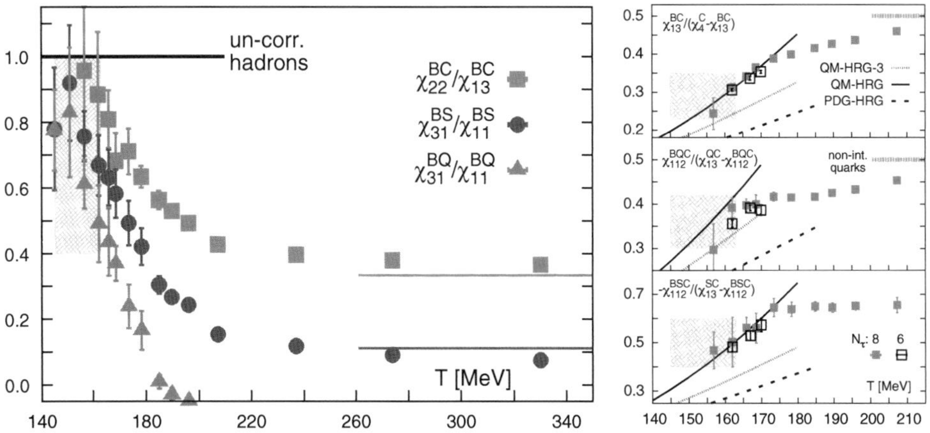

Fig. 25. (Color online) (Left) Ratios between cumulants of correlations of net baryon number with net charm, net strangeness and net electric charge fluctuations.[152] Deviations of these ratios from unity in the crossover region (shaded band) suggest that the onset of the melting/deconfinement of open charm hadrons as well as open strange hadrons and hadrons with light up/down quarks starts in the same temperature range (see text for details). (Right) Thermodynamic contributions of all charmed baryons (top), all charged charmed baryons (middle) and all strange charmed baryons (bottom) relative to that of corresponding charmed mesons.[152] The dashed lines (PDG-HRG) are predictions for an uncorrelated hadron gas using only the PDG states.[31] The solid lines (QM-HRG) are similar HRG predictions including also the states predicted by the quark model.[203,204] The dotted lines (QM-HRG-3) are the same QM predictions, but only including states having masses less than 3 GeV.

mesons and strange-charmed baryons to strange-charmed mesons, respectively. As can be seen from Fig. 25 (right), results for these ratios at low temperatures are well described by a hadron gas that uses open charm resonances obtained in quark model calculations (QM-HRG) while, not surprisingly, it differs strongly from a simple hadron resonance gas based only on the very few experimentally known charmed hadrons (PDG-HRG). This observation is similar to that made for strange hadron fluctuations in Sec. 5 (right panel of Fig. 18). These observations clearly provide evidence for contributions from experimentally yet unobserved charmed/strange hadrons to charm and strangeness fluctuations in the vicinity of the crossover transition temperature. In additions, the agreement between lattice QCD data and QM-HRG model calculations below T_c and the onset of deviations just in the transition region also suggests that open charm hadrons start to dissociate in the crossover region. This conclusion is consistent with that drawn from the temperature dependence of the ratio in the left panel of Fig. 25, which has the advantage of being independent of details of the charmed hadron spectrum.

7.2. *Heavy quarkonia*

Inside the QGP the interaction between a heavy quark–anti-quark pair gets weakened due to the screening effects of the intervening deconfined colored medium.

This telltale signature of the presence of a color deconfined medium is expected to be manifested through melting of heavy quarkonium states,[18] i.e., bound states of a quark–anti-quark pair. Melting of heavy quarkonium states in the QGP is also expected to follow a distinctive sequential pattern with the smallest, most tightly bound quarkonium state surviving up to the highest temperature, effectively serving as a 'thermometer' to probe the temperature of the medium created at RHIC (for a recent review see Ref. 205). Since quarkonia do not carry flavor quantum numbers, unlike the open heavy flavor hadrons, their melting cannot be accessed in lattice QCD calculations through quantum number correlation based studies as discussed in Sec. 7.1. Currently, at least three approaches are being actively pursued to study thermal modifications of heavy quarkonia on the lattice. The first approach is to use potential models with heavy quark potentials computed on the lattice. The second approach relies on the extraction of quarkonium spectral functions from Euclidean temporal correlators. The third approach involves spatial correlation functions of quarkonia and the study of their in-medium screening properties. Here, we summarize the recent progresses and developments concerning these three approaches. More details about earlier lattice results can also be found in Refs. 206–205.

7.2.1. *Heavy quark potential*

Properties of heavy quarkonia in the vacuum have been successfully described by the potential model approach. In this approach, the interaction between a heavy quark pair forming the quarkonium is described by an instantaneous potential.[208,209] Due to its success at zero temperature, the potential model approach has also been applied at nonzero temperature, by making the potential between the heavy quarks temperature dependent.[210] The temperature dependent potentials used in these calculations are based either on model calculations or on finite temperature lattice QCD results for the free energy and the internal energy. However, justification for the use of these potentials in a Schrödinger equation is mostly based on phenomenological arguments and has no rigorous connections to the real-time evolution of heavy quarkonia described by the Schrödinger equation.

Recently, a non-relativistic effective theory approach at nonzero temperature, relying on a particular hierarchy of the relevant scales, has been developed.[211–213] Generally by integrating out the hard energy scale, i.e., the rest mass of the heavy quark m, Non-Relativistic QCD (NRQCD) effective theory[214,215] is obtained, and by further integrating out the soft scale, i.e., the typical momentum exchange between the bounded quarks of $\mathcal{O}(mv)$, so-called potential non-relativistic QCD (pNRQCD), is obtained.[216] In pNRQCD a heavy quark bound state can be described by a two-point function satisfying a Schrödinger equation,[211–213]

$$i\partial_t D^{>}_{NR}(t,\vec{r}) = \left(-\frac{1}{m}\nabla_r^2 + V(t,r)\right) D^{>}_{NR}(t,\vec{r}), \tag{65}$$

where $D^{>}_{NR}(t,\vec{r})$ is the real-time forward heavy quark pair correlation function in the non-relativistic limit. The potential $V(t,r)$ in such a Schrödinger equation is well defined and turns out to be complex valued.[211–213] Its real part reflects color Debye screening effects while the imaginary part is related to Landau damping, i.e., the scattering of quarks with the constituents of the medium and the absorption of gluons from the medium via singlet-octet transition.[213] It has been shown in Ref. 217 that in the limit $m \to \infty$ the leading part of the potential $V(t,r)$ can be obtained as

$$V(r) = \lim_{t\to\infty} \frac{i\partial_t W(t,r)}{W(t,r)}, \tag{66}$$

where $W(t,r)$ is the real-time thermal Wilson loop. Through a Fourier transformation one obtains,

$$W(t,r) = \int_{-\infty}^{+\infty} d\omega\, \rho(\omega,r)\, e^{-i\omega t}, \tag{67}$$

where $\rho(\omega,r)$ is the spectral function of the real-time Wilson loop. The above equation can be analytically continued to the Euclidean time, giving

$$W(\tau,r) = \int_{-\infty}^{+\infty} d\omega\, \rho(\omega,r)\, e^{-\omega\tau}. \tag{68}$$

In Eq. (68) the Wilson loop, $W(\tau,r)$ is the usual Euclidean-time Wilson loop which can be obtained from lattice QCD calculations. As if there exists a well defined lowest lying peak structure in the spectral function $\rho(\omega,r)$ it would dominate the dynamics in the Wilson loop in the late time limit thus giving to the complete information of the heavy quark potential as shown in Eq. (66). However, as argued in Ref. 218 the late time dynamics is not completely separated from the early time non-potential physics and to take these effects into account a skewed Lorentzian function including the early time contribution for $\rho(\omega,r)$ is required to fit the Wilson loop. The position and width of this skewed Lorentzian peak are connected to the real and the imaginary parts of the potential, respectively. Thus, by extracting the spectral function, $\rho(\omega,r)$, from the Euclidean-time Wilson loop,[h] $W(\tau,r)$, obtained from lattice QCD calculations and by subsequently fitting this $\rho(\omega,r)$ with a skewed Lorentzian ansatz the real and the imaginary part of the heavy quark potential, $V(r)$, can be obtained.

This procedure has been tested using the Wilson loop calculated using hard thermal loop perturbation theory[219] as well as for the pure gauge theory.[220] Figure 26 shows the real (left) and imaginary (right) parts of the heavy quark potential in the pure gluonic medium.[220] The real part of the heavy quark potential turns out to be close to the heavy quark singlet free energy, defined through the

[h]As illustrated in Ref. 219 in order to avoid the cusp divergence and the large noise to signal ratio in computations of Wilson loops, some other observables, e.g., Wilson line correlation functions defined in the Coulomb gauge, can be a substitution and are actually used in lattice calculations.

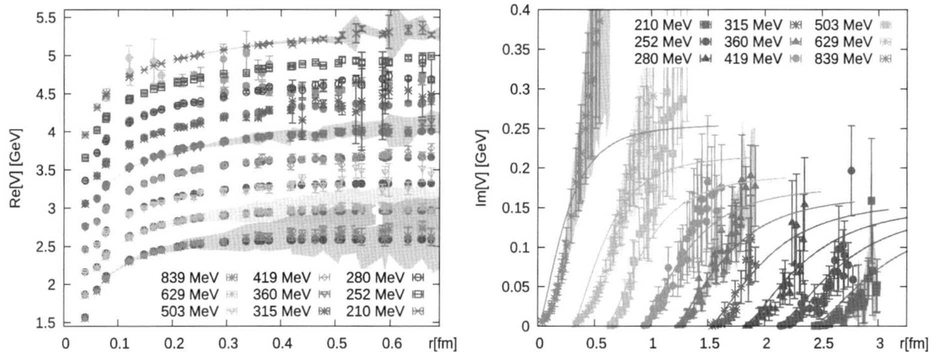

Fig. 26. (Color online) Real (left) and imaginary (right) parts of the heavy quark potential calculated in quenched lattice QCD.[220] Values are shifted for better visibility. Gray circles denote color singlet free energies and solid lines are leading order hard thermal loop results.

spatial correlation function of the Polyakov loop and its conjugate,[221] at all temperatures. At high temperatures the imaginary part of the heavy quark potential is close to the leading order hard thermal loop result, but at low temperatures it lies below hard thermal loop result.

Using the same procedure the real part of the heavy quark potential has also been extracted in (2+1)-flavor QCD[220] on lattices with temporal extent of only $N_\tau = 12$. Of course, on such small lattices it is hardly possible to extract reliable information on the imaginary part of the heavy quark potential. Nonetheless, similar to the quenched limit result the real part of the heavy quark potential is found to be close to the singlet free energies. However, string breaking is not observed at $T < T_c \approx 174\,\mathrm{MeV}$ up to ~1.2 fm. This could be due to the large pion mass ($m_\pi \approx 300\,\mathrm{MeV}$) employed in this exploratory QCD calculation.

The heavy quark potential has also been extracted from the Wilson line correlation function calculated on $48^3 \times 12$ lattices at $T = 250\,\mathrm{MeV}$ and $305\,\mathrm{MeV}$ using a fit ansatz motivated by hard thermal loop calculations.[222] These calculations used HISQ action with quark masses that would correspond to $m_\pi \approx 160\,\mathrm{MeV}$ in the continuum limit. In this calculation the real part of the potential was found to be equal or larger than the singlet free energies at these two temperatures and it always was smaller than the zero temperature potential. Moreover, the imaginary part is found to be of similar size as predicted by leading order hard thermal loop perturbation theory.[222]

7.2.2. *Spectral functions of quarkonia*

The spectral function of quarkonium can be extracted from lattice QCD calculations of the two-point correlation functions of the heavy quark currents in the Euclidean time

$$G(\tau, T) = \int d^3\vec{x} \langle (\bar{q}(\tau, \vec{x})\, \Gamma\, q(\tau, \vec{x}))(\bar{q}(0, \vec{0})\, \Gamma\, q(0, \vec{0}))^\dagger \rangle, \tag{69}$$

where q represents the heavy quark field. Γ are Dirac matrices defining the spin structure of quarkonium, in particular $\Gamma = 1, \gamma_5, \gamma_\mu, \gamma_5\gamma_\mu$ corresponds to the scalar, pseudo-scalar, vector and axial-vector quarkonium states, respectively. Signatures of medium modification and melting of quarkonium are reflected in the structure of the spectral density $\rho(\omega, T)$. It can be obtained from the quarkonium correlator in the Euclidean-time

$$G(\tau, T) = \int_0^\infty \frac{d\omega}{2\pi} \rho(\omega, T) \frac{\cosh(\omega(\tau - 1/2T))}{\sinh(\omega/2T)}. \tag{70}$$

While $\rho(\omega, T)$ is a continuous function of the frequency ω over an infinite range, $G(\tau, T)$ is calculable on the lattice only at a finite number of discrete points along the temporal direction due to the finiteness of the lattice. As has been discussed in Sec. 6, inverting Eq. (70) to get spectral function $\rho(\omega, T)$ from the correlation function $G(\tau, T)$ is a typical ill-posed problem and a suitable inversion method is needed to solve the problem. The commonly used inversion methods for the extractions of quarkonium spectral functions is the Maximum Entropy Method (MEM).[153] Variants of this have recently been suggested which include a modified MEM with an extended search space[223,224] and a Bayesian method which is analogous to MEM but uses a different prior distribution.[225]

The charmonium spectral functions at finite temperature have been studied extensively for quenched lattice QCD.[227–231] Recently quenched lattice QCD calculations[226] have been performed using lattices with large temporal extents, leading to quite reliable extractions of the charmonium spectral functions. Figure 27 illustrates the spectral functions of J/Ψ (left) and η_c (right) for various temperatures. These studies suggest that all charmonium states are dissociated for $T \gtrsim 1.5T_c$ in a gluonic plasma.[226] Investigations on charmonium spectral functions in lattice QCD calculations with dynamical quarks lead to consistent conclusions.[232,233]

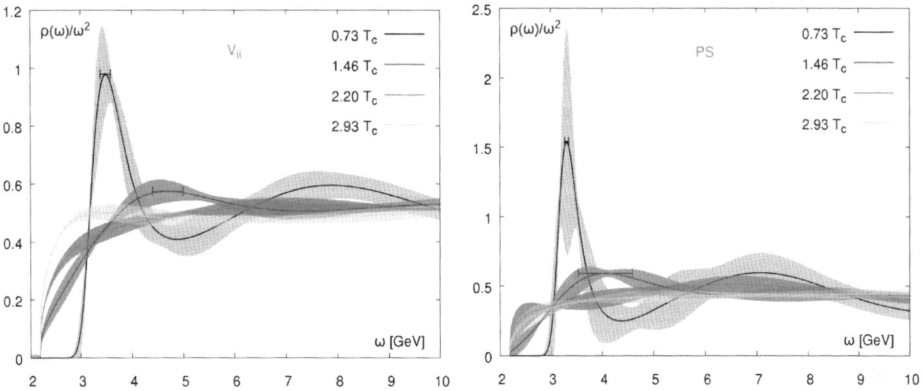

Fig. 27. (Color online) Temperature dependence of J/Ψ (left) and η_c (right) spectral functions obtained from quenched lattice QCD calculations.[226]

Lattice QCD calculations for bottom quarks are technically challenging since present computational limitations do not permit calculations with sufficiently fine lattice spacings that are much smaller than the inverse of the heavy bottom quark mass. Large discretization errors arise from the large heavy quark bare mass in units of the lattice spacing, $m_b a$. Nonetheless, although it is computationally very demanding relativistic bottomonium correlations functions have been computed recently in quenched QCD using isotropic lattices with $N_\sigma = 192$ and large temporal extent,[235] $N_\tau = 96$ and 48 at the same value of the lattice spacing, $a \simeq 0.01\,\text{fm}$, i.e., $m_b a \simeq 0.2$. The temporal lattice sizes correspond to temperatures $T = 0.7T_c$ and $1.4T_c$. In this calculation it has been found that at $1.4T_c$ only the bottomonium correlation function in the vector channel show significantly smaller thermal modifications than the charmonium correlation function in the same channel.[235]

To circumvent the problem of large cut-off effects in calculations with bottom quarks without performing calculations at extremely small lattice spacings, one needs to adapt a lattice discretization scheme for heavy quarks that is capable to describe heavy quark physics in some chosen kinematic regime also at moderately small values of the cut-off. One such approach is Non-Relativistic QCD (NRQCD)[236,237] where the heavy quark mass is assumed to be much larger than the inverse lattice spacing, but the momentum dependence of the heavy quark energy is included in the non-relativistic limit. NRQCD does not possess a proper continuum limit because the radiative corrections to coefficients of the NRQCD Lagrangian diverge as $m_b a \to 0$. However, since the non-relativistic quark field is not compactified along the Euclidean-time direction, the relation between Euclidean-time quarkonium correlation functions and the corresponding spectral functions becomes simpler,

$$G(\tau, T) = \int_0^\infty \frac{d\omega}{2\pi}\, \rho(\omega, T)\, e^{-\omega\tau}, \tag{71}$$

giving rise to a temperature independent integration kernel, $\exp(-\omega\tau)$. Moreover, the transport peak is also absent in the NRQCD spectral function, which makes its determination much easier.

Treating the bottom quark within the NRQCD framework the fate of bottomonium states has been studied on anisotropic lattices using $N_f = 2$ and $2 + 1$ of improved Wilson quarks with $m_\pi \approx 400\,\text{MeV}$ and $m_\pi \approx 500\,\text{MeV}$, respectively.[234,238–241] Figure 28 shows the temperature dependence of the Υ (top) and χ_{b1} (bottom) spectral functions extracted using the MEM.[234] This indicates that the S-wave ground state Υ survives up to $1.9T_c$ while the P-wave ground state χ_{b1} melts just above T_c.[234] However, a recent lattice QCD study[242] that uses a different Bayesian method suggested in Ref. 225 for the extraction of the spectral functions finds that the P-wave ground state χ_{b1} may survive up to $1.6T_c$. This analysis, however, has been performed on lattices with a rather small temporal extents, $N_\tau = 12$.

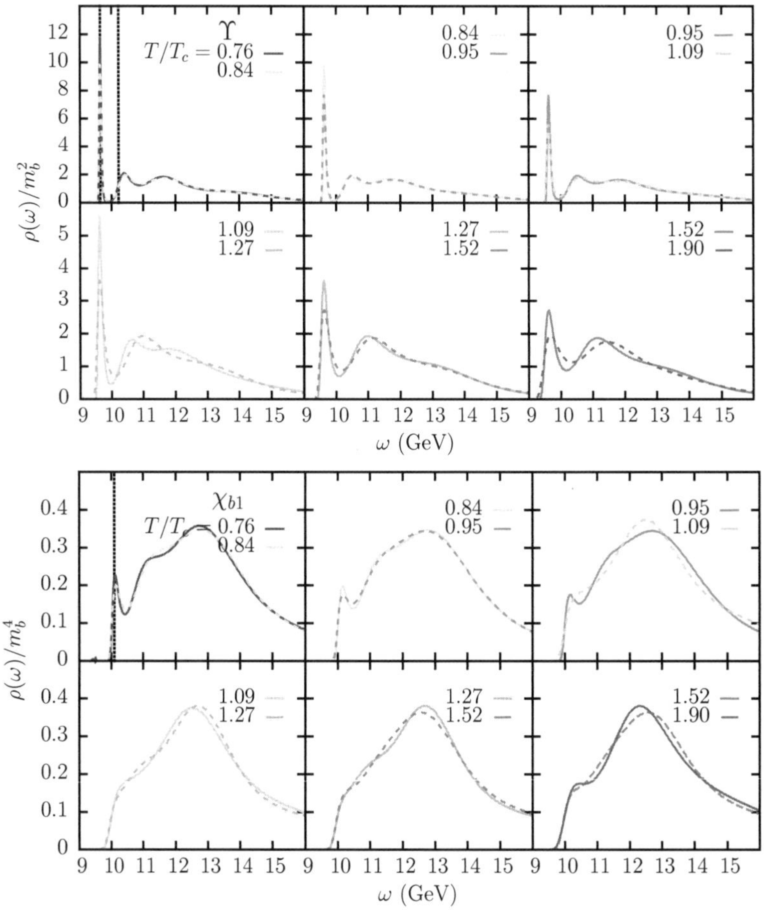

Fig. 28. (Color online) Temperature dependence of spectral functions for the Υ (top) and χ_{b1} states (bottom). Results are obtained by using a lattice discretized form of NRQCD.[234]

7.2.3. *Spatial correlation functions of quarkonia*

Similarly as the temporal correlation functions spatial quarkonium correlation functions are defined as the sum of heavy quark current–current correlation functions instead over a 'funny' space of τ, x and y

$$G(z,T) = \int_0^{1/T} d\tau \int dxdy \langle (\bar{q}(\tau,\vec{x})\,\Gamma\, q(\tau,\vec{x}))(\bar{q}(0,\vec{0})\,\Gamma\, q(0,\vec{0}))^\dagger \rangle, \tag{72}$$

are also related to the quarkonium spectral functions as follows

$$G(z,T) = \int_{-\infty}^{\infty} \frac{dp_z}{2\pi}\, e^{ip_z z} \int_0^{\infty} \frac{d\omega}{2\pi}\, \frac{\rho(\omega,p_z,T)}{\omega}. \tag{73}$$

The spatial correlation functions, however, are sensitive also to the momentum dependence of spectral functions which is reflected in an additional integration over the momentum p_z along the z-direction. Despite of its more complicated connection to the spectral function compared to that of temporal correlators (see Eq. (70)), the spatial correlation function has some distinct advantages. Since the spatial separation is not limited by the inverse temperature, the spatial correlation function can be studied at separations larger than $1/T$ and thus is more sensitive to in-medium modifications of quarkonia. Furthermore, in contrast to the temporal correlation functions, the spatial correlation functions can be directly compared to the corresponding vacuum correlation function to quantify modifications of the in-medium spectral function as the entire temperature dependence of spatial correlation functions emerges from the temperature dependence of the spectral function (Eq. (73)), while in the case of temporal correlation functions (Eq. (70)) this temperature dependence is folded with the temperature dependence of the kernel, $\cosh(\omega(\tau - 1/2T))/\sinh(\omega/2T)$.

While the relation between spectral functions and spatial meson correlators is more involved, in some limiting cases it becomes more intuitive. At large distances the spatial correlation functions decay exponentially, $G(z,T) \sim \exp(-M(T)z)$, characterized by the screening mass M. At low temperatures this exponential drop of a mesonic correlator projecting on a given quantum number channel Γ is controlled by the bound state with smallest mass in this channel. The spectral function then is dominated by this state,

$$\rho(\omega, p_z, T) \sim \delta(\omega^2 - p_z^2 - m_0^2), \tag{74}$$

where M becomes equal to the pole mass m_0 of the meson. At high temperatures, when the mesonic excitations have completely dissolved and become uncorrelated, the spatial meson correlation functions describe the propagation of a free quark anti-quark pair and the screening mass is given by[243]

$$M_{\text{free}} = \sqrt{m_{q_1}^2 + (\pi T)^2} + \sqrt{m_{q_2}^2 + (\pi T)^2}, \tag{75}$$

where m_{q_1} and m_{q_2} are the masses of the quark and anti-quark that form the meson. This form of the screening mass in the non-interacting limit is a consequence of the fermionic (anti-periodic) boundary condition along the Euclidean-time direction, which leads to the appearance of a smallest nonzero Matsubara frequency, πT, in the quark and anti-quark propagators. As the bosonic meson state is dissolved in the non-interacting limit the screening mass reflects two independently propagating fermionic degrees of freedom. Thus, the transition between these two limiting values for the screening mass can be used as an indicator for the thermal modification and eventual dissociation of charmonia and open charm mesons.

A first analysis of spatial correlation functions of charmonia was performed using staggered fermions with the p4 action.[245] Recently, a detailed comparative study of the spatial correlation functions of charmonia and open charm mesons was performed using the HISQ action.[244] Figure 29 shows ratios of the spatial correlation functions for the D_s^* (top left) and the J/Ψ (top right) at different temperatures to the corresponding zero temperature results. As discussed before,

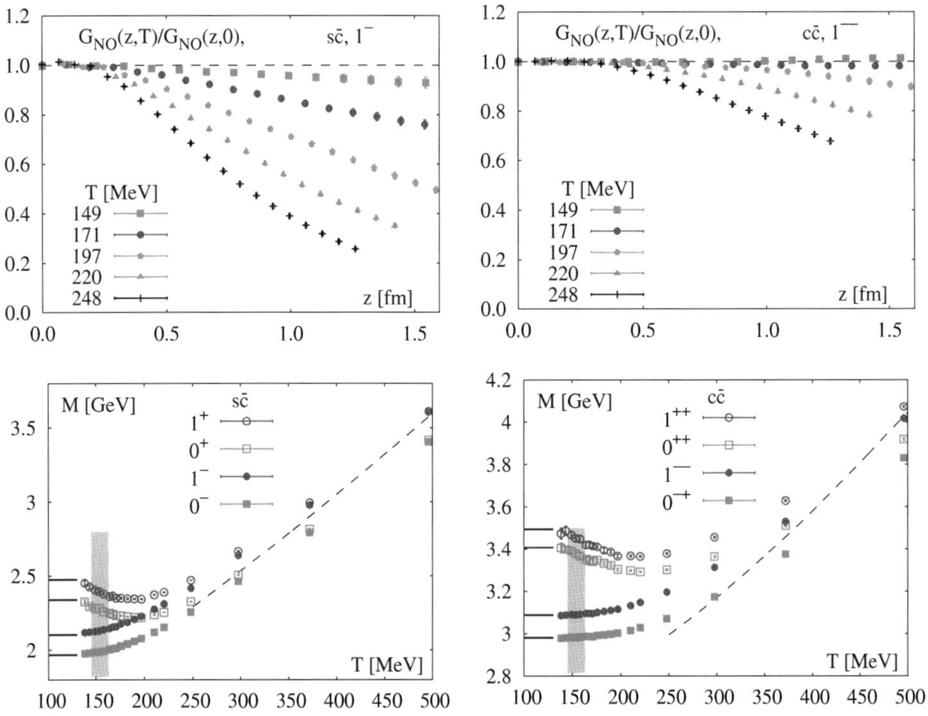

Fig. 29. (Color online) (Top) Temperature dependence of ratios of spatial correlation functions to those at zero temperature for D_s^* (top left) and J/Ψ (top right).[244] (Bottom) Temperature dependence of screening masses of ground state $s\bar{c}$ mesons (left) D_{s1}^* (1^+), D_{s0}^* (0^+), D_s^* (1^-) and D_s (0^-) and ground state charmonia (right) χ_{c1} (1^{++}), χ_{c0} (0^{++}), J/Ψ (1^{--}) and η_c (0^{-+}).[244] The horizontal solid lines denote the corresponding zero-temperature meson masses and the dashed line depicts the non-interacting theory limit of a freely propagating quark–anti-quark pair. The yellow band denotes the chiral crossover temperature region, i.e., $T_c = 154 \pm 9$ MeV.

such a ratio can directly probe the thermal modifications of the spectral functions. If there is no change in the meson spectral functions, these ratios will be equal to unity and deviations from unity will indicate in-medium modification of the meson spectral functions at nonzero temperature. As can be seen, even for $T \sim T_c \approx 154$ MeV the ratio for the open charm sector (Fig. 29 (top left)) shows significant deviation from unity. This is completely consistent with the previously discussed quantum number correlation based studies in Sec. 7.1 which concluded that open charm mesons start to melt in the vicinity of T_c.[152] On the other hand, for the charmonia the ratio of nonzero to zero temperature correlation functions remains very close to unity for $T \lesssim 170$ MeV and significant deviations, comparable to the open charm sector at $T \sim T_c$, only show up for $T \gtrsim 200$ MeV (Fig. 29 (top right)). This indicates that up to $T \simeq 1.1T_c$ J/Ψ exists almost like a vacuum state and significant thermal modifications appears only for $T \gtrsim 1.3T_c$. The same physics is also reflected in temperature dependence of the screening masses of open charm mesons and charmonia as shown in Fig. 29 (bottom). The screening masses of all open charm mesons (bottom left) start to deviate from unity significantly already in the chiral crossover region. The J/Ψ and η_c screening masses (bottom right), however, remain to be almost identical to their vacuum pole masses up to $T \simeq 1.1T_c$. The screening masses for χ_{c0} and χ_{c1} deviate from their vacuum masses even below T_c indicating significant thermal modifications of these states. At higher temperatures, i.e., $T \gtrsim 2T_c$ the screening masses in all channels increase linearly with temperature. This rise agrees well with that of a freely propagating quark–anti-quark pair.

8. QCD in External Magnetic Fields

In recent past it has been realized that large magnetic fields created during the early stages of heavy ion collisions may give rise to fascinating observable effects induced through the coupling between the magnetic field and the chiral anomaly.[20,21] This observation also motivated a plethora of activities including lattice QCD studies of hot-dense strong-interaction matter under the influence of external magnetic fields. This section presents a very brief summary of these lattice QCD studies. More comprehensive reviews on this topic can be found in Refs. 246–249.

An external magnetic field along the z-direction, $\vec{B} = B\hat{z}$, can be induced by choosing electromagnetic gauge fields: $A_{\hat{y}} = Bx$ and $A_{\hat{x}} = A_{\hat{z}} = A_{\hat{\tau}} = 0$. On the lattice this can be implemented simply by multiplying the $SU(3)$ gauge field variables $U_{n,\hat{\mu}}$ with the corresponding $U(1)$ phase factors: $u_{n,\hat{y}} = e^{ia^2 qBn_x}$ and $u_{n,\hat{x}} = u_{n,\hat{z}} = u_{n,\hat{\tau}} = 1$. Here, q is the (electric) charge of a quark and $n = (n_x, n_y, n_z, n_\tau)$ denotes a lattice site, with $n_{x,y,z} = 1, \ldots, N_\sigma$ and $n_\tau = 1, \ldots, N_\tau$. This choice ensures that the magnetic flux, $a^2 B$, is constant through all the plaquettes in the x–y plane, except at the boundary $(N_\sigma, n_y, n_z, n_\tau)$ owing to the periodic boundary condition on the gauge fields along the spatial directions. Thus, to preserve the smoothness of the external magnetic field across the boundary and the gauge invariance of the fermion action the $U(1)$ factor on the boundary links must be modified: $u_{(N_\sigma, n_y, n_z, n_\tau),\hat{x}} = e^{-ia^2 qN_\sigma Bn_y}$. By making such a choice one pays the price that the magnetic flux becomes quantized, $a^2 B = 2\pi n_B/(qN_\sigma^2)$ with $n_B = 0, \ldots, N_\sigma^2$. Thus, at a given temperature T and on a finite volume there is a minimal value for the constant, external magnetic field $B = 2\pi T^2 N_\tau^2/(qN_\sigma^2)$. On the other hand, for given temporal extent, N_τ, of the lattice the maximal value of the magnetic field achieved in this case is $B = 2\pi T^2 N_\tau^2/q$. Most importantly, since the external magnetic field is induced by multiplying the $SU(3)$ gauge field variables with purely imaginary phase factors the fermion determinant remains real and positive, allowing direct Monte Carlo sampling.

The QCD crossover transition in an external magnetic field has been extensively studied using lattice QCD. In contrast to the earlier effective model based predictions and the results of first lattice QCD calculations[250,251] using unimproved staggered fermions with heavy pion masses, the QCD crossover temperature T_c turns out to decease with increasing magnetic field.[252–254] Figure 30 (left) illustrates the

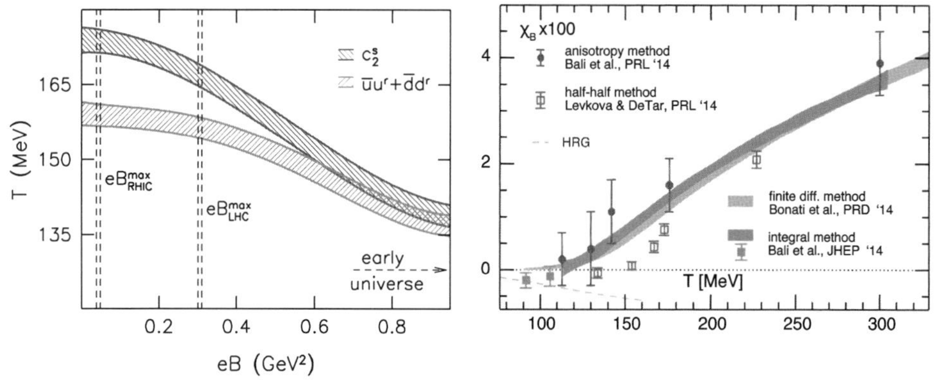

Fig. 30. (Color online) (Left) The QCD crossover temperature as a function of the magnetic field obtained from the inflection point of the quark condensate (red band) and strange quark susceptibility (blue band).[252] (Right) The magnetic susceptibility χ_B, i.e., the correction to the pressure proportional to quadratic power of the magnetic field, as a function of temperature.[257]

decrease of T_c as function of the magnetic field.[252] The reduction of T_c follows from the reduction of the chiral condensate with increasing B, a phenomenon known as the inverse magnetic catalysis, for the relevant intermediate temperature range. However, for low as well as high enough temperatures the chiral condensate shows the expected magnetic catalysis, i.e., an increase with increasing B.[254–256]

The response of the hot strong-interaction matter to an external magnetic field has also been investigated through lattice QCD calculations of the magnetic susceptibility. The magnetic susceptibility is the second derivative of the QCD partition function with respect to the external magnetic field B and, thus, gives the order B^2 corrections to the pressure. Since the standard approach leads to discrete, quantized values for B on the lattice, unlike for the susceptibilities of the chemical potentials, taking direct derivatives of the QCD partition function with respect to B is not straightforward. Lattice calculations of the magnetic susceptibility have been performed using three different approaches: (i) by computing the free energy difference through an integration over the varying magnetic field;[258,259,259] (ii) through computations of the magnetization via pressure anisotropies parallel and orthogonal to the direction of the magnetic field;[260,261] (iii) by adopting a non-uniform external field such that the net magnetic flux across the x–y plane of the whole lattice is zero, resulting in a non-quantized magnetic field and allowing for a direct computation of the derivatives of the partition function with respect to B.[262] As shown in Fig. 30 (right), all these different approaches produce more or less consistent results. The QGP phase turns out to be paramagnetic. However, for low enough temperatures QCD is weakly diamagnetic, in accordance with the hadron resonance gas model.[263] These results also suggests that for $T \lesssim 300\,\text{MeV}$ the order B^2 magnetic contribution to the QCD pressure is at best a few percent for a magnetic field of the order of 10^{15} T, which is relevant for heavy ion collision experiments.[21]

Lattice QCD studies have also been carried out to investigate the chiral magnetic effect,[21] namely the phenomenon of (electric) charge separation along the direction of the magnetic field in presence of chiral anomaly induced topological fluctuations. Various different avenues have been pursued by studying: (i) charge fluctuations on given topological background;[264] (ii) enhanced fluctuations of the electric current along the magnetic field,[265] (iii) dependence of the electric current on a given axial chemical potential,[265] (iv) correlations of the electric polarization and the topological charge.[266] In all cases, the chiral magnetic effect has been observed, however, in a suppressed magnitude compared to the model based expectations. Obviously, in lattice QCD studies of the chiral magnetic effect, which is closely related to the chiral anomaly of QCD, it would be of great interest to have lattice QCD calculations perfomed with a chiral fermion formulation. Fully dynamical calculations with DWF or overlap fermions at present do not yet exist.

9. Summary

In this review we have discussed selected topics in finite temperature QCD that have a direct link to the ongoing experimental study of the phase diagram of strong-interaction matter and the properties of matter formed in heavy ion collisions. We naturally focused on topics to which lattice QCD calculations can contribute and have lead or may lead to definite, quantitative answers in the future.

Lattice QCD calculations at finite temperature started more than 30 years ago with simulations of pure $SU(N)$ gauge theories on computers that delivered less than one Mega-Flop/s peak-performance. Nowadays these calculations are performed with dynamical quarks and the correct mass spectrum of QCD on computers that reach a peak-performance of more than ten Peta-Flop/s. The field thus could utilize computing resources with a peak-performance that increased by more than ten orders of magnitude, or doubled the speed almost every year. Although we know quite well from the physics of the hadron gas that an exponential rise cannot continue for ever, we still do not seem to have reached the critical point and we may also in the future expect to gain further inside into the physics of strong-interaction matter through numerical lattice QCD calculations.

During the last years lattice QCD calculations have accomplished two long-standing goals: (i) the pseudo-critical temperature of strong-interaction matter with physical light and strange quark masses has been determined at vanishing net-baryon number, $T_c = (154 \pm 9)\,\mathrm{MeV}$, and (ii) the equation of state has been calculated in a wide temperature range. It now is frequently used in the hydrodynamic modeling of heavy ion collisions. The extension of these results to nonzero chemical potential is well controlled at least in leading order Taylor expansion of the pressure in terms of baryon, electric charge and strangeness chemical potentials and we will soon have results for bulk thermodynamics in a range of baryon chemical potentials $\mu_B/T \lesssim 3$, which will cover most of the parameter space that can be explored in heavy ion experiments at RHIC and the LHC. These Taylor expansions also provide specific information on fluctuations of and correlations among conserved charges. Lattice QCD calculations of these observables have reached a stage where guidance can be provided to the experimental search for the critical end point and, with some care, a comparison between experimental results for measured proton number, strangeness and, in particular, electric charge fluctuations on the one hand and lattice QCD calculations on the other hand starts to become possible.

A major focus in the ongoing experimental heavy ion program is to explore in detail the properties of matter formed in heavy ion collisions. The analysis of transport coefficients, thermal masses and screening lengths does play an important role in this effort. Lattice QCD calculations of these quantities are possible, although in many cases they are complicated as an analytic continuation from Euclidean time to Minkowski time is needed. For this reason most of the existing calculations have been performed in the quenched approximation where fermionic degrees of freedom are taken into account only as valence quarks and the influence of virtual quark loops on the gluonic background is neglected. Still these calculations prepare the ground for future fully dynamical calculations of transport properties of strong-interaction matter, for which in some cases exploratory work has already started.

There remain many obvious, open questions in the study of the phase diagram of strong-interaction matter and the properties of the strongly interacting medium created in heavy ion collisions which can be addressed in future lattice QCD calculations and which will profit from future increasing computing resources as well as algorithmic advances. Calculations at nonzero baryon number density and the search for the critical point, the calculation of transport properties with fully dynamical, light quarks as well as calculations of thermodynamic quantities using chiral fermion discretization schemes are among the most demanding problems which hopefully can be addressed in the coming years with improved techniques and resources.

Acknowledgements

This review would not have been possible without the input from many of our colleagues, in particular, Prasad Hegde, Olaf Kaczmarek, Christian Schmidt and Mathias Wagner, with whom we could collaborate over many years in the HotQCD and Bielefeld–BNL–CCNU collaborations. This work has been supported in part through contract DE-SC0012704 with the U.S. Department of Energy, the BMBF under grant 05P12PBCTA and the GSI BILAER grant.

References

1. R. Hagedorn, *Nuovo Cim. Suppl.* **3** (1965) 147–186.
2. G. Baym, *Physica A* **96** (1979) 131.
3. T. Celik, F. Karsch and H. Satz, *Phys. Lett. B* **97** (1980) 128–130.
4. K. Redlich and H. Satz, arXiv:1501.07523 (2015).
5. H. D. Politzer, *Phys. Rev. Lett.* **30** (1973) 1346–1349.

6. D. J. Gross and F. Wilczek, *Phys. Rev. Lett.* **30** (1973) 1343–1346.
7. N. Cabibbo and G. Parisi, *Phys. Lett. B* **59** (1975) 67–69.
8. J. C. Collins and M. Perry, *Phys. Rev. Lett.* **34** (1975) 1353.
9. E. V. Shuryak, *Phys. Rept.* **61** (1980) 71–158.
10. R. D. Pisarski and F. Wilczek, *Phys. Rev. D* **29** (1984) 338–341.
11. K. Rajagopal and F. Wilczek, *hep-ph/0011333.* (2000).
12. K. Fukushima and T. Hatsuda, *Rept. Prog. Phys.* **74** (2011) 014001.
13. K. G. Wilson, *Phys. Rev. D* **10** (1974) 2445–2459.
14. M. Creutz, *Phys. Rev. D* **21** (1980) 2308–2315.
15. L. D. McLerran and B. Svetitsky, *Phys. Lett. B* **98** (1981) 195.
16. J. Kuti, J. Polonyi and K. Szlachanyi, *Phys. Lett. B* **98** (1981) 199.
17. J. Engels, F. Karsch, H. Satz and I. Montvay, *Phys. Lett. B* **101** (1981) 89.
18. T. Matsui and H. Satz, *Phys. Lett. B* **178** (1986) 416.
19. R. Rapp and J. Wambach, *Adv. Nucl. Phys.* **25** (2000) 1.
20. D. E. Kharzeev, K. Landsteiner, A. Schmitt and H.-U. Yee, *Lect. Notes Phys.* **871** (2013) 1–11.
21. D. E. Kharzeev, L. D. McLerran and H. J. Warringa, *Nucl. Phys. A* **803** (2008) 227–253.
22. A. D. Linde, *Phys. Lett. B* **96** (1980) 289.
23. E. Braaten and R. D. Pisarski, *Nucl. Phys. B* **337** (1990) 569.
24. N. Haque, A. Bandyopadhyay, J. O. Andersen, M. G. Mustafa, M. Strickland *et al.*, *JHEP* **1405** (2014) 027.
25. E. Braaten and A. Nieto, *Phys. Rev. D* **51** (1995) 6990–7006.
26. A. Hietanen, K. Kajantie, M. Laine, K. Rummukainen and Y. Schroder, *Phys. Rev. D* **79** (2009) 045018.
27. I. Montvay and G. Muenster, *Quantum Fields on a Lattice* (Cambridge Monographs on Mathematical Physics, 2010).
28. H. J. Rothe, *Lattice Gauge Theories: An Introduction (Fourth Edition)*. (World Scientific Lecture Notes in Physics, 2012).
29. T. Degrand and C. Detar, *Lattice Methods for Quantum Chromodynamics* (World Scientific, 2006).
30. C. Gattringer and C. Lang, *Quantum Chromodynamics on the Lattice-An Introductory Presentation* (Springer, 2010).
31. K. Olive *et al.*, *Chin. Phys. C* **38** (2014) 090001.
32. P. Hasenfratz and F. Karsch, *Phys. Lett. B* **125** (1983) 308.
33. R. Gavai, *Phys. Rev. D* **32** (1985) 519.
34. G. Aarts, F. A. James, E. Seiler and I.-O. Stamatescu, *Eur. Phys. J. C* **71** (2011) 1756.
35. G. Aarts, L. Bongiovanni, E. Seiler, D. Sexty and I.-O. Stamatescu, *Eur. Phys. J. A* **49** (2013) 89.
36. D. Sexty, *PoS.* **Lattice2015** (2014) 016.
37. M. Cristoforetti, F. Di Renzo and L. Scorzato, *Phys. Rev. D* **86** (2012) 074506.
38. H. Fujii, D. Honda, M. Kato, Y. Kikukawa, S. Komatsu *et al.*, *JHEP* **1310** (2013) 147.
39. M. Cristoforetti, F. Di Renzo, G. Eruzzi, A. Mukherjee, C. Schmidt *et al.*, *Phys. Rev. D* **89** (2014) 114505.
40. R. V. Gavai and S. Gupta, *Phys. Rev. D* **68** (2003) 034506.
41. C. Allton, S. Ejiri, S. Hands, O. Kaczmarek, F. Karsch *et al.*, *Phys. Rev. D* **68** (2003) 014507.
42. K. Symanzik, *Nucl. Phys. B* **226** (1983) 187.

43. K. Symanzik, *Nucl. Phys. B* **226** (1983) 205.
44. H. B. Nielsen and M. Ninomiya, *Phys. Lett. B* **105** (1981) 219.
45. J. B. Kogut and L. Susskind, *Phys. Rev. D* **11** (1975) 395–408.
46. D. B. Kaplan, *Phys. Lett. B* **288** (1992) 342–347.
47. Y. Shamir, *Nucl. Phys. B* **406** (1993) 90–106.
48. Y. Shamir, *Phys. Lett. B* **305** (1993) 357–365.
49. V. Furman and Y. Shamir, *Nucl. Phys. B* **439** (1995) 54–78.
50. H. Neuberger, *Phys. Lett. B* **427** (1998) 353–355.
51. H. Neuberger, *Phys. Lett. B* **417** (1998) 141–144.
52. B. Sheikholeslami and R. Wohlert, *Nucl. Phys. B* **259** (1985) 572.
53. E. Follana, Q. Mason, C. Davies, K. Hornbostel, G. P. Lepage, J. Shigemitsu, H. Trottier and K. Wong, *Phys. Rev. D* **75** (2007) 054502.
54. C. Morningstar and M. J. Peardon, *Phys. Rev. D* **69** (2004) 054501.
55. U. M. Heller, F. Karsch and B. Sturm, *Phys. Rev. D* **60** (1999) 114502.
56. D. J. Antonio *et al.*, *Phys. Rev. D* **77** (2008) 014509.
57. D. Renfrew, T. Blum, N. Christ, R. Mawhinney and P. Vranas, *PoS.* **LATTICE2008** (2008) 048.
58. P. H. Ginsparg and K. G. Wilson, *Phys. Rev. D* **25** (1982) 2649.
59. M. I. Buchoff, M. Cheng, N. H. Christ, H. T. Ding, C. Jung *et al.*, *Phys. Rev.D* **89** (2014) 054514.
60. G. Cossu, S. Aoki, H. Fukaya, S. Hashimoto, T. Kaneko *et al.*, *Phys. Rev. D* **87** (2013) 114514.
61. T. Bhattacharya, M. I. Buchoff, N. H. Christ, H.-T. Ding, R. Gupta *et al.*, *Phys. Rev. Lett.* **113** (2014) 082001.
62. V. Dick, F. Karsch, E. Laermann, S. Mukherjee and S. Sharma, arXiv:1502.06190 (2015).
63. H. Saito *et al.*, *Phys. Rev. D* **84** (2011) 054502.
64. H. Saito, S. Ejiri, S. Aoki, K. Kanaya, Y. Nakagawa *et al.*, *Phys. Rev. D* **89** (2014) 034507.
65. F. Karsch, E. Laermann and C. Schmidt, *Phys. Lett. B* **520** (2001) 41–49.
66. P. de Forcrand and O. Philipsen, *JHEP* **0701** (2007) 077.
67. G. Endrodi, Z. Fodor, S. Katz and K. Szabo, *PoS.* **LAT2007** (2007) 182.
68. H.-T. Ding, A. Bazavov, P. Hegde, F. Karsch, S. Mukherjee *et al.*, *PoS.* **LATTICE2011** (2011) 191.
69. A. Butti, A. Pelissetto and E. Vicari, *JHEP* **0308** (2003) 029.
70. A. Pelissetto and E. Vicari, *Phys. Rev. D* **88** (2013) 105018.
71. M. Grahl and D. H. Rischke, *Phys. Rev. D* **88** (2013) 056014.
72. Y. Aoki, G. Endrodi, Z. Fodor, S. Katz and K. Szabo, *Nature* **443** (2006) 675–678.
73. E. K. Riedel and F. J. Wegner, *Phys. Rev. Lett.* **29** (1972) 349–352. http://link.aps.org/
74. J. Engels, S. Holtmann, T. Mendes and T. Schulze, *Phys. Lett. B* **514** (2001) 299–308.
75. J. Engels, L. Fromme and M. Seniuch, *Nucl. Phys. B* **675** (2003) 533–554.
76. S. Ejiri, F. Karsch, E. Laermann, C. Miao, S. Mukherjee *et al.*, *Phys. Rev. D* **80** (2009) 094505.
77. C. Bonati, P. de Forcrand, M. D'Elia, O. Philipsen and F. Sanfilippo, *Phys. Rev. D* **90** (2014) 074030.
78. A. Bazavov, T. Bhattacharya, M. Cheng, C. DeTar, H. Ding *et al.*, *Phys. Rev. D* **85** (2012) 054503.
79. Y. Aoki, Z. Fodor, S. Katz and K. Szabo, *Phys. Lett. B* **643** (2006) 46–54.

80. Y. Aoki, S. Borsanyi, S. Durr, Z. Fodor, S. D. Katz *et al.*, *JHEP* **0906** (2009) 088.
81. S. Borsanyi *et al.*, *JHEP* **1009** (2010) 073.
82. O. Kaczmarek, F. Karsch, E. Laermann, C. Miao, S. Mukherjee *et al.*, *Phys. Rev. D* **83** (2011) 014504.
83. G. Endrodi, Z. Fodor, S. Katz and K. Szabo, *JHEP* **1104** (2011) 001.
84. P. Cea, L. Cosmai and A. Papa, *Phys. Rev. D* **89** (2014) 074512.
85. C. Bonati, M. D'Elia, M. Mariti, M. Mesiti, F. Negro *et al.*, *Phys. Rev. D* **90** (2014) 114025.
86. A. Andronic, P. Braun-Munzinger, K. Redlich and J. Stachel, *J. Phys. G* **38** (2011) 124081.
87. Z. Fodor and S. Katz, *JHEP* **0404** (2004) 050.
88. S. Datta, R. V. Gavai and S. Gupta, *PoS.* **LATTICE2013** (2014) 202.
89. S. Ejiri, *Phys. Rev. D* **73** (2006) 054502.
90. P. de Forcrand and O. Philipsen, *Nucl. Phys. B* **642** (2002) 290–306.
91. X.-Y. Jin, Y. Kuramashi, Y. Nakamura, S. Takeda and A. Ukawa, *1504.00113.* (2015).
92. C. Allton, M. Doring, S. Ejiri, S. Hands, O. Kaczmarek *et al.*, *Phys. Rev. D* **71** (2005) 054508.
93. R. Gavai and S. Gupta, *Phys. Rev. D* **78** (2008) 114503.
94. O. Kaczmarek, F. Karsch, P. Petreczky and F. Zantow, *Phys. Lett. B* **543** (2002) 41–47.
95. L. D. McLerran and B. Svetitsky, *Phys. Rev. D* **24** (1981) 450.
96. A. Bazavov and P. Petreczky, *Phys. Rev. D* **87** (2013) 094505.
97. R. Bellwied, S. Borsanyi, Z. Fodor, S. D. Katz and C. Ratti, *Phys. Rev. Lett.* **111** (2013) 202302.
98. A. Bazavov, H. T. Ding, P. Hegde, O. Kaczmarek, F. Karsch *et al.*, *Phys. Rev. Lett.* **111** (2013) 082301.
99. V. Koch, A. Majumder and J. Randrup, *Phys. Rev. Lett.* **95** (2005) 182301.
100. S. Ejiri, F. Karsch and K. Redlich, *Phys. Lett. B* **633** (2006) 275–282.
101. S. L. Adler, *Phys. Rev.* **177** (1969) 2426–2438.
102. J. Bell and R. Jackiw, *Nuovo Cim. A* **60** (1969) 47–61.
103. G. 't Hooft, *Phys. Rev. Lett.* **37** (1976) 8–11.
104. D. J. Gross, R. D. Pisarski and L. G. Yaffe, *Rev. Mod. Phys.* **53** (1981) 43.
105. E. V. Shuryak, *Comments Nucl. Part. Phys.* **21** (1994) 235–248.
106. M. Cheng, S. Datta, A. Francis, J. van der Heide, C. Jung *et al.*, *Eur. Phys. J. C* **71** (2011) 1564.
107. S. R. Sharpe, *PoS.* **LAT2006** (2006) 022.
108. G. C. Donald, C. T. Davies, E. Follana and A. S. Kronfeld, *Phys. Rev. D* **84** (2011) 054504.
109. A. Bazavov *et al.*, *Phys. Rev. D* **86** (2012) 094503.
110. A. Tomiya, G. Cossu, H. Fukaya, S. Hashimoto and J. Noaki, arXiv:1412.7306 (2014).
111. T.-W. Chiu, W.-P. Chen, Y.-C. Chen, H.-Y. Chou and T.-H. Hsieh, *PoS.* **LATTICE2013** (2014) 165.
112. F. Karsch, *Lect. Notes Phys.* **583** (2002) 209–249.
113. S. Borsanyi, Z. Fodor, C. Hoelbling, S. D. Katz, S. Krieg *et al.*, *Phys. Lett. B* **730** (2014) 99–104.
114. A. Bazavov *et al.*, *Phys. Rev. D* **90** (2014) 094503.
115. A. Majumder and B. Muller, *Phys. Rev. Lett.* **105** (2010) 252002.
116. A. Bazavov, H. T. Ding, P. Hegde, O. Kaczmarek, F. Karsch *et al.*, *Phys. Rev. Lett.* **113** (2014) 072001.

117. R. V. Gavai and S. Gupta, *Phys. Rev. D* **64** (2001) 074506.
118. C. Allton, S. Ejiri, S. Hands, O. Kaczmarek, F. Karsch *et al.*, *Phys. Rev. D* **66** (2002) 074507.
119. S. Ejiri, F. Karsch, E. Laermann and C. Schmidt, *Phys. Rev. D* **73** (2006) 054506.
120. S. Borsanyi, Z. Fodor, S. D. Katz, S. Krieg, C. Ratti *et al.*, *JHEP* **1201** (2012) 138.
121. A. Bazavov *et al.*, *Phys. Rev. D* **86** (2012) 034509.
122. S. Borsanyi, G. Endrodi, Z. Fodor, S. Katz, S. Krieg *et al.*, *JHEP* **1208** (2012) 053.
123. S. Borsanyi, Z. Fodor, S. Katz, S. Krieg, C. Ratti *et al.*, *Phys. Rev. Lett.* **111** (2013) 062005.
124. P. Hegde, arXiv:1412.6727 (2014).
125. M. Laine and Y. Schroder, *Phys. Rev. D* **73** (2006) 085009.
126. M. Stephanov, *Phys. Rev. Lett.* **102** (2009) 032301.
127. M. Asakawa, S. Ejiri and M. Kitazawa, *Phys. Rev. Lett.* **103** (2009) 262301.
128. B. Friman, F. Karsch, K. Redlich and V. Skokov, *Eur. Phys. J. C* **71** (2011) 1694.
129. M. Stephanov, *Phys. Rev. Lett.* **107** (2011) 052301.
130. M. A. Stephanov, K. Rajagopal and E. V. Shuryak, *Phys. Rev. Lett.* **81** (1998) 4816–4819.
131. M. A. Stephanov, K. Rajagopal and E. V. Shuryak, *Phys. Rev. D* **60** (1999) 114028.
132. S. Jeon and V. Koch, arXiv:hep-ph/0304012 (2003).
133. Y. Hatta and M. Stephanov, *Phys. Rev. Lett.* **91** (2003) 102003.
134. C. Athanasiou, K. Rajagopal and M. Stephanov, *Phys. Rev. D* **82** (2010) 074008.
135. S. Collaboration, https://drupal.star.bnl.gov/STAR/files/BES_WPII_ver6.9_Cover.pdf.
136. M. Aggarwal *et al.*, *Phys. Rev. Lett.* **105** (2010) 022302.
137. L. Adamczyk *et al.*, *Phys. Rev. Lett.* **112** (2014) 032302.
138. L. Adamczyk *et al.*, *Phys. Rev. Lett.* **113** (2014) 092301.
139. N. R. Sahoo, *J. Phys. Conf. Ser.* **535** (2014) 012007.
140. X. Luo, *PoS.* **CPOD2014** (2014) 019.
141. J. T. Mitchell, *Nucl. Phys. A* **904–905** (2013) 903c–906c.
142. A. Bazavov, H. Ding, P. Hegde, O. Kaczmarek, F. Karsch *et al.*, *Phys. Rev. Lett.* **109** (2012) 192302.
143. M. Kitazawa and M. Asakawa, *Phys. Rev. C* **86** (2012) 024904.
144. A. Bzdak and V. Koch, *Phys. Rev. C* **86** (2012) 044904.
145. A. Bzdak and V. Koch, *Phys. Rev. C* **91** (2015) 027901.
146. F. Karsch, *Central Eur. J. Phys.* **10** (2012) 1234–1237.
147. S. Mukherjee and M. Wagner, *PoS.* **CPOD2013** (2013) 039.
148. S. Borsanyi, Z. Fodor, S. Katz, S. Krieg, C. Ratti *et al.*, *Phys. Rev. Lett.* **113** (2014) 052301.
149. S. Mukherjee, *PoS.* **CPOD2014** (2014) 005.
150. S. Capstick and N. Isgur, *Phys. Rev. D* **34** (1986) 2809.
151. D. Ebert, R. Faustov and V. Galkin, *Phys. Rev. D* **79** (2009) 114029.
152. A. Bazavov, H.-T. Ding, P. Hegde, O. Kaczmarek, F. Karsch *et al.*, *Phys. Lett. B* **737** (2014) 210–215.
153. M. Asakawa, T. Hatsuda and Y. Nakahara, *Prog. Part. Nucl. Phys.* **46** (2001) 459–508.
154. R. Rapp and H. van Hees, arXiv:0803.0901 (2008).
155. H. B. Meyer, *Eur. Phys. J. A* **47** (2011) 86.
156. F. Karsch, E. Laermann, P. Petreczky, S. Stickan and I. Wetzorke, *Phys. Lett. B* **530** (2002) 147–152.

157. Y. Nakahara, M. Asakawa and T. Hatsuda, *Phys. Rev. D* **60** (1999) 091503.
158. G. Aarts, C. Allton, J. Foley, S. Hands and S. Kim, *Phys. Rev. Lett.* **99** (2007) 022002.
159. H.-T. Ding, A. Francis, O. Kaczmarek, F. Karsch, E. Laermann *et al.*, *Phys. Rev. D* **83** (2011) 034504.
160. O. Kaczmarek, E. Laermann, M. Müller, F. Karsch, H. Ding *et al.*, *PoS.* **ConfinementX** (2012) 185.
161. O. Kaczmarek and M. Müller, *PoS.* **LATTICE2013** (2014) 175.
162. H.-T. Ding, O. Kaczmarek and F. Meyer, arXiv:1412.5869 (2014).
163. S. Gupta, *Phys. Lett. B* **597** (2004) 57–62.
164. B. B. Brandt, A. Francis, H. B. Meyer and H. Wittig, *JHEP* **1303** (2013) 100.
165. G. Aarts, C. Allton, A. Amato, P. Giudice, S. Hands *et al.*, arXiv:1412.6411 (2014).
166. A. Amato, G. Aarts, C. Allton, P. Giudice, S. Hands *et al.*, *Phys. Rev. Lett.* **111** (2013) 172001.
167. D. Fernandez-Fraile and A. Gomez Nicola, *Phys. Rev. D* **73** (2006) 045025.
168. A. Adare *et al.*, *Phys. Rev. Lett.* **98** (2007) 172301.
169. B. Abelev *et al.*, *Phys. Rev. Lett.* **98** (2007) 192301.
170. B. Abelev *et al.*, *JHEP* **1209** (2012) 112.
171. M. He, R. J. Fries and R. Rapp, arXiv:1409.4539 (2014).
172. S. Cao, Y. Huang, G.-Y. Qin and S. A. Bass, arXiv:1404.3139 (2014).
173. H. Ding, A. Francis, O. Kaczmarek, F. Karsch, H. Satz *et al.*, *J. Phys. G* **38** (2011) 124070.
174. S. Caron-Huot and G. D. Moore, *Phys. Rev. Lett.* **100** (2008) 052301.
175. P. Kovtun, D. T. Son and A. O. Starinets, *JHEP* **0310** (2003) 064.
176. O. Kaczmarek, arXiv:1409.3724 (2014).
177. D. Banerjee, S. Datta, R. Gavai and P. Majumdar, *Phys. Rev. D* **85** (2012) 014510.
178. S. Caron-Huot, M. Laine and G. D. Moore, *JHEP* **0904** (2009) 053.
179. J. Casalderrey-Solana and D. Teaney, *Phys. Rev. D* **74** (2006) 085012.
180. H. B. Meyer, *Phys. Rev. D* **76** (2007) 101701.
181. M. Luscher and P. Weisz, *JHEP* **0109** (2001) 010.
182. G. Parisi, R. Petronzio and F. Rapuano, *Phys. Lett. B* **128** (1983) 418.
183. P. De Forcrand and C. Roiesnel, *Phys. Lett. B* **151** (1985) 77–80.
184. G. D. Moore and D. Teaney, *Phys. Rev. C* **71** (2005) 064904.
185. F. Karsch and H. Wyld, *Phys. Rev. D* **35** (1987) 2518.
186. H. B. Meyer, *JHEP* **0401** (2004) 030.
187. A. Nakamura and S. Sakai, *Phys. Rev. Lett.* **94** (2005) 072305.
188. H. B. Meyer, *Phys. Rev. Lett.* **100** (2008) 162001.
189. F. Karsch, D. Kharzeev and K. Tuchin, *Phys. Lett. B* **663** (2008) 217–221.
190. D. Kharzeev and K. Tuchin, *JHEP* **0809** (2008) 093.
191. P. Romatschke and D. T. Son, *Phys. Rev. D* **80** (2009) 065021.
192. P. Romatschke, *Class. Quant. Grav.* **27** (2010) 025006.
193. G. D. Moore and K. A. Sohrabi, *Phys. Rev. Lett.* **106** (2011) 122302.
194. G. D. Moore and K. A. Sohrabi, *JHEP* **1211** (2012) 148.
195. G. Denicol, H. Niemi, E. Molnar and D. Rischke, *Phys. Rev. D* **85** (2012) 114047.
196. R. Baier, P. Romatschke, D. T. Son, A. O. Starinets and M. A. Stephanov, *JHEP* **0804** (2008) 100.
197. Y. Kohno, M. Asakawa and M. Kitazawa, *Phys. Rev. D* **89** (2014) 054508.
198. O. Philipsen and C. Schäfer, *JHEP* **1402** (2014) 003.
199. P. Romatschke, *Int. J. Mod. Phys. E* **19** (2010) 1–53.
200. A. Adil and I. Vitev, *Phys. Lett. B* **649** (2007) 139–146.

201. R. Sharma, I. Vitev and B.-W. Zhang, *Phys. Rev. C* **80** (2009) 054902.
202. M. He, R. J. Fries and R. Rapp, *Phys. Rev. C* **86** (2012) 014903.
203. D. Ebert, R. Faustov and V. Galkin, *Eur. Phys. J. C* **66** (2010) 197–206.
204. D. Ebert, R. Faustov and V. Galkin, *Phys. Rev. D* **84** (2011) 014025.
205. A. Mocsy, P. Petreczky and M. Strickland, *Int. J. Mod. Phys. A* **28** (2013) 1340012.
206. A. Bazavov, P. Petreczky and A. Velytsky, *0904.1748.* (2009).
207. N. Brambilla, S. Eidelman, B. Heltsley, R. Vogt, G. Bodwin *et al.*, *Eur. Phys. J. C* **71** (2011) 1534.
208. E. Eichten, K. Gottfried, T. Kinoshita, K. Lane and T.-M. Yan, *Phys. Rev. D* **17** (1978) 3090.
209. E. Eichten, K. Gottfried, T. Kinoshita, K. Lane and T.-M. Yan, *Phys. Rev. D* **21** (1980) 203.
210. A. Mocsy, *Eur. Phys. J. C* **61** (2009) 705–710.
211. M. Laine, O. Philipsen, P. Romatschke and M. Tassler, *JHEP* **0703** (2007) 054.
212. A. Beraudo, J.-P. Blaizot and C. Ratti, *Nucl. Phys. A* **806** (2008) 312–338.
213. N. Brambilla, J. Ghiglieri, A. Vairo and P. Petreczky, *Phys. Rev. D* **78** (2008) 014017.
214. W. Caswell and G. Lepage, *Phys. Lett. B* **167** (1986) 437.
215. G. T. Bodwin, E. Braaten and G. P. Lepage, *Phys. Rev. D* **51** (1995) 1125–1171.
216. N. Brambilla, A. Pineda, J. Soto and A. Vairo, *Nucl. Phys. B* **566** (2000) 275.
217. A. Rothkopf, T. Hatsuda and S. Sasaki, *Phys. Rev. Lett.* **108** (2012) 162001.
218. Y. Burnier and A. Rothkopf, *Phys. Rev. D* **86** (2012) 051503.
219. Y. Burnier and A. Rothkopf, *Phys. Rev. D* **87** (2013) 114019.
220. Y. Burnier, O. Kaczmarek and A. Rothkopf, *1410.2546.* (2014).
221. S. Nadkarni, *Phys. Rev. D* **34** (1986) 3904.
222. A. Bazavov, Y. Burnier and P. Petreczky, arXiv:1404.4267 (2014).
223. A. Rothkopf, *J. Comput. Phys.* **238** (2013) 106–114.
224. A. Rothkopf, *PoS.* **LATTICE2012** (2012) 100.
225. Y. Burnier and A. Rothkopf, *Phys. Rev. Lett.* **111** (2013) 182003.
226. H. Ding, A. Francis, O. Kaczmarek, F. Karsch, H. Satz *et al.*, *Phys. Rev. D* **86** (2012) 014509.
227. M. Asakawa and T. Hatsuda, *Phys. Rev. Lett.* **92** (2004) 012001.
228. S. Datta, F. Karsch, P. Petreczky and I. Wetzorke, *Phys. Rev. D* **69** (2004) 094507.
229. A. Jakovac, P. Petreczky, K. Petrov and A. Velytsky, *Phys. Rev. D* **75** (2007) 014506.
230. H. Iida, T. Doi, N. Ishii, H. Suganuma and K. Tsumura, *Phys. Rev. D* **74** (2006) 074502.
231. H. Ohno *et al.*, *Phys. Rev. D* **84** (2011) 094504.
232. G. Aarts, C. Allton, M. B. Oktay, M. Peardon and J.-I. Skullerud, *Phys. Rev. D* **76** (2007) 094513.
233. S. Borsanyi, S. Dürr, Z. Fodor, C. Hoelbling, S. D. Katz *et al.*, *JHEP* **1404** (2014) 132.
234. G. Aarts, C. Allton, T. Harris, S. Kim, M. P. Lombardo *et al.*, *JHEP* **1407** (2014) 097.
235. H. Ohno, H. T. Ding and O. Kaczmarek, arXiv:1412.6594 (2014).
236. B. Thacker and G. P. Lepage, *Phys. Rev. D* **43** (1991) 196–208.
237. G. P. Lepage, L. Magnea, C. Nakhleh, U. Magnea and K. Hornbostel, *Phys. Rev. D* **46** (1992) 4052–4067.
238. G. Aarts, S. Kim, M. Lombardo, M. Oktay, S. Ryan *et al.*, *Phys. Rev. Lett.* **106** (2011) 061602.
239. G. Aarts, C. Allton, S. Kim, M. Lombardo, S. Ryan *et al.*, *JHEP* **1312** (2013) 064.

240. G. Aarts, C. Allton, S. Kim, M. P. Lombardo, M. B. Oktay *et al.*, *JHEP* **1303** (2013) 084.
241. G. Aarts, C. Allton, S. Kim, M. Lombardo, M. Oktay *et al.*, *JHEP* **1111** (2011) 103.
242. S. Kim, P. Petreczky and A. Rothkopf, arXiv:1409.3630 (2014).
243. W. Florkowski and B. L. Friman, *Z. Phys. A* **347** (1994) 271–276.
244. A. Bazavov, F. Karsch, Y. Maezawa, S. Mukherjee and P. Petreczky, arXiv:1411.3018 (2014).
245. F. Karsch, E. Laermann, S. Mukherjee and P. Petreczky, *Phys. Rev. D* **85** (2012) 114501.
246. M. D'Elia, *Lect. Notes Phys.* (2013).
247. A. Yamamoto, *Lect. Notes Phys.* **871** (2013) 387–397.
248. G. Endrodi, arXiv:1410.8028 (2014).
249. M. D'Elia, arXiv:1502.06047 (2015).
250. M. D'Elia, S. Mukherjee and F. Sanfilippo, *Phys. Rev. D* **82** (2010) 051501.
251. E.-M. Ilgenfritz, M. Kalinowski, M. Muller-Preussker, B. Petersson and A. Schreiber, *Phys. Rev. D* **85** (2012) 114504.
252. G. Bali, F. Bruckmann, G. Endrodi, Z. Fodor, S. Katz *et al.*, *JHEP* **1202** (2012) 044.
253. V. Bornyakov, P. Buividovich, N. Cundy, O. Kochetkov and A. Schäfer, *Phys. Rev. D* **90** (2014) 034501.
254. E. M. Ilgenfritz, M. Muller-Preussker, B. Petersson and A. Schreiber, *Phys. Rev. D* **89** (2014) 054512.
255. M. D'Elia and F. Negro, *Phys. Rev. D* **83** (2011) 114028.
256. G. Bali, F. Bruckmann, G. Endrodi, Z. Fodor, S. Katz *et al.*, *Phys. Rev. D* **86** (2012) 071502.
257. G. Bali, F. Bruckmann, G. Endrödi, S. Katz and A. Schäfer, *JHEP* **1408** (2014) 177.
258. C. Bonati, M. D'Elia, M. Mariti, F. Negro and F. Sanfilippo, *Phys. Rev. Lett.* **111** (2013) 182001.
259. C. Bonati, M. D'Elia, M. Mariti, F. Negro and F. Sanfilippo, *Phys. Rev. D* **89** (2014) 054506.
260. G. Bali, F. Bruckmann, G. Endrodi, F. Gruber and A. Schaefer, *JHEP* **1304** (2013) 130.
261. G. Bali, F. Bruckmann, G. Endrodi and A. Schafer, *Phys. Rev. Lett.* **112** (2014) 042301.
262. L. Levkova and C. DeTar, *Phys. Rev. Lett.* **112** (2014) 012002.
263. G. Endrödi, *JHEP* **1304** (2013) 023.
264. M. Abramczyk, T. Blum, G. Petropoulos and R. Zhou, *PoS.* **LAT2009** (2009) 181.
265. P. Buividovich, M. Chernodub, E. Luschevskaya and M. Polikarpov, *Phys. Rev. D* **80** (2009) 054503.
266. G. Bali, F. Bruckmann, G. Endrödi, Z. Fodor, S. Katz *et al.*, *JHEP* **1404** (2014) 129.

Initial State and Thermalization in the Color Glass Condensate Framework

François Gelis
Institut de Physique Théorique,
CEA / DSM / Saclay,
91191 Gif sur Yvette cedex, France
francois.gelis@cea.fr

In this review, I present the description of the early stages of heavy ion collisions at high energy in the Color Glass Condensate framework, from the pre-collision high energy nuclear wave function to the point where hydrodynamics may start becoming applicable.

1. Introduction

Heavy ion collisions pose a challenge for Quantum Chromodynamics (QCD) because a comprehensive description of these collisions involves a mix of hard short distance phenomena, and non-perturbative long distance soft physics. Even the aspects of these collisions where a hard scale may justify a weak coupling approach are not perturbative in the naive sense of a strict loop expansion. Indeed, resummations are often required even though the coupling is small, usually because the bosonic constituents of the system have a large occupation number, that may compensate the smallness of the coupling.

One of the areas where a weak coupling QCD approach is expected to be most effective is the description of the early stages of a heavy ion collision. The term "early stages" usually comprises the description of the relevant degrees of freedom in the wave functions of the two projectiles prior to their collision, the interactions that happen during the very brief duration of the collision itself, and the subsequent evolution of the produced gluons and quarks shortly after the collision. Roughly speaking, the temperature (or the fourth root of the energy density if the system is not yet thermalized) of the system can serve as a measure of the applicability of weak coupling techniques, since this scale sets the value of the running coupling constant.

A lot of progress has been made in the last 20 years in understanding how to apply QCD to these collisions. The starting point was the realization that a hard scale naturally emerges from the non-linear interactions among the gluons when their density is large,[1–3] as is the case in the wave function of high energy hadrons or nuclei. Thus, the bulk of particle production in high energy heavy ion collisions is amenable to a weak coupling treatment. The formalism for doing this — known as the Color Glass Condensate (CGC) — was progressively established and refined during this period, and has by now reached a mature state allowing quantitative and systematic calculations.

An outstanding problem, that has not yet reached a satisfactory state of understanding, is the transition from the CGC description to a more macroscopic description such as hydrodynamics. The main question is to explain, within the CGC framework, why an hydrodynamical description is possible in the first place. In other words, why does the system produced in a heavy ion collision flow as well as it seems to do? A satisfactory matching between the CGC and hydrodynamics implies that there should be a certain range of time in which the two descriptions predict the same evolution. At the moment, we are not there yet, even though considerable progress has been made in the past years.

After a brief account of why heavy ion collisions are interesting from the point of view of QCD (Sec. 2), the rest of this review follows the time-line of a collision. We recall the main physical ideas behind the parton model in Sec. 3, and we describe *gluon saturation* in Sec. 4. Section 5 is devoted to the Color Glass Condensate, the QCD-based effective theory for the saturated regime, and in Sec. 6 we show how to apply it in order to make leading order calculations in heavy ion collisions. Next-to-Leading order contributions are considered in Sec. 7, as well as the scale evolution of the gluon distribution in the projectiles via the JIMWLK equation. In Sec. 8, we first introduce the main issues and puzzles posed by attempts to match CGC calculations and hydrodynamics. In Sec. 9, we discuss the instabilities that exist in the solutions of the classical Yang–Mills equations, and their disastrous consequences for fixed loop-order CGC predictions beyond leading order. A resummation that cures these pathologies is presented in Sec. 10, leading to a scheme known as the *Classical Statistical Approximation.* We also present alternative derivations of this approximation scheme in order to make connections with other approaches. We present in Sec. 11 the results obtained by using this approximation in the context of heavy ion collisions, using different types of initial conditions. A discussion of some known shortcomings of this approximation is presented in Sec. 12.

2. Heavy Ion Collisions

2.1. *Reminder on QCD*

Although they occupy only a tiny fraction of the volume of atoms, the atomic nuclei make up for most of the mass of ordinary matter. The protons and neutrons that are contained in nuclei each contain three valence quarks, that give them their quantum numbers. However, these valence quarks account only for a small part of the nucleon mass. Most of it comes from binding energy, i.e. from the cloud of

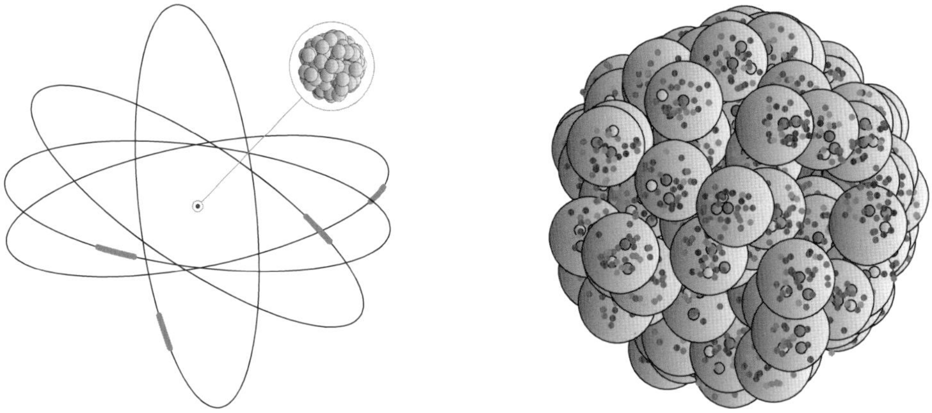

Fig. 1. (Color online) Atoms and nuclei.

gluons and virtual quark–antiquark pairs that surrounds the valence quarks. This predominance of binding energy in the mass of hadrons reflects a crucial property of the force which is responsible of the cohesion of hadrons and nuclei: this force becomes strong on distance scales comparable to the proton size, around 10^{-15} m. On the other hand, the measurements of structure functions in deep inelastic scattering experiments, first performed at SLAC in the 1960's, can be understood if one assumes that this force becomes weak on distance scales that are much smaller than the proton size.

The combination of these two properties paved the way to the development of Quantum Chromodynamics (QCD), as the microscopic theory that governs the interactions between quarks and gluons. On the surface, QCD is a gauge theory that resembles very much Quantum Electrodynamics. The matter degrees of freedom are spin 1/2 quarks, that interact by the exchange of vector particles, the gluons,

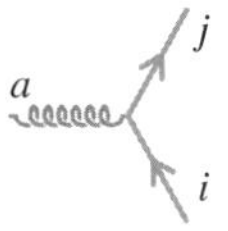

$\sim \; g\,(t^a)_{ij}$
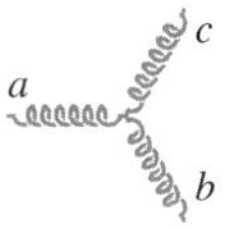

$\sim \; g\,(T^a)_{bc}$

(1)

The quark–gluon coupling in QCD is very similar to the electron photon coupling in QED, except that it has more "structure" since it involves a matrix of the fundamental representation of SU(3), t^a_{ij}. In this object, the index a (running from 1 to 8, the dimension of the SU(3) algebra) is the *color charge* of the gluon, and the indices i and j (running from 1 to 3, the dimension of the matrices in the fundamental representation of SU(3)) are the color charges of the incoming and outgoing quarks. The fact that the gluons themselves carry a color charge is the essential difference between QCD and QED, since it leads to novel interactions vertices that involve only gluons. These new interactions are a requirement of gauge symmetry, and can be derived from the following gauge invariant Lagrangian:

$$\mathcal{L} = -\frac{1}{4}F^2 + \sum_f \overline{\psi}_f(i\not{D} - m_f)\psi_f. \tag{2}$$

At the classical level, the only free parameters in QCD are the quark masses m_f and a coupling constant g. In the quantized theory, the coupling is usually traded for a scale[a] Λ_{QCD} that emerges from the renormalization of the coupling. This new scale arises in the running of the coupling constant $\alpha_s \equiv g^2/(4\pi)$. At one loop, this is given by[4–8]

$$\alpha_s(E) = \frac{2\pi N_c}{(11N_c - 2N_f)\log(E/\Lambda_{\text{QCD}})}, \tag{3}$$

where E is the energy scale, N_c the number of colors and N_f the number of quark flavors. The main difference compared to QED, due to the self-interactions of the gluons, is the fact that the coupling becomes smaller at short distances as shown in Fig. 2, a property known as *asymptotic freedom*.

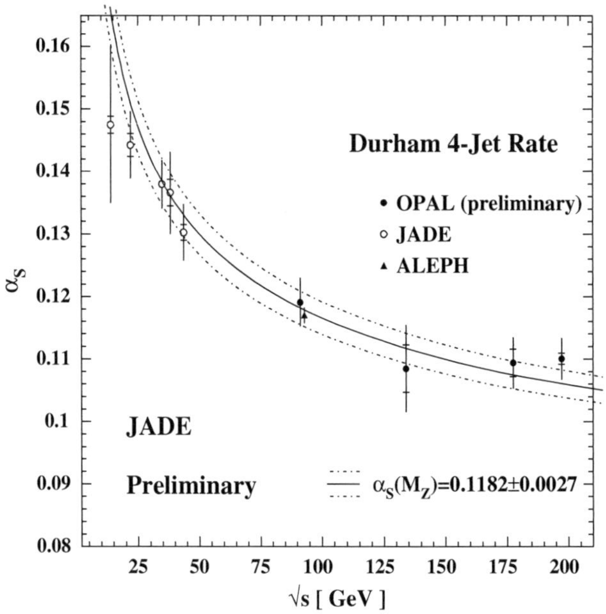

Fig. 2. Running coupling in QCD.

[a]Note that, in the absence of quarks (or with only massless quarks), QCD is scale invariant at the classical level. Loop corrections induce a breaking of this scale invariance, which is the reason for the appearance of Λ_{QCD} in the quantized theory.

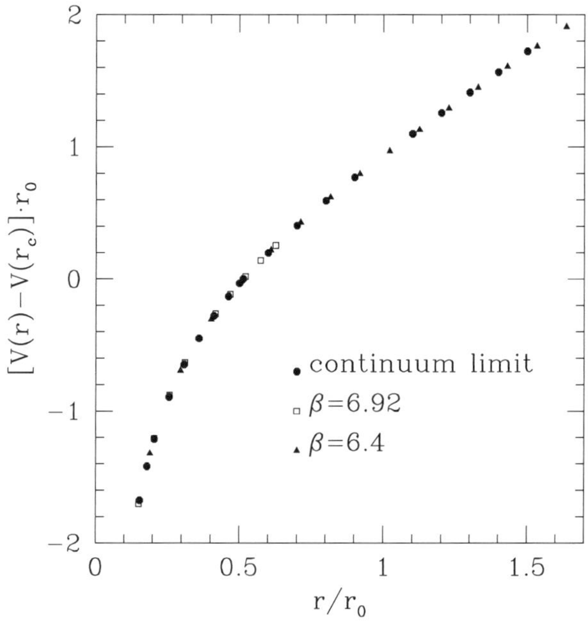

Fig. 3. Coulomb potential of a heavy quark and antiquark pair, from lattice QCD.

A related property of QCD is the long distance behavior of the interaction potential between a quark and an antiquark. This can be calculated numerically in lattice simulations for heavy quarks (that are therefore static). This potential, shown in Fig. 3, behaves as a standard $1/r$ Coulomb potential at short distance, but increases linearly at large distance, in sharp contrast with electromagnetic interactions. This leads to *color confinement*, that is the fact that free color charges cannot exist in Nature. Quarks only appear in color singlet bound states called hadrons, made of 3 quarks (baryons) or quark–antiquark pairs (mesons). The spectrum of these bound states can in principle be determined from first principles from the QCD Lagrangian, and it depends only on the quark masses and on the QCD scale $\Lambda_{\rm QCD}$. However, this dependence is non-perturbative and lattice simulations are the only way to perform these calculations. Presently, lattice calculations can reproduce the spectrum of light hadrons with an accuracy of the order of 5%, as illustrated in Fig. 4.

The QCD running coupling shown in Fig. 2 can also be viewed with a different perspective: it suggests that if one squeezes many hadrons in a small volume, then the average inter-quark distance will be small and their interactions will be weak. In such a situation, the quarks would not be confined into individual hadrons, and would instead form a plasma made of deconfined quarks and gluons. This idea is substantiated by lattice calculations of the QCD partition function as a function of temperature, that indicate a rapid increase of the number of effective degrees

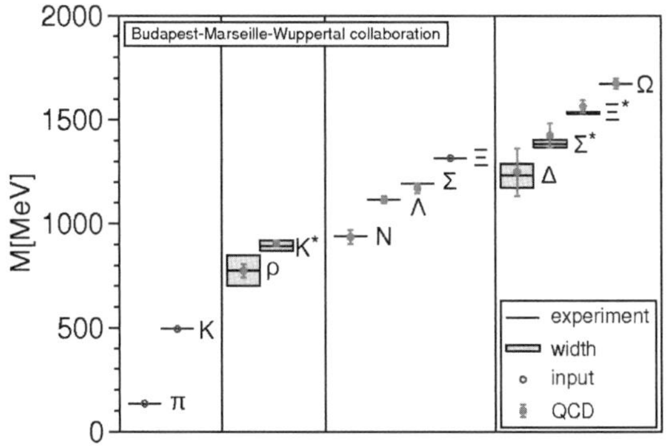

Fig. 4. (Color online) Hadron spectrum from lattice QCD.

of freedom at a temperature around 160 MeV.[b] This suggests that the relevant degrees of freedom are no longer the color singlet light hadrons (pions, kaons, ...) and have been replaced by quarks and gluons (that are more numerous because of the uncovered color degree of freedom).

2.2. *Heavy ion collisions*

Experimentally, the conditions of such a transition can be realized by colliding heavy nuclei at high energy. Such experiments are presently being performed by the RHIC (gold nuclei collided at 200 GeV) and by the LHC (lead nuclei collided at 5.5 TeV). Just after the impact of the two nuclei, the energy density reaches values that are more than ten times the normal nuclear matter density, well above the energy density at the deconfinement transition inferred from lattice calculations. Such a collision, whose total duration is of the order of 10 fm/c, can be divided

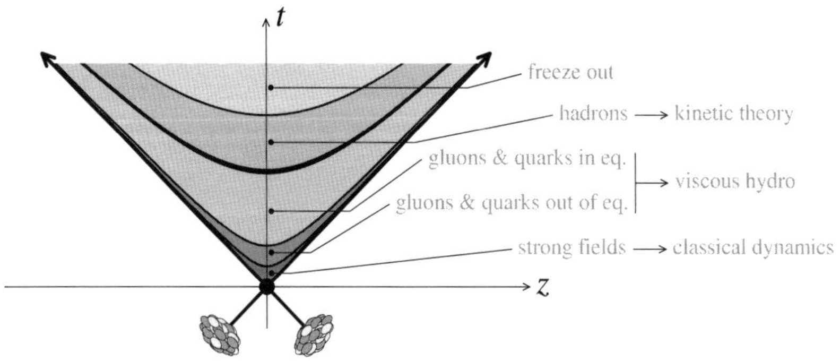

Fig. 5. (Color online) Successive stages of a heavy ion collision.

[b]This is for QCD with 3 light quark flavors. The transition temperature is higher for pure glue QCD.

into several stages, as shown in Fig. 5. In this figure, we have also indicated what kind of tool one may employ for each of these stages. It turns out that macroscopic descriptions such as relativistic hydrodynamics are quite successful at describing the bulk evolution of the system. Somewhat surprisingly, the matter produced in these collisions seems to behave almost like a perfect fluid with close to no viscosity. A very small viscosity suggests that this matter is not the siege of strong dissipative processes that would rearrange its microscopic degrees of freedom.

In these lectures, we will be primarily interested in the beginning of the collision, up to the point where a hydrodynamical description may become plausible. We will adopt a weak coupling perspective,[c] and we will try to follow a heavy ion collision in a description which is as close as possible to QCD. Indeed, in collisions at very high energy, the initial energy density is so large that the early stages of such a collision should be amenable to a weak coupling description, thanks to the asymptotic freedom of QCD. Note that a small viscosity, such that could explain the success of hydrodynamics, is more naturally obtained in the strong coupling limit because the viscosity is inversely proportional to the scattering cross-section of the quarks and gluons. However, it is also be possible to get strong interactions at weak coupling, provided that the occupation number is inversely proportional to the coupling g^2. In this case, the coupling disappears from the scattering rate, and the system has many of the features of a strongly coupled system.

3. Parton Model

3.1. *Kinematics*

As discussed earlier, free quarks and gluons do not exist in normal nuclear matter. Instead they are confined into color singlet bound states, whose spectrum depends non-perturbatively on the parameters of the QCD Lagrangian. The same is true of the energy levels of a nucleus: they could in principle be derived from the underlying QCD dynamics, but this is even more complicated than in the case of light hadrons and at the moment far out of reach of lattice computations.

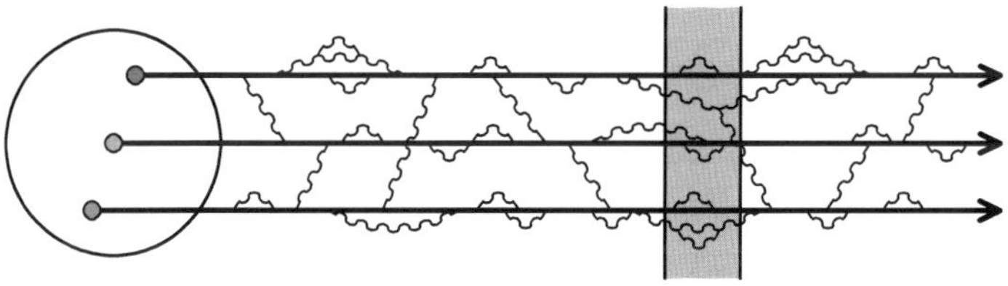

Fig. 6. (Color online) Dynamics of the constituents inside a slow nucleon.

[c]In the discussion on gluon saturation, we will see that the emergence of the *saturation scale*, that increases with the collision energy, justifies this assumption.

Does this mean that we should give up any hope of using QCD to describe collisions between such objects? Fortunately, the answer is no, *for collisions at sufficiently high energy.* The kinematics of these collisions is the key to overcome this difficulty. Let us consider first a nucleon at low energy (i.e. when the nucleon is almost at rest in the observer's frame), shown in Fig. 6. In this cartoon, the thick lines represent the three valence quarks, and the horizontal axis represents time. Only gluon constituents are shown, not the sea quarks. In such a frame, the valence quarks orbit with a period comparable to the proton size (they are ultrarelativistic). These quarks exchange gluons that provide the binding force, which also happens on scales of the order of the proton size. Moreover, the quarks and gluons can briefly fluctuate: for instance, a quark can temporarily become a quark+gluon state. The lifetime of these virtual states can be anything smaller than the proton size.[d] When studying reactions involving hadrons, one should compare the typical timescale of the collision (shown as a blue strip in the figure) with the timescales of the internal dynamics of the nucleon. In collisions involving low energy hadrons, the hadron has a complicated internal dynamics on timescales comparable to the duration of the collision, which makes these collisions untractable in perturbative QCD.

Contrast this with what happens in a collision at very high energy. Although scattering amplitudes are boost invariant and may be discussed in any frame, it is convenient to imagine that we do not change the momentum of one of the hadrons, and that all the energy increase is achieved by boosting the second hadron. This is illustrated in Fig. 7. The blue strip, unchanged compared to the low energy case, may be viewed as the size of the first hadron, that we did not boost. All the changes are in the internal dynamics of the second hadron, whose timescales are now stretched by Lorentz time dilation. The gluon exchanges between the valence quarks are now happening over timescales that are much longer than the duration of the collision, which means that the constituents of the nucleon can be viewed as free during the collision. The same happens to the fluctuations of the constituents. The lifetime of these virtual states is increased much beyond the collision timescales,

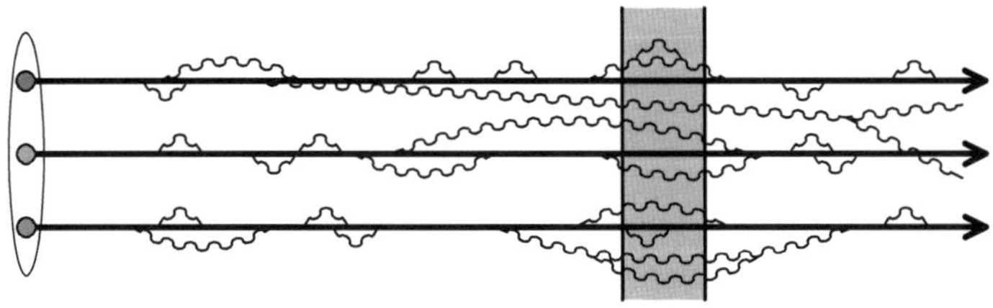

Fig. 7. (Color online) Dynamics of the constituents inside a highly boosted nucleon.

[d]But since QCD is a renormalizable theory, the physics of the strong interactions at hadronic energy scales does not depend on what happens on much higher energy scales. Therefore, these short lived fluctuations have essentially no relevance in hadronic physics.

making these off-shell constituents undistinguishable from on-shell particles.[e] Since there are fluctuations at arbitrary small timescales in a nucleon at rest, increasing the energy will uncover more and more of these fluctuations. These simple kinematical considerations are the essence of the *parton model*, that approximates a high energy nucleon or nucleus as a collection of quasi-free constituents (called partons), and whose density grows with energy.

3.2. *Factorization*

From this discussion, it seems that a QCD description of high energy collisions between hadrons may be feasible, provided we can provide "snapshots" of their partonic content at the time of the collision. What information is necessary in this snapshot is not completely obvious at this point, and may vary depending on the observable one intends to calculate, but for instance one may think of the following:

- flavor and color of each parton,
- transverse position and longitudinal momentum.

Of course, these properties of a hadron cannot be known event by event, which means that at best a probabilistic description may be achieved, that would allow to compute expectation values for event averaged observables. However, the possibility of describing hadronic collisions with only a probabilistic partonic description of the incoming hadrons is highly non-trivial, because it is an approximation that amounts to discarding certain quantum interferences. Without doing any approximation, the transition probability from a pair of hadrons $h_1 h_2$ to some final state X is obtained by summing all the relevant reaction channels before squaring the amplitude,

$$\begin{matrix}\text{transition probability}\\ \text{from hadrons to X}\end{matrix} \equiv \left|\sum \begin{matrix}\text{Amplitudes}\\ h_1 h_2 \to X\end{matrix}\right|^2 .$$

In contrast, the parton model as described above would express the transition probability as follows,

$$\begin{matrix}\text{transition probability}\\ \text{from hadrons to X}\end{matrix} \equiv \sum_{\substack{\text{partons}\\ \{q,g\}}} \begin{matrix}\text{probability to find}\\ \{q,g\} \text{ in } \{h_1,h_2\}\end{matrix} \otimes \left|\sum \begin{matrix}\text{Amplitudes}\\ \{q,g\} \to X\end{matrix}\right|^2$$

which is clearly an approximation of the previous formula. This approximation is called *initial state factorization*. Roughly speaking, the physical motivation for such a factorization is that the neglected terms are interferences between a hard process that occurs on the timescale of the collision and a process internal to one of the projectiles, happening on much longer timescales. The vast separation in

[e]The concept of on-shell or off-shell particles depends on the duration of the measurement. The only way to know that a particle is exactly on-shell is to perform an infinitely long measurement. Indeed, an off-shell particle may be viewed as a particle of momentum $\boldsymbol{p}$ whose energy differs from the on-shell energy $E_{\boldsymbol{p}}$ (given by the dispersion relation). A measurement that lasts Δt can only resolve energy differences of order $1/\Delta t$ or larger.

their timescales is what makes the corresponding interference small. At a more formal level, this factorization can be established in QCD, with various degrees of sophistication[f] depending on the observable.

3.3. *Single parton distributions*

The most developed framework for this type of factorization is the DGLAP formalism, in which one describes the incoming hadrons by *single parton distributions.* These distributions depend on the hadron and on the parton under consideration, on the fraction x of longitudinal momentum carried by the parton, and on a momentum scale Q that can be viewed as the inverse of the transverse spatial resolution with which the hadron is probed. In Fig. 8, the single parton distributions of a proton, extracted from deep inelastic scattering data, are shown at a fixed resolution scale Q. Although these distributions are non-perturbative and cannot be computed from the QCD Lagrangian,[g] QCD predicts how they change if one increases the

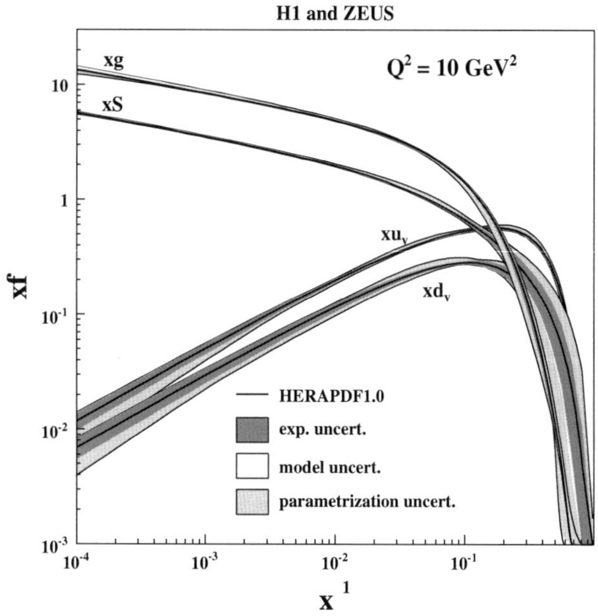

Fig. 8. (Color online) Parton distributions of a proton, at the resolution scale $Q^2 = 10\,\text{GeV}^2$. From Ref. 13.

[f]The weakest of these *factorization theorems* are based on *leading log factorization*, where the two formulas are shown to be equivalent for an infinite series of terms of the form $(\alpha_s \log(Q))^n$ (where Q is some hard scale), but not for terms of the form $\alpha_s(\alpha_s \log(Q))^n$. *Next-to-leading log factorization* extends the proof of this equivalence to include terms in $\alpha_s(\alpha_s \log(Q))^n$, and so on. *All-orders factorization theorems*[10–12] prove that the two formulas are equivalent up to terms that decrease as inverse powers of the hard scale.

[g]Lattice QCD can be used to evaluate the first few Mellin moments of parton distributions, since they are given by expectation values of local operators that can be evaluated in the Euclidean theory.

resolution scale Q, via the *DGLAP equation*.[14–17] From this figure, one sees that the valence quark distribution is predominant at large values of the momentum fraction $x \gtrsim 0.1$, and is totally negligible at small x. At any value $x \lesssim 0.1$, the gluons are the dominant species of partons, and their density increases like a power of $1/x$ when $x \to 0$. The sea quarks follow the trend set by the gluons, but with a suppression factor of order α_s since they are produced by the process $g \to q\bar{q}$.

4. Gluon Saturation

4.1. *Dense regime of QCD*

Since the DGLAP factorization framework is based solely on the single parton distributions, it is expected to become insufficient at large parton densities. The problem that will arise in this regime is illustrated in Fig. 9, that shows side to side a typical scattering process in the dilute (left) and dense (right) regimes. In the dilute situation, the incoming hadrons are "mostly empty", and hard scatterings are rare processes. Moreover, reactions involving more than one parton in each projectile are extremely rare (their rate scales as the square of the probability to find a parton). But when the parton density is large, processes initiated by multiple partons become more likely to happen. Moreover, the partons that lie nearby in phase-space may be correlated, and therefore multiparton distributions cannot be inferred simply from single parton distributions. A framework that would enable one to calculate these processes should provide information about multiparton distributions in hadrons and nuclei, and thus should go beyond the DGLAP framework. Moreover, when the parton density becomes of the order of the inverse coupling $1/g^2$, a strongly interacting regime — called *gluon saturation*[1–3] — is reached, where an infinite series of Feynman graphs contribute at each order in g^2.

A hint of the fact that the small x saturation regime is qualitatively different from the dilute regime appears when plotting the deep inelastic scattering cross-section slightly differently. This cross-section depends on two Lorentz invariant quantities, x and the 4-momentum squared Q^2 of the photon exchanged in the scattering. However, when plotted against the combination $x^{0.32}Q^2$, this data appears to line up on a unique curve (see Fig. 10). This scaling signals the emergence of an x dependent momentum scale, that behaves roughly as $Q_s^2(x) \sim x^{-0.32}$.

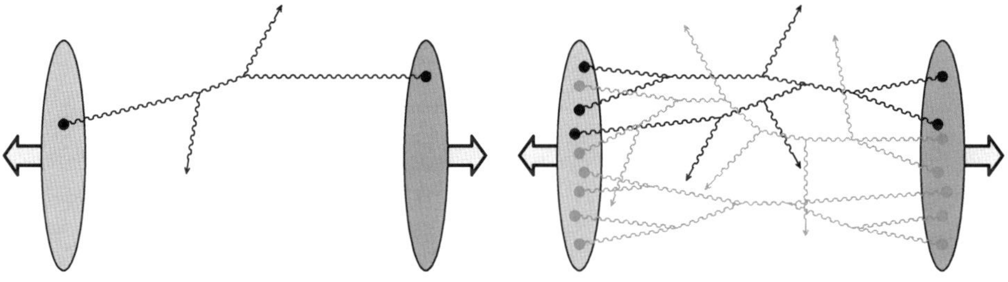

Fig. 9. (Color online) Typical partonic processes in a collision between dilute (left) and dense (right) projectiles.

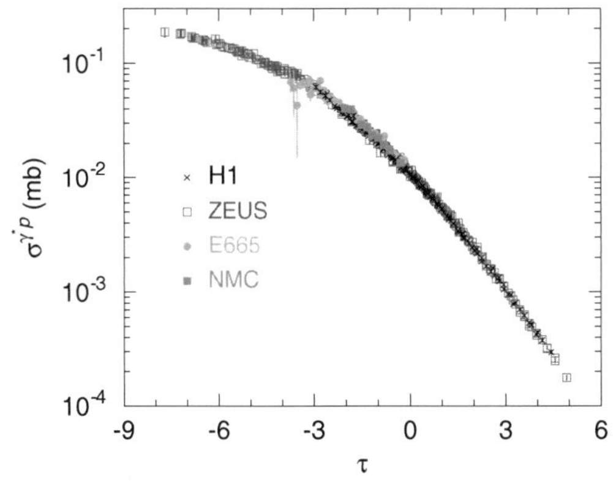

Fig. 10. (Color online) Geometrical scaling in the DIS cross-section at small x. The horizontal axis represents the variable $\tau \equiv x^{0.32}Q^2$.

This scale, known as the *saturation momentum*, appears as a consequence of the non-linear interactions among the gluons, that become important at high density.

4.2. *Saturation condition*

To understand the onset of gluon saturation, it is instructive to go back to the dilute regime at large x. In this situation, a hadron appears as a loose collection of a few partons. When the hadron is progressively boosted, these partons radiate more gluons by bremsstrahlung,[h] as illustrated in the top panel of Fig. 11. As long as the density remains low enough, these cascades of gluons develop independently and the evolution[i] of the hadron structure is governed by the linear *BFKL equation*.[18,19] Since these additional gluons are contained within the geometrical volume of the

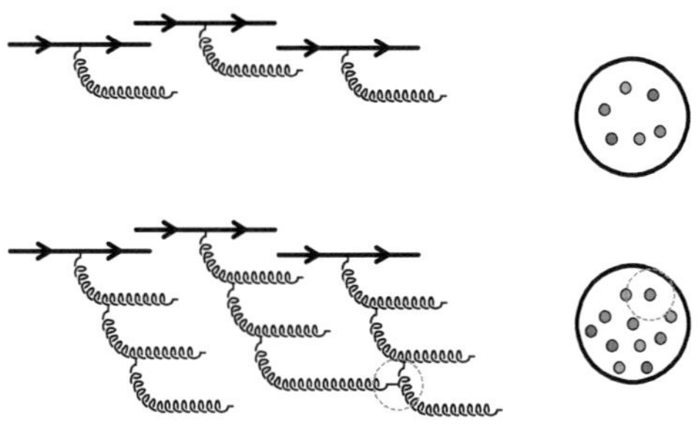

Fig. 11. (Color online) Gluon cascades in the small x evolution of a hadron. Top: dilute regime. Bottom: onset of the recombination corrections.

[h] As discussed before, these gluons are not truly on-shell, but can be viewed as real gluons if the lifetime of the quantum fluctuation that gave them birth is longer than the observation time.
[i] Although this terminology is commonly used, it is somewhat of a misnomer, since the hadron does not truly "evolve". It is the observer's view of the hadron content that changes as the observer's frame is increasingly boosted with respect to the hadron.

hadron, their density increases rapidly. At some point, their wave functions start to overlap and their interactions are no longer negligible. Gluons from two different cascades can recombine, which tames the growth of the gluon density. Moreover, this recombination process makes the x evolution of the gluon distribution non-linear.

Before going into a more quantitative description of gluon saturation, it is easy to derive a simple criterion for the onset of saturation. Gluon recombination becomes likely when the product of the numbers of gluons per unit area with the cross-section for recombining two gluons into one becomes larger than one,

$$\underbrace{\alpha_s Q^{-2}}_{\sigma_{gg\to g}} \times \underbrace{A^{-2/3} xG(x,Q^2)}_{\text{surface density}} \geq 1. \tag{4}$$

This condition can be rearranged in order to obtain an inequality on Q:

$$Q^2 \leq \underbrace{Q_s^2 \equiv \frac{\alpha_s xG(x,Q_s^2)}{A^{2/3}}}_{\text{saturation momentum}} \sim A^{1/3} x^{-0.3}. \tag{5}$$

This argument justifies the emergence of the saturation momentum, which characterizes the physics of gluon saturation. Saturation is important when the typical momentum scales in a process are smaller than Q_s. The region where this condition is satisfied is shown in Fig. 12. From the more quantitative plot on the right panel of this figure, a typical value to keep in mind is that Q_s^2 is in the range 2–4 GeV^2 for nuclei at the energy of the LHC. In the saturation domain, non-linear gluon interactions are important, which arguably makes the calculation of such processes more complicated. However, this also has an unexpected positive aspect: Q_s now supersedes all the softer momentum scales in determining the typical momentum of the relevant partons, and thus also controls the running of the coupling. Since Q_s increases when x decreases (i.e. when going at higher energy), this opens an avenue for an *ab initio* weak coupling treatment of multiparton interactions (sometimes called the "underlying event" in other contexts) in high energy hadronic

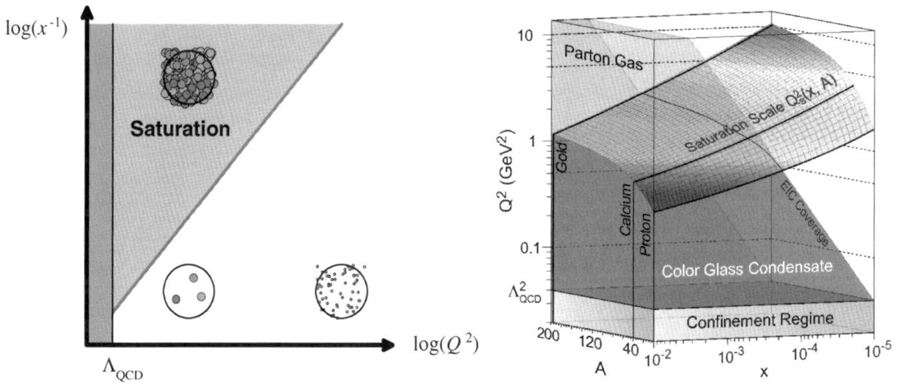

Fig. 12. (Color online) Saturation domain in the x and Q plane. The 3-dimensional plot on the right adds information about the A dependence. From Ref. 20.

scatterings. From Eq. (5), one sees that the saturation momentum also increases with the mass number of nuclei. This implies that, at a given energy, saturation effects are stronger in nucleus–nucleus collisions, for large nuclei.[j] This is important for heavy ion collisions at the RHIC and the LHC, because in these collisions the bulk of particle production is controlled by saturation physics.

5. Color Glass Condensate

5.1. *Degrees of freedom*

The CGC[k] is a QCD-based effective theory whose aim is to describe quantitatively the gluon saturation regime. The CGC exploits the high energy kinematics in order to simplify the description of the non-perturbative valence partons. The main idea was already encountered in the qualitative discussion of the parton model: a high energy hadronic collision is so brief that the fact that the partons are strongly bound by confinement is totally irrelevant. In fact, over such short timescales, the internal motion of the partons inside the hadron appears completely frozen. Thus, one may view the partons as static in the transverse plane, with a large longitudinal momentum.[26–28] For an observer sitting in the center of mass frame of the collision, the only information that matters about these partons is the color current J_a^μ that they carry along the beam direction. The dominant component of this 4-vector is the longitudinal one. In light-cone coordinates, for a hadron moving in the $+z$ direction, it reads

$$J_a^\mu(x) = \rho_a(x^-, \boldsymbol{x}_\perp)\delta^{\mu +}, \tag{6}$$

where the function $\rho_a(x)$ is the density of color charges of the partons. It does not depend on the light-cone "time" x^+ because of time dilation. Moreover, due to the concomitant Lorentz contraction, its x^- dependence is very peaked around $x^- = 0$.

It is important to realize that this drastic simplification cannot be used for all partons: it is applicable only to those partons whose longitudinal momentum (in the observer's frame) is large enough. The partons that have a rapidity close to the observer's rapidity have comparable transverse and longitudinal momenta, and thus cannot be approximated by a longitudinal current. Moreover, for these partons, the Lorentz boost factor that slows down their time evolution is not large and one cannot neglect their dynamics. Therefore, these slower partons must be treated as full fledged quantum fields.

The situation is summarized in Fig. 13: a cutoff $y_{\rm cut}$ must be introduced somewhere between the rapidity of the observer and the rapidity of the hadron under consideration. The partons close to the observer (mostly gluons, at least at leading order in α_s) are described as gauge fields, while those that are close to the projectile are approximated as a static longitudinal color current.

[j]For gold or lead nuclei, the factor $A^{1/3}$ that appears in Q_s^2 is approximately 6.

[k]For more detailed reviews of the Color Glass Condensate, one can consult Refs. 21, 22, 23, 24, 25.

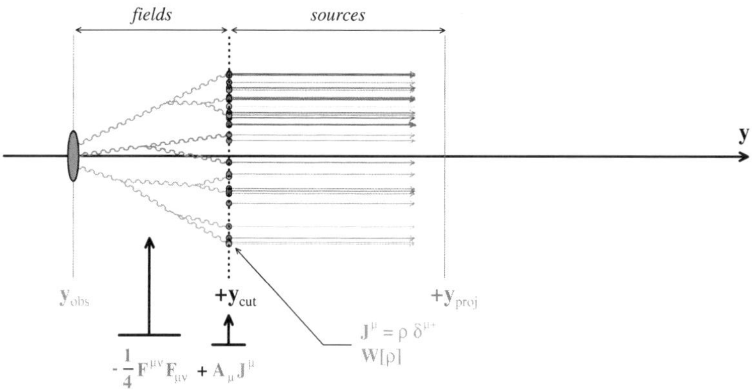

Fig. 13. (Color online) Degrees of freedom in the CGC effective theory.

5.2. *CGC effective theory*

Thanks to the rapidity separation between the slow and the fast degrees of freedom in the CGC, their coupling can be treated as eikonal, i.e. via a term of the form $J^\mu A_\mu$. Therefore, the CGC can be summarized by the following effective action

$$\mathcal{S}_{\rm CGC} = \int d^4x \left(-\frac{1}{4}F_{\mu\nu}F^{\mu\nu} + J^\mu A_\mu\right). \tag{7}$$

(For a collision of two hadrons, the current J^μ is the sum of two terms, one for each hadron.) The function $\rho_a(x^-, \boldsymbol{x}_\perp)$ that appears in the current J^μ reflects the particular arrangement of the fast partons at the time of the collision. It is not a quantity that can be predicted event by event, and one can only have a statistical knowledge of this object. Therefore, the CGC also introduces a probability distribution $W[\rho]$. As we shall see later, all observable quantities must be averaged over all the possible configurations of ρ, according to the distribution $W[\rho]$,

$$\langle \mathcal{O} \rangle = \int [D\rho] W[\rho] \mathcal{O}[\rho]. \tag{8}$$

In words, one should first calculate the observable for an arbitrary configuration of the color charge density ρ (in a collision of two hadrons, there is a ρ_1 and a ρ_2), and then perform a weighted average over all the possible ρ's. The justification of this procedure will be given in the following two sections.

6. CGC at Leading Order

6.1. *Power counting*

So far, we have not assumed anything about the magnitude of the color charge density ρ_a that describes the fast partons in the CGC effective theory. For the CGC to be applicable to the saturated regime, we must allow ρ_a to be as large as the inverse coupling $1/g$. Indeed, the recombinations due to non-linear gluon interactions can stabilize the gluon occupation number at a value of order $1/g^2$, where the gluon splittings and the recombinations balance each other. Since the occupation number is quadratic in the gauge field, such a value corresponds to $\rho \sim g^{-1}$. As we shall see, such a large value of the source ρ simplifies the dynamics by making it classical at leading order, but complicates things by making an infinite set of graphs contribute at each order in g^2. In order to see this, let us first examine the power counting in the CGC effective theory. Consider a generic connected graph,[l] as shown in Fig. 14. For such a graph, one finds that the order of magnitude depends only on the number of produced gluons and the number of loops, via the following formula[29,30]

$$\frac{1}{g^2}\, g^{\#\text{ produced gluons}}\, g^{2(\#\text{ loops})}. \tag{9}$$

The main consequence of assuming that $\rho \sim g^{-1}$ is that this power counting does not depend on the number of sources that are included into the graph. The reason for this is that each additional source (of order g^{-1}) is attached to the rest of the graph by a vertex (of order g), and therefore does not contribute to the overall magnitude of the graph. This also means that, at each order in g^2, one must sum an infinite set of graphs.

For instance, the inclusive gluon spectrum has the following expansion in powers of g^2

$$\frac{dN_1}{d^3\vec{\boldsymbol{p}}} = \frac{1}{g^2}[c_0 + c_1 g^2 + c_2 g^4 + \cdots], \tag{10}$$

where each of the coefficients $c_0, c_1, \ldots$ is itself an infinite series of terms of the form $(g\rho)^n$,

$$c_i \equiv \sum_{n=0}^{\infty} c_{i,n}(g\rho_{1,2})^n. \tag{11}$$

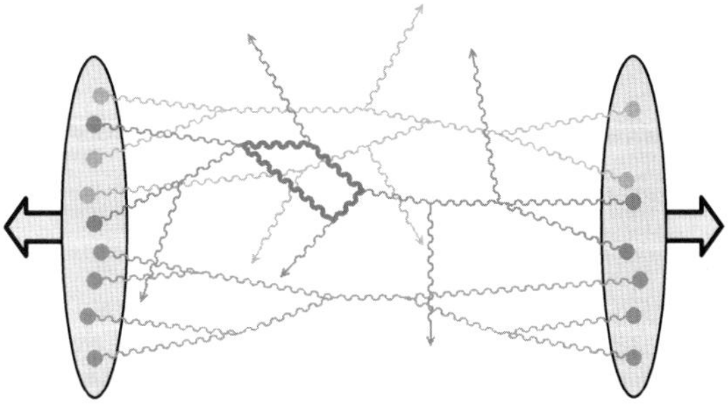

Fig. 14. (Color online) Generic connected graph in the CGC effective theory. The dots represent the sources ρ. The lines terminated by arrows are the gluons produced in the final state.

[l]Typical graphs are not connected: they are made of many disconnected subgraphs. But it is sufficient to discuss the properties of one of these subgraphs.

At this point, we should make an important remark regarding exclusive versus inclusive observables. From the above power counting, we see that the average number of produced gluons in a high energy nucleus–nucleus collision is of order $1/g^2$. If we assume for simplicity that the multiplicity distribution is Poissonian[m], the probability to have a given final state (e.g. a final state with a prescribed number of gluons) is exponentially suppressed by a factor $\exp(-\#/g^2)$. This factor may be viewed as a Sudakov factor that arises from excluding all the other final states. It turns out that these exclusive quantities are very difficult to calculate. In particular, the (disconnected) graphs that involve spectator partons contribute to exclusive observables, because these spectator partons may end up producing the unwanted final states.

In contrast, many simplifications occur in the calculation of inclusive quantities, that involve an average over all possible final states

$$\langle \mathcal{O} \rangle \equiv \sum_{\substack{\text{all final} \\ \text{states } f}} \mathcal{P}(AA \to f)\mathcal{O}[f]. \tag{12}$$

In particular, the high energy factorization results that will be discussed in the next section can only be established for these inclusive quantities, and their proof fails if one tries to generalize it to exclusive quantities.

6.2. *Calculation of inclusive observables*

The definition of Eq. (12) suggests an elementary method for calculating inclusive quantities: compute the exclusive probabilities to end up in a given final state f, and sum over all possible f's. However, it turns out that one can calculate them in a much more effective way without having to perform explicitly the sum over the final states. For this, one should use the *Schwinger–Keldysh formalism*,[31,32] in which this sum is already "built in". Any contribution to Eq. (12) is the product of an amplitude, a complex conjugate amplitude going to the same final state, and the observable evaluated on this final state, as illustrated in Fig. 15. The diagrammatic rules for the amplitude (right of the dotted line) are the usual time-ordered Feynman rules. The propagator is the usual Feynman propagator, which

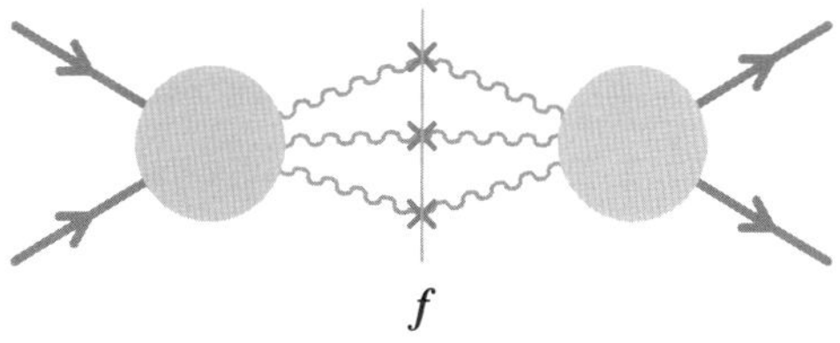

Fig. 15. (Color online) Illustration of the Schwinger–Keldysh formalism.

[m]This is not exactly true in the CGC, but this fact does not change the essence of this argument.

for a (massless) scalar particle reads

$$G^0_{++}(p) = \frac{i}{p^2 + i\epsilon}. \tag{13}$$

For the complex conjugate amplitude (left of the dotted line), one needs the complex conjugate of the vertices and propagators. The propagator is therefore

$$G^0_{--}(p) = \frac{-i}{p^2 - i\epsilon}. \tag{14}$$

Across the dotted line, one must use special propagators that represent the on-shell particles of the final state f,

$$G^0_{+-}(p) = 2\pi\theta(-p^0)\delta(p^2). \tag{15}$$

The Schwinger–Keldysh formalism amounts to the following:

- Draw all the graphs $AA \to AA$ that have a given order in g^2 (the power counting is the same as before, with each loop adding one power of g^2).
- Sum over all the possibilities of assigning the labels $+$ and $-$ to the internal vertices.
- Only connected graphs contribute, because when summed over the $+$ and $-$ labels, the subgraphs that are not attached to the observable vanish.

These rules will automatically provide the sum over final states that was included in the formula (12). Note that when used in this context, the Schwinger–Keldysh formalism is equivalent to Cutkosky's cutting rules,[33,34] that were developed as a tool to compute the imaginary part of transition amplitudes. The superficial description of the Schwinger–Keldysh formalism that we have given here can be made more rigorous by writing the generating functional for its Green's functions. It can be obtained as follows from the generating functional $Z[j]$ of time-ordered perturbation theory:

$$Z[j_+, j_-] = \exp\left[\int d^4x d^4y G^0_{+-}(x,y)\Box_x\Box_y \frac{\delta^2}{\delta j_+(x)\delta j_-(y)}\right] Z[j_+]Z^*[j_-]. \tag{16}$$

This formula makes more obvious the fact that the Schwinger–Keldysh formalism is made of two copies of the ordinary Feynman perturbation theory (one of them complex conjugated), "stitched" together by the on-shell propagators G^0_{+-}.

At this point, we have not really simplified the calculation of Eq. (12). It has just been rephrased in a more systematic language. The simplifications come from noticing the following identities,

$$\begin{aligned} &G_{++} + G_{--} = G_{+-} + G_{-+} \\ &G_{++} - G_{+-} = G_{-+} - G_{--} = G_R \quad \text{(retarded propagator)}. \end{aligned} \tag{17}$$

When using the Schwinger–Keldysh formalism to calculate inclusive observables at leading order, the sum over the $+$ and $-$ indices always generates the particular combinations of propagators that appear in the second of these equations, and these propagators therefore become retarded propagators.

6.3. *Classical equation of motion*

This simplification is particularly dramatic for inclusive observables at leading order. The starting point is a double sum over all possible tree diagrams (they all have the same order in g^2 when the source is $\rho \sim g^{-1}$) and over all the indices $+$ and $-$ of the Schwinger–Keldysh formalism. Thanks to the previous remark, the second sum merely replaces all the propagators by retarded propagators. One is thus left with a sum over all the tree diagrams built with retarded propagators, whose first few terms would be

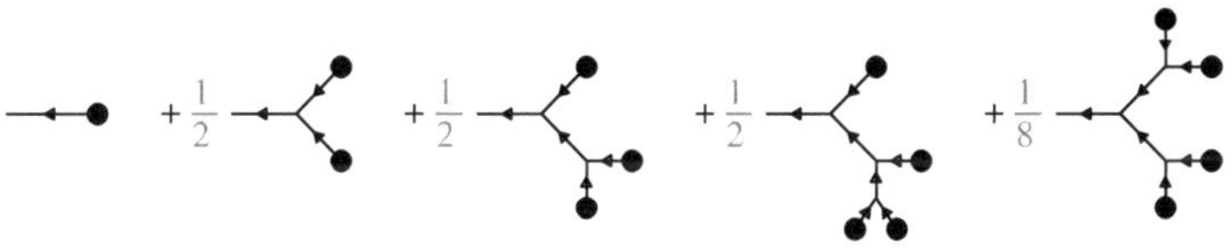

(here for a ϕ^3 scalar field theory). It is then easy to see that this sum is the solution of the classical field equations of motion that vanishes when $x^0 \to -\infty$ (this retarded boundary condition follows from the fact that we are summing tree that are made of retarded propagators[n]). Although in interacting theories the classical equation of motion is a non-linear wave equation, this is a considerable simplification because we have now a problem that can be solved numerically.

The same simplifications work in the case of the CGC: at leading order, it is sufficient to solve the classical Yang–Mills equations with null retarded boundary conditions

$$[D_\mu, F^{\mu\nu}] = \rho_1 \delta^{\nu +} + \rho_2 \delta^{\nu -}, \qquad \lim_{x^0 \to -\infty} A^\mu(x) = 0. \tag{18}$$

Assuming that we have solved this equation, all the inclusive observables at leading order can be expressed in terms of its solution $\mathcal{A}^\mu$. For instance, the single inclusive gluon spectrum is given by

$$\left.\frac{dN_1}{dY\, d^2\vec{\boldsymbol{p}}_\perp}\right|_{\rm LO} = \frac{1}{16\pi^3} \int_{x,y} e^{ip\cdot(x-y)} \square_x \square_y \sum_\lambda \epsilon_\lambda^\mu \epsilon_\lambda^\nu \mathcal{A}_\mu(x) \mathcal{A}_\nu(y), \tag{19}$$

and the inclusive multigluon spectra simply read

$$\left.\frac{dN_n}{d^3\boldsymbol{p}_1 \cdots d^3\boldsymbol{p}_n}\right|_{\rm LO} = \left.\frac{dN_1}{d^3\boldsymbol{p}_1}\right|_{\rm LO} \times \cdots \times \left.\frac{dN_1}{d^3\boldsymbol{p}_n}\right|_{\rm LO}. \tag{20}$$

[n]One can see here why it is important to consider inclusive observables for this simplification to happen. It is the sum over the final states that leads to the sum over the indices $+$ and $-$. Without this sum, one would be left with time-ordered propagators, which would make the boundary conditions of the classical solution untractable.

Similarly, the components of the energy–momentum tensor have simple expressions in terms of the classical chromo-electric and chromo-magnetic fields $\boldsymbol{E}^i$ and $\boldsymbol{B}^i$,

$$T^{00}_{\rm LO} = \frac{1}{2}[\boldsymbol{E}^2 + \boldsymbol{B}^2] \quad T^{0i}_{\rm LO} = [\boldsymbol{E} \times \boldsymbol{B}]^i \tag{21}$$

$$T^{ij}_{\rm LO} = \frac{\delta^{ij}}{2}[\boldsymbol{E}^2 + \boldsymbol{B}^2] - [\boldsymbol{E}^i\boldsymbol{E}^j + \boldsymbol{B}^i\boldsymbol{B}^j]. \tag{22}$$

6.4. *Numerical implementation*

In order to perform this calculation in practice, one should be aware of the following:

(i) High energy collisions are nearly invariant under boosts in the longitudinal direction. This invariance has its simplest manifestation if one uses proper-time ($\tau \equiv \sqrt{2x^+x^-}$) and rapidity ($\eta \equiv \frac{1}{2}\log(x^+/x^-)$) as the coordinates inside the forward light-cone. When written in this system of coordinates, the classical Yang–Mills equations do not depend explicitly on rapidity, and thus become $1+2$ dimensional equations.

(ii) The sources ρ_1 and ρ_2 have support on the light-cone, where they are singular, i.e. proportional respectively to $\delta(x^-)$ and $\delta(x^+)$. These sources divide the space-time in four distinct regions, as shown in the left Fig. 16. The gauge field is identically zero in the region 0, and it can be found analytically in the regions 1 and 2.[35] In the region 3, the best one can do analytically is to obtain the value of the gauge fields and the conjugate electrical fields just above the forward light-cone, at a proper time $\tau = 0^+$[36]:

$$A^i_0 = \alpha^i_1 + \alpha^i_2, \quad E^i_0 = 0, \quad \alpha^i_n = \frac{i}{g}U^\dagger_n \partial^i U_n \quad (n = 1, 2),$$

$$A_{0\eta} = 0, \quad E^\eta_0 = i\frac{g}{2}[\alpha^i_1, \alpha^i_2], \tag{23}$$

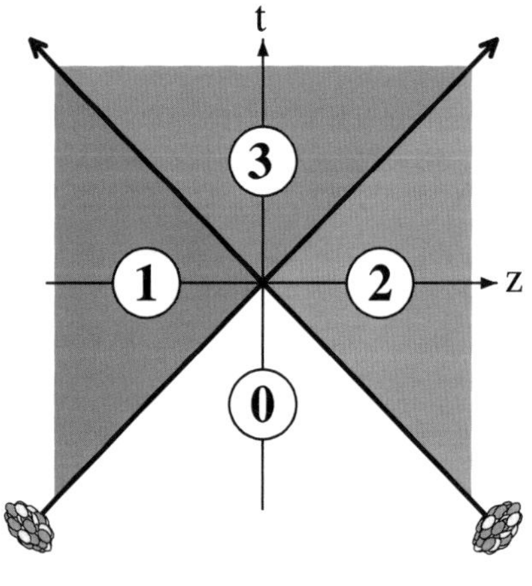

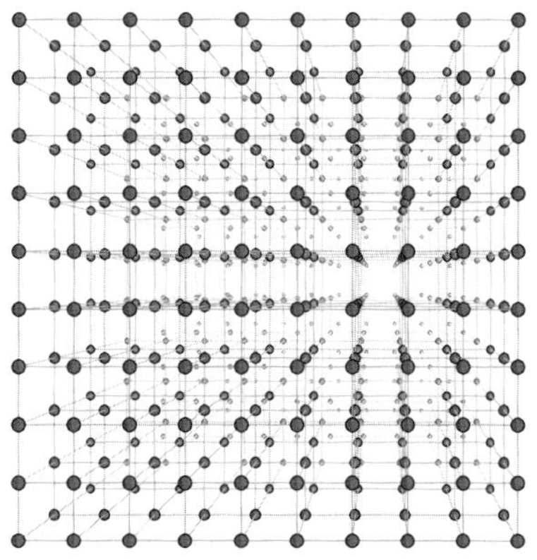

Fig. 16. (Color online) Left: space-time structure of the classical gauge field $\mathcal{A}^\mu$. Right: 3-dimensional cubic lattice.

where the Wilson lines $U_{1,2}(\boldsymbol{x}_\perp)$ read

$$U_1(\boldsymbol{x}_\perp) = \mathrm{P}e^{ig\int dx^- \frac{1}{\nabla_\perp^2}\rho_1(x^-,\boldsymbol{x}_\perp)}. \tag{24}$$

Given these remarks, we need only to solve numerically (1 + 2)-dimensional classical Yang–Mills equations inside the forward light-cone,[37–46] starting with the initial conditions (23). Choosing the time variable τ determines the Hamiltonian of the system, and from there one can determine the Yang–Mills equations in Hamiltonian form. In order to handle them numerically, one must discretize space on a cubic lattice (see Fig. 16(right)), while time remains a continuously varying variable. This is most easily done in the temporal gauge $A^\tau = 0$ (also called the Fock–Schwinger gauge in this context). After having adopted this gauge condition, the problem has a residual gauge invariance, under any gauge transformation that depends only on space. Naive discretizations based on the gauge potentials A^μ are not very adequate because they lead to violations of this residual gauge invariance. Instead, one should adopt Wilson's formulation, in which the gauge potentials are traded in favor of link variables (see Fig. 17(left)), i.e. Wilson lines that span one elementary edge of the lattice

$$U_i(x) \equiv \mathrm{P}\exp ig\int_x^{x+\hat{\imath}} ds A^i(s). \tag{25}$$

Under a residual gauge transformation, these links transform as

$$U_i(x) \to \Omega(x)U_i(x)\Omega^\dagger(x+\hat{\imath}). \tag{26}$$

The electrical fields E^i that appear in Hamilton's equations transform covariantly,

$$E^i(x) \to \Omega(x)E^i(x)\Omega^\dagger(x), \tag{27}$$

and therefore they should live on the nodes of the lattice. In the $A^\tau = 0$ gauge, the Hamiltonian discretized in this fashion reads

$$\begin{aligned}\mathcal{H} &= \sum_{\vec{\boldsymbol{x}};i}\frac{E^i(\boldsymbol{x})E^i(\boldsymbol{x})}{2}\\ &\quad - \frac{6}{g^2}\sum_{\vec{\boldsymbol{x}};ij} 1 - \frac{1}{3}\mathrm{Re}\ \mathrm{Tr}\,(\underbrace{U_i(x)U_j(x+\hat{\imath})U_i^\dagger(x+\hat{\jmath})U_j^\dagger(x)}_{\text{plaquette at the point } \vec{\boldsymbol{x}} \text{ in the } ij \text{ plane}}).\end{aligned} \tag{28}$$

The only combinations of link variables that enter in this formula are *plaquettes* (i.e. the trace of the product of the four link variables that form an elementary

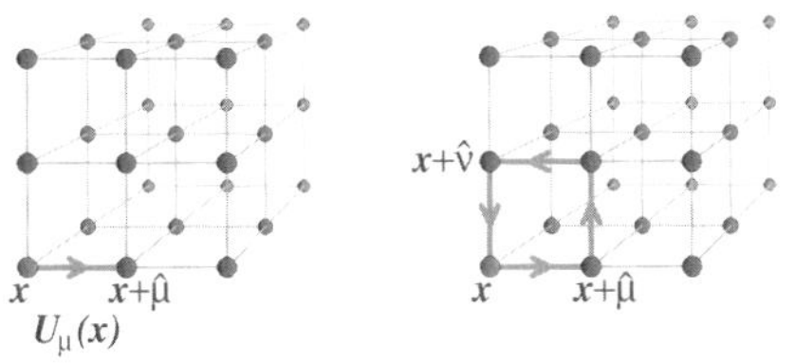

Fig. 17. (Color online) Left: link variable. Right: elementary plaquette variable.

square on the cubic lattice), which are gauge invariant. The Hamilton equations that can be derived from this Hamiltonian form a large (but finite) set of ordinary differential equations, that can be solved numerically by standard methods such as the leapfrog algorithm.

6.5. *Structure of the classical color fields*

At very short times after the collisions ($\tau \ll Q_s^{-1}$), the classical chromo-electric and chromo-magnetic fields are parallel to the collision axis,[47] as illustrated in Fig. 18. By studying how the expectation value of transverse Wilson loops,[48,49]

$$W \equiv \left\langle \mathrm{P}\exp ig\int_\gamma dx^i \mathcal{A}^i \right\rangle , \tag{29}$$

depends on the area enclosed by the loop, one can infer the typical transverse size of these flux tubes. Indeed, one can roughly view the argument of the exponential as the magnetic flux going through the loop. It has been found that W decreases roughly as $\exp(-\# \times \text{Area})$ for areas larger than Q_s^{-2}, which indicates that the

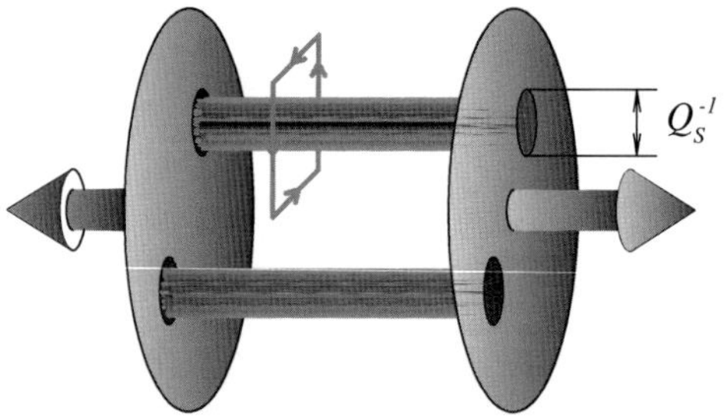

Fig. 18. (Color online) Color flux tubes just after the collision.

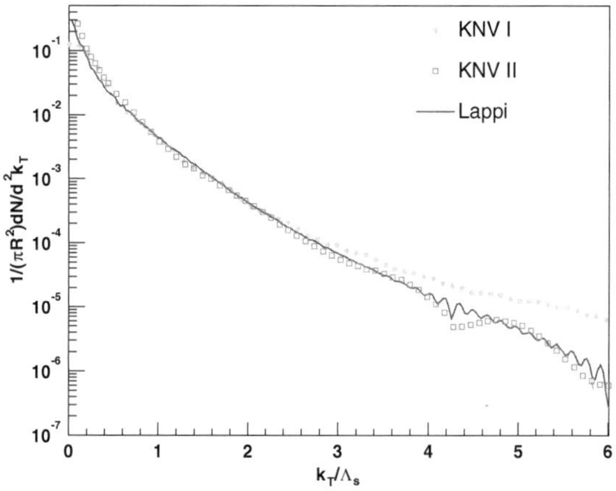

Fig. 19. (Color online) Inclusive gluon spectrum.

fields are not correlated over transverse distances larger than Q_s^{-1}. Q_s^{-1} is therefore the typical radius of these flux tubes.

From the classical gauge fields, one can compute the spectrum of gluons produced in a heavy ion collision (see Fig. 19). At large transverse momentum, the spectrum decreases as $k_\perp^{-4}$. Indeed, when $k_\perp \gg Q_s$, the saturation criterion is not satisfied and one should recover the usual perturbative results of the dilute regime. In contrast, saturation effects are quite large at small transverse momentum, $k_\perp \lesssim Q_s$, where they produce a strong softening of the spectrum, while the dilute result would still give a spectrum that grows as $k_\perp^{-4}$ at small $k_\perp$ (because there is no dimensionful scale other than $k_\perp$ in the dilute calculation). Let us close this section by a remark concerning the energy dependence of the gluon multiplicity, obtained by integrating the gluon spectrum over $k_\perp$. The result is proportional to the transverse area $S_\perp$ multiplied by Q_s^2,

$$N_{\text{gluon}} \sim \frac{S_\perp Q_s^2}{\alpha_s}. \tag{30}$$

From this pocket formula, one sees that the energy dependence of the gluon multiplicity is directly inherited from that of the saturation momentum,

$$N_{\text{gluon}} \sim x^{-0.3} \sim s^{0.15}. \tag{31}$$

Note that there is no contradiction between the fact that the multiplicity grows like a power of the collision energy and the Froissart bound, that tells us that hadronic cross-sections cannot grow faster than $\sigma \sim \log^2(s)$. The difference between the two is due to the fact that these two objects measure very different things. The total cross-section measures the probability that the two projectiles interact. As a probability, its growth is constrained by unitarity. In contrast, the gluon multiplicity measures the "amount of stuff" which is produced in a collision. It is not a probability, and is not bound by the same constraints. More precisely, the Froissart bound is related to the growth of the radius of the "black disk" region with energy (this radius grows like the log of energy). However, even after a certain region has become "black" (and thus its probability of interacting cannot grow anymore, because it has reached the unitarity limit), the number of gluons produced in this region will continue to grow like a power of energy (because the number of gluons per unit area in the incoming projectiles continues to increase like Q_s^2).

7. Next to Leading Order

7.1. *Improved power counting*

From the power counting formula derived earlier, we expected for the gluon spectrum an expansion of the form

$$\frac{dN}{d^3\vec{p}} = \frac{1}{g^2}[c_0 + c_1 g^2 + c_2 g^4 + \cdots], \tag{32}$$

and we have seen in the previous section how to calculate the term c_0/g^2. The following terms, $c_1, c_2 g^2, \ldots$, are given respectively by the sum of 1-loop, 2-loop, etc. diagrams. When calculating loops in the CGC framework, we have to recall that the degrees of freedom have been divided into sources and fields, separated by a cutoff $y_{\rm cut}$ in rapidity. In the integration over the loop momentum, we must use this cutoff in order to prevent the loop momentum to go into the kinematical domain which is described in terms of static sources. Failing to do this would lead to a double counting of the contribution of the modes that lie in this region.

In general, loop diagrams will depend on this cutoff[50]: a graph with n loops can contain up to n powers of $y_{\rm cut}$. Therefore, the coefficients that appear in the g^2 expansion of the gluon spectrum can themselves be expanded in powers of the cutoff as follows:

$$\begin{aligned} c_1 &= c_{10} + c_{11} y_{\rm cut} \\ c_2 &= c_{20} + c_{21} y_{\rm cut} + \underbrace{c_{22} y_{\rm cut}^2}_{\text{Leading Log terms}} \end{aligned} \tag{33}$$

The terms that have the maximal degree in $y_{\rm cut}$, i.e. a degree equal to the number of loops, are called *leading log terms.*[o]

The cutoff $y_{\rm cut}$ was introduced by hand when defining the CGC, as a way of separating the two kinds of degrees of freedom, and it is therefore not a physical parameter. Observables should not depend upon it. As we shall see in this section, the leading log cutoff dependence that arises from loop corrections to observables can be absorbed into a redefinition of the probability distribution $W[\rho]$. This redefinition turns $W[\rho]$ into a cutoff dependent object, but its cutoff dependence is universal, which means that the same distributions can be used for all inclusive observables.

Before we continue with a discussion of the leading log terms, let us also mention the fact that the next-to-leading log corrections are now known in some cases: in the BK equation (a mean field approximation of the JIMWLK equation)[51–54] and also for the JIMWLK equation.[55–57] The running coupling corrections have been used in some phenomenological studies[58–62] where they appear to be quantitatively important.

7.2. *NLO result*

Let us give here a sketch of the proof of this statement. The first step is the derivation of an expression for the NLO correction to inclusive observables. It turns out that there exists a formal relationship between the LO and NLO contributions, valid for any inclusive observable, that reads[63,64]:

$$\mathcal{O}_{\rm NLO} = \left[\frac{1}{2} \int_{\boldsymbol{u},\boldsymbol{v}} \boldsymbol{\Gamma}_2(\boldsymbol{u},\boldsymbol{v}) \mathbb{T}_{\boldsymbol{u}} \mathbb{T}_{v} + \int_{\boldsymbol{u}} \boldsymbol{\alpha}(\boldsymbol{u}) \mathbb{T}_{u} \right] \mathcal{O}_{\rm LO}. \tag{34}$$

[o]The terminology comes from the fact that $y_{\rm cut}$ is the logarithm of a cutoff on the longitudinal momentum.

In this formula, the LO observable $\mathcal{O}_{\rm LO}$ should be viewed as a functional of the initial value of the classical field on some space-like hypersurface (the integrations over the variables $\boldsymbol{u}$ and $\boldsymbol{v}$ are on this surface). The operator $\mathbb{T}_{\boldsymbol{u}}$ is the generator of shifts of this initial condition, in the sense that its exponential translates the initial field $\mathcal{A}_{\rm init}$ in any quantity that can be expressed in terms of the classical field

$$\exp\left[\int_{\boldsymbol{u}} \boldsymbol{\alpha}(\boldsymbol{u})\mathbb{T}_{\boldsymbol{u}}\right]\mathcal{F}[\mathcal{A}_{\rm init}] = \mathcal{F}[\mathcal{A}_{\rm init}+\boldsymbol{\alpha}]. \tag{35}$$

The remarkable property of Eq. (34) is that the coefficient functions $\boldsymbol{\Gamma}_2$ and $\boldsymbol{\alpha}$ are universal: they do not depend on the observable under consideration. Note however that although this formula is valid for all inclusive observables, it is not true for exclusive observables.

7.3. *Classical phase-space formulation of quantum mechanics*

In the formula (34), the operator between the square brackets acts only on the initial value of the classical fields, while the time evolution from the initial time surface to the time where the observable is evaluated remains classical. This is in fact a completely general result in quantum mechanics: at the first order in $\hbar$, the time evolution remains classical and $\hbar$ enters only in the initial condition. Let us make a digression to justify this important point. This is best viewed if one rewrites the evolution equation for the density operator,

$$\frac{\partial\widehat{\rho}_\tau}{\partial\tau} = i\hbar[\widehat{H},\widehat{\rho}_\tau], \tag{36}$$

in terms of the Wigner transforms of $\widehat{\rho}_\tau$ and $\widehat{H}$,

$$\begin{aligned} W_\tau(\boldsymbol{x},\boldsymbol{p}) &\equiv \int ds e^{i\frac{\boldsymbol{p}\cdot\boldsymbol{s}}{\hbar}}\left\langle \boldsymbol{x}+\frac{\boldsymbol{s}}{2}|\widehat{\rho}_\tau|\boldsymbol{x}-\frac{\boldsymbol{s}}{2}\right\rangle \\ \mathcal{H}(\boldsymbol{x},\boldsymbol{p}) &\equiv \int ds e^{i\frac{\boldsymbol{p}\cdot\boldsymbol{s}}{\hbar}}\left\langle \boldsymbol{x}+\frac{\boldsymbol{s}}{2}|\widehat{H}|\boldsymbol{x}-\frac{\boldsymbol{s}}{2}\right\rangle . \end{aligned} \tag{37}$$

(The Wigner transform of $\widehat{H}$ is nothing but the classical Hamiltonian of the system.) Note that in these Wigner transforms, the variables $\boldsymbol{x}$ and $\boldsymbol{p}$ are classical phase-space variables, not operators. It is straightforward to show that Eq. (36) is equivalent to

$$\begin{aligned} \frac{\partial W_\tau}{\partial\tau} &= \mathcal{H}(\boldsymbol{x},\boldsymbol{p})\frac{2}{i\hbar}\sin\left(\frac{i\hbar}{2}(\overleftarrow{\partial}_{\boldsymbol{p}}\overrightarrow{\partial}_{\boldsymbol{x}}-\overleftarrow{\partial}_{\boldsymbol{x}}\overrightarrow{\partial}_{\boldsymbol{p}})\right)W_\tau(\boldsymbol{x},\boldsymbol{p}) \\ &= \underbrace{\{\mathcal{H},W_\tau\}}_{\text{Poisson bracket}} + \mathcal{O}(\hbar^2). \end{aligned} \tag{38}$$

The first line is known as the *Moyal equation.* It is equivalent to the von Neumann equation for $\widehat{\rho}_\tau$, except that it is expressed entirely in terms of classical phase-space variables. This equation makes the classical limit particularly transparent if one expands in powers of $\hbar$ the operator that appears in its right-hand side. As one can see immediately, its zeroth order in $\hbar$ is the usual Poisson bracket. This means that at the order $\hbar^0$, one recovers classical Hamiltonian mechanics. A remarkable feature of the Moyal equation is that it has no term of order $\hbar^1$. This means that at NLO, the time evolution of the system remains purely classical. The only quantum effects at order $\hbar^1$ come via the initial condition, through the fact that the support of the Wigner distribution of a quantum state must have an extension of at least $\hbar$. The first correction to the Moyal equation arises at the order $\hbar^2$, i.e. at NNLO. At this order, one gets quantum corrections both in the initial condition and in the time evolution itself. This discussion also indicates that a formula such as Eq. (34), where only the initial conditions are altered, presumably does not exist beyond NLO.

7.4. *Cutoff dependence*

Equation (34) is very useful in order to extract the cutoff dependence in inclusive observables at NLO, because this cutoff dependence is already present in the operator that acts on $\mathcal{O}_{\rm LO}$. If we keep only the terms that are linear in the cutoff, we have[63–65]

$$\frac{1}{2}\int_{\boldsymbol{u},\boldsymbol{v}} \boldsymbol{\Gamma}_2(\boldsymbol{u},\boldsymbol{v})\mathbb{T}_{\boldsymbol{u}}\mathbb{T}_{\boldsymbol{v}} + \int_{\boldsymbol{u}} \boldsymbol{\alpha}(\boldsymbol{u})\mathbb{T}_{\boldsymbol{u}} = y_{\rm cut}^{+}\mathcal{H}_1 + y_{\rm cut}^{-}\mathcal{H}_2. \tag{39}$$

In this equation, $y_{\rm cut}^{+}$ and $y_{\rm cut}^{-}$ are the cutoffs corresponding to the right and left moving nucleus respectively, and $\mathcal{H}_{1,2}$ are operators known as the *JIMWLK Hamiltonians*[p] of the two nuclei,[66–74]

$$\mathcal{H} \equiv \frac{1}{2}\int_{\vec{\boldsymbol{x}}_\perp,\vec{\boldsymbol{y}}_\perp} \frac{\delta}{\delta\rho_a(\vec{\boldsymbol{x}}_\perp)}\chi_{ab}(\vec{\boldsymbol{x}}_\perp,\vec{\boldsymbol{y}}_\perp)\frac{\delta}{\delta\rho_b(\vec{\boldsymbol{y}}_\perp)}, \tag{40}$$

where

$$\chi_{ab}(\vec{\boldsymbol{x}}_\perp,\vec{\boldsymbol{y}}_\perp) \equiv \frac{\alpha_s}{4\pi^3}\int d^2\vec{\boldsymbol{z}}_\perp \frac{(\vec{\boldsymbol{x}}_\perp-\vec{\boldsymbol{z}}_\perp)\cdot(\vec{\boldsymbol{y}}_\perp-\vec{\boldsymbol{z}}_\perp)}{(\vec{\boldsymbol{x}}_\perp-\vec{\boldsymbol{z}}_\perp)^2(\vec{\boldsymbol{y}}_\perp-\vec{\boldsymbol{z}}_\perp)^2}$$
$$\times\,[(1-\widetilde{U}^\dagger(\vec{\boldsymbol{x}}_\perp)\widetilde{U}(\vec{\boldsymbol{z}}_\perp))(1-\widetilde{U}^\dagger(\vec{\boldsymbol{z}}_\perp)\widetilde{U}(\vec{\boldsymbol{y}}_\perp))]_{ab}. \tag{41}$$

In this equation, $\widetilde{U}$ is a Wilson line in the adjoint representation, constructed from the gauge field A^+ such that $\boldsymbol{\nabla}_\perp^2 A^+ = -\rho$.

7.5. *Factorization*

The formula (39) has two important consequences:

(i) This formula is the *sum* of two terms corresponding to the two nuclei, but there is no cutoff dependent term mixing the two nuclei. This means that the cutoff dependent terms are intrinsic properties of the nuclei prior to their collision, and this is the reason why it is possible to eliminate them by redefining the distributions of the sources $\rho_{1,2}$ of the two projectiles.

[p]If one expands this Hamiltonian at small ρ, one can recover the BFKL equation.[18,19]

(ii) Since the operator in the square brackets in Eq. (34) is the same for all inclusive observables, the cutoff dependence is equally universal. This is the reason why it will be possible to define cutoff dependent distributions $W[\rho_{1,2}]$ such that they cancel the cutoff dependence of all observables.

The property (i), about the absence of mixing of the cutoff dependence between the two nuclei, can be understood simply in terms of causality. This is illustrated in Fig. 20. Indeed, the cutoff dependence arises from the phase-space integration of the soft gluons emitted by bremsstrahlung. Since they are soft, the formation time of these gluons is large: they cannot be emitted during the very brief duration of the collision, so they have to be emitted before the collision. Because the separation between the two nuclei is space-like until the collision, causality forbids any cutoff dependent term that would mix the two nuclei. The property (ii) also follows the same reasoning: gluon emissions that happen before the collision should be the same for all observables measured after the collision.

If we compute observables for fixed configurations ρ_1 and ρ_2 of the color charge densities in the two projectiles, there is no way to get rid of the cutoff dependence. The only way to remove it is to integrate over all the possible configurations of $\rho_{1,2}$. The main ingredient in this manipulation is the fact that the JIMWLK Hamiltonian $\mathcal{H}$ is a self-adjoint operator:

$$\int [D\rho] W(\mathcal{H}\mathcal{O}) = \int [D\rho](\mathcal{H}W)\mathcal{O}. \tag{42}$$

This property can be used to transfer the action of $\mathcal{H}$ from the observable onto the distribution $W[\rho]$. From Eq. (39), one can see that ρ-averaged quantities such as

$$\frac{dN_1}{d^3\vec{\boldsymbol{p}}} \underset{\text{Leading Log}}{=} \int [D\rho_1 D\rho_2] W_1[\rho_1] W_2[\rho_2] \underbrace{\left.\frac{dN_1}{d^3\vec{\boldsymbol{p}}}\right|_{\text{LO}}}_{\text{fixed } \rho_{1,2}} \tag{43}$$

are independent of the cutoff, provided that the distributions $W[\rho]$ themselves depend on the cutoff according to the JIMWLK equation

$$\frac{\partial W}{\partial y} = \mathcal{H}W. \tag{44}$$

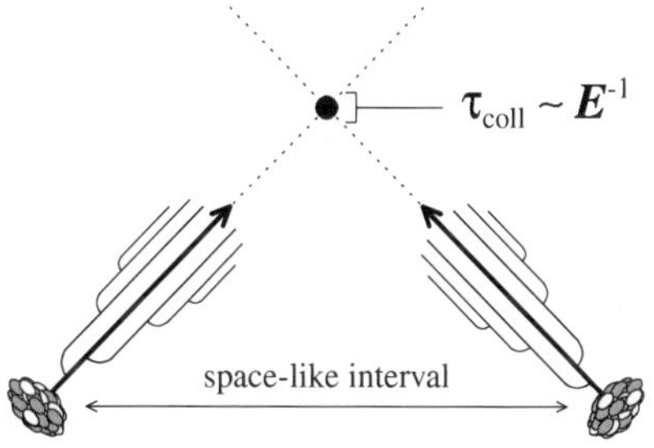

Fig. 20. Causality argument for factorization.

From Eqs. (34) and (39), it is furthermore obvious that the same factorization formula (with the same W's) applies to any inclusive observable.

In Eq. (43), it is the evolution with rapidity of the distributions W that gives the gluon spectrum its rapidity dependence. Indeed, the gluon spectrum for fixed $\rho_{1,2}$ that enters in the integrand is still independent of rapidity. From the JIMWLK equation, one sees that the distributions W evolve significantly for changes of the rapidity of the order of $\Delta y \sim \alpha_s^{-1}$. This factorization result thus predicts that the gluon spectrum is rather flat in rapidity at weak coupling.

7.6. *Ridge correlations*

Equation (43) can be generalized to the inclusive multigluon spectrum. Recalling also Eq. (20), we obtain the following factorization formula:

$$\frac{dN_n}{d^3\vec{\boldsymbol{p}}_1 \cdots d^3\vec{\boldsymbol{p}}_n} \underset{\text{Leading Log}}{=} \int [D\rho_1 D\rho_2]\, W_1[\rho_1]\, W_2[\rho_2]\, \frac{dN_1}{d^3\vec{\boldsymbol{p}}_1} \cdots \frac{dN_1}{d^3\vec{\boldsymbol{p}}_n}\bigg|_{\text{LO}}. \tag{45}$$

This equation, valid at leading log accuracy, shows that at this order the correlations between the produced gluons only originate from correlations of the ρ's of the incoming projectiles, since in the integrand the n gluons still appear completely factorized. Since the relevant rapidity interval for a significant JIMWLK evolution is $\Delta y \sim \alpha_s^{-1}$, this is also the typical rapidity distance over which the produced gluons will be correlated.

This is the basis of an interpretation of the peculiar shape of the 2-hadron correlations observed in heavy ion collisions. This correlation function is represented in Fig. 21, as a function of the relative azimuthal angle and relative rapidity. As one can see, the correlation is very narrow in azimuthal angle, and very elongated

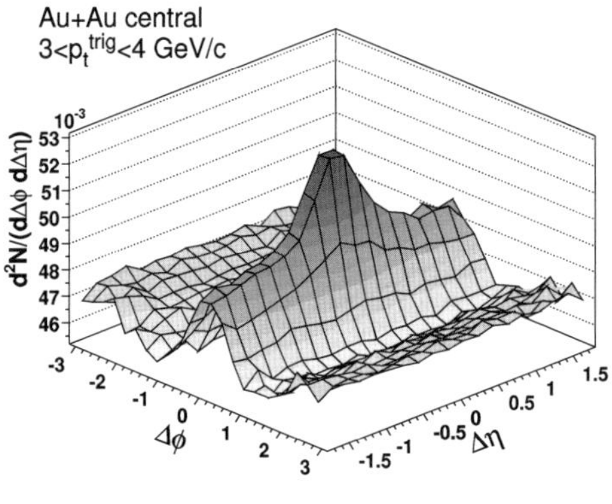

Fig. 21. Two hadron correlation measured in heavy ion collisions.[75]

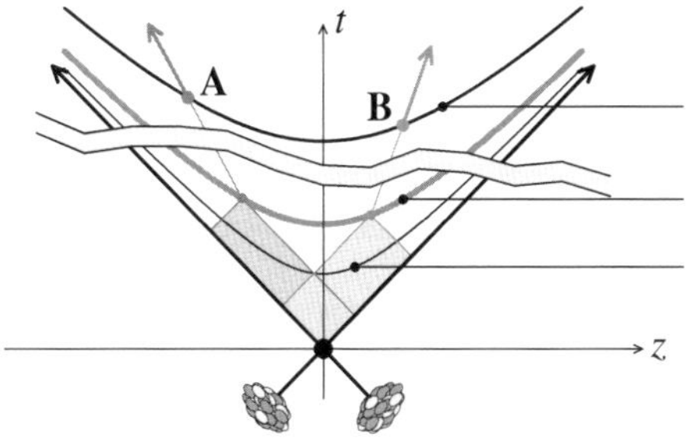

Fig. 22. (Color online) Origin of the rapidity correlations.

in rapidity.[q] Because of causality, the existence of a correlation between particles that are widely separated in rapidity must originate from phenomena that happened very shortly after the collision. This is explained in Fig. 22. Let us consider the time evolution of two particles A and B in reverse, starting from their last interaction on the freeze-out surface. Obviously, by causality, they must come from the light-cones represented respectively in red and green in the figure. A correlation is an event that had an influence on both the particle A and the particle B. It must therefore have happened in the overlap between these two light-cones, that we have represented in blue. One sees clearly that there is a latest time at which this correlation could possibly have been created. From the time of the freeze-out and the rapidity separation of the two particles, it is easy to determine this upper bound of the time,

$$\tau_{\text{correlation}} \le \tau_{\text{freeze out}}\, e^{-|\Delta y|/2}. \tag{46}$$

This bound decreases very rapidly as one increases the rapidity separation Δy. In heavy ion collisions, the order of magnitude of the freeze-out time is 10 fm/c. For instance, a correlation between particles separated in rapidity by $\Delta y = 6$ must be produced before the time 0.5 fm/c, which is well within the regime where the CGC is still applicable.

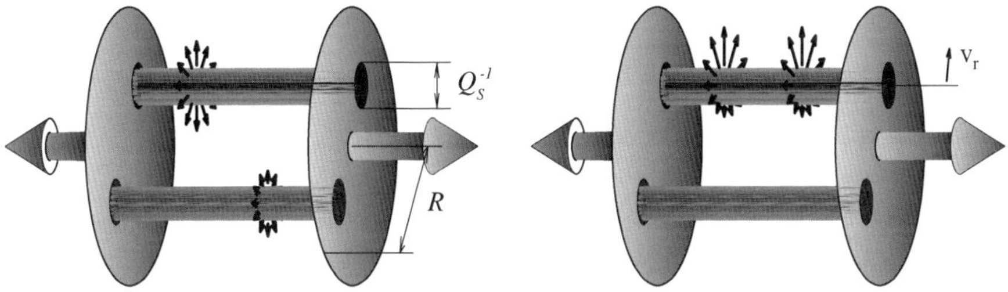

Fig. 23. (Color online) Rapidity correlations from color flux tubes.

[q]The peak in the middle is a jet-like correlation, due to quasi-collinear splittings in the final state.

It is in fact easy to understand qualitatively the main features of the observed correlation from the structure of the classical color fields produced at early times in heavy ion collisions.[76,77] As we have said before, these fields are organized in flux tubes that have a typical transverse size of Q_s^{-1} and that remain coherent over rapidity intervals of order α_s^{-1}. Two gluons emitted from the same tube are correlated if they are produced with a rapidity separation $|\Delta y| \lesssim \alpha_s^{-1}$, but are not correlated if they come from two different flux tubes.[r] From the size of the flux tubes, we conclude that the probability that two particles are correlated scales as $(RQ_s)^{-2}$ where R is the transverse radius of the collision zone.

This explains the existence of a long range correlation in rapidity between pairs of particles, but not why their correlation is peaked in azimuthal angle. The 2-gluon correlation one gets from the CGC color fields is nearly independent of the azimuthal angle, because on average these gluons can be emitted in any transverse direction. However, one should keep in mind that the above causality argument applies only to the correlation in rapidity, not to the correlation in azimuthal angle that can be produced much later. These azimuthal correlations are generated by the radial flow[78,79] that develops subsequently and expels radially the matter produced in the collision. Simple relativistic kinematics indeed shows that if one boosts a 2-particle spectrum independent of azimuth, it becomes peaked around $\Delta\phi = 0$ (the prominence of the peak increases with the velocity of the boost).[76] In Fig. 24, we show a comparison of the strength of the azimuthal correlation in data and in a very basic radial boost model where a unique radial boost velocity is applied to a flat spectrum (the radial velocity is estimated from the slope of $p_\perp$ spectra at small momentum). The centrality dependence, which in this model comes from the increase of the radial velocity with centrality, is in fair qualitative agreement with the measurement.

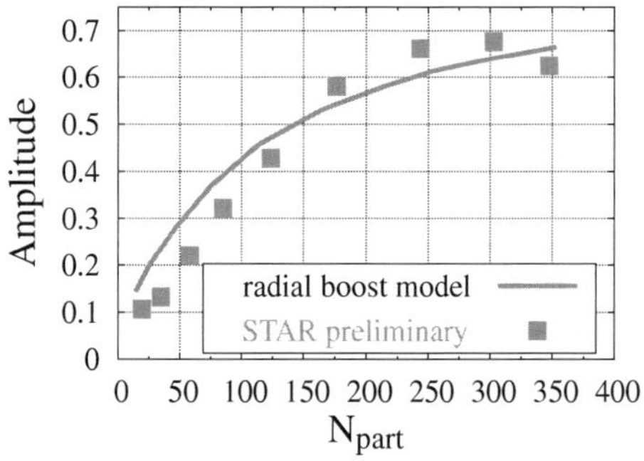

Fig. 24. (Color online) Comparison of the strength of the azimuthal correlation in data and in a simple radial boost model[80] (see also Ref. 81).

[r]The fact that the chromo-electric and chromo-magnetic fields are purely longitudinal at early time does not seem to play any role in this argument. The important properties are their coherence length in rapidity and in the transverse plane.

8. From the CGC to Hydrodynamics

8.1. *Requirements for hydrodynamics*

The CGC provides a self-contained QCD-based framework for describing heavy ion collisions from first principles. It also provides tools for calculating inclusive observables at leading log accuracy, i.e. leading order plus a resummation of all the leading log contributions coming from higher loop diagrams. However, there is some physics that plays an important role in heavy ion collisions but is not easily captured in the CGC. The fact that the produced gluons and quarks will eventually hadronize when their energy density falls below the QCD critical energy density is obviously not present in the CGC (at least, at any fixed loop order).

For this reason, CGC calculations can only describe the early stages of the fireball expansion, and should be later on matched onto another description such as relativistic hydrodynamics.[82–89] Since the components of the energy momentum tensor can be computed in the CGC framework, one could use it as initial data for the hydrodynamical evolution. Firstly, for such a matching to be possible, the CGC must bring the system to a state that hydrodynamics can handle. This means that the transverse and longitudinal pressure should not be too different (in particular, the longitudinal pressure should not be negative), and that the viscous effects (e.g. the ratio η/s of the shear viscosity to the entropy density) should be small.

Moreover, when performing such a matching between two descriptions that use different degrees of freedom, one should be careful to check that the two descriptions are compatible in a certain time window. In other words, there should be some range where the two models predict equivalent results. If this is the case, the precise time τ_0 at which the matching is realized is not important, and it can be varied in this range without any incidence on the final result.[s]

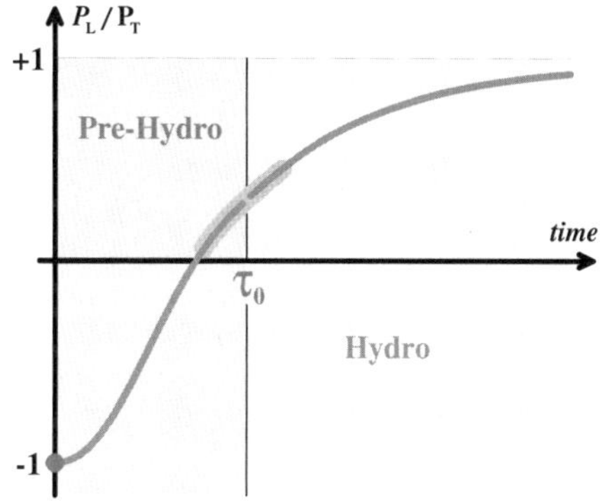

Fig. 25. (Color online) Smooth matching between CGC and hydrodynamics.

[s]This is very similar to the factorization of the source distribution $W[\rho]$ in the calculation of inclusive observables. The independence with respect to the cutoff $y_{\rm cut}$ is possible because the static sources describe the same physics as the gauge fields in a certain range of longitudinal momentum.

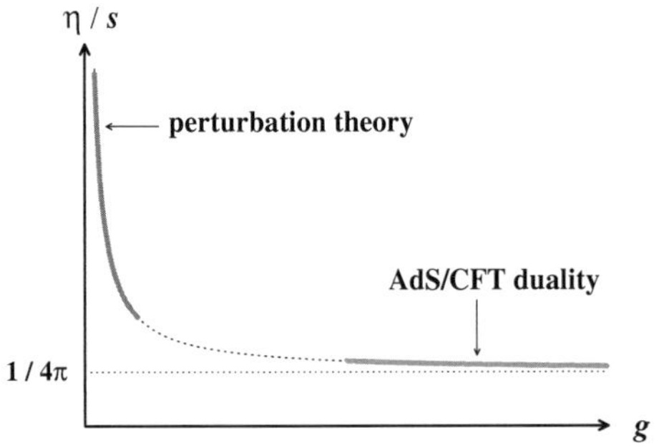

Fig. 26. (Color online) Shear viscosity to entropy ratio in a gauge theory as a function of the coupling.

The typical behavior of the ratio η/s in a gauge theory in equilibrium is shown in Fig. 26. When the coupling is small, this ratio can be calculated in a weak coupling expansion. For QCD, it is given by

$$\frac{\eta}{s} \approx \frac{5.1}{g^4 \ln\left(\frac{2.4}{g}\right)}. \tag{47}$$

This formula shows that η/s is large at weak coupling. At large coupling, this quantity cannot be calculated in QCD, but there is an exact result for a supersymmetric cousin of QCD, the $N = 4$ SUSY Yang–Mills theory[t]: $\eta/s = 1/(4\pi)$.[90,91] From this plot, it seems that only strongly coupled systems can have a small η/s ratio. There is however another possibility to evade this conclusion. Firstly, one should recall the kinetic interpretation of the ratio η/s,

$$\frac{\eta}{s} \sim \frac{\text{mean free path}}{\text{de Broglie wave length}}. \tag{48}$$

In a system where the degrees of freedom have a typical momentum Q, the de Broglie wave length is of order Q^{-1}, while the inverse mean free path is given by

$$(\text{mean free path})^{-1} \sim \underbrace{g^4 Q^{-2}}_{\text{cross-section}} \times \underbrace{\int_{\boldsymbol{k}} f_{\boldsymbol{k}}}_{\text{density}} \underbrace{(1 + f_{\boldsymbol{k}})}_{\text{Bose enhancement}}. \tag{49}$$

The factor $1 + f_{\boldsymbol{k}}$ under the integral is not needed when discussing dilute plasmas, but is important if the occupation number is large. In the CGC, just after the collision of two heavy ions, one has $f_{\boldsymbol{k}} \sim g^{-2}$ up to $k \sim Q$. Therefore, in such a system one has $\eta/s \sim g^0$, which is much smaller than the perturbative result in equilibrium. It seems therefore plausible that the strong color fields produced in heavy ion collisions may flow, not because they are strongly coupled but because they are highly occupied.

[t]This gauge theory is conformal, and is dual to a string theory in an $AdS_5 \times S_5$ background. The large coupling limit of the gauge theory corresponds to the weak coupling limit of the string theory, in which gravity becomes classical and reduces to Einstein's equations.

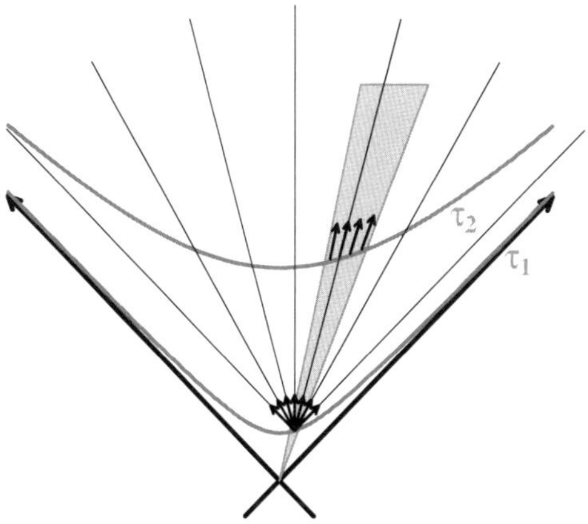

Fig. 27. (Color online) Role of the expansion in decreasing the longitudinal pressure.

8.2. *Expansion and free streaming*

The other main feature of heavy ion collisions is the very rapid expansion of the system in the longitudinal direction, which causes a redshifting of the longitudinal momenta. As illustrated in Fig. 27, the pressure tensor tends to become anisotropic because of this, unless the interactions are strong enough to overcome the expansion. The figure shows what the expansion does on non-interacting particles. Starting from a nearly isotropic distribution of the velocities at the time τ_1, the expansion will "filter out" the particles so that at the time τ_2 only particles with the momentum rapidity $y \approx \eta$ exist at the space-time rapidity η. This means that, in the local comoving frame, the longitudinal pressure is much smaller than the transverse pressure. A large pressure anisotropy is a problem for hydrodynamics. Indeed, the difference between the longitudinal and transverse pressures goes into the viscous terms,[u] and large viscous corrections are a sign that hydrodynamics may incorrectly reproduce the underlying dynamics.

Let us now describe the CGC prediction for the energy momentum tensor, starting with the LO calculation.[v] Immediately after the collision, at $\tau = 0^+$, it is known analytically that the chromo-electric and chromo-magnetic fields are both parallel to the collision axis. This structure of the color fields implies that the energy momentum tensor is diagonal, $T^{\mu\nu} = \mathrm{diag}(\epsilon, P_T, P_T, P_L)$, with[w] $P_T = \epsilon$ and $P_L = -\epsilon$. At later times, the energy–momentum tensor must be determined numerically by solving the classical Yang–Mills equations, and by computing $T^{\mu\nu}$ from the classical gauge fields according to Eqs. (22). The results of this calculation[47,99]

[u]There are now attempts to view hydrodynamics as an expansion around a non-isotropic background. In this formulation, this may be less of a problem.[92–98]

[v]This discussion also applies to the LO result improved by the resummation of the leading log corrections. This does not change the conclusion of this paragraph since this resummation is totally absorbed into the rapidity evolution of the distributions $W[\rho]$.

[w]The negative longitudinal pressure in a longitudinal flux tube is the analogue of a string tension.

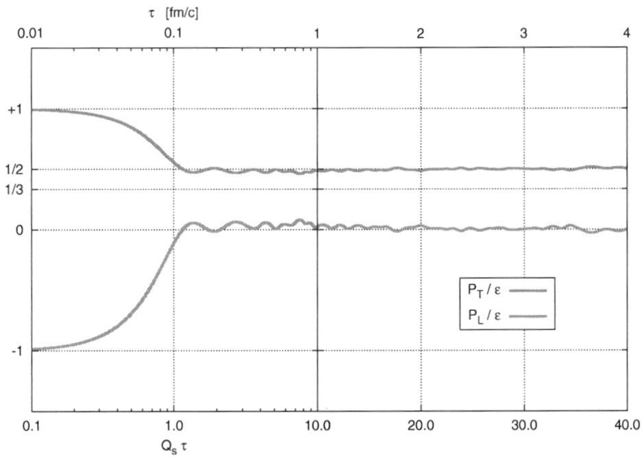

Fig. 28. (Color online) Transverse and longitudinal pressure to energy density ratios, in the CGC at leading order.

are shown in Fig. 28. After starting at -1, the ratio P_L/ϵ increases and becomes mostly positive at a time of order $Q_s\tau \sim 1$. However, this calculation shows that the longitudinal pressure remains at all times much smaller than the transverse one. Thus, the CGC at leading order leads to a situation which is similar to free streaming, where $\epsilon, P_T \sim \tau^{-1}$ and $P_L \ll \tau^{-1}$. This leads to an unsatisfactory matching between the CGC at LO and hydrodynamics,[x] as illustrated in Fig. 29.

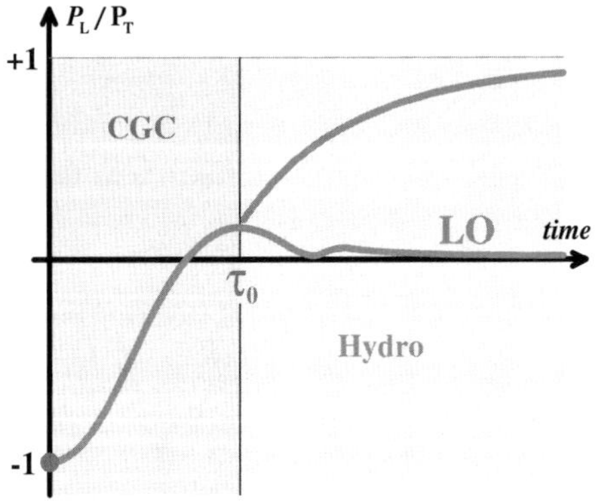

Fig. 29. (Color online) Matching between hydrodynamics and the CGC at LO.

[x]In principle, this matching requires several steps: (1) compute $T^{\mu\nu}(x)$ from CGC, (2) find its time-like eigenvector such that $u_\mu T^{\mu\nu}(x) = \epsilon u^\nu$ (this defines the local rest frame, and the local energy density), (3) compute the pressure from some equation of state $P = f(\epsilon)$, (4) compute the viscous stress tensor as the difference between the full $T^{\mu\nu}$ and the ideal part (obtained from ϵ, P and u^μ). In many calculations using "CGC initial conditions", a simplified procedure is often employed, where one assumes that $u^\mu = (1, \mathbf{0})$: (1) compute $\epsilon = T^{00}$, (2) define $P = f(\epsilon)$, (3) neglect the viscous stress tensor.

Indeed, in hydrodynamics the ratio P_L/P_T increases to approach 1 while it remains near zero in the CGC at LO.

8.3. *A simpler model study case: Thermalization in a box*

Before presenting recent works on isotropization and thermalization in heavy ion collisions, it is interesting to pause a brief moment on the same question for a hypothetical system of gluons enclosed in a fixed volume. In this case, the ultimate outcome is completely clear from the start: the system will eventually thermalize, and since its total energy is conserved one can predict from the start what will be the equilibrium temperature. The only issues are the timescale of the thermalization process, its possible dependence on the nature of the initial condition, and the shape of the gluon distribution at intermediate times before thermalization is complete.

A particularly interesting class of initial conditions are the so called "overoccupied" initial distributions, because they are realized in heavy ion collisions in the CGC framework. The typical CGC-like initial condition has modes occupied up to the saturation momentum Q_s, with a large occupation number of order g^{-2}. It is called overoccupied because it contains too many particles for its energy density:

$$n \sim \frac{Q_s^3}{g^2}, \qquad \epsilon \sim \frac{Q_s^4}{g^2}, \qquad n\epsilon^{-3/4} \sim g^{-1/2} \gg 1, \tag{50}$$

while the dimensionless combination $n\epsilon^{-3/4}$ should be of order 1 in a system with the same energy density at equilibrium. With such an initial condition, the mean free path in the system is parametrically shorter than the thermalization time, which leaves ample time to the system to forget its initial conditions long before reaching thermal equilibrium. With this type of initial condition, the hard scale of the system grows with time according to the law[100–102]

$$\Lambda_{\rm H} \sim Q_s (t/t_0)^{1/7}, \tag{51}$$

and the distribution scales as

$$f(t,p) \sim (t_0/t)^{4/7} f(p/\Lambda_{\rm H}). \tag{52}$$

The thermalization time can be estimated as the time at which the hard scale reaches the expected equilibrium temperature (i.e. the fourth root of the initial energy density), $T_{\rm eq} \sim Q_s g^{-1/2}$. This gives a thermalization that has the following parametric dependence on the coupling,

$$Q_s t_{\rm eq} \sim g^{-7/2}. \tag{53}$$

In a recent work[103] using the effective kinetic theory[y] of Ref. 104, this estimate was made more quantitative and the numerical prefactors computed to give

$$t_{\rm eq} \approx \frac{72}{1+0.12\log(\lambda^{-1})}\frac{1}{\lambda^2 T_{\rm eq}}, \tag{54}$$

where $\lambda \equiv g^2 N_c$.

This type of overoccupied initial condition has also led to speculations about the possibility of forming a Bose–Einstein condensate (BEC) in such systems, that would accommodate the excess of particles.[101] Such a condensate can only be transient, because the number of particles is not conserved in a relativistic quantum field theory. Therefore, the true equilibrium state cannot have a BEC. Whether such a condensate can form as transient phenomenon depends on the magnitude of the initial overoccupation and the rate of the number changing processes. In a scalar ϕ^4 theory, where the rate of the inelastic processes is quite small compared to the elastic one, such a condensate has been observed in a number of numerical simulations.[105,106] The situation is much less clear in QCD, because the inelastic rate is enhanced by soft and collinear divergences. Unsurprisingly, kinetic theory calculations (based on a small scattering angle approximation of the matrix element) neglecting the inelastic processes do observe the formation of a BEC.[107] Extensions of these QCD kinetic theory computations to include inelastic processes[110] suggest that a BEC may still appear after including number changing processes, while other computations do not see any evidence for it.[102,103]

Note that for initially underoccupied[z] initial conditions, the evolution towards equilibrium is quite different. In particular, it is a lot less universal because the mean free path associated with the initial distribution is comparable to or longer than its value at equilibrium. In this situation, the initial hard particles first create a bath of soft particles by bremsstrahlung radiation, which thermalizes quickly. Then the remaining hard particles lose their energy by successive $1 \to 2$ splittings induced by collisions on the soft background.

9. Weibel Instability and Resummation

9.1. *Unstable classical solutions*

There is however a good reason to explore the CGC beyond leading order: namely, the fact that the boost invariant solutions of the classical Yang–Mills equations are unstable.[99,113–121] When their initial condition is modified by a small but rapidity dependent perturbation, the solution diverges from the unperturbed one. This is

[y] This effective theory includes $2 \to 2$ processes dressed by in-medium masses, and effective $1 \to 2$ and $2 \to 1$ processes due to the quasi-collinear splitting of hard gluons by bremsstrahlung. For the latter, one must resum multiple scatterings, due to the Landau Pomeranchuk Migdal effect.

[z] In the bottom up scenario of Ref. 111 (see also Ref. 112), it has been argued that in heavy ion collisions the expansion may turn a CGC-like initial condition into an underoccupied distribution before full thermalization is reached.

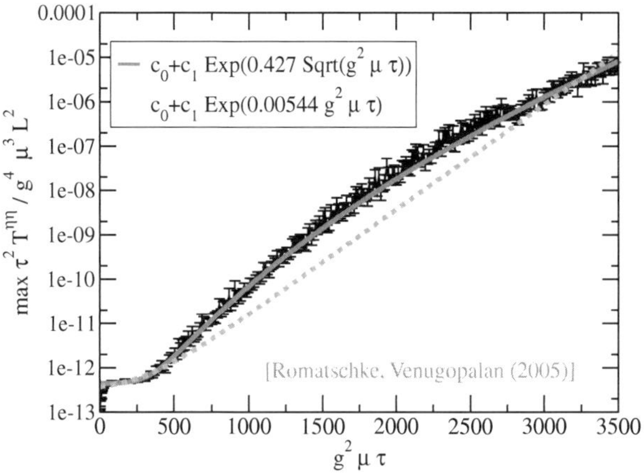

Fig. 30. (Color online) Growth of unstable modes in classical Yang–Mills dynamics.

illustrated in Fig. 30, that shows a component of the energy momentum tensor that should be very small at all times if the perturbation was stable. Instead, it growths like an exponential[aa] $\exp\sqrt{\mu\tau}$ (the characteristic growth rate μ is of the order of the saturation momentum Q_s). These unstable modes in the classical Yang–Mills equations are closely related to the Weibel instability that occurs in anisotropic plasmas in QED and in QCD.[100,112,122–141] More details on how these instabilities of the classical solutions develop can also be found in Refs. 119, 120, 142 and 99.

As we shall see in the next subsection, these instabilities are disastrous for the power counting that we have exposed earlier, where one keeps track only of the powers of g^2. Indeed, terms that have a higher order in g^2 because they arise at a higher loop order may in fact contain time dependent factors that increase exponentially with time. These *secular divergences* mean that fixed loop order calculations are most likely unreliable (we will show an example of this in the next subsection), and that the power counting should be revisited and improved in order to capture the most important among those terms.

On the other hand, the existence of instabilities in the classical solutions of the Yang–Mills equations may play a very important role in the statistical equilibration of the system. If we consider this system from the point of view of its classical phase-space, its initial condition at $\tau = 0^+$ occupies a very compact region in phase-space. Indeed, at LO, the support of its Wigner distribution is a single point, since the LO is purely classical. At NLO, this support is broadened into a region of extension $\hbar$. If the dynamics was completely stable (as in an integrable system for instance, where the number of conservation laws equals the number of degrees of freedom), the size of this support would remain roughly constant in time, and at any subsequent time

[aa]The fact that $\sqrt{\tau}$ appears here instead of τ itself is due to the longitudinal expansion of the system. Because of the expansion, the equation that drives the growth of the perturbations is a Bessel equation instead of an ordinary wave equation.

the system would still be far from statistical equilibrium. In contrast, an unstable dynamics will map this initially compact support into the extended portion of the phase-space allowed by the few conservation laws that remain valid. Measurements performed after such an evolution should bring results that are more in line with microcanonical equilibrium.

9.2. *Pathologies at fixed loop order*

These instabilities force us to reconsider the power counting that was the basis for organizing the expansion in powers of g^2 of inclusive observables. As we have seen before, the 1-loop corrections — that are formally of relative order g^2 — contain leading log terms proportional to the cutoff $y_{\rm cut}$, that can be absorbed into a redefinition of the distributions $W[\rho]$. Because of the instabilities, the 1-loop correction also contain some terms that grow exponentially in time. The best place to see this is via the following expression for the 2-point function $\mathbf{\Gamma}_2(\boldsymbol{u},\boldsymbol{v})$ that enters in the formula (34),

$$\mathbf{\Gamma}_2(\boldsymbol{u},\boldsymbol{v}) = \int \frac{d^3\boldsymbol{k}}{(2\pi)^3 2\omega_{\boldsymbol{k}}} a_{\boldsymbol{k}}(\tau_0,\boldsymbol{u}) a^*_{\boldsymbol{k}}(\tau_0,\boldsymbol{v}), \tag{55}$$

(τ_0 is the initial time surface on which Eq. (34) is expressed) where the functions $a_{\boldsymbol{k}}(\tau,\boldsymbol{x})$ are small perturbations around the classical field encountered at leading order (for Yang–Mills theory, these functions would also carry color, spin and Lorentz indices not written explicitly in Eq. (55)). They obey the equation

$$\begin{aligned} &[\mathcal{D}_\rho \mathcal{D}^\rho \delta^\nu_\mu - \mathcal{D}_\mu \mathcal{D}^\nu + ig\mathcal{F}^\nu_\mu] a^\mu_{\boldsymbol{k}} = 0 \\ &\qquad\qquad \lim_{x^0\to-\infty} a^\mu_{\boldsymbol{k}}(x) = \epsilon^\mu(\boldsymbol{k}) e^{ik\cdot x}, \end{aligned} \tag{56}$$

in which the covariant derivatives contain the LO classical field. The plane wave initial condition used in Eq. (56) for these small perturbations can be understood from the explicit derivation of the formula (34).

The existence of instabilities means that, for some range of $\boldsymbol{k}$, the mode functions $a_{\boldsymbol{k}}$ grow exponentially with time. The naive power counting derived so far implicitly assumed that the function $\mathbf{\Gamma}_2$ is of order g^0, but we never looked at its time dependence. From Eq. (55), it is now obvious that it will become very large if some of the $a_{\boldsymbol{k}}$'s are unstable. Eventually, these exponential factors in $\mathbf{\Gamma}_2$ will compensate the g^2 that comes from the loop, and these terms will be as large as the LO terms. This statement is illustrated in Fig. 31, where we compare the energy density and pressure at LO and LO+NLO in a ϕ^4 scalar field theory,[143] for which the solutions of the classical field equations of motion are also unstable.[bb] This computation shows clearly that the fixed order LO+NLO result cannot be trusted after some time, because it becomes much larger than the LO result (note that this

[bb]The instability in the ϕ^4 theory is of a totally different nature, since it is caused by parametric resonance. However, the detailed mechanism of the instability is not important in this discussion.

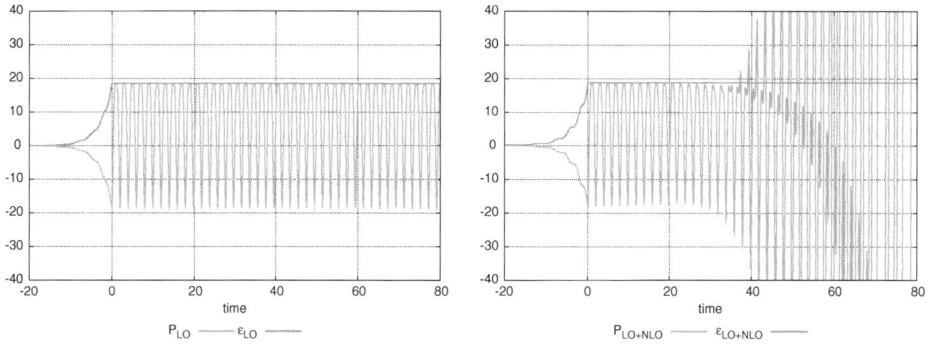

Fig. 31. (Color online) Effect of parametric instabilities on the perturbative expansion of the energy–momentum tensor in a scalar ϕ^4 theory.

happens only for the pressure, because the energy density is protected from this exponential growth by energy conservation).

Although a similar NLO calculation has not been done in the CGC framework, one can guess what would happen. The Weibel instabilities would produce an unbound growth of the longitudinal pressure, and this time the ratio P_L/P_T would be driven to arbitrarily large values exceeding unity. Attempting to match such a NLO CGC initial condition to hydrodynamics would not be better than at LO, as illustrated in Fig. 32.

9.3. *Resummation of the leading terms*

The same problem would in fact occur at any fixed loop order. The only way to improve the situation is to examine each loop order and to resum the most important terms at each order. For this, we need first to modify the power counting rules that we have established earlier, in order to keep track of the unstable modes.

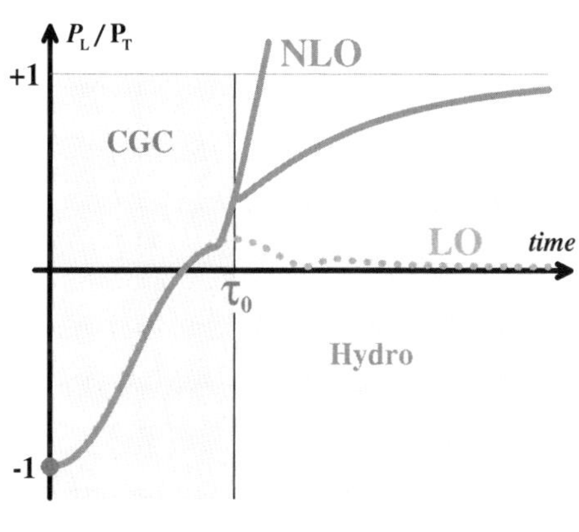

Fig. 32. (Color online) Matching between hydrodynamics and the CGC at NLO.

Let us first examine the graph that contributes at 1-loop, represented in the top-left corner of Fig. 33, in conjunction with the formula (34). In this formula, each of the derivatives with respect to the initial classical field creates a perturbation to this classical field, that we have indicated by green propagators in Fig. 33 (in the top-left graph, only the term proportional to $\mathbf{\Gamma}_2$, that has second derivatives with respect to the initial fields, has been represented — the term with only one derivative has a slower growth with time). The "standard" power counting[cc] would assign a factor 1 to $\mathbf{\Gamma}_2$ and a factor g to each of the derivatives $\mathbb{T}_{\boldsymbol{u}}$ (represented by a blue dot in the graphs). Thus the NLO correction to the energy–momentum tensor is expected to be of order g^0, while the LO is of order g^{-2}. From this diagrammatic representation, it also easy to count the number of perturbations of the classical field. Each of them will develop into a factor of order $\exp(\sqrt{\mu\tau})$ (μ is of order Q_s). Thus, the expansion of $T^{\mu\nu}$ is more accurately written[dd] as:

$$T^{\mu\nu} = c_0 g^{-2} + c_1 g^0 e^{2\sqrt{\mu\tau}} + \cdots , \tag{57}$$

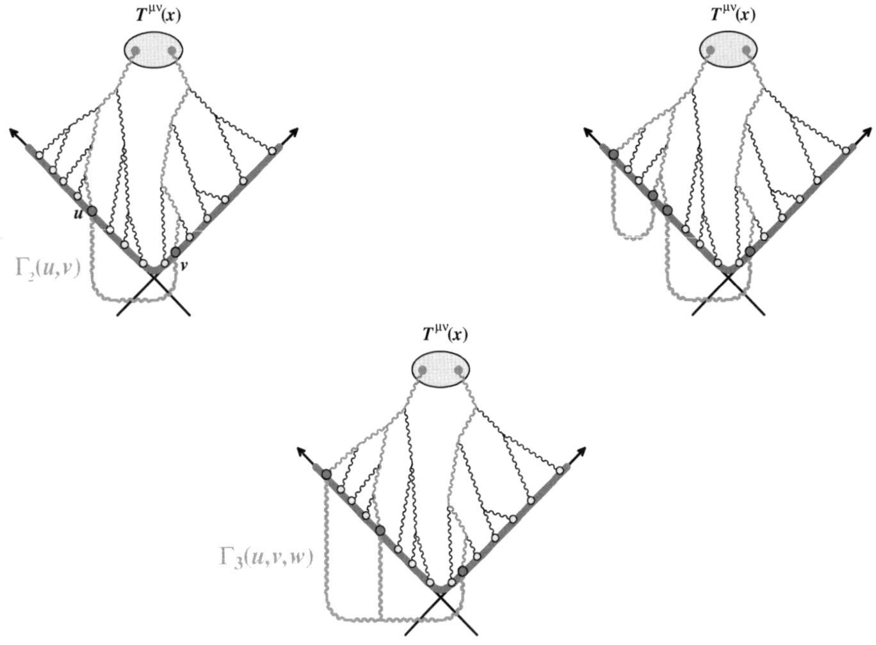

Fig. 33. (Color online) Improved power counting taking into account the growth of the unstable modes.

[cc]For the term in $\boldsymbol{\alpha}\mathbb{T}$ in Eq. (34), the standard counting has a factor g from $\boldsymbol{\alpha}$ and a factor g from the operator $\mathbb{T}$. However, we will not need to consider this term further since it has a subleading growth in time.

[dd]This formula indicates the worst possible behavior. It is possible that some components of the energy–momentum tensor will not be affected by the instability, as was the case in the scalar field theory considered in Fig. 31.

where the coefficients c_0, c_1 do not grow exponentially with time. From this pocket formula, one can deduce at which time the naive loop expansion breaks down. This is the time when the 1-loop result becomes as large as the leading order, i.e.

$$\tau_{\max} \sim \mu^{-1} \log^2(1/g^2). \tag{58}$$

Up to a logarithmic factor, this time is of the order of the inverse saturation momentum.

At two loops, the naive power counting tells us that we should get terms of order g^2 (i.e. g^4 relative to the leading order). However, not all the terms have the same growth in time. In Fig. 33, we have represented two types of 2-loop contributions, in order to illustrate these differences. In the top-right graph, the two loops are the seed of four perturbations to the classical field, while in the bottom graph, only three of these perturbations are created. The latter term will therefore have a subleading behavior in time. Moreover, one sees that the distinguishing feature of the top-right graph is that it can be generated by acting twice with the quadratic part of the operator that appears in Eq. (34). This reasoning extends to all orders. At the n-th loop order, the maximal number of field perturbations that can be seeded on the light-cone is $2n$, and the corresponding graphs are generated by acting n times with this quadratic operator. The sum of all these leading terms can be obtained by

$$T^{\mu\nu}_{\text{resummed}} \equiv \exp\left[\frac{1}{2}\int_{\boldsymbol{u},\boldsymbol{v}} \boldsymbol{\Gamma}_2(\boldsymbol{u},\boldsymbol{v})\mathbb{T}_{\boldsymbol{u}}\mathbb{T}_{\boldsymbol{v}}\right] T^{\mu\nu}_{\text{LO}}. \tag{59}$$

Note that the Taylor coefficients of the exponential correspond precisely to the symmetry factors of graphs such as the top-right diagram of Fig. 33 (the 1/2! of the second Taylor coefficient gives the symmetry factor that corresponds to the freedom of swapping the two $\boldsymbol{\Gamma}_2$'s that are hanging below the light cone).

Moreover, if the 2-point function $\boldsymbol{\Gamma}_2$ used in Eq. (59) is precisely the one that enters in the NLO result (34), then one has

$$T^{\mu\nu}_{\text{resummed}} = T^{\mu\nu}_{\text{LO}} + T^{\mu\nu}_{\text{NLO}} + \cdots \tag{60}$$

In other words, this resummation contains the exact LO and NLO contributions, and a subset of all the higher loop contributions. It is important to keep in mind that, starting at the 2-loop order, it is an approximation which is not equivalent to the complete underlying theory. This will have important consequences that we will discuss later.

10. Classical Statistical Approximation (CSA)

10.1. *Reformulation as a Gaussian average*

At this point, the resummation performed via Eq. (59) is quite formal. Three questions must be addressed: (1) can this formula in terms of functional derivatives be evaluated in a practical way? (2) does Eq. (59) lead to results whose time dependence is bounded? (3) when doing this, are we introducing other pathologies?

In order to answer the first question, one should recall the following identity:

$$e^{\frac{\alpha}{2}\partial_x^2} f(x) = \int_{-\infty}^{+\infty} dz \frac{e^{-z^2/2\alpha}}{\sqrt{2\pi\alpha}} f(x+z). \tag{61}$$

This formula can be established e.g. by applying a Fourier transform to both sides. Although we cast it here in a space of functions of a single variable, this formula can be generalized to operators that are Gaussian in derivatives over a functional space. It enables us to rewrite Eq. (59) as[ee]

$$T^{\mu\nu}_{\rm resummed} = \int [Da] \exp\left[-\frac{1}{2}\int_{\boldsymbol{u},\boldsymbol{v}} a(\boldsymbol{u})\boldsymbol{\Gamma}_2^{-1}(\boldsymbol{u},\boldsymbol{v})a(\boldsymbol{v})\right] T^{\mu\nu}_{\rm LO}[\mathcal{A}_{\rm init}+a]. \tag{62}$$

This resummation procedure, where one averages classical trajectories over an ensemble of initial conditions, is known as the *Classical Statistical Approximation* (CSA).

From this equation, one can easily see that the problem of the unbounded growth of the fluctuations has been cured. Indeed, this resummation has promoted the linearized perturbations[ff] that appear in the NLO contribution into an integral part of the non-linear classical field (the initial condition of the classical field is modified by the perturbation, but its evolution remains fully non-linear). In any theory where the potential prevents the fields from running away to infinity, this guarantees that the resummed quantity will not diverge in time.

10.2. *Practical implementation*

In the form of Eq. (62), the procedure for evaluating the resummed energy–momentum tensor[gg] is quite straightforward:

(1) Determine the 2-point function $\boldsymbol{\Gamma}_2(\boldsymbol{u},\boldsymbol{v})$ that defines the Gaussian fluctuations, for the initial time $Q_s\tau_0$ of interest. This is an initial value problem, whose outcome is uniquely determined by the state of the system at $x^0 = -\infty$, and depends on the history of the system from $x^0 = -\infty$ to $\tau = \tau_0$. This problem is solvable analytically as long as the fluctuations remain weak, $a^\mu \ll Q_s/g$. If they grow larger, the fluctuations start to interact non-linearly, and their spectrum becomes non-Gaussian. To avoid this, the initial time should be chosen such that $Q_s\tau_0 \ll 1$.

(2) Solve the classical Yang–Mills equations from τ_0 to τ_f. In high energy collisions, the problem as a whole is boost invariant, but individual field configurations are now rapidity dependent because of the fluctuating part superimposed to their initial condition. Therefore, unlike in the CGC at LO, the classical Yang–Mills equations must be solved in $3+1$ dimensions.

[ee] Such a Gaussian averaging procedure has also been reached in other approaches.[145–149]

[ff] One would recover the pathological behavior of the NLO result by linearizing the equation of motion for the classical field of initial condition $\mathcal{A}_{\rm init} + a$.

[gg] Although the discussion here is centered on the energy–momentum tensor, the same resummation can be applied to any inclusive quantity.

(3) Do a Monte Carlo sampling of the fluctuating initial conditions.

The setup for doing this is the same as the one described when discussing the CGC at LO, except for the fact that one must keep rapidity explicitly. One must discretize the classical Yang–Mills equations in the system of coordinates $\tau, \eta, \boldsymbol{x}_\perp$. The lattice used in these computations usually represents a sub-volume of the interaction region that expands in the longitudinal direction, as illustrated in Fig. 34. This implies that the lattice spacing in the z coordinate is time dependent. In order to be able to resolve the physically relevant scales at the final time τ_f of the simulation, it is usually necessary to have a larger number of lattice spacings in the longitudinal direction.

10.3. *CSA in quantum mechanics*

The Classical Statistical Approximation has an analogue in ordinary quantum mechanics, which is helpful to understand what approximation is being made when one uses it. The starting point is the transformation that goes from the von Neumann equation (36) for the density operator to the Moyal equation (38) for its Wigner transform. The Moyal equation is still an exact representation of the quantum evolution of the system, although it is expressed in terms of objects that depend on classical phase-space variables. However, we have seen that if one expands the operator in the right-hand side of Eq. (38) to lowest order in $\hbar$, one recovers the classical Poisson bracket. In this approximation, the evolution of the system is thus purely classical. Note however that the initial condition can remain fully quantum in this approximation. For instance, if the system is initially in a pure quantum state $|\psi\rangle$, its initial Wigner distribution would be

$$W_0(\boldsymbol{x}, \boldsymbol{p}) \equiv \int d^3\boldsymbol{r}\, e^{i\boldsymbol{p}\cdot\boldsymbol{r}} \left\langle \boldsymbol{x} + \frac{\boldsymbol{r}}{2} \middle| \psi \right\rangle \left\langle \psi \middle| \boldsymbol{x} - \frac{\boldsymbol{r}}{2} \right\rangle$$
$$= \int d^3\boldsymbol{r}\, e^{i\boldsymbol{p}\cdot\boldsymbol{r}} \psi^* \left(\boldsymbol{x} + \frac{\boldsymbol{r}}{2}\right) \psi \left(\boldsymbol{x} - \frac{\boldsymbol{r}}{2}\right). \tag{63}$$

If the initial state $|\psi\rangle$ is known, then no approximation is needed in the CSA for the initial Wigner distribution. In this analogy, the Gaussian initial Wigner distribution

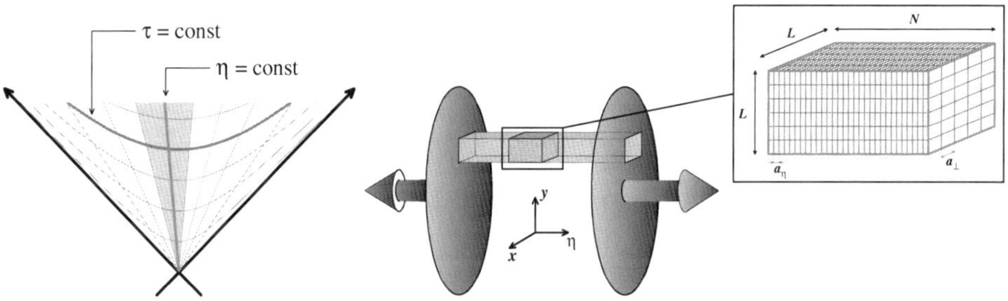

Fig. 34. (Color online) Lattice setup for numerical implementations of the classical statistical method.

that we have obtained in the previous subsection would correspond to starting from a coherent state, i.e. a state whose Wigner distribution is a Gaussian of width $\hbar$ centered on some classical configuration.

10.4. *CSA from the path integral*

It is also instructive to see how the CSA can be derived from the path integral formulation of quantum field theory. Since we aim at calculating the expectation value of observables in a system initialized in a known state, we must use the Schwinger–Keldysh formalism

$$\langle \mathcal{O} \rangle = \int [D\varphi_{\pm}^{(i)}(\boldsymbol{x})] \rho_0(\varphi_+^{(i)}, \varphi_-^{(i)}) \int [D\phi_{\pm}(x)] e^{iS[\phi_+] - iS[\phi_-]} \mathcal{O}(\phi), \tag{64}$$

where the field is doubled into a + component (that represents the evolution in the amplitude) and a − component (that describes the conjugate amplitude). In this formulation, the initial state of the system is represented by a density operator $\widehat{\rho}_0$, whose matrix elements $\rho_0(\varphi_+^{(i)}, \varphi_-^{(i)})$ control the distribution of the initial values of the fields $\phi_{\pm}$. The second integral is restricted to fields whose boundary value at the initial time is

$$\phi_{\pm}(t_0, \boldsymbol{x}) = \varphi_{\pm}^{(i)}(\boldsymbol{x}). \tag{65}$$

Inclusive observables do not put any constraint on the final state, and therefore the fields have no specific boundary condition at $t = +\infty$ in the above path integral, except for $\phi_+ = \phi_-$ when the measurement is done.

From Eq. (64), one should define new fields as the sum and difference of ϕ_+ and ϕ_-,

$$\phi_2 \equiv \frac{\phi_+ + \phi_-}{2}, \quad \phi_1 \equiv \phi_+ - \phi_-. \tag{66}$$

The difference $S[\phi_+] - S[\phi_-]$ is obviously odd in the field ϕ_1, and it is trivial to verify that the term linear in ϕ_1 comes as a prefactor of the classical equation of motion for ϕ_2. In addition, in an interacting theory, there are terms that are cubic in ϕ_1,

$$S[\phi_+] - S[\phi_-] = \phi_1 \cdot \frac{\delta S[\phi_2]}{\delta \phi_2} + \text{terms cubic in } \phi_1. \tag{67}$$

We are seeking an approximation that applies in the regime of strong fields, e.g. when the fields are excited by a large external source like in the CGC framework, which implies that ϕ_2 is large. Since ϕ_+ and ϕ_- are the fields in the amplitude and conjugate amplitude respectively, their difference is a quantum effect whose magnitude is suppressed by $\hbar$. Therefore, in such a situation, we have $\phi_1 \ll \phi_2$, and it is natural to neglect the cubic term in ϕ_1 in the action. After doing this, the field ϕ_1 becomes a Lagrange multiplier for the classical equation of motion for ϕ_2. The evolution of ϕ_2 is now deterministic, and the only remaining fluctuations are those inherited from the average over the initial density matrix $\rho_0(\varphi_+^{(i)}, \varphi_-^{(i)})$. In the Hamiltonian formulation of the classical equation of motion for ϕ_2, the average over $\varphi_+^{(i)}, \varphi_-^{(i)}$ becomes an average over $\varphi_2^{(i)}$ and its conjugate momentum $\Pi_2^{(i)}$, with a distribution obtained as the Wigner transform of ρ_0:

$$W_0[\varphi_2^{(i)}, \Pi_2^{(i)}] \equiv \int [D\chi] e^{i\int \chi\cdot\Pi_2} \rho_0 \left(\varphi_2^{(i)} + \frac{\chi}{2}, \varphi_2^{(i)} - \frac{\chi}{2}\right). \tag{68}$$

10.5. *CSA in perturbation theory*

The path integral derivation of the classical statistical approximation also clarifies what this approximation amounts to in perturbation theory. This knowledge will be useful later when we discuss the non-renormalizability of this approximation. Let us discuss this in the simple framework of a scalar field theory with a ϕ^4 interaction term.

The first step is to express the transformation from the $\phi_\pm$ fields to the $\phi_{1,2}$ fields as a "rotation", in order to obtain the diagrammatic rules in this new basis.[150–154] From the propagators in the $\pm$ basis,

$$\begin{aligned} G^0_{++}(p) &= \frac{i}{p^2 - m^2 + i\epsilon}, & G^0_{--}(p) &= \frac{-i}{p^2 - m^2 - i\epsilon}, \\ G^0_{+-}(p) &= 2\pi\theta(-p^0)\delta(p^2 - m^2), & G^0_{-+}(p) &= 2\pi\theta(p^0)\delta(p^2 - m^2), \end{aligned} \tag{69}$$

one can define a set of new propagators by a linear transformation on the two indices

$$\mathbb{G}^0_{\alpha\beta} \equiv \sum_{\epsilon,\epsilon'=\pm} \Omega_{\alpha\epsilon}\Omega_{\beta\epsilon'} G^0_{\epsilon\epsilon'}, \tag{70}$$

with the following transformation matrix:

$$\Omega_{\alpha\epsilon} \equiv \begin{pmatrix} 1 & -1 \\ 1/2 & 1/2 \end{pmatrix}. \tag{71}$$

The free rotated propagators are

$$\mathbb{G}^0_{\alpha\beta} = \begin{pmatrix} 0 & G^0_A \\ G^0_R & G^0_S \end{pmatrix}, \tag{72}$$

where we define

$$G^0_R = G^0_{++} - G^0_{+-}, \quad G^0_A = G^0_{++} - G^0_{-+}, \quad G^0_S = \frac{1}{2}(G^0_{+-} + G^0_{-+}). \tag{73}$$

(The subscripts R, A and S mean respectively for *retarded*, *advanced* and *symmetric*.) Note that the symmetric propagator depends on the initial state of the system, while the retarded and advanced ones are independent of this initial data (they only reflect how modes propagate in the theory under consideration). Under the rotation of Eq. (71), the vertices are transformed into

$$\begin{aligned} \Gamma_{1111} &= \Gamma_{1122} = \Gamma_{2222} = 0, \\ \Gamma_{1222} &= -ig^2, \quad \Gamma_{1112} = -ig^2/4. \end{aligned} \tag{74}$$

(The ones not listed explicitly here should be obtained by permutations of the indices.) The CSA simply amounts to setting to zero the vertex Γ_{1112} (and its permutations) wherever it appears in the diagrammatic expansion, while all the other Feynman rules are unmodified.

11. Applications of the CSA to Heavy Ion Collisions

The classical statistical approximation has recently been applied to heavy ion collisions in two sets of works, that mostly differ in the nature of the initial conditions that were used.

11.1. *CGC initial conditions*

In a strict application of the CSA to the description of heavy ion collisions in the CGC framework, the Gaussian ensemble of initial conditions arises from the exponentiation of the 1-loop result. In other words, the CGC description of heavy ion collisions is an initial value problem, and the state of the system at $\tau = 0^+$ is therefore prescribed uniquely from the fact that the system was in the vacuum state at $x^0 = -\infty$. The initial Gaussian ensemble of fields is characterized by the following mean values and variance (written here in very sketchy way, without color and Lorentz indices):

$$\begin{aligned} \langle \mathcal{A}^\mu \rangle &= \mathcal{A}^\mu_{\rm LO} \\ \mathbf{\Gamma}_2(\boldsymbol{u}, \boldsymbol{v}) &= \int \frac{d^2\boldsymbol{k}}{(2\pi)^3 2\omega_{\boldsymbol{k}}} a_{\boldsymbol{k}}(\tau_0, \boldsymbol{u}) a^*_{\boldsymbol{k}}(\tau_0, \boldsymbol{v}) \\ [\mathcal{D}_\rho \mathcal{D}^\rho \delta^\nu_\mu - \mathcal{D}_\mu \mathcal{D}^\nu + ig\mathcal{F}^\nu_{\ \mu}] a^\mu_{\boldsymbol{k}} &= 0 \\ \lim_{x^0 \to -\infty} a_{\boldsymbol{k}}(x) &= e^{ik\cdot x}. \end{aligned} \tag{75}$$

The mean value is already known from the LO calculation,[36] given in Eqs. (23). From these equations, one sees that the determination of the variance $\mathbf{\Gamma}_2$ amounts to solving the linearized Yang–Mills equations for all the mode functions $a^\mu_{\boldsymbol{k}}$ (this formula comes from the derivation of the NLO contribution).[155] As long as we stay in a regime where these perturbations are not yet enhanced by instabilities, the variance is small compared to the mean value squared, and the Gaussian distribution of the initial fields is a narrow distribution centered on the LO color fields (see Fig. 35(right)). The mode functions describe how plane waves with a well defined momentum, color and polarization at $x^0 = -\infty$ are distorted while they propagate over the LO color background field $\mathcal{A}^\mu_{\rm LO}$. The linearized Yang–Mills equations must be solved for each mode function, up to the time τ_0 (in the forward light-cone) at which the numerical simulation will start (Fig. 35(left)). Explicit formulas for these mode functions were derived in Ref. 155, in the Fock–Schwinger gauge $A^\tau = 0$ used in these computations. For given quantum numbers: ν (the Fourier conjugate of the

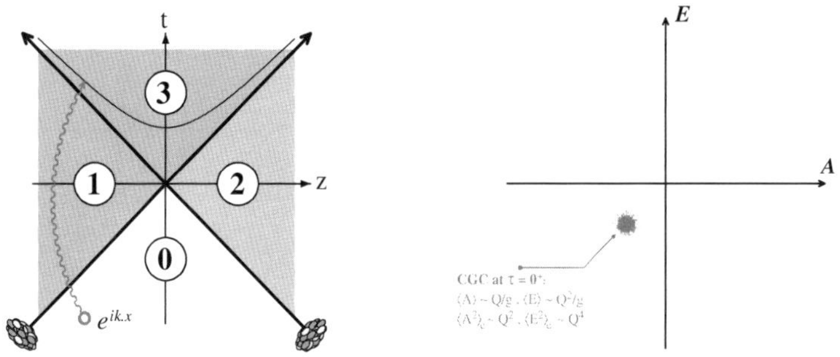

Fig. 35. (Color online) CGC spectrum of initial fluctuations at $Q_s\tau_0 \ll 1$.

rapidity η), $\boldsymbol{k}_\perp, \lambda, c$, the gauge potential and the associated electrical fields read

$$\begin{aligned} a^i &= \beta^{+i} + \beta^{-i} \qquad & a^\eta &= \mathcal{D}^i\left(\frac{\beta^{+i}}{2+i\nu} - \frac{\beta^{-i}}{2-i\nu}\right) \\ e^i &= -i\nu(\beta^{+i} - \beta^{-i}) \quad & e^\eta &= -\mathcal{D}^i(\beta^{+i} - \beta^{-i}), \end{aligned} \tag{76}$$

where

$$\begin{aligned} \beta^{+i} \equiv\, & e^{\frac{\pi\nu}{2}}\Gamma(-i\nu)e^{i\nu\eta}\mathcal{U}_1^\dagger(\boldsymbol{x}_\perp) \\ & \times \int_{\boldsymbol{p}_\perp} e^{i\boldsymbol{p}_\perp\cdot\boldsymbol{x}_\perp}\widetilde{\mathcal{U}}_1(\boldsymbol{p}_\perp + \boldsymbol{k}_\perp)\left(\frac{p_\perp^2\tau}{2k_\perp}\right)^{i\nu}\left(\delta^{ij} - 2\frac{p_\perp^i p_\perp^j}{p_\perp^2}\right)\epsilon_\lambda^j \end{aligned} \tag{77}$$

and

$$\begin{aligned} \beta^{-i} \equiv\, & e^{-\frac{\pi\nu}{2}}\Gamma(i\nu)e^{i\nu\eta}\mathcal{U}_2^\dagger(\boldsymbol{x}_\perp) \\ & \times \int_{\boldsymbol{p}_\perp} e^{i\boldsymbol{p}_\perp\cdot\boldsymbol{x}_\perp}\widetilde{\mathcal{U}}_2(\boldsymbol{p}_\perp + \boldsymbol{k}_\perp)\left(\frac{p_\perp^2\tau}{2k_\perp}\right)^{-i\nu}\left(\delta^{ij} - 2\frac{p_\perp^i p_\perp^j}{p_\perp^2}\right)\epsilon_\lambda^j. \end{aligned} \tag{78}$$

$U_{1,2}$ are the Wilson lines that describe the color source content of the two colliding nuclei (they are defined in terms of $\rho_{1,2}$ in Eq. (24)).

These mode functions have been used in Ref. 156 in order to compute the time evolution of the components of the energy–momentum tensor from CGC initial conditions. The main difficulty and source of uncertainty in these calculation is the subtraction of the ultraviolet divergences. Firstly, since the energy–momentum tensor is a dimension four operator, it picks up a pure vacuum contribution that behaves as the fourth power of the inverse lattice spacing. This contribution can be removed by subtracting the result of a second computation done with the same lattice parameters, but without the colliding nuclei. A large statistics in the Monte Carlo evaluation of the average over the initial fields is required for this subtraction. A second kind of ultraviolet contribution, that depends on the background field, was also observed to affect the energy density and the pressure,[hh] but not the

[hh]This contribution is possibly due to the non-renormalizability of the CSA — see Subsec. 12.1.

transverse pressure. Since at this level of approximation, the energy momentum tensor is traceless:

$$\epsilon = P_L + 2P_T, \tag{79}$$

and obeys Bjorken's law due to energy–momentum conservation,

$$\partial_\tau(\tau\epsilon) + P_L = 0, \tag{80}$$

the only possible form of an ultraviolet sensitive contribution that does not affect P_T is a term $const \times \tau^{-2}$ that affects equally ϵ and P_L. This is indeed what was observed. Lacking a more precise understanding of this term, the constant was fitted in order to subtract this term from ϵ and P_L. After this subtraction, the early time behavior of the ratios P_L/ϵ and P_T/ϵ closely follows the LO up to $Q_s\tau \sim 1$, which is indeed expected since this is before the unstable modes may have affected the time evolution. For a moderately large coupling $g = 0.5$, it was found that the longitudinal pressure increases significantly compared to its value in the LO calculation (see Fig. 36), and now becomes a sizable fraction of the transverse pressure.

11.2. *Particle-like initial conditions*

Alternatively, one may depart somewhat from the CGC framework and use the same approximation scheme with a different Gaussian ensemble of initial conditions. The simplest model of initial conditions one may consider is an ensemble of fields that describe a distribution of free particles.[157,159–161] In this case, the center of the

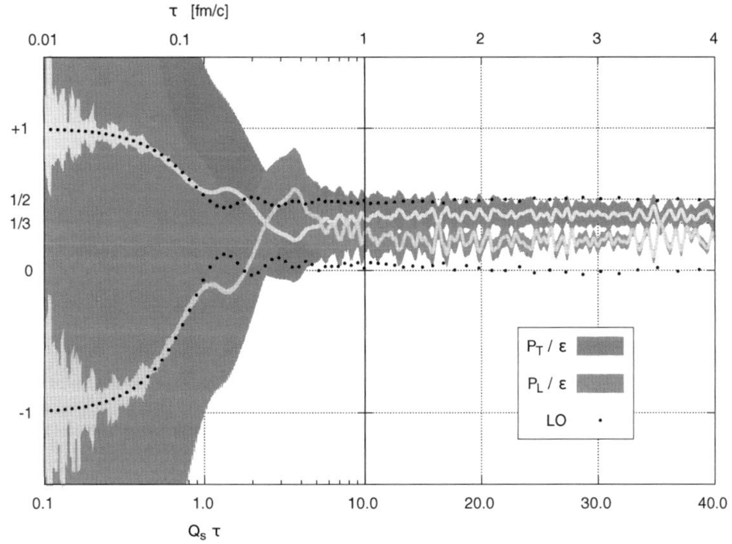

Fig. 36. (Color online) Behavior of P_L/ϵ and P_T/ϵ in the CSA with CGC initial conditions. The upper time scale, in fm/c, is based on a saturation momentum $Q_s = 2\,\text{GeV}$.

Gaussian is $\mathcal{A}^\mu \equiv 0$, and the variance is constructed from the in-vacuum mode functions (derived in[162] for gluons in the Fock–Schwinger gauge) as follows:

$$\langle \mathcal{A}^\mu \rangle = 0, \quad \mathbf{\Gamma}_2(\boldsymbol{u}, \boldsymbol{v}) = \int_{\text{modes } \boldsymbol{k}} f_0(\boldsymbol{k}) a_{\boldsymbol{k}}(\boldsymbol{u}) a^*_{\boldsymbol{k}}(\boldsymbol{v}), \quad a_{\boldsymbol{k}}(x) \equiv e^{ik\cdot x}, \tag{81}$$

where $f_0(\boldsymbol{k})$ is the initial momentum distribution of these particles. It is plausible that the CGC initial state at $\tau_0 = 0^+$ may evolve after some time into an incoherent distribution of gluons such as the one described by Eqs. (81), but this certainly deserves a more thorough study.

The difference and complementarity between the initial conditions in Eqs. (75) and (81) becomes more transparent if one recalls the symmetric 2-point Green's function in the Schwinger–Keldysh formalism in the presence of a bath of particles,

$$G_S(k) = 2\pi \left(\frac{1}{2} + f_0(\boldsymbol{k}) \right) \delta(k^2). \tag{82}$$

In this formula, the term in $f_0(\boldsymbol{k})$ represents an initial particle distribution, while the term in $1/2$ corresponds to pure vacuum quantum fluctuations. It thus becomes clear that the CGC initial conditions at $\tau = 0^+$ can be viewed as vacuum fluctuations that are somewhat altered by the presence of the LO color background field. It is the minimal modification that quantum mechanics can bring to a classical state, promoting it to a (pure) coherent state, while the particle-like initial conditions defined by Eqs. (81) are a mixed state from the point of view of quantum mechanics.

This type of particle-like initial condition was used in Refs. 157–161, in order to study the evolution of a longitudinally system, in Yang–Mills theory and also in the ϕ^4 scalar field theory.[161] Note that, if one chooses an initial distribution proportional to the inverse coupling g^{-2}, then one can completely scale out the coupling from the calculation by an appropriate rescaling of the fields. In this work, the shape of the initial distribution was controlled by two parameters n_0 (that controls the overall normalization) and ξ_0 (that controls the anisotropy).

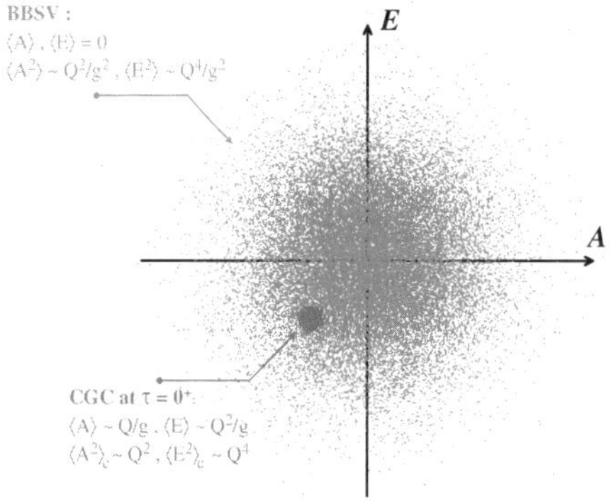

Fig. 37. (Color online) Illustration of the difference between the initial conditions (75) and (81) for the Wigner distribution of the initial fields.

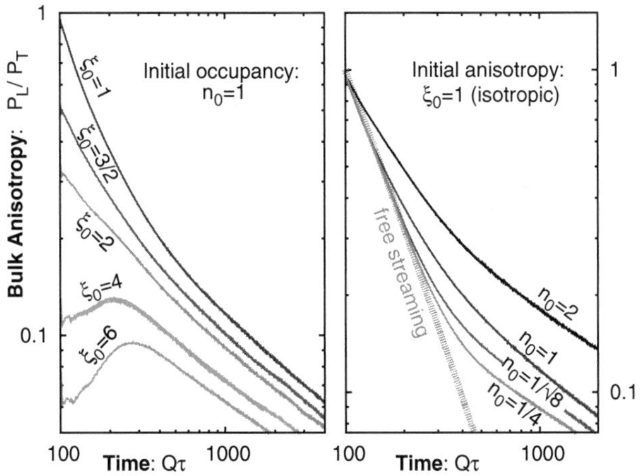

Fig. 38. (Color online) Behavior of P_L/P_T in the CSA with particle-like initial conditions in Yang–Mills theory, for various anisotropies (ξ_0) and densities (n_0) of the initial particle distribution. From Ref. 157.

Some results regarding the behavior of the ratio P_L/P_T are shown in Fig. 38. Despite substantial variations of these parameters, the system was always observed to reach a scaling regime in which on has the following approximate behaviors,

$$f(\boldsymbol{p}) \sim \tau^{-2/3}, \quad p_\perp \sim \tau^0, \quad p_z \sim \tau^{-1/3}, \quad \frac{P_L}{P_T} \sim \tau^{-2/3}. \tag{83}$$

Note that these scaling laws are not the free streaming ones (where $p_z \sim \tau^{-1}$, $f(\boldsymbol{p}) \sim \tau^0$ and $P_{_L}/P_{_T} \sim \tau^{-2}$) which means that this system interacts significantly but not strongly enough in order to overcome the expansion, at least in the classical approximation. The scaling behavior of the longitudinal momentum $p_z \sim \tau^{-1/3}$ can be understood semi-analytically[111] if the elastic scattering rate is dominated by small angle scatterings. Quite surprisingly, the same scaling behavior of p_z was also observed[161] in this approximation in the scalar ϕ^4 theory, despite the fact that the leading order scattering cross-section in this theory is dominated by large angle scatterings.[ii]

In this approximation, this scaling behavior of Eq. (83) will continue forever, and the pressure tensor becomes more and more anisotropic over time. However, since this is a classical approximation, it is know to break at least when the occupation number $f(\boldsymbol{p})$ becomes of order one. This is expected to happen when $Q_s\tau \sim g^{-3}$, which corresponds to the end of the first stage of the bottom-up scenario of Ref. 111. If this scenario — illustrated in Fig. 39 — is confirmed, the isotropization of the system would only begin after this time, while the occupation number in the system is no longer large. The quantum corrections would eventually bring the energy-momentum tensor to a quasi-isotropic shape compatible with

[ii]See Subsec. 12.3 for a discussion of the possible interplay between the classical approximation and large angle scatterings in anisotropic systems.

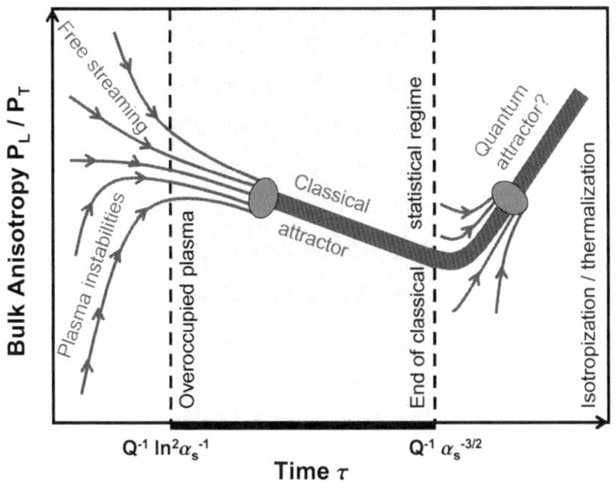

Fig. 39. (Color online) Thermalization/isotropization scenario that emerges from CSA studies with particle-like initial conditions. From Ref. 157.

nearly ideal hydrodynamics. However, what would be more difficult to understand in this scenario is why the shear viscosity to entropy ratio is small. Indeed, in a weak coupling system with occupation numbers of order 1 or below, this ratio is expected to be parametrically large $\eta/s \sim g^{-4}$.

12. Limitations of the Classical Statistical Approximation

The main appeal of the CSA is the fact that it is quite straightforward to implement, (unlike other schemes like the two-particle irreducible approach, that has never been applied to Yang–Mills theory so far), while at the same time staying very close to the dynamics of the gauge fields so that it is a natural extension of the CGC framework. However, it is not without problems and limitations, that we discuss in this section.

12.1. *Ultraviolet divergences and non-renormalizability*

A very important difference between the initial conditions (75) and (81) is the momentum dependence of the spectrum of field fluctuations. In Eqs. (81) the large momentum behavior of the spectrum is controlled by the initial particle distribution $f_0(\boldsymbol{k})$, and therefore it does not extend to infinity if one chooses a $f_0(\boldsymbol{k})$ that has a compact support. In contrast, the vacuum fluctuations that are the source of the field fluctuations in the CGC initial conditions have a spectrum which is flat up to $k = \infty$.

This difference in the spectra at large momentum leads to very different dependences on the ultraviolet cutoff (i.e. the inverse lattice spacing) in numerical implementations of the CSA using these two types of initial conditions. This issue can be phrased as follows: starting from a renormalizable quantum field theory (e.g. Yang–Mills theory, or a ϕ^4 scalar field theory), does the CSA preserve its renormalizability? It turns out that the answer to this question depends on the spectrum of the initial field fluctuations. This can be studied perturbatively by using the retarded-advanced basis, and by using the fact that the CSA amounts to dropping the vertex that has 3 indices of type 1.

From studies performed in the context of quantum field theory at finite temperature,[153,163–166] it has been known for a long time that a particle-like spectrum that falls at least as fast as k^{-1} leads to a super-renormalizable approximation. In this case, it is sufficient to make a finite number of subtractions in order to make predictions that do not depend on the ultraviolet cutoff. This applies directly to initial conditions of type (81), provided that the initial distribution $f_0(\boldsymbol{k})$ falls quickly enough with momentum.

The situation is quite different with the vacuum-like CGC initial conditions, because they have a flat spectrum of fluctuations. By using the perturbative CSA described in Subsec. 10.5, it has been shown in Ref. 154 that the CSA is a non-renormalizable approximation of the underlying quantum field theory when this type of initial condition is used. For instance, for the self-energy Σ_{12} at two loops (in a ϕ^4 scalar field theory), the CSA gives an ultraviolet divergent result

Im 
$$= -\frac{g^4}{1024\pi^3}\left(\Lambda^2_{\text{UV}} - \frac{2}{3}p^2\right). \tag{84}$$

Since this divergence occurs in the imaginary part of a correlator, it cannot be removed by a counterterm added to the action (otherwise that would make the action non-Hermitian). The consequence of this is that one cannot take the continuum limit in computations based on the CSA with vacuum-like initial conditions. This 2-loop divergence in a self-energy has a counterpart in the classical approximation of the Boltzmann equation, as we shall see in the next subsection.

Non-renormalizable graphs can even be found in some 1-loop 4-point functions, such as the following Γ_{1122} function:

$-i\Gamma^{1\ \text{loop}}_{1122} =$ 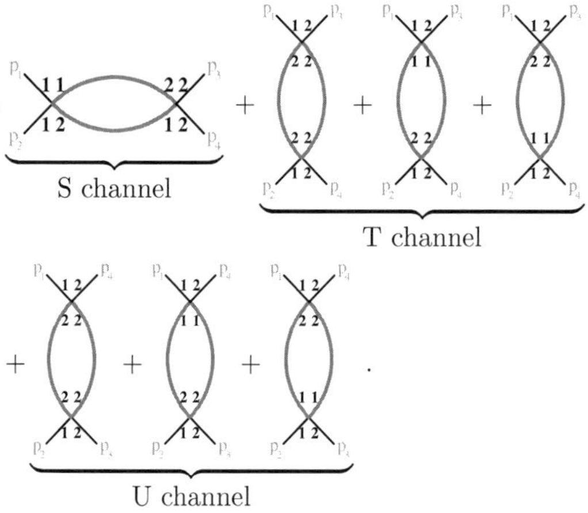

(85)

Here, we have shown all the vertex assignments that would appear in the retarded-advanced basis. Calculating and adding up all these contributions would lead to an ultraviolet finite result, as is expected in a renormalizable theory since the Lagrangian has no $\phi_1^2\phi_2^2$ operator. Note however that some of these graphs contain a 1112 vertex, and would therefore be discarded in the CSA. In this approximation, we only have

$$-i\left[\Gamma_{1122}\right]^{\text{1 loop}}_{\text{CSA}} =$$

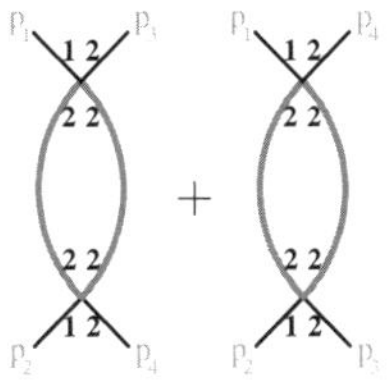

$$, \qquad (86)$$

the result of which is given by

$$-i[\Gamma_{1122}]^{\text{1 loop}}_{\text{CSA}} = -\frac{g^4}{64\pi}\left[\text{sign}(t) + \text{sign}(u) + 2\Lambda_{\text{UV}}\left(\frac{\theta(-t)}{|\boldsymbol{p}_1+\boldsymbol{p}_3|} + \frac{\theta(-u)}{|\boldsymbol{p}_1+\boldsymbol{p}_4|}\right)\right], \qquad (87)$$

with

$$t \equiv (p_1+p_3)^2, \quad u \equiv (p_1+p_4)^2, \qquad (88)$$

the standard Mandelstam variables and where Λ_{UV} is the ultraviolet cutoff on 3-momentum. Despite the zero superficial degree of divergence of these graphs, they contain a linear divergence. Moreover, the coefficient of the divergent terms is non-polynomial in the momenta, implying that it a non-local ultraviolet divergence. Such non-renormalizable contributions will appear at 2-loops and beyond in the expectation value of inclusive quantities like the energy-momentum tensor. Because of them, CSA calculations performed with initial conditions that contain vacuum fluctuations should be performed with an ultraviolet cutoff (i.e. the inverse lattice spacing) which is not too large compared to the physical scales.

12.2. *CSA in kinetic theory*

Assessing the cutoff dependence within the CSA itself is very costly, because it requires repeating the same calculation several times with smaller and smaller lattice spacings.[160] A much cheaper way of understanding the interplay between the classical approximation and the dependence on the ultraviolet cutoff is to consider the same approximation at the level of kinetic theory.[jj] Let us start with the Boltzmann equation, with the collision term expressed in terms of self-energies in the

[jj] Here, we employ kinetic theory as a way to assess some formal aspects of the underlying quantum field theory, like its dependence on the ultraviolet cutoff. A number of works[107–110,167–172] have also used kinetic theory as a tool for studying thermalization in models of heavy ion collisions.

Schwinger–Keldysh formalism:

$$[\partial_t + \boldsymbol{v_p}\cdot\boldsymbol{\nabla}]f(p) = \underbrace{\frac{i}{2\omega_{\boldsymbol p}}\left[f(\boldsymbol p)\Sigma_{-+}(P) - (1+f(\boldsymbol p))\Sigma_{+-}(P)\right]}_{C_{\boldsymbol p}[f]}. \tag{89}$$

In order to perform in kinetic theory the same approximation as in the CSA, one should first rewrite the collision term in the retarded-advanced basis:

$$C_{\boldsymbol p}[f] = \frac{i}{2\omega_{\boldsymbol p}}\left[\boldsymbol\Sigma_{11}(P) + \left(f(\boldsymbol p)+\frac12\right)(\boldsymbol\Sigma_{21}(P) - \boldsymbol\Sigma_{12}(P))\right]. \tag{90}$$

($\boldsymbol\Sigma_{11}$ is imaginary, as well as $\boldsymbol\Sigma_{21}(P) - \boldsymbol\Sigma_{12}(P)$.) At this point, we just need to calculate the self-energies that appear in the right-hand side (at 2-loops if we want $2 \to 2$ collisions) by neglecting the 1112 vertex. The two versions of the CSA, with or without vacuum fluctuations, simply correspond keeping the 1/2 in the factor $1/2 + f(\boldsymbol p)$ that appears in the propagator G_{22} or not.[173–175] It is easy to check that the classical approximation with no vacuum fluctuations amounts to keeping only the terms that are cubic in the particle distribution, while the classical approximation with vacuum fluctuations leads to the collision term

$$\begin{aligned}
&\frac{g^4}{4\omega_{\boldsymbol p}}\int_{\boldsymbol k}\int_{\boldsymbol p'}\int_{\boldsymbol k'}(2\pi)^4\delta(P+K-P'-K')\\
&\quad\times\left[\left(f(\boldsymbol p')+\frac12\right)\left(f(\boldsymbol k')+\frac12\right)(1+f(\boldsymbol p)+f(\boldsymbol k))\right.\\
&\quad\left.-\left(f(\boldsymbol p)+\frac12\right)\left(f(\boldsymbol k)+\frac12\right)(1+f(\boldsymbol p')+f(\boldsymbol k'))\right],
\end{aligned} \tag{91}$$

that has the same cubic and quadratic terms as the exact collision, but also some spurious terms that are linear in the particle distribution.

The fixed points of these "classical" Boltzmann equations are

$$\begin{aligned}
\text{without vacuum fluctuations:}\quad & f(\boldsymbol p) = \frac{T}{\omega_{\boldsymbol p}-\mu}\\
\text{with vacuum fluctuations:}\quad & f(\boldsymbol p) = \frac{T}{\omega_{\boldsymbol p}-\mu} - \frac12,
\end{aligned} \tag{92}$$

($\omega_{\boldsymbol p} \equiv \sqrt{p^2+m^2}$), i.e. respectively the first and the first two terms in the expansion of a Bose–Einstein distribution around low energy.

The parameters T and μ that appear in the asymptotic distributions can be determined from conservation laws. In the Boltzmann equation with only elastic scatterings, both the energy and the particle number are conserved. For the CSA with vacuum fluctuations, these conservation laws lead to

$$n = n_c + \int \frac{d^3\boldsymbol{p}}{(2\pi)^3} \left(\frac{T}{\omega_{\boldsymbol{p}} - \mu} - \frac{1}{2} \right)$$
$$\epsilon = n_c m + \int \frac{d^3\boldsymbol{p}}{(2\pi)^3} \omega_{\boldsymbol{p}} \left(\frac{T}{\omega_{\boldsymbol{p}} - \mu} - \frac{1}{2} \right). \tag{93}$$

In this exercise, we allow the formation of a Bose–Einstein condensate with a particle density n_c, in case one wishes to consider highly populated initial conditions. Note that in any case, there are only two unknowns in these equations: T and μ if there is no condensation, or T and n_c if there is condensation (in which case $\mu = m$). From these two equations, we can determine the two unknown parameters. However, since the integrals are ultraviolet divergent, it is necessary to introduce a cutoff Λ on $|\boldsymbol{p}|$, which turns T, μ and n_c into Λ-dependent quantities.[158] The solution of Eqs. (93) is shown in Fig. 40 (Q is a physical momentum scale that characterizes the initial momentum distribution of the particles), where we have also superimposed results from Ref. 160 obtained by lattice classical statistical simulations. These curves show a very strong dependence on the ultraviolet cutoff when it becomes much larger than the physical scales, as expected given the fact that the CSA with vacuum fluctuations is not renormalizable. On the other hand, there is a region where the cutoff is a few times the physical scale, and where the parameters that characterize the asymptotic distribution are rather insensitive to the cutoff. Therefore, simulations using the CSA with vacuum fluctuations as initial conditions

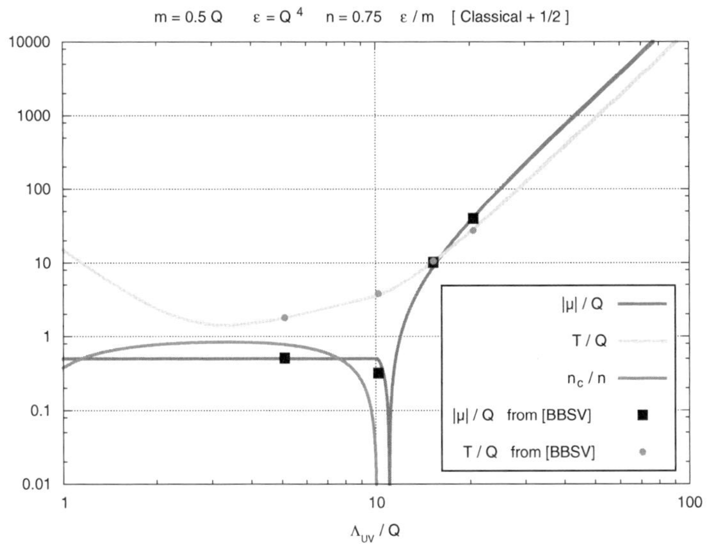

Fig. 40. (Color online) Evolution of T, μ and n_c as a function of the ultraviolet cutoff. The points reproduce the values listed in Fig. 10 of Ref. 160, obtained with a classical statistical field simulation.

should preferably be performed with an ultraviolet cutoff chosen in that range, in order to minimize the sensitivity of the results on the cutoff.

12.3. *Quantum corrections in anisotropic systems*

Note that this non-renormalizability problem arises only when one uses the CSA in conjunction with a spectrum of initial conditions that represents vacuum fluctuations, as in Eqs. (75). When using particle-like initial conditions, such as those described in Eqs. (81), the CSA is ultraviolet finite provided that the initial particle distribution falls faster than $1/k$. Because of this, if one forgets that this type of initial condition is not derived directly from the CGC, this implementation of the CSA may seem better since one does not need to worry about the dependence on the ultraviolet cutoff.

However, this choice of initial conditions in the CSA leads to a different kind of problem when used in studies of isotropization in heavy ion collisions, for the following reason. For the purpose of this argument, let us again reason in terms of the corresponding Boltzmann equation. For $2 \to 2$ collisions, it reads

$$\partial_t f_3 \sim g^4 \int_{124} \cdots [f_1 f_2 (f_3 + f_4) - f_3 f_4 (f_1 + f_2)]$$
$$+ g^4 \int_{124} \cdots [f_1 f_2 - f_3 f_4]. \tag{94}$$

(Here we are tracking the distribution of particles of momentum p_3 — see Fig. 41.) In this equation, we have separated the purely classical terms (in f^3) from the subleading f^2 terms. When only particle-like fluctuations are included, the CSA applied to the Boltzmann equation gives only the f^3 terms, and neglects all the other terms. In contrast, the CSA where vacuum fluctuations are included has the f^3 and f^2 terms, and also some unphysical terms that are linear in f (see Eq. (91)).

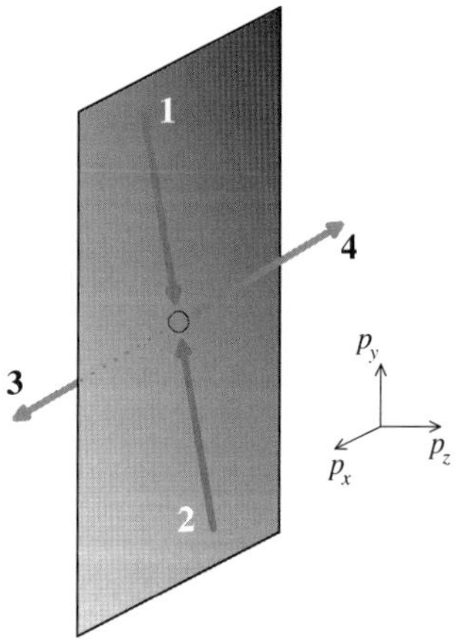

Fig. 41. (Color online) $2 \to 2$ scattering contributing to isotropization.

Consider now a situation where the particle distribution is strongly anisotropic, with a support in p_z which is squeezed compared to the support in $\boldsymbol{p}_\perp$. Starting from a generic initial condition of this form, the unapproximated Boltzmann equation will in general lead to isotropization because two purely transverse particles can be scattered outside of the transverse plane.[kk] But does this still happen in approximations of the Boltzmann equation? In Fig. 41, one has $p_3^z = -p_4^z \neq 0$, and given our assumption about the support of the particle distribution, this means that $f_3 = f_4 \approx 0$. In the collision term of the Boltzmann equation, a number of terms therefore vanish:

$$\partial_t f_3 \sim g^4 \int_{124} \cdots [f_1 f_2 \underbrace{(f_3 + f_4)}_{0} - \underbrace{f_3 f_4}_{0}(f_1 + f_2)] + g^4 \int_{124} \cdots [f_1 f_2 - \underbrace{f_3 f_4}_{0}]. \tag{95}$$

In particular, all the cubic terms are zero. The problem is that these are the only terms that are kept in the CSA with no vacuum fluctuations. The only non-zero term is the term in $f_1 f_2$, which would be present in the CSA only if vacuum initial fluctuations are present. From this discussion, the particle-like initial conditions in the CSA, despite their appeal since they lead to UV finite results, may be inappropriate because they lead to missing the most important contribution to

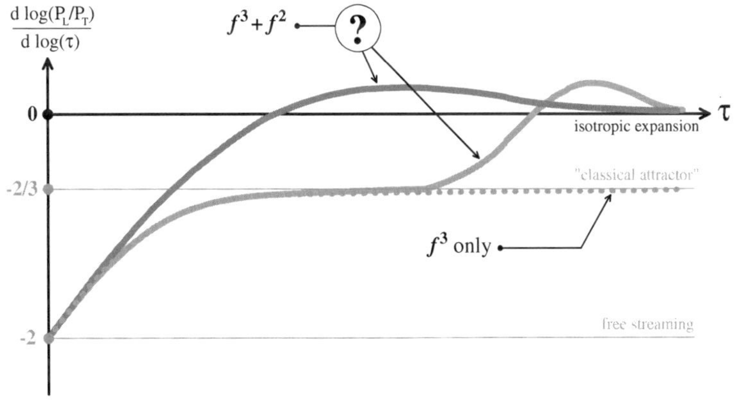

Fig. 42. (Color online) Behavior of the logarithmic derivative of the ratio P_L/P_T. Dotted curve: classical approximation where one keeps only the f^3 terms. Blue curve: full collision term, if there exists a "classical attractor". Red curve: full collision term, if the f^2 terms prevent the classical attractor.

[kk]This is obvious in the ϕ^4 scalar field theory, where the leading term in the two body cross-section is point-like and where scatterings are large angle are dominant. This is also the case in the CGC, thanks to screening. Indeed, since the occupation number is of order g^{-2}, the Debye mass is of order Q_s. Since this is comparable to the typical momentum of the gluons, large angle scatterings are also expected to be the norm.

isotropization. In other words, in an anisotropic system, the classical approximation could break down long before the occupation number becomes of order 1.

This may also have an incidence on the behavior at intermediate times. With only the f^3 terms, the ratio P_L/P_T decreases like $\tau^{-2/3}$ at late times, and therefore the system never isotropizes in this approximation. On the other hand, the system is expected to isotropize eventually with the complete Boltzmann equation (f^3 and f^2 terms), with a ratio $P_L/P_T \sim \tau^0$ at late times. If the f^2 terms are truly negligible over some extended period of time, then the full Boltzmann equation should lead to the blue curve in Fig. 42, in which the system spends some time stuck into a "classical attractor" before eventually leaving it in order to isotropize. In contrast, if the fact that the f^2 terms are always dominant in the tail plays an important role, then one may instead get the red curve. The Boltzmann equation offers an interesting playground in order to test these possibilities, since it can be solved with and without the f^2 terms at comparable computational costs.

13. Summary

A lot of progress has been made in the past 10 years in QCD-based studies of the early stages of heavy ion collisions. The most promising framework for these studies is the Color Glass Condensate, which allows one to systematically include the non-linear effects that prevail when the gluon occupation numbers are large.

The description of the relevant degrees of freedom in the wave function of the incoming nuclei and how they evolve as one varies the energy of the collision is entering a very mature stage, since one now knows this evolution at next-to-leading log accuracy. A lot more work will be needed before these NLL evolution equations can be implemented in phenomenological studies of heavy ion collisions, but this is a very important step towards more quantitative works.

Regarding the collision itself, i.e. how two objects described by means of the CGC interact while they collide, one has now a clearer understanding of how observables can be expressed at leading order in terms of a classical solution of the Yang–Mills equations (and how the inclusive nature of an observable translates into retarded boundary conditions for this classical solution), and how to calculate their next-to-leading order corrections in terms of linearized perturbations around this classical solution.

The recent years have also witnessed a renewed interest in the question of the isotropization and thermalization of the system produced in heavy ion collisions. This issue has indeed been made more pressing by the many phenomenological successes of hydrodynamical models in describing the expansion of the fireball as a nearly perfect fluid, something which is at odds with the LO CGC results. From recent works, there appears to be two tools of choice for these studies: kinetic theory (in the small scattering angle approximation, or the more sophisticated effective kinetic theory of Ref. 104) and the classical statistical approximation, which is more directly related to the CGC framework. Although this problem seems to have now reached a satisfactory level of understanding — both qualitatively and quantitatively– in the simpler case of a system confined in a fixed volume, the situation is at the moment somewhat unclear in the more realistic situation of a longitudinally expanding system, for which two different results have been obtained in the classical statistical approximation with different initial conditions. These initial conditions mostly differ in whether some vacuum quantum fluctuations are included or not, and a prime question to clarify in the near future will be to understand to what extent quantum fluctuations play a role in isotropization and thermalization.

Acknowledgments

This work is supported by the Agence Nationale de la Recherche project 11-BS04-015-01.

References

1. L. V. Gribov, E. M. Levin and M. G. Ryskin, *Phys. Rept.* **100** (1983) 1.
2. A. H. Mueller and J.-W. Qiu, *Nucl. Phys. B* **268** (1986) 427.
3. J. P. Blaizot and A. H. Mueller, *Nucl. Phys. B* **289** (1987) 847.
4. D. J. Gross and F. Wilczek, *Phys. Rev. Lett.* **30** (1973) 1343.
5. D. J. Gross and F. Wilczek, *Phys. Rev. D* **8** (1973) 3633.
6. D. J. Gross and F. Wilczek, *Phys. Rev. D* **9** (1974) 980.
7. H. D. Politzer, *Phys. Rev. Lett.* **30** (1973) 1346.
8. H. D. Politzer, *Phys. Rept.* **14** (1974) 129.
9. R. P. Feynman, *Photon-Hadron Interactions*, Frontiers in Physics (W. A. Benjamin, 1972).
10. J. C. Collins, D. E. Soper and G. Sterman, *Nucl. Phys. B* **250** (1985) 199.
11. J. C. Collins, D. E. Soper and G. Sterman, *Nucl. Phys. B* **261** (1985) 104.
12. J. C. Collins, D. E. Soper and G. Sterman, *Nucl. Phys. B* **263** (1986) 37.
13. H1 and ZEUS Collab. (F. D. Aaron *et al.*), *JHEP* **1001** (2010) 109.
14. G. Altarelli and G. Parisi, *Nucl. Phys. B* **126** (1977) 298.
15. V. N. Gribov and L. N. Lipatov, *Sov. J. Nucl. Phys.* **15** (1972) 438.
16. V. N. Gribov and L. N. Lipatov, *Sov. J. Nucl. Phys.* **15** (1972) 675.
17. Yu. Dokshitzer, *Sov. Phys. JETP* **46** (1977) 641.
18. I. Balitsky and L.N. Lipatov, *Sov. J. Nucl. Phys.* **28** (1978) 822.
19. E. A. Kuraev, L. N. Lipatov and V. S. Fadin, *Sov. Phys. JETP* **45** (1977) 199.
20. A. Deshpande, R. Ent and R. Milner, *CERN Courier*, October 2009.
21. E. Iancu and R. Venugopalan, in *Quark Gluon Plasma 3*, eds. R. C. Hwa and X. N. Wang (World Scientific, 2004), hep-ph/0303204.
22. T. Lappi, *Int. J. Mod. Phys. E* **20** (2011) 1.
23. H. Weigert, *Prog. Part. Nucl. Phys.* **55** (2005) 461.
24. F. Gelis, E. Iancu, J. Jalilian-Marian and R. Venugopalan, *Ann. Rev. Part. Nucl. Sci.* **60** (2010) 463.
25. F. Gelis, *Int. J. Mod. Phys. A* **28** (2013) 1330001.
26. L. D. McLerran and R. Venugopalan, *Phys. Rev. D* **49** (1994) 2233.
27. L. D. McLerran and R. Venugopalan, *Phys. Rev. D* **49** (1994) 3352.
28. L. D. McLerran and R. Venugopalan, *Phys. Rev. D* **50** (1994) 2225.

29. F. Gelis and R. Venugopalan, *Nucl. Phys. A* **776** (2006) 135.
30. F. Gelis and R. Venugopalan, *Nucl. Phys. A* **779** (2006) 177.
31. J. Schwinger, *J. Math. Phys.* **2** (1961) 407.
32. L. V. Keldysh, *Sov. Phys. JETP* **20** (1964) 1018.
33. R. E. Cutkosky, *J. Math. Phys.* **1** (1960) 429.
34. G. 't Hooft and M. J. G. Veltman, *CERN report* 73–9.
35. Yu. V. Kovchegov, *Phys. Rev. D* **54** (1996) 5463.
36. A. Kovner, L. D. McLerran and H. Weigert, *Phys. Rev. D* **52** (1995) 6231.
37. A. Krasnitz and R. Venugopalan, *Nucl. Phys. B* **557** (1999) 237.
38. A. Krasnitz and R. Venugopalan, *Phys. Rev. Lett.* **84** (2000) 4309.
39. A. Krasnitz and R. Venugopalan, *Phys. Rev. Lett.* **86** (2001) 1717.
40. A. Krasnitz, Y. Nara and R. Venugopalan, *Phys. Rev. Lett.* **87** (2001) 192302.
41. A. Krasnitz, Y. Nara and R. Venugopalan, *Nucl. Phys. A* **727** (2003) 427.
42. T. Lappi, *Phys. Rev. C* **67** (2003) 054903.
43. T. Lappi, *Phys. Lett. B* **643** (2006) 11.
44. A. Krasnitz, Y. Nara and R. Venugopalan, *Phys. Lett. B* **554** (2003) 21.
45. A. Krasnitz, Y. Nara and R. Venugopalan, *Nucl. Phys. A* **717** (2003) 268.
46. T. Lappi and R. Venugopalan, *Phys. Rev. C* **74** (2006) 054905.
47. T. Lappi and L. D. McLerran, *Nucl. Phys. A* **772** (2006) 200.
48. A. Dumitru, Y. Nara and E. Petreska, *Phys. Rev. D* **88** (2013) 054016.
49. A. Dumitru, T. Lappi and Y. Nara, *Phys. Lett. B* **734** (2014) 7.
50. A. Ayala, J. Jalilian-Marian, L. D. McLerran and R. Venugopalan, *Phys. Rev. D* **53** (1996) 458.
51. I. Balitsky, *Phys. Rev. D* **75** (2007) 014001.
52. I. Balitsky and G. A. Chirilli, *Phys. Rev. D* **77** (2008) 014019.
53. Yu. V. Kovchegov and H. Weigert, *Nucl. Phys. A* **784** (2007) 188.
54. E. Gardi, J. Kuokkanen, K. Rummukainen and H. Weigert, *Nucl. Phys. A* **784** (2007) 282.
55. A. V. Grabovsky, *JHEP* **1309** (2013) 141.
56. A. Kovner, M. Lublinsky and Y. Mulian, *Phys. Rev. D* **89** (2014) 061704.
57. A. Kovner, M. Lublinsky and Y. Mulian, arXiv:1405.0418.
58. J. L. Albacete, N. Armesto, J. G. Milhano, C. A. Salgado and U. A. Wiedemann, *Phys. Rev. D* **71** (2005) 014003.
59. J. L. Albacete and Y. Kovchegov, *Phys. Rev. D* **75** (2007) 125021.
60. J. L. Albacete, N. Armesto, J. G. Milhano and C. A. Salgado, *Phys. Rev. D* **80** (2009) 034031.
61. J. L. Albacete, N. Armesto, J. G. Milhano and C. A. Salgado, arXiv:0906.2721.
62. T. Lappi and H. Mäntysaari, *Eur. Phys. J. C* **73** (2013) 2307.
63. F. Gelis, T. Lappi and R. Venugopalan, *Phys. Rev. D* **78** (2008) 054019.
64. F. Gelis, T. Lappi and R. Venugopalan, *Phys. Rev. D* **78** (2008) 054020.
65. F. Gelis, T. Lappi and R. Venugopalan, *Phys. Rev. D* **79** (2009) 094017.
66. I. Balitsky, *Nucl. Phys. B* **463** (1996) 99.
67. J. Jalilian-Marian, A. Kovner, L. D. McLerran and H. Weigert, *Phys. Rev. D* **55** (1997) 5414.
68. J. Jalilian-Marian, A. Kovner, A. Leonidov and H. Weigert, *Nucl. Phys. B* **504** (1997) 415.
69. J. Jalilian-Marian, A. Kovner, A. Leonidov and H. Weigert, *Phys. Rev. D* **59** (1998) 014014.
70. J. Jalilian-Marian, A. Kovner, A. Leonidov and H. Weigert, *Phys. Rev. D* **59** (1999) 034007.

71. J. Jalilian-Marian, A. Kovner, A. Leonidov and H. Weigert, *Phys. Rev. D* **59** (1999) 099903.
72. E. Iancu, A. Leonidov and L. D. McLerran, *Nucl. Phys. A* **692** (2001) 583.
73. E. Iancu, A. Leonidov and L. D. McLerran, *Phys. Lett. B* **510** (2001) 133.
74. E. Ferreiro, E. Iancu, A. Leonidov and L. D. McLerran, *Nucl. Phys. A* **703** (2002) 489.
75. STAR Collab. (B. I. Abelev *et al.*), *Phys. Rev. C* **80** (2009) 064912.
76. A. Dumitru, F. Gelis, L. McLerran and R. Venugopalan, *Nucl. Phys. A* **810** (2008) 91.
77. T. Lappi, S. Srednyak and R. Venugopalan, *JHEP* **1001** (2010) 066.
78. S. A. Voloshin, *Phys. Lett. B* **632** (2006) 490.
79. C. A. Pruneau, S. Gavin and S. A. Voloshin, *Nucl. Phys. A* **802** (2008) 107.
80. S. Gavin, L. McLerran and G. Moschelli, *Phys. Rev. C* **79** (2009) 051902.
81. E. V. Shuryak, *Phys. Rev. C* **76** (2007) 047901.
82. STAR Collab. (J. Adams *et al.*), *Nucl. Phys. A* **757** (2005) 102.
83. PHENIX Collab. (K. Adcox *et al.*), *Nucl. Phys. A* **757** (2005) 184.
84. BRAHMS Collab. (I. Arsene *et al.*), *Nucl. Phys. A* **757** (2005) 1.
85. PHOBOS Collab. (B. B. Back *et al.*), *Nucl. Phys. A* **757** (2005) 28.
86. P. Huovinen and P. V. Ruuskanen, *Ann. Rev. Nucl. Part. Sci.* **56** (2006) 163.
87. P. Romatschke, *Int. J. Mod. Phys. E* **19** (2010) 1.
88. D. Teaney, *Prog. Part. Nucl. Phys.* **62** (2009) 451.
89. P. Romatschke and U. Romatschke, *Phys. Rev. Lett.* **99** (2007) 172301.
90. G. Policastro, D. T. Son and A. O. Starinets, *Phys. Rev. Lett.* **87** (2001) 081601.
91. G. Policastro, D. T. Son and A. O. Starinets, *JHEP* **0209** (2002) 043.
92. M. Martinez and M. Strickland, *Nucl. Phys. A* **856** (2011) 68.
93. M. Martinez, R. Ryblewski and M. Strickland, *Phys. Rev. C* **85** (2012) 064913.
94. W. Florkowski, R. Maj, R. Ryblewski and M. Strickland, *Phys. Rev. C* **87** (2013) 034914.
95. W. Florkowski, R. Ryblewski and M. Strickland, *Nucl. Phys. A* **916** (2013) 249.
96. D. Bazow, U. W. Heinz and M. Strickland, arXiv:1311.6720.
97. M. Strickland, *Nucl. Phys. A* **926** (2014) 92.
98. M. Nopoush, R. Ryblewski and M. Strickland, *Phys. Rev. C* **90** (2014) 014908.
99. K. Fukushima and F. Gelis, *Nucl. Phys. A* **874** (2012) 108.
100. A. Kurkela and G. D. Moore, *JHEP* **1112** (2011) 044.
101. J. P. Blaizot, F. Gelis, J. Liao, L. McLerran and R. Venugopalan, *Nucl. Phys. A* **873** (2012) 68.
102. M. C. Abraao York, A. Kurkela, E. Lu and G. D. Moore, *Phys. Rev. D* **89** (2014) 074036.
103. A. Kurkela and E. Lu, *Phys. Rev. Lett.* **113** (2014) 182301.
104. P. Arnold, G. D. Moore and L. G. Yaffe, *JHEP* **0301** (2003) 030.
105. T. Epelbaum and F. Gelis, *Nucl. Phys. A* **872** (2011) 210.
106. J. Berges and D. Sexty, *Phys. Rev. Lett.* **108** (2012) 161601.
107. J. P. Blaizot, J. Liao and L. D. McLerran, *Nucl. Phys. A* **920** (2013) 58.
108. F. Scardina, D. Perricone, S. Plumari, M. Ruggieri and V. Greco, arXiv:1408.1313.
109. Z. Xu, K. Zhou, P. Zhuang and C. Greiner, *Phys. Rev. Lett.* **114** (2015) 18.
110. X. G. Huang and J. Liao, *Int. J. Mod. Phys. E* **23** (2014) 1430003.
111. R. Baier, A. H. Mueller, D. Schiff and D. Son, *Phys. Lett. B* **502** (2001) 51.
112. A. Kurkela and G. D. Moore, *JHEP* **1111** (2011) 120.
113. P. Romatschke and R. Venugopalan, *Phys. Rev. Lett.* **96** (2006) 062302.
114. P. Romatschke and R. Venugopalan, *Eur. Phys. J. A* **29** (2006) 71.

115. P. Romatschke and R. Venugopalan, *Phys. Rev. D* **74** (2006) 045011.
116. T. S. Biro, C. Gong, B. Muller and A. Trayanov, *Int. J. Mod. Phys.* **C 5** (1994) 113.
117. U. W. Heinz, C. R. Hu, S. Leupold, S. G. Matinyan and B. Muller, *Phys. Rev. D* **55** (1997) 2464.
118. J. Bolte, B. Müller and A. Schäfer, *Phys. Rev. D* **61** (2000) 054506.
119. H. Fujii and K. Itakura, *Nucl. Phys. A* **809** (2008) 88.
120. H. Fujii, K. Itakura and A. Iwazaki, *Nucl. Phys. A* **828** (2009) 178.
121. T. Kunihiro, B. Muller, A. Ohnishi, A. Schafer, T. T. Takahashi and A. Yamamoto, *Phys. Rev. D* **82** (2010) 114015.
122. S. Mrowczynski, *Phys. Lett. B* **314** (1993) 118.
123. S. Mrowczynski, *Phys. Lett. B* **393** (1997) 26.
124. A. K. Rebhan, P. Romatschke and M. Strickland, *Phys. Rev. Lett.* **94** (2005) 102303.
125. A. K. Rebhan, P. Romatschke and M. Strickland, *JHEP* **0509** (2005) 041.
126. S. Mrowczynski, A. Rebhan and M. Strickland, *Phys. Rev. D* **70** (2004) 025004.
127. P. Romatschke and M. Strickland, *Phys. Rev. D* **68** (2003) 036004.
128. P. Romatschke and M. Strickland, *Phys. Rev. D* **70** (2004) 116006.
129. A. K. Rebhan and D. Steineder, *Phys. Rev. D* **81** (2010) 085044.
130. A. K. Rebhan, M. Strickland and M. Attems, *Phys. Rev. D* **78** (2008) 045023.
131. P. Arnold, J. Lenaghan and G. D. Moore, *JHEP* **0308** (2003) 002.
132. P. Arnold, J. Lenaghan, G. D. Moore and L. G. Yaffe, *Phys. Rev. Lett.* **94** (2005) 072302.
133. P. Arnold and G. D. Moore, *Phys. Rev. D* **73** (2006) 025013.
134. P. Arnold and G. D. Moore, *Phys. Rev. D* **73** (2006) 025006.
135. P. Arnold and G. D. Moore, *Phys. Rev. D* **76** (2007) 045009.
136. P. Arnold, G. D. Moore and L. G. Yaffe, *Phys. Rev. D* **72** (2005) 054003.
137. A. Dumitru, Y. Nara and M. Strickland, *Phys. Rev. D* **75** (2007) 025016.
138. D. Bodeker and K. Rummukainen, *JHEP* **0707** (2007) 022.
139. J. Berges, D. Gelfand, S. Scheffler and D. Sexty, *Phys. Lett. B* **677** (2009) 210.
140. J. Berges, S. Scheffler and D. Sexty, *Phys. Rev. D* **77** (2008) 034504.
141. M. Attems, A. Rebhan and M. Strickland, *Phys. Rev. D* **87** (2013) 025010.
142. K. Fukushima, *Phys. Rev. C* **75** (2007) 021902; Erratum: arXiv:0711.2634.
143. K. Dusling, T. Epelbaum, F. Gelis and R. Venugopalan, *Nucl. Phys. A* **850** (2011) 69.
144. P. B. Greene, L. Kofman, A. D. Linde and A. A. Starobinsky, *Phys. Rev. D* **56** (1997) 6175.
145. D. Polarski and A. A. Starobinsky, *Class. Quant. Grav.* **13** (1996) 377.
146. D. T. Son, hep-ph/9601377.
147. S. Yu. Khlebnikov and I. I. Tkachev, *Phys. Rev. Lett.* **77** (1996) 219.
148. R. Micha and I. I. Tkachev, *Phys. Rev. D* **70** (2004) 043538.
149. K. Fukushima, F. Gelis and L. McLerran, *Nucl. Phys. A* **786** (2007) 107.
150. P. Aurenche and T. Becherrawy, *Nucl. Phys. B* **379** (1992) 259.
151. M. A. van Eijck, R. Kobes and Ch. G. van Weert, *Phys. Rev. D* **50** (1994) 4097.
152. F. Gelis, *Nucl. Phys. B* **508** (1997) 483.
153. G. Aarts and J. Smit, *Nucl. Phys. B* **511** (1998) 451.
154. T. Epelbaum, F. Gelis and B. Wu, *Phys. Rev. D* **90** (2014) 6.
155. T. Epelbaum and F. Gelis, *Phys. Rev. D* **88** (2013) 085015.
156. T. Epelbaum and F. Gelis, *Phys. Rev. Lett.* **111** (2013) 232301.
157. J. Berges, K. Boguslavski, S. Schlichting and R. Venugopalan, *Phys. Rev. D* **89** (2014) 074011.
158. T. Epelbaum, F. Gelis, N. Tanji and B. Wu, *Phys. Rev. D* **90** (2014) 12.

159. J. Berges, K. Boguslavski, S. Schlichting and R. Venugopalan, *Phys. Rev. D* **89** (2014) 114007.
160. J. Berges, K. Boguslavski, S. Schlichting and R. Venugopalan, *JHEP* **1405** (2014) 054.
161. J. Berges, K. Boguslavski, S. Schlichting and R. Venugopalan, *Phys. Rev. Lett.* **114** (2015) 6.
162. K. Dusling, F. Gelis and R. Venugopalan, *Nucl. Phys. A* **872** (2011) 161.
163. D. Bodeker, L. D. McLerran and A. Smilga, *Phys. Rev. D* **52** (1995) 4675.
164. P. Arnold, D. T. Son and L. G. Yaffe, *Phys. Rev. D* **55** (1997) 6264.
165. G. Aarts and J. Smit, *Phys. Lett. B* **393** (1997) 395.
166. G. Aarts, B. J. Nauta and C. G. van Weert, *Phys. Rev. D* **61** (2000) 105002.
167. M. Ruggieri, F. Scardina, S. Plumari and V. Greco, *Phys. Rev. C* **89** (2014) 054914.
168. A. Puglisi, S. Plumari and V. Greco, arXiv:1407.2559.
169. A. Puglisi, S. Plumari and V. Greco, arXiv:1408.7043.
170. J. Uphoff, O. Fochler, F. Senzel, C. Wesp, Z. Xu and C. Greiner, arXiv:1401.1364.
171. O. Fochler, J. Uphoff, Z. Xu and C. Greiner, *Phys. Rev. D* **88** (2013) 014018.
172. M. Greif, I. Bouras, Z. Xu and C. Greiner, *Phys. Rev. D* **90** (2014) 094014.
173. A. H. Mueller and D. T. Son, *Phys. Lett. B* **582** (2004) 279.
174. S. Jeon, *Phys. Rev. C* **72** (2005) 014907.
175. V. Mathieu, A. H. Mueller and D. N. Triantafyllopoulos, *Eur. Phys. J. C* **74** (2014) 2873.

Introduction to Hydrodynamics

Sangyong Jeon* and Ulrich Heinz†

*Department of Physics, McGill University,
3600 University Street, Montreal, QC, H3A-2T8, Canada

†Department of Physics, The Ohio State University,
Columbus, OH 43210, USA

Hydrodynamics has been successful in providing a good description of the bulk dynamics in ultra-relativistic heavy ion collisions. In this brief review, we provide basics of the theory of viscous hydrodynamics. Topics covered include derivation of the 2nd order viscous hydrodynamics from the linear response theory and kinetic theory, viscous anisotropic hydrodynamics, and numerical implementation of relativistic hydrodynamics.

1. Introduction

Application of hydrodynamics in high-energy physics has a long and illustrious history starting from L. D. Landau's seminal paper.[1] In its history of more than half a century, many papers have been written on a broad spectrum of topics, too numerous to list them all here. In this review, our emphasis will be more on the basics of the theory of hydrodynamics than to report on the current phenomenological status, of which several excellent reviews already exist (for instance, see Refs. 2–4).

Recent ultra-relativistic heavy ion collision experiments at the Super Proton Synchrotron (SPS), Relativistic Heavy Ion Collider (RHIC) and the Large Hadron Collider (LHC) have demonstrated beyond any doubt that Quark–Gluon Plasma (QGP) is being created in these collisions. Unfortunately, direct access to the QGP properties such as the temperature, equation of state, transport coefficients, etc. is not very feasible. The only experimentally accessible information is contained in the spectra of the final state particles. To connect them to the QGP properties such as above, one must use theoretical models. It would be wonderful to have an analytic or a numerical method that can calculate evolution of heavy ion collisions from first principles. But this microscopic, non-equilibrium, many-body QCD problem is currently intractable. What is tractable is the coarse-grained collective motion of the system as a fluid after the local thermal equilibrium is established. Since the properties of QGP we are after are mostly (local) equilibrium properties, it is natural that the dynamics of collective motion — hydrodynamics — is an integral part of the theoretical modelling.

What has been exciting and interesting in QGP research is the close discourse between the experiment and theory. In elementary particle experiments, perturbative QCD is being tested with amazing successes. More and more precise perturbative QCD calculations prove to describe experimental data more and more accurately. In contrast, QGP research is much more dynamic. For instance, before the discovery of QGP, theoretical expectation was that QGP would be a weakly coupled plasma of quarks and gluons based partially on the fact that the QGP properties seem to approach about 80% of the Stefan–Boltzmann limit rather quickly on lattice, around $2T_c$.[5] But almost from the day-1 of RHIC operation, strong elliptic flow quickly proved this initial expectation not very viable. QGP around the transition temperature turned out to be *the* most strongly coupled many-body system ever observed. Soon after, the authors of Ref. 6 used string theory techniques to calculate the infinite coupling limit of the shear viscosity and came up with a surprising result that the limit is small, but has a non-zero lower bound. Subsequent Hydrodynamic calculations then demonstrated the importance of small but finite shear viscosity in understanding the RHIC flow data.[7–9] Comparing the ensuing LHC predictions with the LHC data now confirmed the expectation that as the temperature increases, shear viscosity of QGP should also increase.[10–13]

All these connections between exciting theoretical developments and experiments cannot be made without hydrodynamics. More recently, the systems created in the highest multiplicity proton–proton collisions and proton–nucleus collisions were also seen to exhibit strong collective behavior.[14–17] This is deeply puzzling as the size of the system ought to be too small to behave collectively. It is hoped that more thorough investigation of the possible origin of the collectivity in such small systems can illuminate the inner workings of the QGP formation greatly.[18]

As mentioned in the beginning, the aim of this review is the introduction of the theory of the hydrodynamics in ultra-relativistic heavy ion collisions. This actually entails a large number of disciplines in addition to the relativistic fluid dynamics. Our plan for this review is as follows. Section 2 contains the basic concepts of hydrodynamics and their definitions. In Sec. 3, second-order viscous hydrodynamics is derived from a very general linearly response theory of conserved currents. Section 4 discusses how coarse-graining of kinetic theory can result in more general form of viscous hydrodynamics. In Sec. 5, various numerical techniques needed to implement relativistic viscous hydrodynamics in ultra-relativistic heavy ion collisions are discussed. We conclude in Sec. 6.

2. Hydrodynamic Form of the Stress–Energy Tensor and the Net Baryon Number Current

Hydrodynamics is all about flow of conserved quantities. In this review, we strictly deal with relativistic hydrodynamics. Therefore, unlike the non-relativistic case, mass is a part of the energy budget. In the Minkowski coordinates, conservation laws in their local form are

$$\partial_\mu T^{\mu\nu} = 0,$$
$$\partial_\mu J_i^\mu = 0, \tag{1}$$

where $T^{\mu\nu}$ is the stress–energy tensor and the roman letter i on the current J_i^μ labels any other conserved charges such as the net baryon number, net electric charge, etc. For the bulk evolution in relativistic heavy ion collisions, usually only the net baryon number current, J_B^μ, is considered. If needed, additional electric current and the strangeness current can be easily accommodated.

Using the divergence theorem, the integral form of the conservation laws read

$$\frac{d}{dt}\int_V d^3x\, T^{0\nu} = -\int_{\partial V} dS_i T^{i\nu},$$
$$\frac{d}{dt}\int_V d^3x\, J_B^0 = -\int_{\partial V} dS_i J_B^i, \tag{2}$$

where on the right-hand side the integration is over the boundary of the volume V assuming that the size and the shape of the volume is independent of time.

This form admits a very physical interpretation that the rate of change of the conserved quantity in a fixed volume equals the net current entering the volume. Hence, the dynamics of conserved quantities are governed by the dynamics that governs the currents. In essence, hydrodynamics is all about the dynamics of the currents.

Hydrodynamics is useful because it is a coarse-grained theory. When a system contains too many particles, it becomes difficult to follow microscopic details of the system. When the system contains sufficiently many particles, the system again starts to admit analytic studies because thermodynamic concepts start to apply, in particular the static equilibrium. A system in static equilibrium is characterized by only few quantities such as the temperature, collective velocity and chemical potential. These control the energy density, momentum density and charge density, respectively. The price to pay for this simplification is that questions on short time scale phenomena or short length scale phenomena cannot be answered any longer.

The systems we would like to study, the ultra-relativistic heavy ion collisions, do contain a large number of particles, but they certainly are not static. In fact, they will never actually reach the state of static equilibrium. Nevertheless, if one is interested only in the coarse-grained collective motion of the system, the concept of *local equilibrium* may still apply provided that the expansion rate is much slower than the microscopic interaction rate. If one considers a macroscopically small but microscopically large fluid cell around a position $\mathbf{x}$ at a given time t, then within a macroscopically short but microscopically long time scale, the time averages should approach the static equilibrium values according to the ergodicity hypothesis of statistical mechanics. More details on the length and the time scale analysis can be found in Sec. 4.1.

When the local equilibrium is reached, it becomes meaningful to describe the system with the local temperature $T(t, \mathbf{x})$, the collective velocity $u^\mu(t, \mathbf{x})$ of the fluid cell and the local chemical potential $\mu_B(t, \mathbf{x})$. One can then study dynamics of only those few thermodynamic quantities. Since T, u^μ, μ_B are basically the Lagrange multipliers to fix the average energy, momentum and net charge, it is natural that we turn to the conservation laws for their dynamics. The goal of hydrodynamics is then to study collective motion of a system using the conservation laws with only the *statistical* inputs from the underlying microscopic theory.

In an dynamically evolving fluid, the concept of the local rest frame is essential in order to apply the concept of local equilibrium. However defining the collective velocity of a fluid cell turned out to be quite intricate. This is not so simple even for the simplest system composed of a single kind of non-interacting particles in the non-relativistic setting. One kind of collective velocity comes from the average momentum

$$\mathbf{u}_p = \frac{\sum_i m\mathbf{v}_i}{\sum_i m} \tag{3}$$

or from the mass current. Here the sum is over all particles in the given fluid cell. Another comes from the energy–weighted average

$$\mathbf{u}_E = \frac{\sum_i (m\mathbf{v}_i^2/2)\mathbf{v}_i}{\sum_i (m\mathbf{v}_i^2/2)} \tag{4}$$

or from the energy current. If the particles in the system have an additional conserved charge, B, and the net charge is non-zero, then one can define yet another collective velocity by performing the charge-weighted average

$$\mathbf{u}_B = \frac{\sum_i B_i\mathbf{v}_i}{\sum_i B_i}. \tag{5}$$

Here B_i is the conserved charge of the ith particle.

For a state in static equilibrium, all three collective velocities above coincide because they must all vanish. However, they do not necessarily coincide to an observer moving with a uniform speed $-\mathbf{u}_O$. The mass weighted average velocity and the charge weighted average velocity are both $\mathbf{u}_O$. But the energy weighted average velocity

$$\mathbf{u}'_E = \frac{\sum_i (m(\mathbf{v}_i + \mathbf{u}_O)^2/2)(\mathbf{v}_i + \mathbf{u}_O)}{\sum_i m(\mathbf{v}_i + \mathbf{u}_O)^2/2} \neq \mathbf{u}_O \tag{6}$$

clearly does not coincide with $\mathbf{u}_O$. There is no unambiguous choice of the flow velocity even in this simple case. One must choose among these options what will be regarded as the flow velocity.

In the relativistic setting, mass is a part of the energy. Hence, there are only two options for choosing the flow velocity: The energy current or the net baryon current. One must choose one of these velocity options in order to decompose $T^{\mu\nu}$ and J_B^μ into a useful hydrodynamic form. The net baryon number is relatively small in ultra-relativistic heavy ion collisions. Furthermore, the flow observables we are interested in are mostly patterns in energy and momentum distributions. Therefore, there is no real benefit to choose the net baryon number collective velocity as long as the heavy ion analysis is concerned.

Choosing to follow the energy current,[a] the flow velocity for the energy current is defined by the eigenvalue problem

$$T^\mu_\nu u^\nu = \varepsilon u^\mu \tag{7}$$

where $T^\mu_\nu = T^{\mu\alpha} g_{\alpha\nu}$ and the flow vector is normalized to $u^\mu u_\mu = g_{\mu\nu} u^\mu u^\nu = 1$. We use $g_{\mu\nu} = \mathrm{diag}(1, -1, -1, -1)$ throughout this review. Note that while $T^{\mu\nu}$ is a symmetric matrix, T^μ_ν is no longer a symmetric matrix. Therefore, there is no guarantee that the eigenvalues are real. But any physically consistent system should admit a positive eigenvalue ε and the associated time-like real eigenvector u^μ.

Decomposing $T^{\mu\nu}$ using ε and u^μ, one gets

$$T^{\mu\nu} = \varepsilon u^\mu u^\nu + \frac{1}{3}(T^\alpha_\alpha - \varepsilon)\Delta^{\mu\nu} + \pi^{\mu\nu} \tag{8}$$

where the local 3-metric $\Delta^{\mu\nu}$ is given by

$$\Delta^{\mu\nu} = g^{\mu\nu} - u^\mu u^\nu. \tag{9}$$

The residual shear tensor $\pi^{\mu\nu}$ is symmetric $\pi^{\mu\nu} = \pi^{\nu\mu}$, transverse $\pi^{\mu\nu} u_\nu = 0$ and traceless $\pi^{\mu\nu} g_{\mu\nu} = 0$. Hence altogether the expression (8) has the required 10 independent degrees of freedom; the local energy density ε, the local fluid velocity $\mathbf{u}$, the trace T^α_α and the residual shear tensor $\pi^{\mu\nu}$. The net baryon current is

$$J_B^\mu = J_{B,\mathrm{id}}^\mu + V_B^\mu \tag{10}$$

where $J_{B,\mathrm{id}}^\mu = \rho_B u^\mu$ is the ideal fluid part of the current and V_B^μ is a space-like vector satisfying the transversality condition $u_\mu V_B^\mu = 0$. It has the required 4 independent degrees of freedom; the local net baryon density ρ_B and the residual vector $\mathbf{V}_B$.

So far we did not use any thermodynamic information. We will do so now to re-write the trace T^α_α in a more physical form. A static medium at rest has

[a]This choice of frame is often referred to as the Landau–Lifshitz frame. If one choose to follow the charge current, it is referred to as the Eckart frame.

$T_{\rm eq}^{\mu\nu} = {\rm diag}(\varepsilon, P, P, P)$ where P is the pressure. Therefore, the trace should contain the equilibrium piece $g_{\mu\nu}T_{\rm eq}^{\mu\nu} = \varepsilon - 3P$. Furthermore, the thermodynamic identities

$$dP = sdT + \rho_B d\mu_B, \tag{11}$$

$$ds = \frac{1}{T}d\varepsilon - \frac{\mu_B}{T}d\rho_B, \tag{12}$$

where s is the entropy density, indicate that the pressure P is a function of the temperature T and the baryon chemical potential μ_B, and they are in turn function of the energy density and the net baryon density. Hence we must be able to find P as a function of ε and ρ_B:

$$P = P(\varepsilon, \rho_B). \tag{13}$$

This relationship is known as the equation of state. Writing $T_\alpha^\alpha = \varepsilon - 3(P + \Pi)$, the stress–energy tensor then becomes

$$T^{\mu\nu} = T_{\rm id}^{\mu\nu} - \Pi\Delta^{\mu\nu} + \pi^{\mu\nu}. \tag{14}$$

where $T_{\rm id}^{\mu\nu} = \varepsilon u^\mu u^\nu - P(\varepsilon, \rho_B)\Delta^{\mu\nu}$ is the ideal fluid stress–energy tensor.

From the arguments presented above, it should be clear that the residual scalar term Π and the tensor term $\pi^{\mu\nu}$ as well as the vector V_B^μ in the baryon current must vanish in the static equilibrium limit. As these quantities represent deviation from equilibrium, the size of these terms will depend on how fast the local equilibrium is achieved. If local equilibration is instantaneous in the macroscopic time scale, then fluid cells will always be in strict local equilibrium and hence $\Pi = \pi^{\mu\nu} = V_B^\mu = 0$. Microscopically, this would happen if scattering cross-section is large so that the mean free path is much shorter than any macroscopic time scale or length scale (c.f. Sec. 4.1). If this is the case, the number of unknowns $(\varepsilon, \mathbf{u}, \rho_B)$ matches the 5 conservation laws and one has the ideal hydrodynamics whose dynamics is completely specified by

$$\partial_\mu(\varepsilon u^\mu u^\nu - P(\varepsilon, \rho_B)\Delta^{\mu\nu}) = 0 \tag{15}$$

and

$$\partial_\mu(\rho_B u^\mu) = 0. \tag{16}$$

The information on the underlying system is only in the equation of state $P(\varepsilon, \rho_B)$.

3. Hydrodynamics from Linear Response Theory

3.1. *Linear response theory*

In realistic systems, the approach to the local equilibrium is never going to be infinitely fast. Therefore, Π, $\pi^{\mu\nu}$ and V_B^μ cannot simply be set to vanish, although if the system is conformal one strictly has $\Pi = 0$. This is because the trace must vanish $T_\alpha^\alpha = 0$ in a conformal system. When any of these quantities are non-zero, the system is out of equilibrium. Therefore, local entropy must increase. The evolution equations of these quantities are then necessarily of the dissipative type. In this and following sections, we use linear response theory to obtain such equations for dissipative hydrodynamics. In Sec. 4, kinetic theory approaches that can go beyond the near-equilibrium restriction of the linear response theory is discussed.

To gain more insights on the behavior of the dissipative quantities, we start with a system very slightly out of equilibrium at $t = 0$. We then consider how the system approaches the equilibrium. This is the realm of the linear response theory. Full analysis of the quantum linear response theory can be found in any number of standard text books (for example, see Ref. 19). Here we will only go over the main ideas.

Suppose that at the remote past, $t = -T$, the density operator had the equilibrium form $\hat{\rho}_0 = e^{-\beta \hat{H}_0}/Z_0$ where $\hat{H}_0$ is the system Hamiltonian and $Z_0 = \mathrm{Tr} e^{-\beta \hat{H}_0}$ is the partition function. Then a force term is adiabatically turned on $f(t, x) = \theta(-t) f(x) e^{\epsilon t}$ at an infinitesimally slow rate ϵ. At $t = 0$, the force term is turned off and at this point the system is out of equilibrium. The full time-dependent Hamiltonian for this process is

$$\hat{H}(t) = \hat{H}_0 - \int d^3x\, \hat{A}(x) f(t, \mathbf{x}). \tag{17}$$

Here $\hat{A}$ represents a set of Hermitian operators for the conserved quantities we are interested in. Namely, $T^{0\mu}$ and ρ_B. Hence, the expressions presented below are in general matrix expressions.

Treating $\delta \hat{H} = -\int d^3x f(t, \mathbf{x}) \hat{A}(\mathbf{x})$ as a perturbation, the formal first-order solution of the quantum Liouville equation $i\partial_t \hat{\rho}(t) = [\hat{H}(t), \hat{\rho}(t)]$ is given by

$$\delta \hat{\rho}_H(t) = -i \int_{-T}^{t} dt' [\delta \hat{H}_H(t'), \hat{\rho}_0(0)] \tag{18}$$

where $\delta \hat{\rho}(t) = \hat{\rho}(t) - \hat{\rho}_0$ and the subscript H denotes Heisenberg picture operators

$$\hat{O}_H(t) = e^{i\hat{H}_0 t} \hat{O} e^{-i\hat{H}_0 t}. \tag{19}$$

Using $\delta \hat{H}$ from Eq. (17), the deviation of an observable A from the equilibrium value is found to be

$$\delta \langle \hat{A}(t, \mathbf{x}) \rangle = \int d^4x'\, G_R^{AA}(t - t', \mathbf{x} - \mathbf{x}') \theta(-t') e^{\epsilon t'} f(\mathbf{x}') \tag{20}$$

where $t > 0$ and

$$G_R^{AA}(t - t', \mathbf{x} - \mathbf{x}') = i\theta(t - t') \mathrm{Tr}(\hat{\rho}_0 [\hat{A}_H(t, \mathbf{x}), \hat{A}_H(t', \mathbf{x}')]) \tag{21}$$

is the retarded response function. We also took the $-T \to -\infty$ limit.

Suppose that one can find an operator $\hat{D}_A$ for which G_R^{AA} is the generalized retarded Green function,

$$\hat{D}_A G_R^{AA}(t - t', \mathbf{x} - \mathbf{x}') = \hat{d}_A \delta(t - t') \delta(\mathbf{x} - \mathbf{x}'), \tag{22}$$

where $\hat{d}_A$ can contain a finite number of derivatives. For $t > 0$, t and t' can never be the same in Eq. (20). Hence $\delta \langle \hat{A} \rangle$ satisfies the evolution equation

$$\hat{D}_A \delta \langle \hat{A}(t, \mathbf{x}) \rangle = 0 \tag{23}$$

for $t > 0$. Therefore finding the pole structure of the response function is equivalent to finding the evolution equation.[20,21] We will use this to find hydrodynamic equations in the following sections.

For further analysis, it is useful to define the spectral density

$$\rho^{AA}(\omega, \mathbf{k}) = \int d^4x \, e^{i\omega t - i\mathbf{k}\cdot\mathbf{x}} \langle [\hat{A}_H(t, \mathbf{x}), \hat{A}_H^\dagger(0)] \rangle. \tag{24}$$

Using the thermal average $\langle(\cdots)\rangle = \mathrm{Tr}\hat{\rho}_0(\cdots)$, it is not hard to show (for instance, see Ref. 19)

$$\rho^{AA}(\omega, \mathbf{k}) = \frac{1}{Z_0} \sum_{m,n} (e^{-\beta E_n} - e^{-\beta E_m})(2\pi)^4 \delta(k - p_m + p_n) |\langle n|\hat{A}|m\rangle|^2 \tag{25}$$

where $k = (\omega, \mathbf{k})$ and $|m\rangle$ is the simultaneous eigenstate of the system Hamiltonian $\hat{H}_0$ and the total momentum $\hat{\mathbf{P}}$ of the system with the eigenvalue $p_m = (E_m, \mathbf{p}_m)$. From this expression, we can derive

$$\rho^{AA}(-\omega, -\mathbf{k}) = -\rho^{AA}(\omega, \mathbf{k}) \tag{26}$$

by exchanging m and n. When the underlying equilibrium system is isotropic, then $\rho^{AA}(\omega, \mathbf{k})$ must be a function of $|\mathbf{k}|$ only. Hence the spectral density is an odd function of ω, $\rho^{AA}(-\omega, \mathbf{k}) = -\rho^{AA}(\omega, \mathbf{k})$. We will use this property often in later sections to parametrize the analytic structure of the response functions.

In terms of ρ^{AA}, the retarded correlator is

$$G_R^{AA}(\omega, \mathbf{k}) = \int \frac{d\omega'}{2\pi} \frac{\rho^{AA}(\omega', \mathbf{k})}{\omega' - \omega - i\epsilon}. \tag{27}$$

The imaginary part of the retarded correlator directly gives the spectral density

$$\mathrm{Im}\, G_R^{AA}(\omega, \mathbf{k}) = \frac{1}{2}\rho^{AA}(\omega, \mathbf{k}). \tag{28}$$

It is also useful to know that the Euclidean correlator is given by (for derivations, see, for instance, Ref. 19)

$$G_E^{AA}(\omega_n, \mathbf{k}) = \int \frac{d\omega'}{2\pi} \frac{\rho^{AA}(\omega', \mathbf{q})}{\omega' - i\omega_n} \tag{29}$$

where $\omega_n = 2\pi n T$ is the Matsubara frequency. One important fact we will often use in the following sections is that the $\omega \to 0$ limit

$$\begin{aligned} G_R^{AA}(0, \mathbf{k}) = G_E^{AA}(0, \mathbf{k}) &= \int \frac{d\omega'}{2\pi} \frac{\rho^{AA}(\omega, \mathbf{k})}{\omega'} \\ &= \frac{1}{Z_0} \sum_{m,n} (e^{-\beta E_n} - e^{-\beta E_m}) \\ &\quad \times \frac{1}{E_m - E_n} (2\pi)^3 \delta(\mathbf{k} - \mathbf{p}_m + \mathbf{p}_n) |\langle n|\hat{A}|m\rangle|^2 \end{aligned} \tag{30}$$

is real and positive and function only of the magnitude of $\mathbf{k}$.[b]

[b]To see this, just exchange m and n.

3.2. *Baryon density diffusion*

From the previous section, it is clear that the analytic structure of the response function determines the evolution of small disturbances. In this and the following sections, we show that this fact combined with the conservation laws is powerful enough to produce dissipative hydrodynamic equations.

The analysis in this section and those up to Sec. 3.5 closely follow the unpublished note by Laurence G. Yaffe (private communication, see also Refs. 22 and 23) in a simplified form. Additional discussion on the second-order formulation of dissipative hydrodynamics is given in Sec. 3.6.

We start with the net baryon number conservation which is the simplest to examine. Suppose we set up a system where only the net baryon density $\langle \rho_B(0, \mathbf{x}) \rangle$ is non-uniform at $t = 0$. The perturbing Hamiltonian is

$$\delta \hat{H}(t) = - \int d^3x \, \hat{\rho}_B(\mathbf{x}) \, e^{\epsilon t} \mu_B(\mathbf{x}) \tag{31}$$

which results in the following linear response in the mixed space of t and the wave vector $\mathbf{k}$,

$$\delta \langle \hat{\rho}_B(t, \mathbf{k}) \rangle = \mu_B(\mathbf{k}) \int_{-\infty}^{\infty} dt' \, \theta(-t') e^{\epsilon t'} G_R^{00}(t - t', \mathbf{k}) \tag{32}$$

for $t > 0$.

Applying the current conservation, $\partial_t \rho_B = -\nabla \cdot \mathbf{J}_B$, to the retarded correlation functions $G_R^{\mu\nu}(t, \mathbf{x}) = i\theta(t) \langle [\hat{J}_B^\mu(t, \mathbf{x}), \hat{J}_B^\nu(0)] \rangle$ results in the following relationships between them in the frequency–wave vector space

$$\omega G_R^{00}(\omega, \mathbf{k}) = k_i G_R^{0i}(\omega, \mathbf{k}), \tag{33}$$

$$\omega G_R^{0j}(\omega, \mathbf{k}) = k_i G_R^{ij}(\omega, \mathbf{k}) \tag{34}$$

provided that $[J_B^0(0, \mathbf{x}), J_B^\nu(0, \mathbf{x}')] = 0$. The underlying isotropic equilibrium permits decomposition into transverse and the longitudinal parts

$$G_R^{ij}(\omega, \mathbf{k}) = \hat{k}^i \hat{k}^j G_L(\omega, \mathbf{k}^2) + \hat{\delta}^{ij} G_T(\omega, \mathbf{k}) \tag{35}$$

where $\hat{\mathbf{k}} = \mathbf{k}/|\mathbf{k}|$ is the unit vector and $\hat{\delta}^{ij} = \delta^{ij} - \hat{k}^i \hat{k}^j$ is the transverse projector. Combining Eqs. (33) and (34) gives

$$\omega^2 G_R^{00}(\omega, \mathbf{k}) = \mathbf{k}^2 G_L(\omega, \mathbf{k}). \tag{36}$$

What we are interested in is the behavior of $G_L(\omega, \mathbf{k})$.

Consider the small ω limit of G_L first. From Eq. (30), we know that

$$g_{00}(\mathbf{k}) = G_R^{00}(0, \mathbf{k}) = G_E^{00}(0, \mathbf{k}) \tag{37}$$

is real and positive. The small ω limit of G_L can be then expressed as

$$G_L(\omega, \mathbf{k}) \approx \frac{\omega^2 g_{00}(\mathbf{k})}{\mathbf{k}^2}. \tag{38}$$

Now consider taking the $\mathbf{k} \to 0$ limit with a fixed $\omega \neq 0$. The retarded density–density correlation function in this limit is

$$\begin{aligned} G_R^{00}(\omega, 0) &= i \int_0^\infty dt\, e^{-i\omega t} \int d^3x \langle [\hat{\rho}_B(t, \mathbf{x}), \hat{\rho}_B(0)] \rangle \\ &= i \int_0^\infty dt\, e^{-i\omega t} \langle [\hat{Q}_B, \hat{\rho}_B(0)] \rangle = 0 \end{aligned} \tag{39}$$

where we used the fact that $\hat{Q}_B = \int d^3x\, \hat{\rho}_B(t, \mathbf{x})$ is the net baryon number operator. Since the net baryon number is conserved, $\hat{Q}_B$ is independent of time. In particular, it can be evaluated at $t = 0$. Therefore, the commutator vanishes. This then indicates that $G_L(\omega, \mathbf{k})$ is well behaved in the zero $|\mathbf{k}|$ limit with $\omega \neq 0$. Consequently, the $\omega \to 0$ limit and the $\mathbf{k} \to 0$ limit do not commute, indicating the presence of a massless pole. We also know that the imaginary part of G_R^{00} (the spectral density ρ^{00}) has to be an odd function of ω since isotropy in space demands that it be a function only of $\mathbf{k}^2$ (c.f. Eq. (26)).

The most general form of G_L consistent with the above conditions is

$$G_L(\omega, \mathbf{k}) = \frac{\omega^2 (g_{00}(\mathbf{k}) + i\omega A(\omega, \mathbf{k}))}{\mathbf{k}^2 - i\omega / D(\omega, \mathbf{k}) - \omega^2 B(\omega, \mathbf{k})}. \tag{40}$$

The functions $D(\omega, \mathbf{k})$, $A(\omega, \mathbf{k})$ and $B(\omega, \mathbf{k})$ are all of the form

$$D(\omega, \mathbf{k}) = D_R(\omega, \mathbf{k}) - i\omega D_I(\omega, \mathbf{k}) \tag{41}$$

where $D_R(\omega, \mathbf{k})$ and $D_I(\omega, \mathbf{k})$ are real-valued even functions of ω and $\mathbf{k}$. The real parts D_R and B_R must have a non-zero limit as $\omega \to 0$ and $\mathbf{k} \to 0$. All other parts of A, B and D must have finite limits as $\omega \to 0$ and $\mathbf{k} \to 0$. In the small ω and $\mathbf{k}$ limit, the response function becomes

$$G_R^{00}(\omega, \mathbf{k}) \approx \frac{D\mathbf{k}^2 g_{00}(0)}{-i\omega + D\mathbf{k}^2} \tag{42}$$

where we defined the diffusion constant $D = D_R(0, 0)$.

The pole structure of G_R^{00} dictates that in the small ω and $|\mathbf{k}|$ limit, $\delta\rho_B(t, \mathbf{x}) = \delta\langle \hat{\rho}_B(t, \mathbf{x}) \rangle$ obeys the diffusion equation

$$\partial_t \delta\rho_B = D\nabla^2 \delta\rho_B. \tag{43}$$

This is our first example of a dissipative hydrodynamic equation. The conservation law, current algebra, thermodynamic stability, and the general analytic structure of the correlation functions are all the ingredients one needs to get this diffusion equation for baryon density. Hence diffusion is a very general phenomenon whenever there is a conserved current. Microscopic dynamics only enters through the value of the diffusion constant.

If we now go back to the conservation equation

$$\partial_t \delta\rho_B = -\partial_i \delta J_B^i, \tag{44}$$

we can see that the diffusion equation above is equivalent to the constitutive relationship

$$\delta J^i = V_B^i = D\partial^i \delta\rho_B \tag{45}$$

valid in the fluid cell rest frame. In the more general frame boosted by u^μ, it becomes

$$V_B^\mu = D\Delta^{\mu\nu}\partial_\nu \rho_B, \tag{46}$$

where again $\Delta^{\mu\nu} = g^{\mu\nu} - u^\mu u^\nu$ is the local 3-metric.

The diffusion constant can be calculated by taking the appropriate limits of G_L,

$$\lim_{\omega\to 0}\lim_{\mathbf{k}\to 0}\frac{1}{\omega}\mathrm{Im}\, G_L(\omega,\mathbf{k}) = Dg_{00}(0), \tag{47}$$

which is our first example of the Kubo formula that relates an *equilibrium* correlation function to a dissipative coefficient. As the response function is singular in the small ω and small $\mathbf{k}$ limit, the order of limits in the Kubo formula is important. The $\mathbf{k} \to 0$ limit must be taken first. Since the transport coefficients are defined as Lorentz scalars, Eq. (47) is to be evaluated using the underlying microscopic theory such as thermal QCD in the rest frame of the equilibrium system.

3.3. *Stress–energy tensor correlation functions*

To carry out the linear response analysis of the energy–momentum currents and their hydrodynamic and dissipative behavior, one need to know the Ward identity among the correlation functions. Defining the retarded correlation functions of $\hat{T}^{\mu\nu}$ turned out to be not so straightforward due to the fact that the conserved quantities $\hat{P}^\mu = \int d^3x\, \hat{T}^{0\mu}$ are also the generators of the space-time evolution. Hence, in general the equal time commutators of $\hat{T}^{0\mu}$ are non-zero unlike the net baryon current case.

To begin the analysis, consider the static partition function given by

$$Z_E[g_E] = \int \mathcal{D}\phi e^{-S_E[\phi,g_E]} \tag{48}$$

where

$$S_E[\phi, g_E] = \int d^3x \int_0^\beta d\tau \sqrt{g_E}\,\mathcal{L}(\phi, g_E) \tag{49}$$

is the Euclidean action and τ is the imaginary time. Here, ϕ denotes the collection of field variables and g_E is the Euclidean metric.

Using the Hilbert definition of the stress–energy tensor density, we have

$$\langle T^{\mu\nu}(x)\rangle_E = -2\frac{\delta}{\delta g^E_{\mu\nu}(x)} \ln Z_E[g_E]. \tag{50}$$

The two point functions are given by

$$\begin{aligned}\bar{G}_R^{\alpha\beta,\mu\nu}(x_E, y_E) &= \langle \mathcal{T}_\tau T^{\mu\nu}(x_E) T^{\alpha\beta}(y_E)\rangle_E \\ &= 4\frac{\delta^2}{\delta g^E_{\alpha\beta}(y_E)\delta g^E_{\mu\nu}(x_E)} \ln Z_E[g_E]\end{aligned} \tag{51}$$

where $x_E = (\tau, \mathbf{x})$ and $\mathcal{T}_\tau$ is the time ordering operator in τ. For the tensor density $T^{\mu\nu}$, the covariant conservation law is

$$\partial_\mu\langle T^{\mu\nu}\rangle_E + \Gamma^\nu_{\sigma\rho}\langle T^{\sigma\rho}\rangle_E = 0 \tag{52}$$

where $\Gamma^\nu_{\sigma\rho} = \frac{1}{2}g_E^{\mu\nu}(g^E_{\sigma\mu,\rho} + g^E_{\rho\mu,\sigma} - g^E_{\sigma\rho,\mu})$ is the Christoffel symbol.

By differentiating Eq. (52) once more with respect to $g^E_{\alpha\beta}(y)$, we obtain the Ward identity among the Euclidean correlation functions in the flat space,[22,23] where $g_E^{\mu\nu} = \delta^{\mu\nu}$,

$$0 = k^E_\alpha(\bar{G}_E^{\alpha\beta,\mu\nu}(k_E) + \delta^{\beta\mu}\langle T^{\alpha\nu}\rangle + \delta^{\beta\nu}\langle T^{\alpha\mu}\rangle - \delta^{\alpha\beta}\langle T^{\mu\nu}\rangle). \tag{53}$$

Here $k_E^\alpha = (\omega_n, \mathbf{k})$ and $\omega_n = 2\pi nT$ is the Matsubara frequency.

To obtain the real time correlation functions, we perform the analytic continuation. From Eqs. (27) and (29) one can see that the analytically continuation

$$\omega_n \to -ik^0 + \epsilon \tag{54}$$

changes the Euclidean correlation function to the real-time retarded correlation function. This also means that each time index of $T^{\mu\nu}$ gets a factor of $(-i)$. Since our Minkowski metric is mostly negative, going from the Euclidean metric to the mostly negative Minkowski metric means $\delta_{\mu\nu} \to -g_{\mu\nu}$. The real-time version of Eq. (53) is then

$$0 = k_\alpha(\bar{G}_R^{\alpha\beta,\mu\nu}(k) - g^{\beta\mu}\langle T^{\alpha\nu}\rangle - g^{\beta\nu}\langle T^{\alpha\mu}\rangle + g^{\alpha\beta}\langle T^{\mu\nu}\rangle) \tag{55}$$

where $k^\mu = (\omega, \mathbf{k})$.

The presence of the single stress–energy tensor average terms in Eq. (55) implies that the correlation function $\bar{G}_R^{\alpha\beta,\mu\nu}(x, x')$ is not the same as

$$G_R^{\mu\nu,\alpha\beta}(x, x') = i\theta(t)\langle[\hat{T}^{\mu\nu}(t, \mathbf{x}), \hat{T}^{\alpha\beta}(t', \mathbf{x}')]\rangle_{\text{eq}}, \tag{56}$$

but differs by terms containing $\delta(x - x')$ (as well as the contact terms containing spatial derivatives of $\delta(x - x')$[24]). As the response function in the linear response theory, these delta-function terms do not matter since t and $t' < t$ can never be the same.

Defining the real-time correlation function by the analytic continuation of the Euclidean correlation function enables us to gain the following important relationship between the two

$$\bar{G}_E^{\mu\nu,\alpha\beta}(0,\mathbf{k}) = \bar{G}_R^{\mu\nu,\alpha\beta}(0,\mathbf{k}) \tag{57}$$

which has a well-defined limit (μ and ν here are not summed)

$$\lim_{\mathbf{k}\to 0} \bar{G}_R^{\mu\nu,\mu\nu}(0,\mathbf{k}) = \lim_{\mathbf{k}\to 0} \bar{G}_E^{\mu\nu,\mu\nu}(0,\mathbf{k}) > 0. \tag{58}$$

3.4. *Momentum diffusion and shear viscosity*

In this and the following sections for the bulk viscosity, we will not consider finite net baryon density for simplicity. For analysis with finite μ_B, see Ref. 23.

Suppose that we set up a system where the flow velocity at $t = 0$ has a single non-zero component in the x-direction $u_x(y)$ which depend only on y. In this situation, two layers of the fluid at y and at $y + \Delta y$ have different fluid velocities in the orthogonal x-direction. In ideal hydrodynamics, this difference is maintained because there is no dissipation. In a normal fluid, however, particle diffusion between the two layers will eventually make them move with the same equilibrated speed. How fast two layers equilibrate depends on the size of the scattering mean free path, which in turn determines the diffusion constant, or the shear viscosity.

To set up a shear flow, let the perturbing Hamiltonian be

$$\delta\hat{H}(t) = -\int d^3x\, e^{\epsilon t}\hat{T}^{x0}(t,\mathbf{x})\beta_x(y). \tag{59}$$

The corresponding linear response is

$$\delta\langle T^{x0}(t,k_y)\rangle = \beta_x(k_y)\int_{-\infty}^{\infty} dt'\theta(-t')e^{\epsilon t'}\bar{G}_R^{x0,x0}(t-t',k_y) \tag{60}$$

for $t > 0$.

In Eq. (55), $\bar{G}_R^{x0,x0}$ appears in the following sequences when $\mathbf{k} = (0, k_y, 0)$

$$\omega(\bar{G}_R^{x0,x0}(\omega,k_y) + \varepsilon) = k_y\bar{G}_R^{x0,xy}(\omega,k_y), \tag{61}$$

$$\omega\bar{G}_R^{x0,xy}(\omega,k_y) = k_y(\bar{G}_R^{xy,xy}(\omega,k_y) + P). \tag{62}$$

Combined, these become

$$\bar{G}_R^{xy,xy}(\omega,k_y) + P = \frac{\omega^2}{k_y^2}(\bar{G}_R^{x0,x0}(\omega,k_y) + \varepsilon). \tag{63}$$

Except the extra P and ε, the structure of these equations is exactly the same as the baryon current case. The following analysis is therefore a repeat of that case.

In the $\omega \to 0$ limit, $\bar{G}_R^{x0,x0}(\omega,k_y)$ must have a well defined limit since it is a thermodynamic quantity. Furthermore, the imaginary part of $\bar{G}_R^{xy,xy}(\omega,k_y)$ must be an odd function of ω. As in the baryon current case, the $k_y \to 0$ limit of the correlation functions must be well-behaved. Thus, we can parametrize $\bar{G}_R^{xy,xy}$ as

$$\bar{G}_R^{xy,xy}(\omega, k_y) = \frac{\omega^2(\varepsilon + g_T(k_y) + i\omega A_T(\omega, k_y))}{k_y^2 - i\omega/D_T(\omega, k_y) - \omega^2 B_T(\omega, k_y)} - P \tag{64}$$

and

$$\bar{G}_R^{x0,x0}(\omega, k_y) = \frac{k_y^2(\varepsilon + g_T(k_y) + i\omega A_T(\omega, k_y))}{k_y^2 - i\omega/D_T(\omega, k_y) - \omega^2 B_T(\omega, k_y)} - \varepsilon \tag{65}$$

where $g_T(k_y) = \bar{G}_R^{x0,x0}(0, k_y)$. Here the functions A_T, B_T and D_T all have the form

$$D_T(\omega, k_y) = D_T^R(\omega, k_y) - i\omega D_T^I(\omega, k_y) \tag{66}$$

where $D_T^R(\omega, k_y)$ and $D_T^I(\omega, k_y)$ are real-valued even functions of ω and k_y. The real parts D_T^R and B_T^R must have a non-zero limit as $\omega \to 0$ and $k_y \to 0$. All other parts of A_T, B_T and D_T must have finite limits as $\omega \to 0$ and $k_y \to 0$.

In the configuration space, the constant $-\varepsilon$ term becomes $-\varepsilon\delta^4(x - x')$. In Eq. (60), this δ-function term does not contribute. Hence, in the small ω and k_y limit, the evolution of T^{x0} is determined by $i\omega = D_T k_y^2$ or

$$(\partial_t - D_T\partial_y^2)T^{x0}(t, y) = 0 \tag{67}$$

where we defined the momentum diffusion constant $D_T = D_T^R(0, 0)$. This is our second dissipative hydrodynamic equation. The diffusion equation combined with the conservation law implies the constitutive relationship

$$T^{xy}(t, y) = D_T\partial^y T^{x0}(t, y) = \eta\partial^y u^x \tag{68}$$

valid in the local rest frame. Here $\eta = D_T(\varepsilon + P)$ is the shear viscosity. It is clear from Eq. (67) that D_T has the physical interpretation of the diffusion constant for momentum diffusion.

Recognizing Eq. (68) to be a part of the spin-2 component of the second rank tensor, we can generalize this result to

$$\pi_{\rm NS}^{ij}(t, \mathbf{x}) = \eta\left(\partial^i u^j + \partial^j u^i - \frac{2g^{ij}}{3}\partial_l u^l\right) \tag{69}$$

again in the rest frame of the fluid cell. In the moving frame, this becomes

$$\pi_{\rm NS}^{\mu\nu}(t, \mathbf{x}) = 2\eta\Delta^{\mu\nu}_{\alpha\beta}\partial^\alpha u^\beta \equiv 2\eta\sigma^{\mu\nu} \tag{70}$$

where $\sigma^{\mu\nu}$ is the velocity shear tensor and $\Delta^{\mu\nu}_{\alpha\beta}$ is the spin-2 projector defined by

$$\Delta^{\mu\nu}_{\alpha\beta} = \frac{1}{2}\left(\Delta^\mu_\alpha\Delta^\nu_\beta + \Delta^\nu_\alpha\Delta^\mu_\beta - \frac{2}{3}\Delta^{\mu\nu}\Delta_{\alpha\beta}\right). \tag{71}$$

Here the label NS indicates that this is the Navier–Stokes form of the shear tensor. The Kubo formula for the shear viscosity is

$$\lim_{\omega\to 0}\lim_{k_y\to 0}\frac{1}{\omega}{\rm Im}\,\bar{G}_R^{xy,xy}(\omega, k_y) = D_T(\varepsilon + P) = \eta \tag{72}$$

where we used the fact that $g_T(0) = P$ which can be determined from Eq. (55).

The Kubo formula for the shear viscosity η can be also expressed in terms of the full shear-tensor correlation function[20,21]

$$\eta = \lim_{\omega\to 0}\lim_{\mathbf{k}\to 0}\frac{1}{10\omega}\mathrm{Im}\bar{G}_R^{\pi_{ij},\pi_{ij}}(\omega,\mathbf{k}) \tag{73}$$

where the shear-tensor is given by

$$\pi_{ij} = T_{ij} - (\delta_{ij}/3)T_k^k. \tag{74}$$

3.5. *Sound propagation and bulk viscosity*

So far, only the diffusion type of hydrodynamic flow is discussed which are not the main bulk excitation. To get the main excitation which must also include the ideal hydrodynamics part, one needs to look at the disturbance in the energy density. This bulk excitation, of course, is the sound wave.

Suppose we perturb the energy density with

$$\delta\hat{H}(t) = -\int d^3x e^{\epsilon t}\,\hat{T}^{00}(t,\mathbf{x})\beta_0(\mathbf{x}). \tag{75}$$

The linear response is then

$$\delta\langle T^{00}(t,\mathbf{k})\rangle = \beta_0(\mathbf{k})\int_{-\infty}^{\infty} dt'\theta(-t')e^{\epsilon t'}\bar{G}_R^{00,00}(t-t',\mathbf{k}). \tag{76}$$

Applying the conservation law to each index of $\bar{G}_R^{\mu\nu,\alpha\beta}$ in Eq. (55), we get

$$\omega^4\bar{G}_R^{00,00}(\omega,\mathbf{k}) = \omega^4\varepsilon - \omega^2\mathbf{k}^2(\varepsilon+P) + \mathbf{k}^4\bar{G}_L(\omega,\mathbf{k}) \tag{77}$$

where

$$\mathbf{k}^4\bar{G}_L(\omega,\mathbf{k}) = k_ik_jk_lk_m(\bar{G}_R^{ij,lm}(\omega,\mathbf{k}) + P(\delta^{il}\delta^{jm}+\delta^{im}\delta^{jl}-\delta^{ij}\delta^{lm})). \tag{78}$$

In the small ω limit, Eq. (77) gives

$$\bar{G}_L(\omega,\mathbf{k}) \approx \frac{\omega^2}{\mathbf{k}^2}(\varepsilon+P) + \frac{\omega^4}{\mathbf{k}^4}(\bar{G}_R^{00,00}(0,\mathbf{k}) - \varepsilon). \tag{79}$$

We also know that the imaginary part of $\bar{G}_L$ must be an odd function of ω and the correlation functions are well-behaved in the $\mathbf{k}\to 0$ limit. The most general form consistent with these conditions is

$$\bar{G}_L(\omega,\mathbf{k}) = \frac{\omega^2(\varepsilon+P+i\omega^3Q(\omega,\mathbf{k}))}{\mathbf{k}^2-\omega^2/Z(\omega,\mathbf{k})+i\omega^3R(\omega,\mathbf{k})}. \tag{80}$$

Here $Z(\omega,\mathbf{k})$, $Q(\omega,\mathbf{k})$ and $R(\omega,\mathbf{k})$ all have the form

$$Z(\omega,\mathbf{k}) = Z_R(\omega,\mathbf{k}) - i\omega Z_I(\omega,\mathbf{k}) \tag{81}$$

where $Z_R(\omega,\mathbf{k})$ and $Z_I(\omega,\mathbf{k})$ are real-valued even functions of ω and $\mathbf{k}$. The real parts Z_R and R_R must have non-zero limits as $\omega\to 0$ and $\mathbf{k}\to 0$. All other parts of Z, Q and R must have finite limits as $\omega \to 0$ and $\mathbf{k} \to 0$. Matching the small ω limit (79) demands that

$$Z_R(0, \mathbf{k}) = \frac{\varepsilon + P}{\bar{G}_R^{00,00}(0, \mathbf{k}) - \varepsilon}. \tag{82}$$

Up to the quadratic terms in ω and $\mathbf{k}$, the poles of $\bar{G}_L(\omega, \mathbf{k})$ for small ω and $|\mathbf{k}|$ are determined by

$$\omega^2 - Z_R(0,0)\mathbf{k}^2 + i\omega Z_I(0,0)\mathbf{k}^2 = 0. \tag{83}$$

This has the structure of the dispersion relationship of a damped sound wave. Hence, in the small ω and the small $|\mathbf{k}|$ limit, $Z_R(0,0) = v_s^2$ is the speed of sound squared and $Z_I(0,0)$ is the sound damping coefficient.

To relate $Z_I(0,0)$ to the shear and the bulk viscosities, let us consider the constitutive relationships once again. From the shear part, we already have the spin-2 part of the stress tensor in the fluid cell rest frame

$$\pi_{\mathrm{NS}}^{ij}(t, \mathbf{x}) = D_T \left(\partial^i T^{j0} + \partial^j T^{i0} - \frac{2g^{ij}}{3} \partial_l T^{l0} \right). \tag{84}$$

To this we add a spin-0 part $-\gamma g^{ij} \partial_t \varepsilon = \gamma g^{ij} \partial_l T^{l0}$ to get

$$\delta T^{ij}(t, \mathbf{x}) = D_T \left(\partial^i T^{j0} + \partial^j T^{i0} - \frac{2g^{ij}}{3} \partial_l T^{l0} \right) + \gamma g^{ij} \partial_l T^{l0}. \tag{85}$$

The energy conservation law in the local rest frame[c] now becomes using Eq. (14) with the dissipative part given by Eq. (85)

$$\begin{aligned} 0 &= \partial_\mu \partial_\nu T^{\mu\nu} \\ &= \partial_t^2 \varepsilon - \nabla^2 P - D_T \frac{4}{3} \nabla^2 \partial_t \varepsilon - \gamma \nabla^2 \partial_t \varepsilon \\ &\to (-\omega^2 + v_s^2 \mathbf{k}^2 - i(4D_T/3 + \gamma)\mathbf{k}^2 \omega)\delta\varepsilon \end{aligned} \tag{86}$$

where we used $g^{ij} \partial_i \partial_j = -\nabla^2$ and $\partial_t \varepsilon = -\partial_l T^{l0}$. Comparing with Eq. (83), one can identify $v_s^2 = \partial P / \partial \varepsilon = Z_R(0,0)$, and

$$Z_I(0,0) = \Gamma = 4D_T/3 + \gamma \tag{87}$$

as the sound attenuation constant. Since we have already identified $D_T = \eta/(\varepsilon + P)$, this allows us to identify $\gamma = \zeta/(\varepsilon + P)$ where ζ is the bulk viscosity. In the fluid cell rest frame, the added term corresponds to the constitutive relationship $\Pi_{\mathrm{NS}} = -\zeta \partial_i u^i$. In the general frame, this becomes

$$\Pi_{\mathrm{NS}} = -\zeta \partial_\mu u^\mu. \tag{88}$$

Again, the label NS indicates that this is the Navier–Stokes form of the bulk pressure. The minus sign in Eq. (88) makes sense since the effective pressure $P + \Pi$

[c]In the local rest frame, $\mathbf{u}(t, \mathbf{x}) = 0$, but $\partial_i u_j \neq 0$.

should be less than the equilibrium pressure when the fluid is expanding (positive $\partial_i u^i$).

We have so far identified $Z_R(0,0)$ and $Z_I(0,0)$ as the speed of sound squared and the sound attenuation coefficient. The role of $R(\omega, \mathbf{k})$ is still to be identified. In the Kubo formula for the attenuation coefficient, $R_R(0,0)$ appears as

$$\lim_{\omega\to 0}\lim_{\mathbf{k}\to 0} \mathrm{Im}\, \frac{\bar{G}_L(\omega, \mathbf{k})}{\omega} = (\varepsilon + P)(Z_I(0,0) - Z_R(0,0)^2\, R_R(0,0)) \tag{89}$$

which does not allow one to identify the right-hand side with Γ if $R_R(0,0) \neq 0$. Actually, the left-hand side of Eq. (89) does yield Γ. It is just that we have not been consistent in power counting since the wave equation Eq. (86) does not contain $O(\omega^3)$ term while the pole of $\bar{G}_L$ does. One may consider this discrepancy as the first sign of the trouble with the first-order constitutive relationship Eq. (85).

3.6. *Second-order viscous hydrodynamics*

Let us consider the consequence of having the first-order constitutive relationship more closely. The diffusion equation with a source S

$$(\partial_t - D_T \nabla^2)\delta n(t, \mathbf{x}) = S(t, \mathbf{x}) \tag{90}$$

has the solution

$$\delta n(x) = \int d^4x' G_R(x - x') S(x'). \tag{91}$$

Here the retarded Green function is

$$G_R(x - x') = \theta(t - t') \frac{e^{-\frac{|\mathbf{x}-\mathbf{x}'|^2}{4D(t-t')}}}{8(\pi D(t - t'))^{3/2}}. \tag{92}$$

If one has a point source, $S(x') = N_0 \delta(x')$, then $\delta n(t, \mathbf{x}) = N_0 G_R(t, \mathbf{x})$. At $t = 0$, the space is empty except at the origin. But at any time after that, there is non-zero δn everywhere. This is clearly acausal.

On the other hand, the solution of the sound equation

$$-(\partial_t^2 - v_s^2 \nabla^2)\delta\epsilon(x) = S(x) \tag{93}$$

for a point source $S(x) = \Lambda_0 \delta(x)$ is

$$\delta\epsilon(t, \mathbf{x}) = \Lambda_0 \theta(t) \frac{1}{4\pi} \frac{\delta(|\mathbf{x}| - v_s t)}{|\mathbf{x}|}. \tag{94}$$

This is causal since the disturbance only moves with the speed of sound.

The origin of acausality in diffusion is the mismatch between the number of time derivatives and the number of spatial derivatives in the diffusion equation.

The diffusive dispersion relationship $\omega = -iD\mathbf{k}^2$ gives the group velocity

$$\frac{\partial\omega}{\partial\mathbf{k}} = -2iD\mathbf{k} \tag{95}$$

which becomes large in the large $\mathbf{k}$ limit. This problem can be remedied if one replaces the constitutive equation, $J^i = D\partial^i n$ with a relaxation type equation

$$\partial_t J^i = -\frac{1}{\tau_R}(J^i - D\partial^i n) \tag{96}$$

then the conservation law becomes

$$\begin{aligned}\partial_t^2 n &= -\partial_i\partial_t J^i \\ &= -\frac{1}{\tau_R}\partial_t n + \frac{D}{\tau_R}\nabla^2 n.\end{aligned} \tag{97}$$

For large k where we previously had a problem, we now have

$$\omega^2 \approx v_R^2 k^2 \tag{98}$$

with the propagation speed $v_R = \sqrt{D/\tau_R}$.

This type of relaxation equation was actually anticipated already: Up to the second-order in ω, the poles of the density–density correlator in Eq. (40) are determined by

$$D\mathbf{k}^2 - i\omega - \omega^2 DB = 0. \tag{99}$$

Comparing, we see that

$$B = \tau_R/D. \tag{100}$$

For the viscous stress–energy tensor components, the following relaxation equations apply in the local rest frame

$$\left(\partial_t + \frac{1}{\tau_\pi}\right)\pi^{ij} = \frac{1}{\tau_\pi}\pi^{ij}_{\rm NS}, \tag{101}$$

$$\left(\partial_t + \frac{1}{\tau_\Pi}\right)\Pi = \frac{1}{\tau_\Pi}\Pi_{\rm NS} \tag{102}$$

where

$$\pi^{lm} = T^{lm} - \frac{g^{lm}}{3}T^k_k \tag{103}$$

is the traceless part of the stress tensor and

$$\Pi = -\left(\frac{1}{3}T^k_k + P\right) \tag{104}$$

is the bulk pressure. They are not to be identified with the Navier–Stokes forms (84) and (88). Rather, they will relax to the Navier–Stokes forms.

To see if the acausality in the momentum diffusion is cured, we start with the momentum conservation

$$\partial_t T^{k0} = -\partial_l T^{kl} \tag{105}$$

Applying the curl gives

$$\partial_t \pi_T^i = -\epsilon_{ijk}\partial_j \partial_l \pi^{kl} \tag{106}$$

where we defined $\pi_T^i = \epsilon_{ijk}\partial_j T^{k0}$. Applying $(\partial_t + 1/\tau_\pi)$ and using Eq. (84) yields

$$0 = (\tau_\pi \partial_t^2 + \partial_t - D_T \nabla^2)\pi_T^i. \tag{107}$$

As long as $D_T/\tau_\pi < 1$, this is now causal.

For the sound modes, we start with the conservation law in the local rest frame

$$\partial_t^2 \varepsilon = \nabla^2 P + \partial_l \partial_m \pi^{lm} - \nabla^2 \Pi. \tag{108}$$

Applying $(\tau_\pi \partial_t + 1)(\tau_\Pi \partial_t + 1)$ to Eq. (108) and using Eqs. (101) and (102), one obtains the following dispersion relation for the bulk mode propagation (in this case $\delta\varepsilon$):

$$\begin{aligned} 0 &= \tau_\pi \tau_\Pi \omega^4 - \tau_\pi \tau_\Pi v_s^2 \omega^2 \mathbf{k}^2 - \tau_\Pi \frac{4D_T}{3}\mathbf{k}^2\omega^2 - \tau_\pi \gamma \mathbf{k}^2 \omega^2 \\ &\quad - \omega^2 + v_s^2 \mathbf{k}^2 - i\left(\frac{4D_T}{3} + \gamma + v_s^2(\tau_\pi + \tau_\Pi)\right)\mathbf{k}^2\omega + i\omega^3(\tau_\pi + \tau_\Pi). \end{aligned} \tag{109}$$

Comparing with the small ω and small $|\mathbf{k}|$ expansion of the denominator in Eq. (80), we can identify $Z_R(0,0) = v_s^2$,

$$Z_I(0,0) = \frac{4D_T}{3} + \gamma + v_s^2(\tau_\pi + \tau_\Pi) \tag{110}$$

and

$$R_R(0,0) = (\tau_\pi + \tau_\Pi)/v_s^2. \tag{111}$$

The Kubo formula for the damping constant now makes more sense

$$\begin{aligned} \lim_{\omega\to 0}\lim_{\mathbf{k}\to 0} \mathrm{Im}\,\frac{\bar{G}_L(\omega,\mathbf{k})}{\omega} &= (\varepsilon + P)(Z_I(0,0) - Z_R(0,0)^2\, R_R(0,0)) \\ &= (\varepsilon + P)\left(\frac{4D_T}{3} + \gamma\right) \\ &= \frac{4\eta}{3} + \zeta \end{aligned} \tag{112}$$

and for the bulk viscosity only,

$$\zeta = \lim_{\omega\to 0}\lim_{\mathbf{k}\to 0} \frac{1}{\omega}\left(\mathrm{Im}\,\bar{G}_L(\omega,\mathbf{k}) - \frac{4}{3}\mathrm{Im}\,\bar{G}_R^{xy,xy}(\omega,k_y)\right). \tag{113}$$

The Kubo formula for the bulk viscosity ζ is also available in terms of the pressure–pressure correlation function[20,21]

$$\zeta = \lim_{\omega\to 0}\lim_{\mathbf{k}\to 0} \frac{1}{\omega}\mathrm{Im}\,G_R^{PP}(\omega,\mathbf{k}). \tag{114}$$

Using the fact that the correlation functions of T^{00} vanishes in the $\mathbf{k} \to 0$ limit, one can use in place of P the trace $T^{\mu}_{\mu}/3$ or the combination $P - v_s^2\varepsilon$ to make it more explicit that the bulk viscosity is non-zero only if the conformal symmetry is broken. The Kubo formulas for the relaxation times τ_π and τ_Π are not simple to determine in this analysis. Simple Kubo formulas for τ_π has been worked out in Refs. 25 and 26 as

$$\eta\tau_\pi = -\lim_{\omega\to 0}\lim_{\mathbf{k}\to 0}\frac{1}{2}\mathrm{Re}\,\partial_\omega^2 G_R^{xy,xy}(\omega,\mathbf{k}) \tag{115}$$

although a simple Kubo formula for τ_Π is still to be found.

In the dispersion relation Eq. (109), the fourth-order terms of $O(\omega^4)$ and $O(\omega^2\mathbf{k}^2)$ are present but $O(\mathbf{k}^4)$ terms are not. This may seem unsatisfactory since there is no reason why this term should be small compared to the other two fourth-order terms when $\omega \sim v_s|\mathbf{k}|$. However, one should recall that hydrodynamics is valid only in the long wavelength and small frequency limits. From this point of view, the fourth-order terms are not so important. They become, however, significant when the equations are solved numerically that can include short wavelength excitations. Fortunately, this Israel–Stewart form of second-order hydrodynamics[27] (comprising of Eqs. (96), (101) and (102)) is shown to be stable in Refs. 28 and 29.

If one wants to include $O(\mathbf{k}^4)$ terms, then one can modify the relaxation equations (101) and (102) to include second derivatives of T^{ij}. However, doing so not only generates $O(\mathbf{k}^4)$ terms, but it also generates (incomplete) terms involving 5 and 6 factors of ω and $\mathbf{k}$. Since higher order terms begin to matter at large ω and $|\mathbf{k}|$, having higher and higher order of frequency and momentum (or derivatives in the configuration space) does not guarantee that the numerical solution in this limit becomes more and more faithful to the real spectrum. One just needs to be careful not to interpret high frequency and momentum modes as physical.

As for the calculation of the viscosities, full leading order perturbative QCD results for both the shear viscosity and the bulk viscosity have been obtained in Refs. 30–32 using the Kubo formulas illustrated above. QCD is an asymptotically free theory.[33–35] In principle, it admits a perturbative expansion only when the energy scale exceeds at least a few GeV. Since the typical QGP energy scale is less than 1 GeV, the strong coupling is not small. Phenomenologically, we must have $\alpha_S \approx 0.3$.[36,37] This value may look weak, but the gauge coupling itself $g = \sqrt{4\pi\alpha_S} \approx 2$ is not so small. In the perturbative many-body QCD, g (or $g/2\pi$) is the expansion parameter not α_S.[38–41] Therefore although the analysis performed in Refs. 30–32 are nothing short of tour de force, having $g \approx 2$ makes numerical values obtained in perturbation theory not too reliable. At this point, reliable first principle calculations at large g can only be performed on numerical lattice in Euclidean space. Lattice QCD can straightforwardly compute static properties such as the equation of state. However, calculations of dynamic properties such as the viscosities become more complicated as they involve estimating real continuous functions from a finite set of discrete Euclidean data. Nonetheless, a great deal has been accomplished in computing the properties of QGP through lattice QCD calculations[42–46] as well as through effective models such as the hadron resonance gas model (HR)[47,48] and the AdS/CFT correspondence.[6,49–51]

The purpose of this section has been to show that the hydrodynamics is very general. No matter what the system is, there usually is a regime where hydrodynamics is in some way applicable as long as there exist "macroscopically small but microscopically large" length and time scales. For more detailed analysis of the length and time scales and also for demonstrating more general structure of hydrodynamic equations, we now turn to the kinetic theory.

4. Hydrodynamics from Kinetic Theory

4.1. *Length scales and validity of hydrodynamic approximations*

Kinetic theory describes a medium microscopically, by following the evolution of the phase-space distribution function $f(x,p)$, a Lorentz scalar that describes the probability of finding a particle with four-momentum p^μ at space-time position x^ν. Classical kinetic theory assumes that the particle momenta are on-shell, $p^2 = m^2$, which requires the system to be sufficiently dilute and the mean free paths sufficiently long to ignore collisional broadening effects on the spectral function $\rho(p) = 2\pi\delta(p^2 - m^2)$ that defines the particles' propagator. The defining equation of classical kinetic theory is the Boltzmann equation,

$$p^\mu \partial_\mu f(x,p) = C(x,p), \tag{116}$$

where $C(x,p)$ is the collision term in which the strength of the interaction enters through their scattering cross-sections. Especially for massless degrees of freedom, its detailed form can be quite complicated.[52] A popular simplification of the collision term is the relaxation time approximation (RTA)[d]

$$C(x,p) = \frac{p^\mu u_\mu(x)}{\tau_{\rm rel}(x,p)}\,[f_{\rm eq}(x,p) - f(x,p)], \tag{117}$$

where the relaxation time $\tau_{\rm rel}$ in general depends on position through the local density and can also depend on the local rest frame energy of the particles (indicated by the p-dependence). Classical kinetic theory is valid if this relaxation time $\tau_{\rm rel}$, and the associated mean free path $\lambda_{\rm mfp} = \langle (p/E)\tau_{\rm rel}\rangle$, are sufficiently large. In other words, interactions among the constituents must be weak.

Hydrodynamics is valid if the system is close enough to thermal equilibrium that its local momentum distribution (and therefore its macroscopic fields, such as particle and energy density and pressure, which can all be expressed as moments of the local momentum distribution) can be characterized by a small number of thermodynamic and transport parameters, such as temperature, chemical potential, shear and bulk viscosity, etc. This requires efficient interactions among the constituents

[d]The Boltzmann equation with this RTA-approximated collision term is known as the Anderson–Witting equation.

of the medium because otherwise any kind of macroscopic dynamics involving local expansion or compression or shear of the fluid will drive its local momentum distribution away from its near-equilibrium form. Hydrodynamics works best for systems made of strongly interacting constituents.

Does this mean that the validity of kinetic theory and hydrodynamics are mutually exclusive? Not necessarily. To gain clarity consider a relativistic system of (almost) massless degrees of freedom. It can be characterized by three length scales, two microscopic and one macroscopic one:

- the thermal wavelength $\lambda_{\rm th} \sim 1/T$;
- the mean free path $\lambda_{\rm mfp} \sim 1/(\langle\sigma v\rangle n)$ where $\langle\sigma v\rangle$ is the momentum-averaged transport cross-section times the relative speed (≈ 1 in units of c) of the colliding objects, and n is the density of scatterers;
- the length scale $L_{\rm hydro}$ over which macroscopic fluid dynamical variables vary; it can be defined in many ways that give quantitatively different but similar order of magnitude results: $L^{-1}_{\rm hydro} \sim \partial_\mu u^\mu \sim |\partial_\mu e|/e$, etc.

The ratio between the two microscopic scales characterizes the magnitude of the transport coefficients η (shear viscosity), ζ (bulk viscosity), and κ (heat conductivity):

$$\frac{\lambda_{\rm mfp}}{\lambda_{\rm th}} \sim \frac{1}{\langle\sigma\rangle n}\,\frac{1}{\lambda_{\rm th}} \sim \frac{1}{\langle\sigma\rangle\lambda_{\rm th}}\,\frac{1}{s} \sim \frac{\eta}{s},\ \frac{\zeta}{s},\ \frac{T\kappa}{s}, \tag{118}$$

where we used $\eta,\ \zeta,\ T\kappa \sim 1/(\langle\sigma\rangle\lambda_{\rm th}) \sim \lambda_{\rm mfp}T^4$ and the entropy density $s \simeq 4n \sim T^3$ for a near-thermalized system of particle number density n for massless degrees of freedom.

In terms of the two microscopic length scales, we can define three regimes of microscopic dynamics:

(1) **Dilute gas regime**:

$$\frac{\lambda_{\rm mfp}}{\lambda_{\rm th}} \sim \frac{\eta}{s} \gg 1 \Leftrightarrow \langle\sigma\rangle \ll \lambda_{\rm th}^2 \sim \frac{1}{T^2}. \tag{119}$$

This is the *weak-coupling regime* where the microscopic system dynamics can be described in terms of on-shell quasi-particles and many-body correlations are suppressed. In this regime the Boltzmann equation applies.

(2) **Dense gas regime**:

$$\frac{\lambda_{\rm mfp}}{\lambda_{\rm th}} \sim \frac{\eta}{s} \sim 1 \Leftrightarrow \langle\sigma\rangle \sim \lambda_{\rm th}^2. \tag{120}$$

In this case interactions happen on the scale $\lambda_{\rm th}$. We call this the *moderate coupling regime* where the microscopic system dynamics must be described by off-shell quasiparticles (whose spectral functions have a finite collisional width) and many-body correlation effects are non-negligible. Here the Boltzmann equation must be replaced by a quantum kinetic approach based on Wigner distributions, and the BBGKY hierarchy of coupled equations for the N-body distribution functions can no longer be efficiently truncated.

(3) **Liquid regime**:

$$\frac{\lambda_{\mathrm{mfp}}}{\lambda_{\mathrm{th}}} \sim \frac{\eta}{s} \ll 1 \Leftrightarrow \langle\sigma\rangle \gg \lambda_{\mathrm{th}}^2. \tag{121}$$

This is the *strong-coupling regime* where the system has no well-defined quasi-particles and no valid kinetic theory description.

To judge the validity of a macroscopic hydrodynamic approach we compare the microscopic to the macroscopic length scales. To simplify the discussion, let us agree on using the inverse of the scalar expansion rate $\theta = \partial_\mu u^\mu$ to represent the macroscopic length scale L_{hydro}.[e] The figure of merit controlling the validity of a fluid dynamic picture is the *Knudsen number*:

$$\mathrm{Kn} = \lambda_{\mathrm{mfp}} \cdot \theta \sim \frac{\eta}{s}\lambda_{\mathrm{th}} \cdot \theta \sim \frac{\eta}{sT} \cdot \theta \sim \theta\tau_{\mathrm{rel}}. \tag{122}$$

The Knudsen number is the small parameter that controls the convergence of the expansion in gradients of thermodynamic quantities that underlies the derivation of hydrodynamics as an effective theory for the long-distance dynamics of a general quantum field theory.[53] Again, we can use it to define three regimes:

(1) **Ideal fluid dynamics:**

$$\mathrm{Kn} \approx 0 \Leftrightarrow \frac{\eta}{s} \approx 0 \text{ or } \theta \approx 0 \text{ such that } \theta\tau_{\mathrm{rel}} \approx 0. \tag{123}$$

(2) **Viscous fluid dynamics:**

$$\mathrm{Kn} \lesssim 1 \Leftrightarrow \frac{\eta}{s} \text{ or } \theta \text{ small such that } \theta\tau_{\mathrm{rel}} \lesssim 1. \tag{124}$$

(3) **Hydrodynamics breaks down:**

$$\mathrm{Kn} \gg 1 \Leftrightarrow \frac{\eta}{s} \text{ or } \theta \text{ large such that } \theta\tau_{\mathrm{rel}} \gg 1. \tag{125}$$

In high-energy heavy-ion collisions, the initial energy deposition occurs in an approximately boost-invariant fashion along the beam direction, leading to an expansion rate θ that diverges like $1/\tau$ for small time τ after impact.[f] On the other hand we now know that the quantum mechanical uncertainty relation places a lower bound $\eta/s \gtrsim 1/(4\pi)$ on any system, even at infinitely strong coupling.[6,54,55] Therefore, hydrodynamics is inapplicable during the earliest stage of a heavy-ion collision. At the end of a heavy-ion collision, the mean free paths of hadrons become

[e]The shorter L_{hydro}, the faster the system is driven away from local equilibrium. The scalar expansion rate directly drives the bulk viscous pressure Π. It is parametrically of the same order as the shear tensor $\sigma^{\mu\nu} = \Delta^{\mu\nu}_{\alpha\beta}\partial^\alpha u^\beta \equiv \nabla^{\langle\mu}u^{\nu\rangle}$ defined in Eq. (70) that drives the shear viscous pressure $\pi^{\mu\nu}$ and as the diffusion force $I^\mu = \nabla^\mu(\mu/T)$ associated with space-time gradients of conserved charge densities that drives the heat flow V^μ.

[f]Here, $\tau = \sqrt{t^2 - z^2}$ is the longitudinal proper time and the boost-invariance refers to independence of the space-time rapidity $\eta = \tanh^{-1}(z/t)$. For more details, see Sec. 5.1.

large compared to the Hubble radius $\sim 1/\theta$ of the expanding fireball, and hydrodynamics breaks down again. This process is called *kinetic decoupling*. Between the early pre-equilibrium and the final decoupling stage stretches an extended period of applicability of viscous fluid dynamics. The most important factor ensuring this is the strong collective coupling of the quark–gluon plasma (QGP) phase which is characterized by a small specific shear viscosity $\eta/s \sim (2-3)/(4\pi)$.[56–59]

Note that the validity of hydrodynamics does not rely directly on η/s being small, only on $(\eta/s)\cdot(\theta/T)$ being small. So, strictly speaking, strong coupling is not required for hydrodynamics to be valid. Only in extreme situations, such as heavy-ion collisions which are characterized by extreme expansion rates, does hydrodynamics require very strong coupling. In this case, hydrodynamics is applicable even though classical kinetic theory is not, because very strongly coupled quantum field theories do not allow a description in terms of on-shell quasi-particles. It is generally believed that the very earliest stage of a heavy-ion collision has no well-defined quasiparticles at all and is better described by a theory of classical or quantum fields than by a (quantum) kinetic approach. On the other hand, weakly coupled system, with very large values of η/s in which the applicability of (even classical) kinetic theory is ensured, can still be describable macroscopically through fluid dynamics if they are sufficiently homogeneous and expand slowly. In this case the smallness of θ can compensate for the largeness of η/s, resulting in a small Knudsen number. Systems with a large η/s but a small product $(\eta/s)\cdot(\theta/T)$ admit simultaneous microscopic classical kinetic and macroscopic hydrodynamic descriptions. In the following subsections we will study such systems to derive the macroscopic hydrodynamic equations from the microscopic kinetic theory. Hydrodynamics being an effective long-distance theory, the form of the resulting equations does not rely on the validity of the underlying kinetic theory (although the values for the transport coefficients do); they can therefore also be applied to a strongly-coupled liquid such as the QGP.

4.2. *Ideal fluid dynamics*

We define p-moments of the distribution function weighted with some momentum observable $O(p)$ by

$$\langle O(p)\rangle \equiv \int_p O(p)f(x,p) \equiv g\int \frac{d^3p}{(2\pi)^3p^0}O(p)f(x,p) \tag{126}$$

(g is a degeneracy factor) and $p^0 = E_p = \sqrt{m^2+\mathbf{p}^2}$. The particle number current and energy momentum tensor are then written as

$$j^\mu = \langle p^\mu\rangle, \quad T^{\mu\nu} = \langle p^\mu p^\nu\rangle. \tag{127}$$

Usually there is more than one particle species in the system, and the conserved baryon charge current J_B^μ and energy–momentum tensor $T^{\mu\nu}$ are given in terms of linear combinations of $\langle p^\mu \rangle_i$ and $\langle p^\mu p^\nu \rangle_i$ where the subscript i labels the particle species whose distribution function is $f_i(x,p)$:

$$J_B^\mu = \sum_i b_i j_i^\mu = \sum_i b_i \langle p^\mu \rangle_i, \quad T^{\mu\nu} = \sum_i T_i^{\mu\nu} = \sum_i \langle p^\mu p^\nu \rangle_i; \tag{128}$$

here b_i is the baryon charge carried by each particle of species i. For simplicity, we restrict the following discussion to a single particle species.

The particle number current and energy–momentum tensor take their ideal fluid dynamical form

$$j_{\rm id}^\mu = n u^\mu, \quad T_{\rm id}^{\mu\nu} = \varepsilon u^\mu u^\nu - P \Delta^{\mu\nu}, \tag{129}$$

where the spatial projector in the local rest frame (LRF) $\Delta^{\mu\nu}$ is given in Eq. (9), if we assume that the system is locally momentum isotropic:

$$f(x,p) = f_{\rm iso}(x,p) \equiv f_{\rm iso}\left(\frac{p^\mu u_\mu(x) - \mu(x)}{T(x)}\right). \tag{130}$$

The local equilibrium distribution

$$f_{\rm eq}(\zeta) = \frac{1}{e^\zeta + a}, \tag{131}$$

where $\zeta \equiv (p^\mu u_\mu(x) - \mu(x))/T(x)$ and $a = 1, -1, 0$ for Fermi–Dirac, Bose–Einstein, and classical Boltzmann statistics, respectively, is a special form of $f_{\rm iso}(x,p)$. It is defined as the distribution for which the collision term $C(x,p)$ in the Boltzmann equation (116) vanishes. Note that the ideal fluid decomposition (129) does not require chemical equilibrium, i.e., it holds for arbitrary values of the chemical potential $\mu(x)$, nor does it require complete thermal equilibrium, i.e., $f_{\rm iso}$ is not required to depend on its argument exponentially as is the case for the equilibrium distribution (131). If the dependence is non-exponential, the collision term in the Boltzmann equation is non-zero, but its p^μ-moment still vanishes, $\int_p p^\mu C = 0$, due to energy–momentum conservation.

The ideal hydrodynamic equations follow by inserting the ideal fluid decomposition (129) into the conservation laws Eq. (1):

$$\dot{n} = -n\theta, \quad \dot{\varepsilon} = -(\varepsilon + P)\theta, \quad \dot{u}^\nu = \frac{\nabla^\mu P}{\varepsilon + P} = \frac{c_s^2}{1 + c_s^2} \frac{\nabla^\mu \varepsilon}{\varepsilon}, \tag{132}$$

where the very last expression assumes an EOS of type $P = c_s^2 \varepsilon$. $\dot{F}$ denotes the LRF time derivative of a function F, $\dot{F} \equiv D_\tau F \equiv u^\mu \partial_\mu F$, and $\nabla^\mu = \Delta^{\mu\nu}\partial_\nu$ the spatial gradient in the LRF. Thus, $\partial_\mu = u^\mu D_\tau + \nabla^\mu$.[g]

Equations (132) can be solved numerically for the local particle density $n(x)$, energy density $\varepsilon(x)$, and flow velocity $u^\mu(x)$, with the temperature $T(x)$, chemical potential $\mu(x)$ and pressure $P(x)$ following from the equation of state (EOS) of the fluid. Local deviations from chemical equilibrium result in a non-equilibrium

[g]Note that in curvilinear coordinates or curved space-times, the partial derivative ∂_μ must be replaced by the covariant derivative d_μ.

value of the local chemical potential $\mu(x)$ and a non-zero right-hand side in the current conservation equation for j^μ. Deviations from thermal equilibrium (while preserving local isotropy) must be accounted for by a non-equilibrium pressure in the EOS $P(\varepsilon, n)$. In both cases, the conservation laws, Eqs. (1), lead to a non-vanishing entropy production rate $\partial_\mu S^\mu \sim 1/\tau_{\rm rel} \neq 0$.

4.3. *Viscous fluid dynamics*

4.3.1. *Navier–Stokes (NS), Israel–Stewart (IS) and Denicol–Niemi–Molnar–Rischke (DNMR) theory* (vHydro)

Israel–Stewart (IS) and Denicol–Niemi–Molnar–Rischke (DNMR) second-order viscous fluid dynamics[27] are obtained by using in (127) for $f(x,p)$ the ansatz

$$f(x,p) = f_{\rm iso}\left(\frac{p^\mu u_\mu(x) - \mu(x)}{T(x)}\right) + \delta f(x,p). \tag{133}$$

The correction δf describes the deviation of the solution $f(x,p)$ of the Boltzmann equation from local momentum isotropy. It is supposed to be "small", in a sense that will become clearer below, and will thus be treated perturbatively.

Most authors set $f_{\rm iso} = f_{\rm eq}$, i.e., they expand around a local equilibrium state. To obtain the correct form of the hydrodynamic equations this is not necessary; only the form of the equation of state $P(\varepsilon, n)$ and the values of the transport coefficients depend on this choice. We, too, will make this choice for simplicity, but emphasize that under certain conditions the perturbative treatment of δf may be better justified if the leading-order distribution $f_{\rm iso}$ is not assumed to be thermal.

For later convenience we decompose p^μ into its temporal and spatial components in the LRF:

$$p^\mu = (u^\mu u^\nu + \Delta^{\mu\nu})p^\mu = \bar{E}_p u^\mu + p^{\langle\mu\rangle} \tag{134}$$

where $\bar{E}_p \equiv u_\mu p^\mu$ and $p^{\langle\mu\rangle} \equiv \Delta^{\mu\nu} p_\nu$ are the energy and spatial momentum components in the LRF. Then

$$n = \langle \bar{E}_p \rangle, \quad \varepsilon = \langle \bar{E}_p^2 \rangle. \tag{135}$$

The decomposition (133) is made unique by Landau matching: First, define the LRF by solving the eigenvalue equation (7) with the constraint $u^\mu u_\mu = 1$ which selects among the four eigenvectors of $T^{\mu\nu}$ the timelike one. Equation (7) fixes the flow vector $u^\mu(x)$ and the LRF energy density. Next, we fix $T(x)$ and $\mu(x)$ by demanding that δf gives no contribution to the local energy and baryon density:

$$\langle \bar{E}_p \rangle_\delta = \langle \bar{E}_p^2 \rangle_\delta = 0. \tag{136}$$

Inserting (133) into (127) we find the general decomposition

$$j^\mu = j^\mu_{\rm id} + V^\mu, \quad T^{\mu\nu} = T^{\mu\nu}_{\rm id} - \Pi\Delta^{\mu\nu} + \pi^{\mu\nu}, \tag{137}$$

with a non-zero number flow in the LRF,

$$V^\mu = \langle p^{\langle\mu\rangle}\rangle_\delta, \tag{138}$$

a bulk viscous pressure

$$\Pi = -\frac{1}{3}\langle p^{\langle\alpha\rangle} p_{\langle\alpha\rangle}\rangle_\delta, \tag{139}$$

and a shear stress

$$\pi^{\mu\nu} = \langle p^{\langle\mu} p^{\nu\rangle}\rangle_\delta, \tag{140}$$

where $\langle\cdots\rangle_\delta$ indicates moments taken with the deviation δf from $f_{\rm iso}$. In the last equation we introduced the notation

$$A^{\langle\mu\nu\rangle} \equiv \Delta^{\mu\nu}_{\alpha\beta} A^{\alpha\beta}, \tag{141}$$

where $\Delta^{\mu\nu}_{\alpha\beta}$ is the spin-2 projector introduced in Eq. (71) denoting the traceless and transverse (to u^μ) part of a tensor $A^{\mu\nu}$. The shear stress tensor $\pi^{\mu\nu} = T^{\langle\mu\nu\rangle}$ thus has 5 independent components while V^μ, which is also orthogonal to u^μ by construction, has 3 independent components.

Using the viscous hydrodynamic decomposition (137) in the conservation laws $\partial_\mu T^{\mu\nu} = 0$ and $\partial_\mu j^\mu = 0$, we obtain the vHydro viscous hydrodynamic evolution equations

$$\begin{aligned}
\dot{n} &= -n\theta - \nabla_\mu V^\mu, \\
\dot{\varepsilon} &= -(\varepsilon + P + \Pi)\theta + \pi_{\mu\nu}\sigma^{\mu\nu}, \\
(\varepsilon + P + \Pi)\dot{u}^\mu &= \nabla^\mu(P + \Pi) - \Delta^{\mu\nu}\nabla^\sigma \pi_{\nu\sigma} + \pi^{\mu\nu}\dot{u}_\nu,
\end{aligned} \tag{142}$$

where $\sigma^{\mu\nu} = \nabla^{\langle\mu} u^{\nu\rangle}$ is the velocity shear tensor introduced in Eq. (70). They differ from the ideal fluid dynamical equations (132) by additional source terms arising from the dissipative flows. Altogether, the deviation δf has introduced $(3 + 1 + 5) = 9$ additional dissipative flow degrees of freedom for which additional evolution equations are needed. These cannot be obtained from the macroscopic conservation laws but require input from the microscopic dynamics. In a system that is initially in local equilibrium, the deviation δf is caused by the dynamical response of the system to gradients in the thermodynamic and flow variables. The forces that drive this deviation can be classified by their Lorentz structure as a scalar, a vector and a tensor force:

$$\begin{aligned}
\text{scalar force:}\quad \theta &= \partial_\mu u^\mu \text{ (scalar expansion rate);} \\
\text{vector force:}\quad I^\mu &= \nabla^\mu \left(\frac{\mu}{T}\right) \text{ (fugacity gradient);} \\
\text{symmetric tensor force:}\quad \sigma^{\mu\nu} &= \nabla^{\langle\mu} u^{\nu\rangle} \text{ (velocity shear tensor);} \\
\text{antisymmetric tensor force:}\quad \omega^{\mu\nu} &= \frac{1}{2}(\nabla^\mu u^\nu - \nabla^\nu u^\mu) \text{ (vorticity tensor).}
\end{aligned} \tag{143}$$

These forces generate dissipative flows, the scalar bulk viscous pressure Π, the heat flow vector V^μ, and the shear stress tensor $\pi^{\mu\nu}$.[h] The strength of the forces (143) driving the system away from local equilibrium is characterized by the Knudsen number. The system response can be characterized by inverse Reynolds numbers associated with the dissipative flows[60]:

$$\mathrm{R}_\Pi^{-1} = \frac{|\Pi|}{P}, \quad \mathrm{R}_V^{-1} = \frac{\sqrt{-I_\mu I^\mu}}{P}, \quad \mathrm{R}_\pi^{-1} = \frac{\sqrt{\pi_{\mu\nu}\pi^{\mu\nu}}}{P}. \tag{144}$$

Due to the time delay $\tau_{\rm rel}$ between the action of the force and the system response, built into the collision term of the Boltzmann equation, the inverse Reynolds numbers are not necessarily of the same order as the Knudsen number: For example, an initially small bulk viscous pressure can remain small, due to critical slowing down, as the system passes through a phase transition, even though the bulk viscosity becomes large during the transition.[61] Conversely, strong deviations from local equilibrium during a rapidly expanding pre-equilibrium stage in heavy-ion collisions can lead to large initial values for the dissipative flows, and a slow equilibration rate may cause them to stay large for a while even though the expansion rate decreases with longitudinal proper time as $1/\tau$. Deviations from equilibrium, and the accuracy of their description by viscous fluid dynamics, are therefore controlled by a combination of Knudsen and inverse Reynolds numbers.[60]

The 9 equations of motion describing the relaxation of the 9 dissipative flow components are controlled by microscopic physics, encoded in the collision term on the right-hand side of the Boltzmann equation, and can be derived from approximate solutions of that equation. This was first done almost 50 years ago by Israel and Stewart in Ref. 27, but when the problem was recently revisited it was found[25,60,62,63] that the relaxation equations take a much more general form than originally derived. Specifically, Denicol *et al.*[60] found the following general structure:

$$\begin{aligned}
\tau_\Pi \dot{\Pi} + \Pi &= -\zeta\theta + \mathcal{J} + \mathcal{K} + \mathcal{R}, \\
\tau_V \Delta^{\mu\nu}\dot{V}_\nu + V^\mu &= \kappa I^\mu + \mathcal{J}^\mu + \mathcal{K}^\mu + \mathcal{R}^\mu, \\
\tau_\pi \Delta^{\mu\nu}_{\alpha\beta}\dot{\pi}^{\alpha\beta} + \pi^{\mu\nu} &= 2\eta\sigma^{\mu\nu} + \mathcal{J}^{\mu\nu} + \mathcal{K}^{\mu\nu} + \mathcal{R}^{\mu\nu}.
\end{aligned} \tag{145}$$

Here all calligraphic terms are of second-order in combined powers of the Knudsen and inverse Reynolds numbers. $\mathcal{J}$ terms contain products of factors that are each of first-order in the Knudsen and inverse Reynolds numbers; $\mathcal{K}$ terms are second-order in Knudsen number, and $\mathcal{R}$ terms are second-order in inverse Reynolds numbers.

In relaxation time approximation, with an energy-independent relaxation time $\tau_{\rm rel}$, the relaxation times for the dissipative flows all agree with each other: $\tau_\Pi = \tau_V = \tau_\pi = \tau_{\rm rel}$.[63] The same does not hold for more general forms of the collision

[h] The energy–momentum tensor is symmetric, so dissipative flows have no antisymmetric tensor contribution, but the antisymmetric vorticity tensor couples to the other dissipative forces and flows at second-order in the Knudsen and inverse Reynolds numbers.

term. If we set in Eqs. (145) the relaxation times and all other second-order terms to zero, we obtain the equations of **relativistic NS theory**:

$$\Pi = -\zeta\theta, \quad V^\mu = \kappa I^\mu, \quad \pi^{\mu\nu} = 2\eta\sigma^{\mu\nu}. \tag{146}$$

The relaxation equations (145) have solutions that, for sufficiently small expansion rates (see below), approach asymptotically (at times $\tau \gg \tau_{\Pi,V,\pi}$) the NS values (146). However, plugging the NS solutions (146) directly back into the decompositions (137) and using them in the conservation laws (1) leads to viscous hydrodynamic equations of motion that are acausal and numerically unstable.[64,65] The physical reason for this is the instantaneous response of the dissipative flows to the dissipative forces encoded in Eqs. (146) which violates causality. A causal and numerically stable implementation of viscous fluid dynamics must account for the time delay between cause and effect of dissipative phenomena and therefore be by necessity of second-order in Knudsen and Reynolds numbers.

The first relativistic causal second-order theory of viscous fluid dynamics was IS theory.[27] It amounts to dropping the $\mathcal{K}$ and $\mathcal{R}$ terms in (145) and replacing (for massless particles) $\mathcal{J} \to -\frac{4}{3}\tau_\Pi\theta\Pi$, $\mathcal{J}^\mu \to -\tau_V\theta V^\mu$, and $\mathcal{J}^{\mu\nu} \to -\frac{4}{3}\tau_\pi\theta\pi^{\mu\nu}$. The importance of keeping specifically these second-order $\mathcal{J}$-terms for the preservation of conformal invariance in a system of massless degrees of freedom was stressed by Baier *et al.*[25]

For conformal systems the resulting IS relaxation equations can be written in the form[66]

$$\begin{aligned}
\dot{\Pi} &= -\frac{1}{\tau'_\Pi}(\Pi + \zeta'\theta),\\
\Delta^{\mu\nu}\dot{V}_\nu &= -\frac{1}{\tau'_V}(V^\mu - \kappa' I^\mu),\\
\Delta^{\mu\nu}_{\alpha\beta}\dot{\pi}^{\alpha\beta} &= -\frac{1}{\tau'_\pi}(\pi^{\mu\nu} - 2\eta'\sigma^{\mu\nu}),
\end{aligned} \tag{147}$$

with effective transport coefficients and relaxation times that are modified by the scalar expansion rate as follows:

$$\zeta' = \frac{\zeta}{1+\gamma_\Pi}, \quad \kappa' = \frac{\kappa}{1+\gamma_V}, \quad \eta' = \frac{\eta}{1+\gamma_\pi}, \quad \tau'_i = \frac{\tau_i}{1+\gamma_i} \quad (i = \Pi,\, V,\, \pi), \tag{148}$$

where $\gamma_i = \frac{4}{3}\theta\tau_i$ for $i = \Pi, \pi$ and $\gamma_i = \theta\tau_i$ for $i = V$. These relaxation equations describe an asymptotic approach of the dissipative flows to effective NS values that, for a positive scalar expansion rate θ, are reduced relative to their first-order values (146) by a factor $1+\gamma_i$, while the effective rate of approach to this effective NS limit is sped up by the same factor. γ_i involves the product of the scalar expansion rate and the respective relaxation time. So, compared to a more slowly expanding system, a rapidly expanding system with the same microscopic scattering cross-sections is characterized by lower effective viscosities and shorter effective relaxation times.[66,67]

When all the second-order terms are kept, the DNMR equations become quite complicated. The $\mathcal{J}$, $\mathcal{K}$ and $\mathcal{R}$ terms listed by DNMR[60] add up to 16 terms for Π, 18 terms for V^μ, and 19 terms for $\pi^{\mu\nu}$, each with its own transport coefficient. While many of these transport coefficients have been computed for massless theories at both weak and strong coupling (a subject too rich to be fully reviewed here, see papers by Denicol *et al.*,[60,63,68] Moore *et al.*[69,70] and Noronha *et al.*[71] for lists of references to relevant work), quite a few are still unknown. Furthermore, the QGP is neither weakly nor strongly enough coupled, nor is it sufficiently conformally symmetric for any of these calculations to be quantitatively reliable for heavy-ion collisions. For these reasons, many practical applications of viscous fluid dynamics employ phenomenological values for the transport coefficients, and work studying which terms need to be kept and which might be of lesser importance is still ongoing.[68]

One important non-linear coupling mechanism that enters at second-order are bulk-shear couplings where shear stress drives a bulk viscous pressure and vice versa.[60,63,68,72] Heavy-ion collisions are characterized by initially very large differences between the longitudinal and transverse expansion rates that cause large shear stress. The latter, in turn, creates a bulk viscous pressure via bulk-shear coupling that can dominate over the one generated *à la* NS by the scalar expansion rate[72] and may actually be able to flip its sign. This should be taken into account in phonemenological applications of viscous hydrodynamics to heavy-ion collisions.

4.3.2. *Anisotropic hydrodynamics* (aHydro)

The dissipative flows are given by moments of the deviation $\delta f(x,p)$ of the distribution function from local equilibrium, and their relaxation equations are derived from the Boltzmann equation using approximations that, in one way or another, assume that δf is small. However, systems featuring strongly anisotropic expansion, such as the early evolution stage of the fireballs created in ultra-relativistic heavy-ion collisions, generate strong local momentum anisotropies: the width of the LRF momentum distribution along a certain direction is inversely proportional to the local expansion rate in that direction, and this momentum-space distortion growth with the magnitude of the shear viscosity. In viscous hydrodynamics, where we expand f around a locally isotropic LO distribution (see Eq. (133)), this local momentum anisotropy must be absorbed entirely by δf, making δf large and rendering the approximations used for calculating the evolution of the dissipative flow generated by δf unreliable. Indeed, even for moderate specific shear viscosities $\eta/s \sim 5-10$ the (negative) longitudinal component of the viscous shear pressure can become so large in IS theory that it overwhelms the thermal pressure, resulting in a negative total pressure along the beam direction — which, according to the kinetic definition $P_L = \frac{1}{3}\langle p_z^2 \rangle$, should never happen.

Anisotropic hydrodynamics[73,74] is based on the idea to account already in the LO distribution for the local momentum anisotropy resulting from anisotropic expansion, by parametrizing[75]

$$f(x,p) = f_{\mathrm{RS}}(x,p) \equiv f_{\mathrm{iso}}\left(\frac{\sqrt{p_\mu \Xi^{\mu\nu}(x) p_\nu} - \tilde{\mu}(x)}{\Lambda(x)}\right), \tag{149}$$

where $\Xi^{\mu\nu}(x) = u^\mu(x)u^\nu(x) + \xi(x) z^\mu(x) z^\nu(x)$, $z^\mu(x)$ being a unit vector in longitudinal z direction in the LRF. The subscript RS refers to Romatschke and Strickland who are the authors of Ref. 75. This distribution is characterized by 3 flow parameters $u^\mu(x)$ and three "thermodynamic" parameters: the "transverse temperature" $\Lambda(x)$, the effective chemical potential $\tilde{\mu}(x)$, and the momentum-anisotropy parameter $\xi(x)$. Inserting (149) into (127) yields the aHydro decomposition

$$j^\mu_{\mathrm{RS}} = n_{\mathrm{RS}} u^\mu, \quad T^{\mu\nu}_{\mathrm{RS}} = \varepsilon_{\mathrm{RS}} u^\mu u^\nu - P_T \Delta^{\mu\nu} + (P_L - P_T) z^\mu z^\nu, \tag{150}$$

$$n_{\mathrm{RS}} = \langle E \rangle_{\mathrm{RS}} = \mathcal{R}_0(\xi)\, n_{\mathrm{iso}}(\Lambda, \tilde{\mu}), \quad \varepsilon_{\mathrm{RS}} = \langle E^2 \rangle_{\mathrm{RS}} = \mathcal{R}(\xi)\, \varepsilon_{\mathrm{iso}}(\Lambda, \tilde{\mu}),$$

$$P_{T,L} = \langle p^2_{T,L} \rangle_{\mathrm{RS}} = \mathcal{R}_{T,L}(\xi) P_{\mathrm{iso}}(\Lambda, \tilde{\mu}). \tag{151}$$

For massless systems, the local momentum anisotropy effects factor out via the $\mathcal{R}(\xi)$-functions[73]:

$$\mathcal{R}_0(\xi) = \frac{1}{\sqrt{1+\xi}}, \qquad \mathcal{R}(\xi) = \frac{1}{2}\left(\frac{1}{1+\xi} + \frac{\arctan\sqrt{\xi}}{\sqrt{\xi}}\right),$$
$$\mathcal{R}_\perp(\xi) = \frac{3}{2\xi}\left(\frac{1+(\xi^2-1)\mathcal{R}(\xi)}{\xi+1}\right), \quad \mathcal{R}_L(\xi) = \frac{3}{\xi}\left(\frac{(\xi+1)\mathcal{R}(\xi)-1}{\xi+1}\right). \tag{152}$$

The isotropic pressure is obtained from a locally isotropic equation of state $P_{\mathrm{iso}}(\Lambda, \tilde{\mu}) = P_{\mathrm{iso}}(\varepsilon_{\mathrm{iso}}(\Lambda, \tilde{\mu}), n_{\mathrm{iso}}(\Lambda, \tilde{\mu}))$. For massless noninteracting partons, $P_{\mathrm{iso}}(\Lambda, \tilde{\mu}) = \frac{1}{3}\varepsilon_{\mathrm{iso}}(\Lambda, \tilde{\mu})$ independent of chemical composition. To compare with ideal and IS viscous hydrodynamics, we need to assign the locally anisotropic system an appropriate temperature $T(x) = T(\xi(x), \Lambda(x), \tilde{\mu}(x))$ and chemical potential $\mu(x) = \mu(\xi(x), \Lambda(x), \tilde{\mu}(x))$, thinking of $f_{\mathrm{RS}}(\xi, \Lambda)$ as an expansion around the locally isotropic distribution $f_{\mathrm{iso}}(T)$. For this we impose the generalized Landau matching conditions $\varepsilon_{\mathrm{RS}}(\xi, \Lambda, \tilde{\mu}) = \varepsilon_{\mathrm{iso}}(T, \mu)$ and For example, using an exponential (Boltzmann) function for f_{iso} with $\mu = \tilde{\mu} = 0$, one finds $T = \Lambda \mathcal{R}^{1/4}(\xi)$. With this matching we can write

$$T^{\mu\nu}_{\mathrm{RS}} = T^{\mu\nu}_{\mathrm{id}} - (\Delta P + \Pi_{\mathrm{RS}})\Delta^{\mu\nu} + \pi^{\mu\nu}_{\mathrm{RS}}, \tag{153}$$

$$\Delta P + \Pi_{\mathrm{RS}} = -\frac{1}{3}\int_p p_\alpha \Delta^{\alpha\beta} p_\beta (f_{\mathrm{RS}} - f_{\mathrm{iso}}) \quad (= 0 \text{ for } m = 0), \tag{154}$$

$$\pi^{\mu\nu}_{\mathrm{RS}} = \int_p p^{\langle\mu} p^{\nu\rangle} (f_{\mathrm{RS}} - f_{\mathrm{iso}}) = (P_T - P_L)\frac{x^\mu x^\nu + y^\mu y^\nu - 2 z^\mu z^\nu}{3}. \tag{155}$$

We see that $\pi^{\mu\nu}_{\mathrm{RS}}$ has only one independent component, $P_T - P_L$, so aHydro leaves 4 of the 5 components of $\pi^{\mu\nu}$ unaccounted for. For massless particles we have $(P_T - P_L)/P_{\rm iso}(\varepsilon) = \mathcal{R}_T(\xi) - \mathcal{R}_L(\xi)$, so the equation of motion for $\pi^{\mu\nu}_{\rm RS}$ can be replaced by one for ξ. For (2+1)-dimensional expansion with longitudinal boost-invariance these equations can be found and were solved numerically by Martinez *et al.*[76] For $m \neq 0$ we need an additional "anisotropic EOS" for $(\Delta P/P_{\rm iso}) \equiv (2P_T + P_L)/(3P_{\rm iso}) - 1$, in order to separate ΔP from the viscous bulk pressure Π.

4.3.3. *Viscous anisotropic hydrodynamics* (vaHydro)

As explained above, aHydro[73,74] accounts only for one (albeit largest) of the five independent components of the shear stress tensor $\pi^{\mu\nu}$. It can therefore not be used to compute the viscous suppression of elliptic flow which is sensitive to e.g., $\pi^{xx} - \pi^{yy}$. On the other hand, since the four remaining components of the shear stress tensor never become as large as the longitudinal/transverse pressure difference (with smooth initial density profiles they start out as zero, and with fluctuating initial conditions they are initially small), they can be treated "perturbatively" *à la* Israel and Stewart, without running into problems even at early times. Combining the non-perturbative dynamics of $P_L - P_T$ via aHydro with a perturbative treatment of the remaining viscous stress terms $\tilde{\pi}^{\mu\nu}$ *à la* IS defines the vaHydro scheme.[77] vaHydro is expected to perform better than both IS theory and aHydro during all evolution stages.

The vaHydro equations are obtained by generalizing the ansatz (149) to include arbitrary (but small) corrections to the spheroidally deformed $f_{\rm RS}(x,p)$:

$$\begin{aligned} f(x,p) &= f_{\rm RS}(x,p) + \delta\tilde{f}(x,p) \\ &= f_{\rm iso}\left(\frac{\sqrt{p_\mu \Xi^{\mu\nu}(x) p_\nu} - \tilde{\mu}(x)}{\Lambda(x)}\right) + \delta\tilde{f}(x,p). \end{aligned} \tag{156}$$

The parameters Λ and $\tilde{\mu}$ in (156) are Landau-matched as before, i.e., by requiring $\langle E\rangle_{\tilde{\delta}} = \langle E^2\rangle_{\tilde{\delta}} = 0$. To fix the value of the deformation parameter ξ one demands that $\delta\tilde{f}$ does not contribute to the pressure anisotropy $P_T - P_L$; this requires $(x_\mu x_\nu + y_\mu y_\nu - 2z_\mu z_\nu)\langle p^{\langle\mu}p^{\nu\rangle}\rangle_{\tilde{\delta}} = 0$. Then, upon inserting (156) into (127), we obtain the vaHydro decomposition

$$j^\mu = j^\mu_{\rm RS} + \tilde{V}^\mu, \quad T^{\mu\nu} = T^{\mu\nu}_{\rm RS} - \tilde{\Pi}\Delta^{\mu\nu} + \tilde{\pi}^{\mu\nu}, \tag{157}$$

with

$$\tilde{V}^\mu = \langle p^{\langle\mu\rangle}\rangle_{\tilde{\delta}}, \quad \tilde{\Pi} = -\tfrac{1}{3}\langle p^{\langle\alpha\rangle} p_{\langle\alpha\rangle}\rangle_{\tilde{\delta}}, \quad \tilde{\pi}^{\mu\nu} = \langle p^{\langle\mu}p^{\nu\rangle}\rangle_{\tilde{\delta}}, \tag{158}$$

subject to the constraints

$$u_\mu \tilde{\pi}^{\mu\nu} = \tilde{\pi}^{\mu\nu} u_\nu = (x_\mu x_\nu + y_\mu y_\nu - 2z_\mu z_\nu)\tilde{\pi}^{\mu\nu} = \tilde{\pi}^\mu_\mu = 0. \tag{159}$$

Clearly, the additional shear stress $\tilde{\pi}^{\mu\nu}$ arising from $\delta\tilde{f}$ has only 4 degrees of freedom.

The strategy in vaHydro is now to solve hydrodynamic equations for aHydro[76] (which treat $P_T - P_L$ non-perturbatively) with added source terms describing the residual viscous flows arising from $\delta\tilde{f}$, together with IS-like "perturbative" equations of motion for $\tilde{\Pi}$, $\tilde{V}^\mu$, and $\tilde{\pi}^{\mu\nu}$. The hydrodynamic equations are obtained by using the decomposition (157) in the conservation laws (1). The evolution equations for the dissipative flows $\tilde{\Pi}$, $\tilde{V}^\mu$, and $\tilde{\pi}^{\mu\nu}$ are derived by generalizing the DNMR[60] procedure to an expansion of the distribution function around the spheroidally deformed $f_{\rm RS}$ in (149), using the 14-moment approximation. These equations are lengthy; for massless systems undergoing $(2+1)$-dimensional expansion with longitudinal boost invariance they were derived by Bazow *et al.*[77] Generalizations to massive systems and full $(3+1)$-dimensional expansion are in progress. We give their simplified form for $(0+1)$-dimensional expansion in the next subsection. Especially at early times $\delta\tilde{f}$ is much smaller than δf, since the largest part of δf is already accounted for by the momentum deformation in (149). The inverse Reynolds number $\tilde{\rm R}_\pi^{-1} = \sqrt{\tilde{\pi}^{\mu\nu}\tilde{\pi}_{\mu\nu}}/\mathcal{P}_{\rm iso}$ associated with the residual shear stress $\tilde{\pi}^{\mu\nu}$ is therefore strongly reduced compared to the one associated with $\pi^{\mu\nu}$, significantly improving the range of applicability of vaHydro relative to standard second-order viscous hydrodynamics.

4.4. *Testing different hydrodynamic approximations*

For $(0+1)$-dimensional longitudinally boost-invariant expansion of a transversally homogeneous system, the Boltzmann equation can be solved exactly in the relaxation time approximation (RTA), both for massless[78–80] and massive particles.[81,82] More recently, an exact solution of this equation was also found for massless systems undergoing $(1+1)$-dimensional expansion,[83,84] with a boost invariant longitudinal and azimuthally symmetric transverse velocity profile discovered by Gubser ("Gubser flow")[85,86] as the result of imposing a particular conformal symmetry ("Gubser symmetry") on the flow. These exact solutions of the kinetic theory can be used to test various hydrodynamic approximation schemes, by imposing the symmetry of the exact solution also on the hydrodynamic solution, solving both with identical initial conditions, and comparing the predictions of both approaches for the evolution of macroscopic observables.[77,79,80,83,84,87]

We will here use the $(0+1)$-dimensional case to test the vaHydro approach.[77] This illustrates the procedure and the kind of conclusions one can draw from such a comparison. Setting homogeneous initial conditions in r and space-time rapidity η_s and zero transverse flow, $\tilde{\pi}^{\mu\nu}$ reduces to a single non-vanishing component $\tilde{\pi}$: $\tilde{\pi}^{\mu\nu} = \mathrm{diag}(0, -\tilde{\pi}/2, -\tilde{\pi}/2, \tilde{\pi})$ at $z = 0$. The factorization $n_{\rm RS}(\xi,\Lambda) = \mathcal{R}_0(\xi) n_{\rm iso}(\Lambda)$ etc. are used to obtain equations of motion for $\dot{\xi}$, $\dot{\Lambda}$, $\dot{\tilde{\pi}}$:[77]

$$\frac{\dot{\xi}}{1+\xi} - 6\frac{\dot{\Lambda}}{\Lambda} = \frac{2}{\tau} + \frac{2}{\tau_{\rm rel}}\left(1 - \sqrt{1+\xi}\,\mathcal{R}^{3/4}(\xi)\right),$$

$$\mathcal{R}'(\xi)\dot{\xi} + 4\mathcal{R}(\xi)\frac{\dot{\Lambda}}{\Lambda} = -\left(\mathcal{R}(\xi) + \frac{1}{3}\mathcal{R}_L(\xi)\right)\frac{1}{\tau} + \frac{\tilde{\pi}}{\varepsilon_{\rm iso}(\Lambda)\tau},$$

$$\begin{aligned}\dot{\tilde{\pi}} = &-\frac{1}{\tau_{\rm rel}}[(\mathcal{R}(\xi) - \mathcal{R}_{\rm L}(\xi))P_{\rm iso}(\Lambda) + \tilde{\pi}] - \lambda(\xi)\frac{\tilde{\pi}}{\tau} \\ &+ 12\left[\frac{\dot{\Lambda}}{3\Lambda}(\mathcal{R}_{\rm L}(\xi) - \mathcal{R}(\xi)) + \left(\frac{1+\xi}{\tau} - \frac{\dot{\xi}}{2}\right) \right. \\ &\left. \times \left(\mathcal{R}^{zzzz}_{-1}(\xi) - \frac{1}{3}\mathcal{R}^{zz}_{1}(\xi)\right)\right] P_{\rm iso}(\Lambda).\end{aligned} \tag{160}$$

$\tau_{\rm rel}$ and the ratio of shear viscosity η to entropy density s, η/s, are related by $\tau_{\rm rel} = 5\eta/(sT) = 5\eta/(\mathcal{R}^{1/4}(\xi)s\Lambda)$. The numerical solution of these equations[77] can be compared with the exact solution of the Boltzmann equation,[79] and also with the other hydrodynamic approximation schemes discussed above, plus a third-order viscous hydrodynamic approximation.[88] As an example, we show in Fig. 1 the entropy production (measured by the increase in particle number $\tau n(\tau)$) between start and end of the dynamical evolution from an initial temperature of 600 MeV to a final one of 150 MeV. For this extreme (0+1)-dimensional scenario, where the difference between longitudinal and transverse expansion rates is maximal, VAHYDRO is seen to reproduce the exact solution almost perfectly, dramatically outperforming all other hydrodynamic approximations. In particular, it should be noted that only the AHYDRO and VAHYDRO approximations are able to correctly reproduce entropy production (or rather the lack thereof) in both the extreme strong coupling (ideal

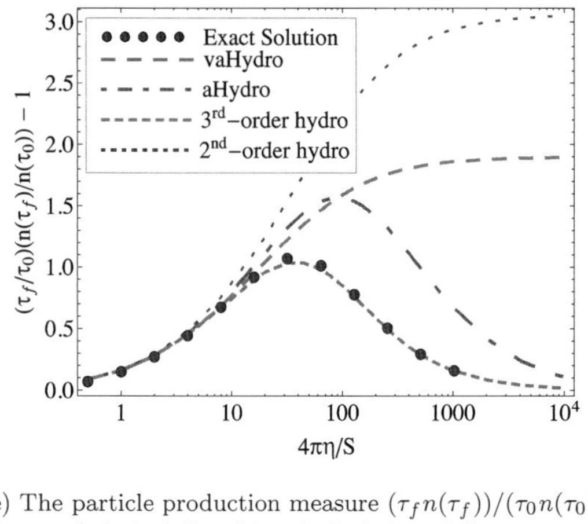

Fig. 1. (Color online) The particle production measure $(\tau_f n(\tau_f))/(\tau_0 n(\tau_0)) - 1$ as a function of $4\pi\eta/s$. The black points, red dashed line, blue dashed–dotted line, green dashed line, and purple dotted line correspond to the exact solution of the Boltzmann equation, VAHYDRO, AHYDRO, third-order viscous hydrodynamics,[88] and DNMR second-order viscous hydrodynamics,[60] respectively. The initial conditions are $T_0 = 600$ MeV, $\xi_0 = 0$, and $\tilde{\pi}_0 = 0$ at $\tau_0 = 0.25$ fm/c. The freeze-out temperature was taken to be $T_f = 150$ MeV.

fluid dynamics, $\tau_{\rm real} = \eta/s = 0$) and the extreme weak coupling (free-streaming particles, no collisions, $\tau_{\rm real}$, $\eta/s = \to \infty$) limits of the microscopic dynamics. vHYDRO schemes based on an expansion around a locally isotropic equilibrium distribution cannot reproduce the constraint that entropy production should vanish as collisions cease; these schemes break down for large η/s values.

Similar comparisons have been done for the massive (0+1)-dimensional case[81,82,89] and for the (1+1)-dimensional Gubser flow. In all cases one finds the following hierarchy of hydrodynamic approximations, when listed in order of improving accuracy in their descriptions of the moments of the exactly known microscopic dynamics: first-order viscous hydrodynamics (NS theory), second-order IS theory, second-order DNMR theory, third-order viscous hydrodynamics *à la* Jaiswal, aHYDRO, and vaHYDRO. In view of the increasing sophistication of these approximation schemes, as discussed in the preceding subsections, this ordering is not surprising, and some variant of vaHYDRO is likely to become the standard hydrodynamic modelling framework in the future. At the moment, howcvcr, only vHYDRO and aHYDRO have been implemented numerically for (2+1)-dimensional and (3+1)-dimensional expansion which do not rely on simplifying assumptions such as longitudinal boost-invariance and azimuthal symmetry. The fireballs created in heavy-ion collisions are not azimuthally symmetric, and experiments tell us that they feature characteristic anisotropic flow patterns that could never arise from an azimuthally symmetric initial condition. Longitudinal boost-invariance is not a good approximation either for particles emitted at large forward and backward rapidities, and it becomes worse when going to lower energies. Therefore, much effort is presently being expended into developing (2+1)-dimensional and (3+1)-dimensional implementations of the aHYDRO and vaHYDRO schemes.

5. Numerical Implementation of Hydrodynamics

5.1. *Need for τ and η*

The hydrodynamic simulations of the ultra-relativistic heavy ion collisions is best implemented in the hyperbolic coordinate system[90] (also known as the Milne coordinate system) where instead of t (the laboratory time) and z (the beam direction), one uses the longitudinal proper time

$$\tau = \sqrt{t^2 - z^2} \tag{161}$$

and the space-time rapidity

$$\eta = \tanh^{-1}(z/t). \tag{162}$$

Equivalently, $t = \tau \cosh \eta$ and $z = \tau \sinh \eta$.

One reason this is useful is that as shown in Fig. 2, the evolution of the ultra-relativistic heavy ion collisions can occur only in the forward light cone bounded by the light cone axes, equivalently $\tau = 0$. More physically, suppose two identical systems were created at $t = 0$ and $z = 0$ by the initial collision of the two nuclei.

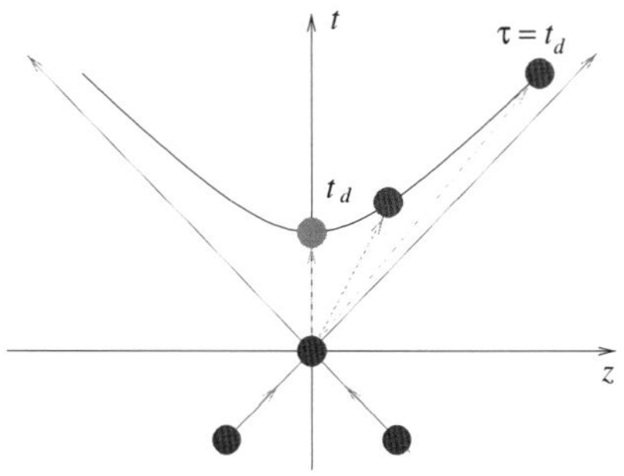

Fig. 2. (Color online) A schematic diagram showing evolution of fireballs with differing v_z after the collision of two heavy ions at $t = 0$ and $z = 0$.

Further suppose that one of the two has the collective velocity v in the z direction, but the other one is at rest. In this case, due to the time dilation, the same stage of the evolution will be reached when the lab time is at t_d for the system at rest and at $t_d/\sqrt{1-v^2}$ for the moving system. In relativistic systems, these two Minkowski times can be very much different even though the two system are at the same stage in their respective evolution. However, in terms of the proper time both are at the same $\tau = t_d$ since τ is nothing but the local rest frame time. Hence, it is very natural that we should use the $\tau - \eta$ coordinate system when there is a strong longitudinal flow of matter. In the ultra-relativistic heavy ion collisions, this is the case due to the original beam momenta of the projectile and the target.

For numerical implementation of hydrodynamics, one first needs to formulate the conservation laws in this coordinate system. For this, we need to know the transformation law between $\tau - \eta$ and $t - z$. We can start with the derivatives

$$\begin{aligned}\partial_\tau &= \frac{\partial t}{\partial \tau}\partial_t + \frac{\partial z}{\partial \tau}\partial_z \\ &= \cosh\eta\partial_t + \sinh\eta\partial_z \end{aligned} \tag{163}$$

and

$$\begin{aligned}\partial_\eta &= \frac{\partial t}{\partial \eta}\partial_t + \frac{\partial z}{\partial \eta}\partial_z \\ &= \tau\sinh\eta\partial_t + \tau\cosh\eta\partial_z \end{aligned} \tag{164}$$

which can be summarized as Lorentz transformations

$$\tilde{\partial}_a = \Lambda_a^\mu \partial_\mu \quad \text{and} \quad \partial_\mu = \Lambda_\mu^a \tilde{\partial}_a \tag{165}$$

where $\tilde{\partial}_a = (\partial_\tau, \nabla_\perp, (1/\tau)\partial_\eta)$ and

$$\Lambda_\mu^a = \begin{pmatrix} \cosh\eta & 0 & 0 & -\sinh\eta \\ 0 & 1 & 0 & 0 \\ 0 & 0 & 1 & 0 \\ -\sinh\eta & 0 & 0 & \cosh\eta \end{pmatrix}. \tag{166}$$

The inverse transform Λ_a^μ is obtained by substituting η with $-\eta$. From now on, we will use the first letters of the roman alphabet $(a, b, \ldots)$ to represent the components in the Milne space as above.

To figure out the transformation of the conservation law, we apply Eq. (165) to

$$\begin{aligned}\partial_t j^0 + \partial_z j^3 &= \cosh\eta\partial_\tau j^0 - \frac{\sinh\eta}{\tau}\partial_\eta j^0 + \frac{\cosh\eta}{\tau}\partial_\eta j^3 - \sinh\eta\partial_\tau j^3 \\ &= \frac{1}{\tau}\partial_\tau(\tau\cosh\eta j^0 - \tau\sinh\eta j^3) \\ &\quad - \partial_\eta\left(\frac{1}{\tau}\sinh\eta j^0 - \frac{1}{\tau}\cosh\eta j^3\right).\end{aligned} \tag{167}$$

Defining

$$\begin{aligned}q^\tau &= \cosh\eta\, q^0 - \sinh\eta\, q^3, \\ \tilde{q}^\eta &= \cosh\eta\, q^3 - \sinh\eta\, q^0\end{aligned} \tag{168}$$

and

$$\tilde{q}^a = (q^\tau, \mathbf{q}_\perp, \tilde{q}^\eta), \tag{169}$$

for any 4-vector q^μ, the conservation law becomes

$$0 = \partial_\tau(\tau j^\tau) + \nabla_\perp \cdot (\tau \mathbf{j}_\perp) + \partial_\eta \tilde{j}^\eta = \tilde{\partial}_a(\tau \tilde{j}^a) \tag{170}$$

where $\mathbf{j}_\perp$ only has the x, y components. One can also show

$$D_\tau = u^\mu \partial_\mu = \tilde{u}^a \tilde{\partial}_a. \tag{171}$$

Note that our definition of the η component is different from the curvilinear definition of the η component by a factor of τ. We do this to keep the dimension of the η component the same as the other components. If one uses the curvilinear definition

$$j^\eta = \frac{1}{\tau}\tilde{j}^\eta \tag{172}$$

then the conservation law becomes

$$0 = \partial_\tau(\tau j^\tau) + \nabla_\perp \cdot (\tau \mathbf{j}_\perp) + \partial_\eta(\tau j^\eta) = \partial_a(\tau j^a) \tag{173}$$

where $\partial_a = (\partial_\tau, \nabla_\perp, \partial_\eta)$.

For the energy–momentum conservation, one needs to apply the transformation law 3 times for the each index in $\partial_\mu T^{\mu\nu} = 0$. The algebra is a bit tedious but straightforward.[91,92] The result is

$$\partial_\tau T^{\tau\tau} + \frac{1}{\tau}\partial_\eta T^{\eta\tau} + \partial_v T^{v\tau} = -\frac{1}{\tau}T^{\eta\eta} \tag{174}$$

where we defined

$$\mathcal{T}^{ab} = \tau \Lambda^a_\mu \Lambda^b_\nu T^{\mu\nu}. \tag{175}$$

and the index $v = x, y$. For the $\tau\eta$ component,

$$\partial_\tau \mathcal{T}^{\tau\eta} + \frac{1}{\tau}\partial_\eta \mathcal{T}^{\eta\eta} + \partial_v \mathcal{T}^{v\eta} = -\frac{1}{\tau}\mathcal{T}^{\eta\tau}. \tag{176}$$

Both $\mathcal{T}^{\tau\tau}$ and $\mathcal{T}^{\tau\eta}$ conservation laws contain the geometrical source term. Transverse momentum conservation is simpler:

$$\partial_\tau \mathcal{T}^{\tau v} + \frac{1}{\tau}\partial_\eta \mathcal{T}^{\eta v} + \partial_w \mathcal{T}^{wv} = 0, \tag{177}$$

where v and w are transverse coordinate indices. When testing one's code for the energy–momentum conservation, the following form may be more convenient:

$$\frac{d}{d\tau}\int dx dy d\eta\, \tilde{T}^{\tau\mu} = 0. \tag{178}$$

The half-transformed $\tilde{T}^{a\mu} = \tau\Lambda^a_\nu T^{\nu\mu}$ satisfies the conservation law $\partial_\tau \tilde{T}^{\tau\mu} + \tilde{\partial}_\eta \tilde{T}^{\eta\mu} + \partial_v \tilde{T}^{v\mu} = 0$ without the geometrical source terms exactly like Eq. (170).

For the shear and the bulk evolution equations in Eq. (145) or (147), transformation from the Minkowski space to the $\tau - \eta$ coordinate space is straightforward, but it is a lot more involved algebraically. For more details see Refs. 9, 67 and 93.

5.2. *Numerical solution of conservation equations*

In this section, we will first discuss conservation laws in the Minkowski coordinates where the conservation laws are

$$\partial_\mu T^{\mu\nu} = 0, \quad \partial_\mu J^\mu_B = 0. \tag{179}$$

In the Milne coordinate system, the energy–momentum conservation takes a slightly different form

$$\tilde{\partial}_a \mathcal{T}^{ab} = \mathcal{S}^b, \tag{180}$$

where $\mathcal{S}^b$ is the geometric source term in Eqs. (174) and (176). However, as will be shown shortly, the methods illustrated below can be easily adapted to this case.

For vHydro, the energy–momentum tensor is given in a general reference frame by the decomposition

$$T^{\mu\nu} = T^{\mu\nu}_{\rm id} + \pi^{\mu\nu} - \Pi\Delta^{\mu\nu} \tag{181}$$

where

$$T^{\mu\nu}_{\rm id} = \varepsilon u^\mu u^\nu - P(\varepsilon, \rho_B)\Delta^{\mu\nu} \tag{182}$$

is the ideal fluid part of the tensor. The net baryon current has the form $J^\mu_B = \rho_B u^\mu + V^\mu_B$. The equations that need to be solved are given in Eq. (179), together with the relaxation equations for the dissipative flows, for example the IS equations (147). The first step in solving the hydrodynamic equations is their initialization. Let us assume that some microscopic pre-equilibrium dynamical theory provides us with a baryon current $J_B^\mu(x)$ and energy–momentum tensor $T^{\mu\nu}(x)$ for points x^μ on some Cauchy surface on which we want to initialize the hydrodynamic evolution stage. The following projection steps, to be taken at each point x on that surface, yield the required initial value fields[i]:

First we define the local fluid rest frame by solving the eigenvalue equation $T^\mu_\nu u^\nu = \varepsilon u^\mu$ for its normalized timelike eigenvector u^μ. The associated eigenvalue gives us the LRF energy density ε. The LRF baryon density is obtained from J_B^μ by projecting onto u_μ: $\rho_B = u_\mu J_B^\mu$. The initial heat flow V_B^μ is the component of J_B^μ perpendicular to u^μ: $V^\mu = \Delta^{\mu\nu} J_{B,\nu}$. Now that we know the LRF energy and baryon densities we can compute the LRF pressure P from the equation of state of the fluid, $P(\varepsilon, \rho_B)$. Next, the bulk viscous pressure is obtained from $\Pi = -\frac{1}{3}\Delta_{\mu\nu}T^{\mu\nu} - P$. Finally, the shear stress is obtained as $\pi^{\mu\nu} = T^{\mu\nu} - \varepsilon u^\mu u^\nu + (P+\Pi)\Delta^{\mu\nu}$ or, equivalently, as $\pi^{\mu\nu} = T^{\langle\mu\nu\rangle} \equiv \Delta^{\mu\nu}_{\alpha\beta}T^{\alpha\beta}$.

For a simple illustration of numerical methods that can solve the hydrodynamic equations, let us first consider a single conservation law in one-dimensional. There is no difficult conceptual obstacle in extending this case to the multi-component, multi-dimension cases such as the IS equations. The conservation equation is

$$\partial_t u = -\partial_x j. \tag{183}$$

We need to supplement this equation with a relationship between the density u and the current j. Simplest example is $j = vu$ with a constant speed of propagation v. But in general j is a non-linear function(al) of u. For instance, the ideal part, Eq. (182), is certainly not in this simple form due to the normalization condition $u_0^2 = \sqrt{1-\mathbf{u}^2}$ and also to the presence of the pressure term. In the dissipative cases, the relaxation equation

$$(\partial_t + 1/\tau_R)j = -(D/\tau_R)\partial_x u \tag{184}$$

determines the relationship between j and u. In such cases, the numerical methods discussed in this section needs to be applied in two steps. In the first step, the conservation laws are used to advance the time component of the currents using the methods that will be discussed here. In the second step, the spatial part of the currents needs to be reconstructed from the time components. The relaxation equations also need to be solved separately, although the techniques discussed here can be easily adapted to handle the relaxation equation as well.

For the simple $j = vu$ case with a constant v, the equation becomes

$$\partial_t u = -v\partial_x u. \tag{185}$$

[i]The following paragraph refers to the vHydro decomposition (10, 181). A slightly modified projection method applies for the aHydro and vaHydro decompositions (150), (151) and (157).[77]

This is an advection equation and has a simple solution

$$u(t, x) = f(x - vt). \tag{186}$$

That is, at any given time, the solution is just the translation of the initial profile by vt. Analytically, this is trivial. However, it is remarkable how difficult it can be to maintain the initial profile in numerical solutions. We will often use this as the simplest test case for our algorithms.

To solve the conservation equation numerically, one first needs to discretize time and space. We define

$$u_i^n = u(t_n, x_i) \tag{187}$$

where $t_n = t_0 + nh$ and $x_i = x_0 + ia$ for any function $u(t, x)$ of time and space. Here h is the time step size and a is the spatial cell size. Physically, it is important to have the following properties in the discretization method. First, the method should conserve total u explicitly, that is, $\sum_i u_i^n = \sum_i u_i^{n+1}$ modulo the boundary terms. For this, one requires that the discretized form of the divergence to take the form

$$(\partial_x j)_i^n \to \frac{j_i(u^n) - j_{i-1}(u^n)}{a} \tag{188}$$

where $j_i(u^n)$ is the discretized representation of the current j at x_i ant t_n. One can easily see that in the sum $\sum_{i=0}^{N} (\partial_x j)_i^n$ only the boundary terms would survive. The details of the boundary terms depend on the method. However, as long as the boundaries are far away from the physical region, the boundary terms should be vanishingly small. If the boundary of the space is not too far away from the physical region, then some suitable discrete boundary conditions should be imposed.

The second requirement is simple, yet quite demanding: If $u_i^0 \geq 0$ for all i, then we would like this property to be maintained for any future time. For instance, if u represents the energy density, then it should never become negative.

To illustrate some of these issues, consider again the advection equation $\partial_t u = -v\partial_x u$ with $v > 0$. Let the initial condition be a rectangle: $u_i^0 = u_c$ for $b \leq i \leq f$ and $u_i^0 = 0$ otherwise. This is a prototype of many situations where two smoothly varying regions are joined by a stiff gradient. The simplest discretization method of $\partial_t u = -v\partial_x u$ is the forward-time centered-space (FTCS) method

$$u_i^{n+1} = u_i^n - \frac{vh}{2a}(u_{i+1}^n - u_{i-1}^n) \tag{189}$$

which is correct up to $O(h^2)$ and $O(a^2)$ errors. Since the second term in the right-hand side is supposed to be a correction, we require $|vh/a| < 1$. This is certainly in the form of Eq. (188) and hence conserves the total u.

According to the analytic solution Eq. (186), the space behind the back edge (x_b) of the rectangle should always have $u^n = 0$. However, according to Eq. (189)

the cell right behind the back edge becomes non-zero and negative after the first time step

$$u_{b-1}^1 = -\frac{vh}{2a} u_c. \tag{190}$$

At the same time, the front edge starts to get distorted by the same amount

$$u_f^1 = u_c - u_{b_1}^1 > u_c. \tag{191}$$

At the next time step, u_{b-1}^2 becomes even more negative

$$u_{b-1}^2 = -\frac{vh}{a} u_c + \left(\frac{vh}{2a}\right)^2 u_c \tag{192}$$

while u_f deviates even more from u_c

$$u_f^2 = u_c - u_{b-1}^2 \tag{193}$$

Furthermore at t_2, the next cell in the empty region becomes non-zero $u_{b-2}^2 = (\frac{vh}{2a})^2 u_c$. Clearly, both the profile preservation and the positivity of the solution are grossly violated by this method even though the total u is conserved. In addition, one can easily see that the growth at x_{b-1} will continue, indicating that this method is unconditionally unstable.

In order to cure the negativity problem, one may note that the trouble above mainly comes from the centered nature of the $O(a^2)$ numerical derivative in Eq. (189). Instead, one may try the first-order approximation[j]

$$u_i^{n+1} = u_i^n - \frac{vh}{a}(u_i^n - u_{i-1}^n) \tag{194}$$

In this "up-wind scheme", u_{b-1}^1 trivially vanishes at t_1. In fact, all u_{b-i} for $i \geq 1$ will remain zero for all times. Similarly, one can show that the numerical solution is positive and bounded everywhere as long as $|vh/a| < 1$. If $v < 0$, mirror-image conclusions can be reached if one uses $(\partial_x u)_i \approx (u_{i+1} - u_i)/a$.

Positivity is thus maintained in this up-wind scheme. However, as the system evolves in time, the shape of the solution gets more and more distorted. This is because the first-order difference

$$(u_i^n - u_{i-1}^n)/a = \partial_x u(t_n, x_i) - \frac{a}{2}\partial_x^2 u(t_n, x_i) + O(a^2) \tag{195}$$

is too crude an approximation of the first-order derivative. The second derivative term in Eq. (195) in fact introduces too much artificial (numerical) damping to preserve the shape for long. In effect, the profile at t is the convolution of the initial profile and the Gaussian Green function of the diffusion equation (Eq. (92)) with the diffusion constant given by $D = \frac{a}{2h}$. As one can see in Eq. (92), the width of the Gaussian grows linearly with time. Hence, the initial profile will be smeared out quickly.

[j]This is in fact trivially exact when $a = vh$ since in that case $u_i^{n+1} = u_{i-1}^n$. That is, the whole profile moves by one spatial cell at each time step. However, this would not work for more general currents.

Better discretization methods must keep the positivity preserving nature of the up-wind method and at the same time must have a better approximation for the spatial derivative than the simple first-order difference for shape preservation. To devise better discretization methods more systematically, consider first dividing the space into $N+1$ cells of size a labeled by integers 0 through N. The ith cell starts at $x_{i-1/2} = x_i - a/2$ and ends at $x_{i+1/2} = x_i + a/2$. (See Fig. 3.) Averaging over one spatial cell and integrating over one time step, the conservation law $\partial_t u = -\partial_x j$ becomes

$$\bar{u}_i^{n+1} = \bar{u}_i^n - \frac{1}{a}\int_{t_n}^{t_{n+1}} dt(j(t, x_{i+1/2}) - j(t, x_{i-1/2})) \tag{196}$$

where $\bar{u}_i^n = \bar{u}_i(t_n)$ is the cell average. This is an exact expression which can be used as the basis for further approximation. This type of methods are known as the finite volume methods.

Approximating the time integral using the midpoint rule, we get

$$\bar{u}_i^{n+1} = \bar{u}_i^n - \frac{h}{a}(j(t_{n+1/2}, x_{i+1/2}) - j(t_{n+1/2}, x_{i-1/2})) + O(h^3) \tag{197}$$

where $t_{n+1/2} = t_n + h/2$. There are few things that should be mentioned here. First, the basic quantities to calculate are the cell-averaged values $\bar{u}_i^n$. Second, we need to approximate the function $u(t, x)$ itself from $\bar{u}_i^n$ because the right hand side contains $u(t, x)$ evaluated at $t_{n+1/2}$ and $x_{i\pm 1/2}$. An approximate form of $u(t_n, x)$ is also needed for obtaining $\bar{u}_i^{n+1}$ for the next time step (see below). Therefore, how we approximate $u(t, x)$ and evaluate $j_{i\pm 1/2}^{n+1}$ determine different numerical schemes and the accuracy of the given scheme.

In many schemes, the values at spatial half points are not unique because the interpolating function approximating $u(t, x)$ is usually only piece-wise continuous. For instance, see Figs. 3 and 4. One way to deal with the ambiguity in evaluating

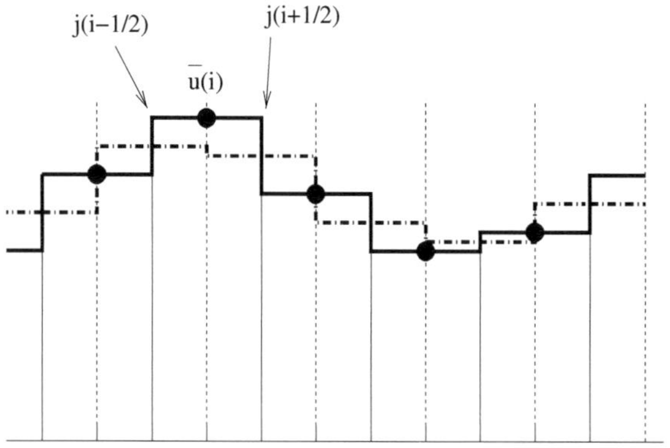

Fig. 3. Spatial grid used in deriving finite volume methods. The solid line is the lowest order histogram function approximation of $u(t, x)$ in the x_i-centered grid. The dot–dash line is the projection of the x_i-centered grid onto the $x_{i+1/2}$-centered grid.

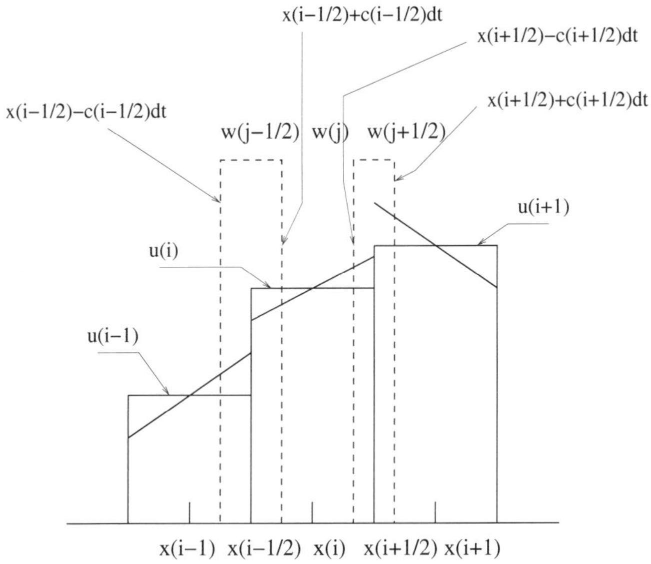

Fig. 4. A schematic view of the cell division used in the Kurganov–Tadmor scheme.

$j(t_{n+1/2}, x_{i\pm 1/2})$ is to just avoid evaluating it at the boundaries. The staggered nature of the space and time indices in Eq. (197) suggests an obvious way to do so. Suppose that the initial data is given for $\bar{u}_i^n$ where the ith grid is centered at x_i and has the spatial interval $[x_{i-1/2}, x_{i+1/2}]$. For the next time step, instead of updating the values of $\bar{u}_i$ within the intervals $[x_{i-1/2}, x_{i+1/2}]$, we update the values of $\bar{u}_{i+1/2}$ within the shifted intervals $[x_i, x_{i+1}]$. Using the lowest order approximation for $u(t, x)$ (the histogram functions in Fig. 3), this yields

$$\bar{u}_{i+1/2}^{n+1} = \frac{\bar{u}_{i+1}^n + \bar{u}_i^n}{2} - \frac{h}{a}(j(u_{i+1}^{n+1/2}) - j(u_i^{n+1/2})) \tag{198}$$

where $(\bar{u}_i^n + \bar{u}_{i+1}^n)/2$ is the average value of $u(t_n, x)$ in the shifted interval $[x_i, x_{i+1}]$ using the histogram function in Fig. 3. For the values of $u_{i+1}^{n+1/2}$ at $t_{n+1/2}$, the forward Euler method

$$u_i^{n+1/2} = \bar{u}_i^n - \frac{h}{2}(\partial_x j)_i^n + O(h^2) \tag{199}$$

is enough since the overall error in the midpoint rule is $O(h^3)$. In this way, we avoid evaluating the current at the cell boundaries. For the next time step, we shift back to the original grid. This staggered method is a slightly generalized form of the Lax–Friedrichs scheme which is second-order accurate in space and third order accurate in time. The original Lax–Friedrichs scheme is only second-order in time because the currents are evaluated at t_n instead of at $t_{n+1/2}$.

Equation (199) involves evaluating numerical derivative of the current j at x_i. This cannot be exactly determined when all one has is data on the average values $\bar{u}_i^n$. For instance, $(j_x)_i^n$ could be the first-order approximations $(j_{i+1}^n - j_i^n)/a$ or

$(j_i^n - j_{i-1}^n)/a$, or the second-order approximation $(j_{i+1}^n - j_{i-1}^n)/2a$, or any other approximate form. In normal situations, choosing a higher order formula should be better than the first-order ones. But this is not always the case when the gradient is stiff.

We have already shown that using the central difference $(j_{i+1}^n - j_{i-1}^n)/2a$ in the forward Euler method (the FTCS method in Eq. (189)) can be disastrous when the gradient is stiff, although it can be safely (and preferably) used in smooth regions. When gradient is stiff, one should instead use the up-wind method Eq. (194) to maintain the positivity. Therefore, an intelligent scheme would choose the derivative according to some approximate measure of the true gradient. One way to do this is to choose the gradient according to the following scheme:

$$(\partial_x u)_i^n = \begin{cases} 0 \text{ if } \bar{u}_i^n < \bar{u}_{i\pm1}^n \text{ or } \bar{u}_i^n > \bar{u}_{i\pm1}^n \\ \text{else } \operatorname{sign}(\bar{u}_{i+1}^n - \bar{u}_i^n) \min\left(\theta \dfrac{|\bar{u}_{i+1}^n - \bar{u}_i^n|}{a}, \dfrac{|\bar{u}_{i+1}^n - \bar{u}_{i-1}^n|}{2}, \theta \dfrac{|\bar{u}_i^n - \bar{u}_{i-1}^n|}{a}\right). \end{cases} \tag{200}$$

The first line indicates that the function has either a maximum or a minimum within the interval. Therefore the slope at x_i can be best approximated by 0. The second line applies when the function is changing monotonically near x_i. The parameter $1 \le \theta < 2$ is there to be slightly more general. This choice of the derivative is called the "generalized minmod flux limiter".

The Lax–Friedrichs scheme represented by Eq. (198) is only $O(a^2)$ accurate because we have used the histogram function as an approximation for $u(t_n, x)$. Needless to say, this is the lowest order approximation. As a result, the Lax–Friedrichs scheme contains too much numerical diffusion to be practically useful. Both of these facts can be easily seen from the Taylor expansion

$$\frac{\bar{u}_i^n + \bar{u}_{i+1}^n}{2} = \bar{u}_{i+1/2}^n + \frac{a^2}{8}\partial_x^2 u_{i+1/2}^n + O(a^4) \tag{201}$$

where the second derivative term is the $O(a^2)$ error term that also causes strong diffusion. In time, this diffusion distorts the solution too much just as in the first-order up-wind method.

To obtain a better approximation, one needs to evaluate $u(t, x)$ more accurately using the average values from the nearest neighbor cells. If one uses the linear interpolant (see Fig. 4)

$$\hat{u}_i^n(x) = \bar{u}_i^n + (\partial_x u)_i^n (x - x_i) \tag{202}$$

for each interval $[x_{i-1/2}, x_{i+1/2}]$, then the scheme should become at least $O(a^3)$. Using this to improve the estimate of $\bar{u}_{i+1/2}^n$ adds a correction term to the lowest-order result Eq. (198)

$$\begin{aligned} \bar{u}_{i+1/2}^{n+1} = \frac{\bar{u}_i^n + \bar{u}_{i+1}^n}{2} &- \frac{h}{a}(j(t_{n+1/2}, x_{i+1}) - j(t_{n+1/2}, x_i)) \\ &- \frac{a}{8}((\partial_x u)_{i+1}^n - (\partial_x u)_i^n). \end{aligned} \tag{203}$$

This staggered approach is the second-order NT (Nessyahu–Tadmor) scheme and works reasonably well.[94] The last term on the right-hand side of Eq. (203) cancels the second derivative term in Eq. (201) so that together they represent $\bar{u}^n_{i+1/2}$ with the $O(a^4)$ error. In other words, the last term in Eq. (203) is the anti-diffusion term that is correcting the large second-order diffusion introduced by the symmetric combination $(\bar{u}^n_i + \bar{u}^n_{i+1})/2$. Therefore, this scheme is $O(a^4)$ in the smooth region.

Why don't we then just replace these terms with $\bar{u}^n_{i+1/2}$? If one does that, it just becomes the FTCS scheme in Eq. (189). The difference, the $O(a^4\partial_x^4 u(x))$ term, provides just enough numerical diffusion so that spurious oscillations do not propagate from the stiff gradient.

The second-order NT algorithm represented by Eq. (203) is related to another often-used finite element method called the SHASTA algorithm. We start with Eq. (203) but replace

$$\frac{\bar{u}^n_i + \bar{u}^b_{i+1}}{2} \approx \bar{u}^n_{i+1/2} + \frac{1}{8}(u_{i+1} - 2u_{i+1/2} + u_i) + O(a^4) \tag{204}$$

so that

$$\begin{aligned}\bar{u}^{n+1}_{i+1/2} = \bar{u}^n_{i+1/2} &- \frac{h}{a}(j(t_{n+1/2}, x_{i+1}) - j(t_{n+1/2}, x_i))\\ &+ \frac{1}{8}(\bar{u}^n_{i+1} - 2\bar{u}^n_{i+1/2} + \bar{u}^n_i) - \frac{a}{8}((\partial_x u)^n_{i+1} - (\partial_x u)^n_i).\end{aligned} \tag{205}$$

At this point, this is no longer a finite volume method even though we will keep using the notation $\bar{u}^n_i$ to represent the value of u at t_n and x_i. It is also not a staggered method any more as the right-hand side contains $\bar{u}^n_{i+1/2}$. Re-labeling the spaces indices $i + 1/2 \to i$ and $i \to i - 1$ so that the new grid size is $a' = a/2$, we get

$$\begin{aligned}\bar{u}^{n+1}_i = \bar{u}^n_i &- \frac{h}{2a}(j(t_{n+1/2}, x_{i+1}) - j(t_{n+1/2}, x_{i-1}))\\ &+ \frac{1}{8}(\bar{u}^n_{i+1} - 2\bar{u}^n_i + \bar{u}^n_{i-1}) - \frac{a}{16}((\partial_x u)^n_{i+1} - (\partial_x u)^n_{i-1})\end{aligned} \tag{206}$$

with the appropriate scaling of the derivatives in the last term and after renaming $a' \to a$. With $u(t_{n+1/2}, x_{i\pm1})$ given by Eqs. (199) and (206) represents the basic SHASTA algorithm.

In practice, Eq. (206) is broken up into two stages to ensure positivity. Specifying $j = vu$, the first transport step is[k]

$$\begin{aligned}w^{n+1}_i &= \bar{u}^n_i + \frac{1}{8}(\bar{u}^n_{i+1} - 2\bar{u}^n_i + \bar{u}^n_{i-1})\\ &\quad - \frac{vh}{2a}\left(\bar{u}^n_{i+1} - \frac{vh}{2}(\partial_x u)^n_{i+1} - \bar{u}^n_{i-1} + \frac{vh}{2}(\partial_x u)^n_{i-1}\right)\end{aligned} \tag{207}$$

[k]In the original SHASTA algorithm, $(\bar{u}^n_{i+1} - 2\bar{u}^n_i + \bar{u}^n_{i-1})$ is used in place of $((\partial_x u)^n_{i+1} - (\partial_x u)^n_{i-1})/2$ in this stage. In practice, as long as $|vh/a|$ is small, this difference does not matter much. But it is crucial that a flux limiter is used in the second anti-diffusion step.

again using Eq. (199). If $\bar{u}_l^n \geq 0$ for all l, then as long as $|vh/a| < 1/4$, w_i^{n+1} is positive.

The second stage is the anti-diffusion step

$$\bar{u}_i^{n+1} = w_i^{n+1} - \frac{a^2}{8}(\partial_x^2 w)_i^{n+1} \tag{208}$$

where $(\partial_x^2 w)_i^{n+1}$ represents numerical estimate of the second derivative at x_i that preserves the positivity. The numerical approximation suggested by the last term in Eq. (206) turned out not to preserve the positivity. The original formulation of the SHASTA algorithm by Boris and Book uses a conservative form

$$(\partial_x^2 w)_i^{n+1} = \frac{1}{a}((w_x)_{i+1}^{n+1} - (w_x)_i^{n+1}) \tag{209}$$

where

$$(w_x)_{i+1}^{n+1} = \begin{cases} 0 \text{ if } \Delta_i^{n+1}, \Delta_{i+1}^{n+1} \text{ and } \Delta_{i+2}^{n+1} \text{ do not all have the same sign.} \\ \text{else } \operatorname{sign}(\Delta_{i+1}^{n+1}) \min(8|\Delta_i^{n+1}|, |\Delta_{i+1}^{n+1}|, 8|\Delta_{i+2}^{n+1}|) \end{cases} \tag{210}$$

with

$$\Delta_{i+1}^{n+1} = (w_{i+1}^{n+1} - w_i^{n+1})/a. \tag{211}$$

This is similar to the minmod flux limiter and maintains the positivity. The SHASTA algorithm is used in many hydrodynamics simulation programs for ultra-relativistic heavy ion collisions[95] including pioneering works in Refs. 96, 91 and also later works in Refs. 67, 97–101.

The second-order NT scheme and the SHASTA scheme in practice work fairly well.[102] However, in these schemes the $h \to 0$ limit cannot be taken since the numerical viscosity behaves like $\sim 1/h$. It would be convenient to be able to take this limit because one can then formulate the discretized problem as a system of coupled ordinary differential equations in time. Many techniques for the ordinary differential equations such as the Runge–Kutta methods then become available to control the accuracy of the time evolution. So far different time integration techniques were not available other than the midpoint rule in Eq. (197).

One way to achieve this is to subdivide the cells as shown in Fig. 4 with the piecewise linear approximation for $u(t,x)$. The size of the cell containing the discontinuity at the half integer point $x_{i+1/2}$ is controlled by the local propagation speed $c_{i+1/2}$. That is, the subcell surrounding $x_{i+1/2}$ is between the left boundary $x_{i+1/2}^l = x_{i+1/2} - c_{i+1/2}h$ and the right boundary $x_{i+1/2}^r = x_{i+1/2} + c_{i+1/2}h$. Then the cells containing the boundaries and the cells not containing the boundaries (between $x_{i-1/2}^r$ to $x_{i+1/2}^l$) are independently evolved. For the subcell containing $x_{i+1/2}$

$$w_{i+1/2}^{n+1} = \frac{\bar{u}_{i+1/2}^l + \bar{u}_{i+1/2}^r}{2} - \frac{1}{2c_{i+1/2}}(j(\bar{u}_{i+1/2}^r) - j(\bar{u}_{i+1/2}^l)) + O(h) \tag{212}$$

which is basically Eq. (198). Here the superscripts r and l means the value of the approximate $u(t_n, x)$ at the boundary points $x_{i+1/2}^r$ and $x_{i+1/2}^l$, respectively. Within the smooth region between $x^r_{i-1/2}$ and $x^l_{i+1/2}$, we get using Eq. (197)

$$w_i^{n+1} = \frac{\bar{u}^l_{i+1/2} + \bar{u}^r_{i-1/2}}{2} - \frac{h}{a}(j(\bar{u}^l_{i+1/2}) - j(\bar{u}^r_{i-1/2})) + O(h^2) \tag{213}$$

where the first term on the right-hand side is obtained by applying the trapezoid rule. The divided cells are then projected onto the original grid using the size of the cells as the weight to get

$$\begin{aligned}\bar{u}_i^{n+1} &= \frac{c_{i-1/2}h}{a} w^{n+1}_{i-1/2} + \frac{c_{i+1/2}h}{a} w^{n+1}_{i+1/2} \\ &\quad + \left(1 - \frac{(c_{i-1/2} + c_{i+1/2})h}{a}\right) w_i^{n+1}.\end{aligned} \tag{214}$$

When the $h \to 0$ limit is taken, this procedure yields

$$\frac{d\bar{u}_i}{dt} = -\frac{H_{i+1/2} - H_{i-1/2}}{a} \tag{215}$$

where

$$H_{i+1/2} = \frac{j(\bar{u}^+_{i+1/2}) + j(\bar{u}^-_{i+1/2})}{2} - c_{i+1/2}\frac{\bar{u}^+_{i+1/2} - \bar{u}^-_{i+1/2}}{2} \tag{216}$$

where now $\bar{u}^+_{i+1/2}$ and $\bar{u}^-_{i+1/2}$ are the values of the piece-wise linear $u(t,x)$ when approaching $x_{i+1/2}$ from the right and from the left, respectively. They are given by

$$\bar{u}^+_{i+1/2} = \bar{u}_{i+1} - (a/2)(\partial_x u)_{i+1}, \tag{217}$$

$$\bar{u}^-_{i+1/2} = \bar{u}_i + (a/2)(\partial_x u)_i \tag{218}$$

again using the minmod flux limiter for the derivatives. This is known as the second-order Kurganov–Tadmor (KT) scheme[103] and it is implemented in the (3+1)-dimensional event-by-event viscous hydrodynamics simulation program Music.[9,92] The numerical viscosity in the smooth regions is known to be $O(a^3\partial_x^4 u)$.

The structure of the KT algorithm, Eqs. (215) and (216), is the same for the lowest-order, second-order and the third-order algorithms. All one needs to do to improve accuracy is to get a better estimate of $\bar{u}^{\pm}_{i+1/2}$. Actually, it is instructive to consider the lowest-order result which uses the histogram function as the approximation of $u(x)$. One then has $\bar{u}^+_{i+1/2} = \bar{u}_{i+1}$ and $\bar{u}^-_{i+1/2} = \bar{u}_i$. It is easy to see in this case that if $j = vu$ and hence $c_{i+1/2} = |v|$, Eq. (215) automatically becomes the up-wind method.

We now have a set of ODE's. What is a good choice of the ODE solver? One of the physical requirement is again the positivity. For instance, suppose u represents particle density. One knows that u can never be negative. One of the schemes that preserves the positivity is the Heun's method which is one of the second-order Runge–Kutta schemes. The equation $du/dt = f$ is numerically solved by following these steps:

$$\bar{u}_j^* = \bar{u}_j^n + hf(t_n, \bar{u}^n), \tag{219}$$

$$\bar{u}_j^{**} = \bar{u}_j^* + hf(t_{n+1}, \bar{u}^*), \tag{220}$$

$$\bar{u}_j^{n+1} = \frac{1}{2}(u_j^n + u_j^{**}). \tag{221}$$

Positivity is maintained at each stage with a suitable choice of the flux limiter such as the minmod flux limiter.

For 3-dimensional (and similarly for 2-dimensional), a simple extension

$$\frac{d}{dt}\bar{u}_{ijk} = -\frac{H^x_{i+1/2,j,k} - H^x_{i-1/2,j,k}}{a_x} - \frac{H^y_{i,j+1/2,k} - H^x_{i,j-1/2,k}}{a_y} - \frac{H^z_{i,j,k+1/2} - H^z_{i,j,k-1/2}}{a_z} \tag{222}$$

works well. In curvilinear coordinate systems such as the Milne coordinates, one may not have the conservation law in the form $\partial_\mu J^\mu = 0$. Instead, it may take the form

$$\partial_a J^a = S \tag{223}$$

where S is the geometrical source term that does not involve derivatives. The KT algorithm can be easily adapted to this case by simply adding the source term on the right-hand side. Namely,

$$\frac{d\bar{u}_i}{dt} = -\frac{H_{i+1/2} - H_{i-1/2}}{a} + S(\bar{u}_i). \tag{224}$$

In implementing the KT scheme, one needs the maximum speed of propagation at $x_{i\pm1/2}$. If there is only one variable u and one current j, then the speed of propagation in the ith direction is

$$c_i = \left|\frac{\partial j_i}{\partial u}\right|. \tag{225}$$

If there are more than one current, then we first define the Jacobian matrix

$$J_i^{ab} = \frac{\partial j_i^a}{\partial u_b} \tag{226}$$

where $i = 1, 2, 3$ is the space index and $a = 1, \ldots, M$ labels the conserved quantities. Therefore we have 3 $M \times M$ matrices. The maximum propagation speed in the ith direction is

$$c_i = \max(|\lambda_1|, \ldots, |\lambda_M|) \tag{227}$$

where λ's are the eigenvalues of the Jacobian J_i^{ab}. When discretizing in time, the original authors of the KT algorithm recommended the time step h to be small enough so that $|\max(c_i)h/a| < 1/8$.

For the explicit form of the c_i for the (3+1)-dimensional ideal hydrodynamics including the net baryon currently used in MUSIC, see Ref. 92. For implementation of the event-by-event (3+1)-dimensional viscous hydrodynamics in MUSIC, see Refs. 9 and 58. Additional valuable information on the algorithms used in solving the VHYDRO equations can be found in Ref. 104. For some comparisons of various schemes discussed in the section, see Ref. 102.

There are many other numerical schemes that are currently in use but we unfortunately did not have space to discuss. These include, but not limited to, a PPM (Piece-wise Parabolic Method) scheme,[105] a Lagrangian scheme where the grid points follow the movement of fluid cells,[106] Riemann solvers,[102,107] and a SPH (smoothed particle hydrodynamics) method.[108] The simple FTCS scheme has also been employed for the viscous hydrodynamics[109,110] for smooth initial conditions.

5.3. *Freeze-out hypersurface and Cooper–Frye formula for particle production*

When a hydrodynamic system is expanding, there comes a time when the system is too dilute to be treated with hydrodynamics (c.f. Sec. 4.1). From this point on, the system is basically a collection of non-interacting particles. In realistic systems, this time is not the same for all fluid cells. As the system starts to expand at τ_0, there are cells at the periphery of the system that are dilute enough to "freezes out" in a very short time. As time elapses, expansion of the system can cause hot and dense matter to flow into the location of those frozen cells. These will eventually freeze-out, too. Therefore, the freeze-out surface volume cannot be a simple 3-dimensional volume. It is a complicated 3-dimensional volume in the 4-dimensional space-time.

To illustrate this point, consider the following Gubser solution of the boost-invariant and azimuthally symmetric ideal-hydrodynamics[85] [1]:

$$T(\tau, r) = \frac{2T_0}{(2q\tau[1 + 2q^2(\tau^2 + r^2) + q^4(\tau^2 - r^2)^2])^{1/3}} \tag{228}$$

where $r^2 = x^2 + y^2$ and $\tau = \sqrt{t^2 - z^2}$. It is a simple matter of taking the τ derivative of $T(\tau, r)$ to see that for $r > r_0$ where $r_0 = \frac{3+\sqrt{5}}{2q}$, there are two values of τ where $\partial_\tau T$ vanishes. Since $\partial_\tau T$ is negative for small τ, this means that $T(\tau, r)$ at fixed $r > r_0$ will have a minimum and then a maximum. This is illustrated in the left panel in Fig. 5. The solid line is for $qr = 1$ which is near the center of the system. The temperature at that position decreases monotonically. At $r = r_0$, there is an inflection point but the behavior is till monotonic. In the $qr = 5$ case, one can clearly see that the temperature decreases at first but a some point it starts to rise again as the pressure pushes hot matter from the central region towards the periphery of the system. Assuming that the freeze-out temperature is between the minimum and the maximum of T/T_0, the position $qr = 5$ will contribute to the final particle spectrum (freeze-out) three times.

[1] Analytic solutions of the ideal hydrodynamics do exist for special cases, but not for general cases.

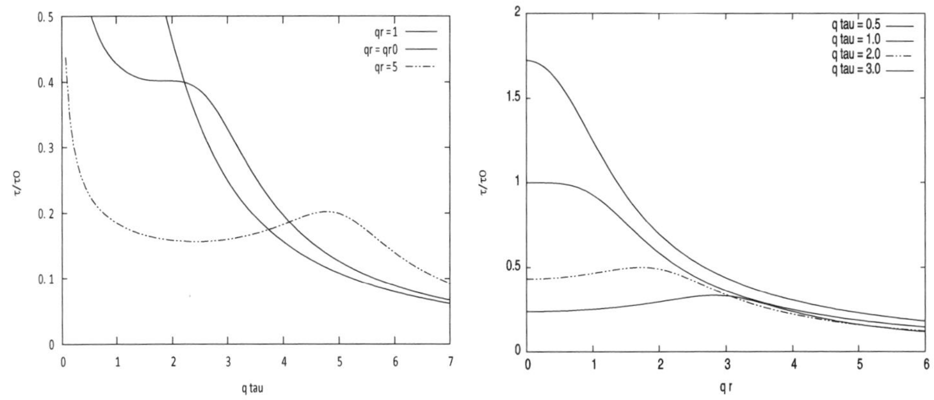

Fig. 5. Plot of $T(\tau, r)/T_0$ for the Gubser solution. Here $qr_0 = (3 + \sqrt{5})/2$. The left panel shows $T(\tau, r)/T_0$ at fixed r's and the right panel shows $T(\tau, r)$ at fixed τ's.

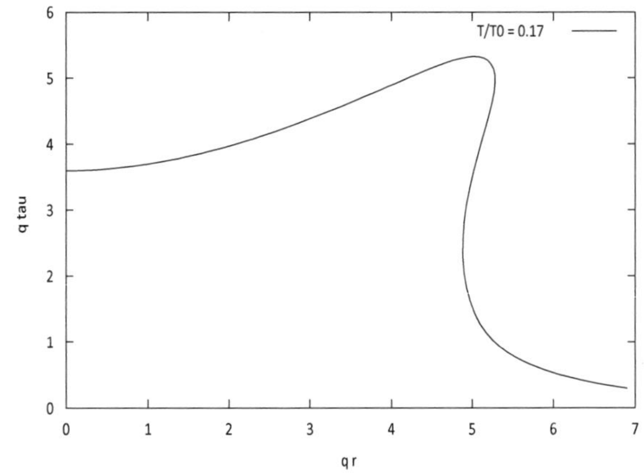

Fig. 6. Plot of the hypersurface with $T(\tau, r)/T_0 = 0.17$ for the Gubser solution.

The plot of isothermal curve with $T(\tau, r)/T_0 = 0.17$ is shown in Fig. 6. One may think of this as representing the "freeze-out" hypersurface in the Gubser solution. The long tail that can be seen above $qr \gtrsim 5$ is unrealistic. In more realistic simulations, the simulation starts at times above the long tail. Nevertheless, the Gubser solution contains much of the features that the more realistic numerical solutions exhibit.[9,111–113]

At the freeze-out hypersurface, the fluid is to be converted to particles according to the local equilibrium condition. In the kinetic theory, the particle number current for the ith particle is given by

$$j_i^\mu(x) = g_i \int \frac{d^3p}{(2\pi)^3 p^0} \, f_i(x, p) \, p^\mu \tag{229}$$

where $f_i(x,p)$ is the phase space density and g_i is the degeneracy. The number of ith particle in a 3-dimensional hypersurface is then given by

$$\int d\sigma_\mu j_i^\mu(x) = g_i \int \frac{d^3p}{(2\pi)^3 p^0} \int d\sigma_\mu\, p^\mu f_i(x,p) \tag{230}$$

where $d\sigma_\mu$ is the hypersurface element. Hence, the momentum spectrum is

$$E_p \frac{dN_i}{d^3p} = \frac{g_i}{(2\pi)^3} \int d\sigma_\mu\, p^\mu f_i(x,p). \tag{231}$$

This is the celebrated Cooper–Frye formula.[114]

Hydrodynamics deals with the energy density and its flow. Experiments measure the momentum distribution of identified particles. Cooper–Frye formula provides the link between them. In ideal hydrodynamics, local equilibrium is strictly maintained. Hence once we find the freeze-out hypersurface (equivalently the energy density via the equation of state), all one has to do is to integrate Eq. (231) over the freeze-out hypersurface with $f_i(x,p) = 1/(e^{(p^\mu u_\mu - \mu_B)/T_{\rm FO}} + a_i)$ where $T_{\rm FO}$ is the freeze-out temperature, $a_i = -1$ for bosonic statistics, $a_i = 1$ for fermionic statistics, and $a_i = 0$ for classical (Boltzmann) statistics.

The freeze-out surface we need to integrate over has the shape shown in Fig. 6 which requires a closer inspection. Suppose all parts of the system reaches the freeze-out temperature at the same Minkowski time $t_{\rm FO}$ and only once. In that case, the freeze-out hypersurface is just the Minkowski spatial volume. The integration element is just then $d\sigma_0 = dxdydz$. All other components vanish. This is simple, but not a realistic scenario in relativistic heavy ion collisions as explained in the previous section.

In the Milne coordinate system, the volume elements in each direction are given by

$$d\sigma_a = (\tau dxdyd\eta, -\tau d\tau dyd\eta, -\tau d\tau dxd\eta, -\tau d\tau dxdy). \tag{232}$$

When the freeze-out hypersurface is specified by the freeze-out proper time $\tau_f(x,y,\eta)$, the surface element on this surface is obtained by replacing $d\tau$ in Eq. (232) with $(\partial\tau_f/\partial x_i)dx_i$:

$$d\sigma_a^f = \left(1, -\frac{\partial\tau_f}{\partial x}, -\frac{\partial\tau_f}{\partial y}, -\frac{\partial\tau_f}{\partial\eta}\right)\tau_f dxdyd\eta. \tag{233}$$

If $d\sigma_a^f$ is time-like, then it is guaranteed that $p^a d\sigma_a^f > 0$. Therefore, Eq. (231) has a well defined interpretation that the particles are flowing out of the hypersurface. If $d\sigma_a^f$ is space-like, then depending on the direction of p^a, $p^a d\sigma_a^f$ could be either positive or negative. Equivalently, particles can be flowing into or out of the hypersurface.

Before we discuss the physical situation when this can happen, first we define the dynamic rapidity $y = \tanh^{-1}(p_z/E)$ so that

$$E = m_T \cosh y, \tag{234}$$

$$p^z = m_T \sinh y \tag{235}$$

with $m_T = \sqrt{m^2 + \mathbf{p}_T^2}$. The Milne space momentum is

$$p^a = (m_T \cosh(y - \eta), \mathbf{p}_T, m_T \sinh(y - \eta)/\tau) \tag{236}$$

which is equivalent to using $\Lambda^a_\mu p^\mu$ from (cf. Eq. (166)) but with an extra factor of $1/\tau$ for the η component to conform with the geometrical definition of the hypersurface element Eq. (233).

To illustrate what happens when $d\sigma^f_a$ is space-like, let us consider the Gubser solution again. In this case, $\tau_f(r)$ is a function of the transverse radius $r = \sqrt{x^2 + y^2}$ only. Since the system is boost-invariant, we can also set $\eta = 0$ without loss of generality. The inner product of the momentum and the volume element is

$$p^a d\sigma^f_a = (m_T \cosh y - (\mathbf{p}_T \cdot \hat{\mathbf{r}})\partial_r \tau_f)\tau_f dx dy d\eta \tag{237}$$

where $\hat{\mathbf{r}}$ is the unit vector in the transverse direction. This can become negative if $\partial\tau_f/\partial r > (m_T/p_T)\cosh y > 1$. From Figs. 5 and 6, one can see that large positive gradient $\partial\tau_f/\partial r$ exists in the region where the temperature reaches $T_{\rm FO}$ the second time. At this time, the temperature is going *up* from the minimum in the left panel of Fig. 5. Therefore, this part of the τ_f is not about the fluid freezing-out. It is rather colder matter being heated up to become the a part of the fluid again.

Before one could use the Cooper–Frye formula, one needs to know the freeze-out hypersurface. This is not a trivial problem because hydrodynamic simulations only provide the freeze-out *space-time points.* The freeze-out hypersurface needs to be reconstructed from these points. In $2 + 1$d, this is not so complicated. But in $3 + 1$-dimensional, it can become very involved. Discussion of this important topic however is beyond the scope of this review. An interested reader should consult Ref. 115.

How do we use the Cooper–Frye formula Eq. (231)? If hydrodynamics is not coupled to the hadronic cascade after-burner, then usually the hypersurface integral in Eq. (231) is performed as it is after the hypersurface is reconstructed. The rationale behind it is that when the cell crosses the freeze-out boundary 3 times, the contributions from the first two times should largely cancel each other. This is physically sound because when the fluid cell crosses the freeze-out surface the second time, the particles that are being heated up again are the remnants of the first crossing. When coupling to the hadronic after-burner, an additional condition such as $p^a d\sigma^f_a > 0$ is usually employed.[115–118]

The presence of non-zero shear tensor $\pi^{\mu\nu}$ and the bulk pressure Π signals non-equilibrium. In this case, the Cooper–Frye formula needs to be modified to take into account the non-equilibrium phase space density. Let

$$f(x,p) = f_{\mathrm{eq}}(x,p) + \delta f(x,p). \tag{238}$$

Then the consistency between the hydrodynamic $T^{\mu\nu}$ and the kinetic theory $T^{\mu\nu}_{\mathrm{Kin}} = \int \frac{d^3p}{(2\pi)^3 p^0} p^\mu p^\nu f$ requires

$$\pi^{\mu\nu} = \int \frac{d^3p}{(2\pi)^3 p^0}\, p^{\langle\mu} p^{\nu\rangle} \delta f(x,p) \tag{239}$$

and

$$\Pi = -\frac{m^2}{3} \int \frac{d^3p}{(2\pi)^3 p^0}\, \delta f(x,p) \tag{240}$$

with

$$p^{\langle\mu} p^{\nu\rangle} = \Delta^{\mu\nu}_{\alpha\beta} p^\alpha p^\beta. \tag{241}$$

These conditions can be satisfied by

$$\begin{aligned} \delta f(x,p) &= (A(\bar{E}_p)\Pi(x,p) \\ &\quad + B(\bar{E}_p) p^{\langle\mu} p^{\nu\rangle} \pi_{\mu\nu}(x,p) + \cdots) f_{\mathrm{eq}}(x,p)(1 + a_i f_{\mathrm{eq}}(x,p)) \end{aligned} \tag{242}$$

where again $\bar{E}_p = p^\mu u_\mu$ is the energy in the local rest frame. Here $A(\bar{E}_p)$ and $B(\bar{E}_p)$ must be consistent with Eqs. (239) and (240), but otherwise arbitrary at this point. In the presence of the net-baryon current Eq. (242) also includes $C(\bar{E}_p)p^{\langle\mu\rangle}q_\mu$ with the constraint on $C(\bar{E}_p)$ given by Eq. (138). Exact forms of $A(\bar{E}_p)$ and $B(\bar{E}_p)$ (and $C(\bar{E}_p)$) depend on the form of the scattering cross-sections in the underlying microscopic system. In Ref. 119, it is argued that for most theories, the $\bar{E}_p$ dependence of A and B should be between 1 and $1/\bar{E}_p$. See also Refs. 120 and 121. In most simulations, the constant ansatz is used.

6. Summary

In this review, we have tried to deliver a more general and pedagogic view of the relativistic hydrodynamics currently used in the study of ultra-relativistic heavy ion collisions. One message we tried to convey to the reader was that hydrodynamics is a very general framework and yet it can describe a vast set of complex phenomena. Especially in QGP studies, hydrodynamics is an indispensable tool that connects the QGP properties to the actual observables.

Another message we tried to convey was that the theory of hydrodynamics is a fascinating subject by itself. As discussed in Secs. 3 and 4, there are still many unsolved problems such as finding Kubo formulas for the second-order transport coefficients and finding the right anisotropic equation of state. In view of the apparent collectivity in the high-multiplicity proton–proton and proton–nucleus collisions at the LHC, the theory of collective motion in small systems is also in urgent need of development. In these systems, thermal fluctuations during the hydrodynamic evolution may not be completely ignored.[122–126] We hope that this review has provided interested readers enough starting points to pursue these and other interesting topics on their own.

In any short review, omission of some important subjects inevitably occurs due to the lack of space. One notable omission in this review is the discussion of the initial state models. As briefly discussed in Sec. 5.2, the initial condition of the hydrodynamic evolution must be given outside of the theory of hydrodynamics. For a meaningful comparison with the experimental data, having the right initial condition including the right energy–momentum fluctuation spectra is crucial. Unfortunately, it is outside the scope of this review and must be left as the subject of a future review.

Acknowledgments

S.J. thanks C. Gale and G. Denicol for many discussions on finer points of hydrodynamics and L.G. Yaffe for generous permission to use some materials from his unpublished note. S.J. is supported in part by the Natural Sciences and Engineering Research Council of Canada. U.H. acknowledges support by the U.S. Department of Energy, Office of Science, Office of Nuclear Physics under Awards No. DE-SC0004286 and (within the framework of the JET Collaboration) DE-SC0004104.

References

1. L. Landau, *Izv. Akad. Nauk Ser. Fiz.* **17** (1953) 51–64.
2. P. F. Kolb and U. W. Heinz, Hydrodynamic description of ultrarelativistic heavy ion collisions, in *Quark Gluon Plasma 3*, eds. R. C. Hwa and X.-N. Wang (World Scientific, 2004), pp. 634–714.
3. P. Huovinen, Hydrodynamical description of collective flow, in *Quark Gluon Plasma 3*, eds. R. C. Hwa and X.-N. Wang (World Scientific, 2004), pp. 600–633.
4. C. Gale, S. Jeon and B. Schenke, *Int. J. Mod. Phys. A* **28** (2013) 1340011.
5. F. Karsch, *Nucl. Phys. A* **698** (2002) 199–208.
6. P. Kovtun, D. T. Son and A. O. Starinets, *Phys. Rev. Lett.* **94** (2005) 111601.
7. H. Song and U. W. Heinz, *Phys. Lett. B* **658** (2008) 279–283.
8. H. Song and U. W. Heinz, *J. Phys. G* **36** (2009) 064033.
9. B. Schenke, S. Jeon and C. Gale, *Phys. Rev. Lett.* **106** (2011) 042301.
10. B. Schenke, S. Jeon and C. Gale, *Phys. Lett. B* **702** (2011) 59–63.
11. H. Song, S. A. Bass and U. Heinz, *Phys. Rev. C* **83** (2011) 054912.
12. H. Petersen, *Phys. Rev. C* **84** (2011) 034912.
13. C. Shen, U. Heinz, P. Huovinen and H. Song, *Phys. Rev. C* **84** (2011) 044903.
14. G. Aad *et al.*, *Phys. Rev. Lett.* **110**(18) (2013) 182302.
15. S. Chatrchyan *et al.*, *Phys. Lett. B* **718** (2013) 795–814.
16. B. Abelev *et al.*, *Phys. Lett. B* **719** (2013) 29–41.
17. A. Adare *et al.*, *Phys. Rev. Lett.* **111**(21) (2013) 212301.
18. I. Kozlov, M. Luzum, G. Denicol, S. Jeon and C. Gale, *Nucl. Phys. A* **931** (2014) 1045.
19. J. Kapusta and C. Gale, *Finite-Temperature Field Theory: Principles and Applications* (Cambridge University Press, 2006).

20. D. Forster, *Hydrodynamic Fluctuations, Broken Symmetry, and Correlation Functions*. Frontiers in Physics: A Lecture Note and Reprint Series (W. A. Benjamin, Advanced Book Program, 1975).
21. L. P. Kadanoff and P. C. Martin, *Annals Phys.* **24** (1963) 419–469.
22. C. P. Herzog, *J. Phys. A* **42** (2009) 343001.
23. P. Kovtun, *J. Phys. A* **45** (2012) 473001.
24. S. Deser and D. Boulware, *J. Math. Phys.* **8** (1967) 1468.
25. R. Baier, P. Romatschke, D. T. Son, A. O. Starinets and M. A. Stephanov, *JHEP* **0804** (2008) 100.
26. G. D. Moore and K. A. Sohrabi, *Phys. Rev. Lett.* **106** (2011) 122302.
27. W. Israel and J. Stewart, *Annals Phys.* **118** (1979) 341–372.
28. G. Denicol, T. Kodama, T. Koide and P. Mota, *J. Phys. G* **35** (2008) 115102.
29. S. Pu, T. Koide and D. H. Rischke, *Phys. Rev. D* **81** (2010) 114039.
30. P. B. Arnold, G. D. Moore and L. G. Yaffe, *JHEP* **0111** (2001) 057.
31. P. B. Arnold, G. D. Moore and L. G. Yaffe, *JHEP* **0305** (2003) 051.
32. P. B. Arnold, C. Dogan and G. D. Moore, *Phys. Rev. D* **74** (2006) 085021.
33. D. Gross and F. Wilczek, *Phys. Rev. D* **8** (1973) 3633–3652.
34. H. D. Politzer, *Phys. Rept.* **14** (1974) 129–180.
35. S. Bethke, *Nucl. Phys. B* (*Proc. Suppl.*) **222–224** (2012) 94–100.
36. B. Schenke, C. Gale and S. Jeon, *Phys. Rev. C* **80** (2009) 054913.
37. P. Tribedy and R. Venugopalan, *Phys. Lett. B* **710** (2012) 125–133.
38. D. J. Gross, R. D. Pisarski and L. G. Yaffe, *Rev. Mod. Phys.* **53** (1981) 43.
39. J. O. Andersen, M. Strickland and N. Su, *JHEP* **1008** (2010) 113.
40. J. O. Andersen, L. E. Leganger, M. Strickland and N. Su, *JHEP* **1108** (2011) 053.
41. N. Haque, M. G. Mustafa and M. Strickland, *Phys. Rev. D* **87**(10) (2013) 105007.
42. A. Ali Khan *et al.*, *Phys. Rev. D* **64** (2001) 074510.
43. M. Cheng, S. Ejiri, P. Hegde, F. Karsch, O. Kaczmarek *et al.*, *Phys. Rev. D* **81** (2010) 054504.
44. S. Borsanyi, Z. Fodor, C. Hoelbling, S. D. Katz, S. Krieg *et al.*, *Phys. Lett. B* **730** (2014) 99–104.
45. H. B. Meyer, *Phys. Rev. Lett.* **100** (2008) 162001.
46. H. B. Meyer, *Phys. Rev. D* **76** (2007) 101701.
47. J. Noronha-Hostler, J. Noronha and C. Greiner, *Phys. Rev. Lett.* **103** (2009) 172302.
48. D. Fernandez-Fraile and A. Gomez Nicola, *Eur. Phys. J. C* **62** (2009) 37–54.
49. A. Buchel, *Phys. Lett. B* **663** (2008) 286–289.
50. S. S. Gubser, S. S. Pufu and F. D. Rocha, *JHEP* **0808** (2008) 085.
51. T. Springer, C. Gale and S. Jeon, *Phys. Rev. D* **82** (2010) 126011.
52. P. B. Arnold, G. D. Moore and L. G. Yaffe, *JHEP* **0301** (2003) 030.
53. S. Dubovsky, L. Hui, A. Nicolis and D. T. Son, *Phys. Rev. D* **85** (2012) 085029.
54. P. Danielewicz and M. Gyulassy, *Phys. Rev. D* **31** (1985) 53–62.
55. G. Policastro, D. T. Son and A. O. Starinets, *Phys. Rev. Lett.* **87** (2001) 081601.
56. P. Romatschke and U. Romatschke, *Phys. Rev. Lett.* **99** (2007) 172301.
57. H. Song, S. A. Bass, U. Heinz, T. Hirano and C. Shen, *Phys. Rev. Lett.* **106** (2011) 192301.
58. C. Gale, S. Jeon, B. Schenke, P. Tribedy and R. Venugopalan, *Phys. Rev. Lett.* **110** (2013) 012302.
59. U. Heinz and R. Snellings, *Ann. Rev. Nucl. Part. Sci.* **63** (2013) 123–151.
60. G. Denicol, H. Niemi, E. Molnar and D. Rischke, *Phys. Rev. D* **85** (2012) 114047.
61. H. Song and U. W. Heinz, *Phys. Rev. C* **81** (2010) 024905.
62. B. Betz, D. Henkel and D. Rischke, *J. Phys. G* **36** (2009) 064029.

63. G. Denicol, S. Jeon and C. Gale, *Phys. Rev. C* **90**(2) (2014) 024912.
64. W. Hiscock and L. Lindblom, *Annals Phys.* **151** (1983) 466–496.
65. W. A. Hiscock and L. Lindblom, *Phys. Rev. D* **31** (1985) 725–733.
66. H. Song and U. W. Heinz, *Phys. Rev. C* **78** (2008) 024902.
67. U. Heinz, H. Song and A. K. Chaudhuri, *Phys. Rev. C* **73** (2006) 034904.
68. E. Molnár, H. Niemi, G. Denicol and D. Rischke, *Phys. Rev. D* **89** (2014) 074010.
69. M. A. York and G. D. Moore, *Phys. Rev. D* **79** (2009) 054011.
70. G. D. Moore and K. A. Sohrabi, *JHEP* **1211** (2012) 148.
71. S. I. Finazzo, R. Rougemont, H. Marrochio and J. Noronha, *JHEP* **1502** (2015) 051 [arXiv:1412.2968 [hep-ph]].
72. G. S. Denicol, W. Florkowski, R. Ryblewski and M. Strickland, *Phys. Rev. C* **90** (2014) 044905.
73. M. Martinez and M. Strickland, *Nucl. Phys. A* **848** (2010) 183–197.
74. W. Florkowski and R. Ryblewski, *Phys. Rev. C* **83** (2011) 034907.
75. P. Romatschke and M. Strickland, *Phys. Rev. D* **68** (2003) 036004.
76. M. Martinez, R. Ryblewski and M. Strickland, *Phys. Rev. C* **85** (2012) 064913.
77. D. Bazow, U. W. Heinz and M. Strickland, *Phys. Rev. C* **90**(5) (2014) 054910.
78. G. Baym, *Phys. Lett. B* **138** (1984) 18–22.
79. W. Florkowski, R. Ryblewski and M. Strickland, *Nucl. Phys. A* **916** (2013) 249–259.
80. W. Florkowski, R. Ryblewski and M. Strickland, *Phys. Rev. C* **88** (2013) 024903.
81. W. Florkowski, E. Maksymiuk, R. Ryblewski and M. Strickland, *Phys. Rev. C* **89**(5) (2014) 054908.
82. W. Florkowski, R. Ryblewski, M. Strickland and L. Tinti, *Phys. Rev. C* **89** (2014) 054909.
83. G. S. Denicol, U. W. Heinz, M. Martinez, J. Noronha and M. Strickland, *Phys. Rev. Lett.* **113** (2014) 202301.
84. G. S. Denicol, U. W. Heinz, M. Martinez, J. Noronha and M. Strickland, *Phys. Rev. D* **90** (2014) 125026.
85. S. S. Gubser, *Phys. Rev. D* **82** (2010) 085027.
86. S. S. Gubser and A. Yarom, *Nucl. Phys. B* **846** (2011) 469–511.
87. M. Nopoush, R. Ryblewski and M. Strickland, Anisotropic hydrodynamics for conformal Gubser flow, *Phys. Rev. D* **91** (2015) 045007 [arXiv:1410.6790 [nucl.th]].
88. A. Jaiswal, *Phys. Rev. C* **88** (2013) 021903.
89. D. Bazow, U. Heinz and M. Martinez, *in preparation.* (2015).
90. J. Bjorken, *Phys. Rev. D* **27** (1983) 140–151.
91. P. F. Kolb, J. Sollfrank and U. W. Heinz, *Phys. Rev. C* **62** (2000) 054909.
92. B. Schenke, S. Jeon and C. Gale, *Phys. Rev. C* **82** (2010) 014903.
93. A. Muronga, *Phys. Rev. C* **69** (2004) 034903.
94. H. Nessyahu and E. Tadmor, *J. Comput. Phys.* **87** (1990) 408–463.
95. V. Schneider, U. Katscher, D. Rischke, B. Waldhauser, J. Maruhn *et al.*, *J. Comput. Phys.* **105** (1993) 92–107.
96. J. Sollfrank, P. Huovinen, M. Kataja, P. Ruuskanen, M. Prakash *et al.*, *Phys. Rev. C* **55** (1997) 392–410.
97. H. Petersen, G.-Y. Qin, S. A. Bass and B. Muller, *Phys. Rev. C* **82** (2010) 041901.
98. I. Karpenko and Y. Sinyukov, *Phys. Rev. C* **81** (2010) 054903.
99. H. Holopainen, H. Niemi and K. J. Eskola, *Phys. Rev. C* **83** (2011) 034901.
100. L. Pang, Q. Wang and X.-N. Wang, *Phys. Rev. C* **86** (2012) 024911.
101. V. Roy and A. Chaudhuri, *Phys. Rev. C* **85** (2012) 024909.
102. Y. Akamatsu, S.-I. Inutsuka, C. Nonaka and M. Takamoto, *J. Comput. Phys.* **256** (2014) 34–54.

103. A. Kurganov and E. Tadmor, *J. Comput. Phys.* **160** (2000) 241–282.
104. C. Shen, Z. Qiu, H. Song, J. Bernhard, S. Bass *et al.*, The iEBE-VISHNU code package for relativistic heavy-ion collisions, arXiv:1409.8164 (2014).
105. T. Hirano and Y. Nara, *PTEP* **2012** (2012) 01A203.
106. C. Nonaka, N. Sasaki, S. Muroya and O. Miyamura, *Nucl. Phys. A* **661** (1999) 353–356.
107. L. Del Zanna, V. Chandra, G. Inghirami, V. Rolando, A. Beraudo *et al.*, *Eur. Phys. J. C* **73** (2013) 2524.
108. C. Aguiar, T. Kodama, T. Osada and Y. Hama, *J. Phys. G* **27** (2001) 75–94.
109. M. Luzum and P. Romatschke, *Phys. Rev. C* **78** (2008) 034915.
110. P. Bozek, *Phys. Rev. C* **85** (2012) 034901.
111. U. W. Heinz and P. F. Kolb, *Phys. Lett. B* **542** (2002) 216–222.
112. H. Song, S. A. Bass and U. Heinz, *Phys. Rev. C* **83** (2011) 024912.
113. H. Holopainen and P. Huovinen, *J. Phys. Conf. Ser.* **389** (2012) 012018.
114. F. Cooper and G. Frye, *Phys. Rev. D* **10** (1974) 186.
115. P. Huovinen and H. Petersen, *Eur. Phys. J. A* **48** (2012) 171.
116. F. Grassi, *Braz. J. Phys.* **35** (2005) 52–69.
117. Y. Sinyukov, S. Akkelin, I. Karpenko and Y. Hama, *Acta Phys. Polon. B* **40** (2009) 1025–1036.
118. D. Oliinychenko, P. Huovinen and H. Petersen, *Phys. Rev. C* **91** (2015) 024906 [arXiv:1411.3912 [nucl-th]].
119. K. Dusling, G. D. Moore and D. Teaney, *Phys. Rev. C* **81** (2010) 034907.
120. G. Denicol and H. Niemi, *Nucl. Phys. A* **904–905** (2013) 369c–372c.
121. J. Noronha-Hostler, G. S. Denicol, J. Noronha, R. P. G. Andrade and F. Grassi, *Phys. Rev. C* **88** (2013) 044916.
122. J. Kapusta, B. Muller and M. Stephanov, *Phys. Rev. C* **85** (2012) 054906.
123. K. Murase and T. Hirano, Relativistic fluctuating hydrodynamics with memory functions and colored noises, arXiv:1304.3243 (2013).
124. T. Hirano, R. Kurita, K. Murase and K. Nagai, *Nucl. Phys. A* (2014).
125. C. Young, J. Kapusta, C. Gale, S. Jeon and B. Schenke, *J. Phys. Conf. Ser.* **535** (2014) 012034.
126. C. Young, J. Kapusta, C. Gale, S. Jeon and B. Schenke, *Phys. Rev. C* **91** (2015) 044901 [arXiv:1407.1077 [nucl-th]].

Early Thermalization, Hydrodynamics and Energy Loss in AdS/CFT

Paul M. Chesler
Department of Physics, Harvard University, Cambridge, MA 02138, USA

Wilke van der Schee
MIT Center for Theoretical Physics, Cambridge, MA 02139, USA

Gauge/gravity duality has provided unprecedented opportunities to study dynamics in certain strongly coupled gauge theories. This review aims to highlight several applications to heavy ion collisions including far-from-equilibrium dynamics, hydrodynamics and jet energy loss at strong coupling.

1. Introduction

Gauge/gravity duality equates certain gauge theories with theories of quantum gravity in one higher dimension.[1] This extra dimension has a natural interpretation as the renormalization group scale, and has led to the name "holographic duality". In the limit where the gauge theory is strongly coupled and has a large number of colors N_c, the dual description reduces to classical supergravity. Hence, challenging strongly coupled quantum dynamics in the gauge theory can be accessed by solving classical partial differential equations. All physics — from microscopic interactions to macroscopic hydrodynamics — is encoded in the dual classical dynamics. Holography can thereby provide systematic and controlled access to strongly coupled dynamics via the solution to the classical partial differential equations. Currently, there is no other theoretical tool capable of accessing strongly coupled real-time dynamics in a controlled setting.

Heavy ion collisions at RHIC and LHC have demonstrated that the produced quark–gluon plasma is strongly coupled.[2] Hence, it is of interest to model heavy ion collisions using holography. However, while there exists many theories with dual gravitational descriptions — some with QCD-like features such as confinement and chiral symmetry breaking[3,4] — the dual description of QCD is not known (if it exists). Nevertheless, given the immense challenge of studying strongly coupled dynamics in QCD it is invaluable to have a model where strongly coupled dynamics can be studied in a controlled systematic setting. This is especially true if the results are valid for a class of strongly coupled gauge theories and hence have some degree of universality.

In this review we shall focus on the simplest theory with a holographic dual: (3 + 1)-dimensional super-Yang–Mills theory with four supersymmetries ($\mathcal{N} = 4$ SYM). In the limit where the SYM is strongly coupled and has a large number of colors N_c, the dual description reduces to classical supergravity in asymptotically five-dimensional Anti-de Sitter spacetime AdS_5[1] (see also Refs. 5 and 6). While the ground state of SYM is very different from QCD — SYM is conformal and contains no particles whereas QCD is confining and has a rich spectrum of hadrons — at temperatures not too far above the deconfinement transition both theories consist of strongly coupled non-Abelian plasmas. It is in this setting where dynamics in SYM can potentially provide lessons for dynamics in QCD.

In what follows we highlight insights obtained via holography in three settings. First, we discuss relativistic hydrodynamics including computations of the shear viscosity at strong coupling and the construction of causal viscous hydrodynamics. We then discuss far-from-equilibrium dynamics and the collisions of shock waves in SYM and their application to early time dynamics in heavy ion collisions. Finally, we give a brief review of progress made in jet quenching and energy loss at strong coupling.

2. First Successes: Viscosity and Relativistic Hydrodynamics

2.1. *Viscosity from black hole horizons*

Gauge/gravity duality provides a very natural way to describe thermal states: they are described by black holes, which are the simplest stationary states in AdS_5 gravity with a notion of temperature. This temperature is simply the black hole's Hawking temperature.[7] The topology of the horizon is inherited from the gauge theory, where a thermal state usually is extended infinitely; this is one major difference from ordinary black holes in our universe, and in fact these black holes in AdS_5 are more appropriately called black branes. Another major difference is the boundary of the AdS_5 spacetime, which can be thought of as the place where the gauge theory lives. This boundary reflects back the thermal radiation, and the black branes are hence in stable thermal equilibrium, as opposed to black holes in flat spacetime, which evaporate away.[a]

It is also possible to construct gradient expansion solutions to Einstein's equations, where the local geometry is approximately that of boosted black brane with slowly varying temperature and boost velocity.[8,9] In this case the corresponding dual gauge theory state is that of a system in local thermal equilibrium with

[a] Also, black holes in our universe tend to have very small temperatures of order of mK, as black holes of higher temperature are hard to form and would quickly evaporate. The thermal stability of black holes in AdS allows any temperature, and in fact the temperature of black holes representing a quark–gluon plasma would have the same very high temperature of order 10^{12} K.

dynamics governed by hydrodynamics. Simply put, in the limit of slowly varying fields, Einstein's equations reduce to hydrodynamics in one less dimension. The connection between slowly varying solutions to Einstein's equations and hydrodynamics is known as the fluid/gravity correspondence. By matching the gradient expansion obtained by solving Einstein's equations to the constitutive relations of hydrodynamics, transport coefficients such as the viscosity can be computed at strong coupling.

At first order in gradients the constitutive relation for the stress tensor $T^{\mu\nu}$ in relativistic neutral hydrodynamics read[10]

$$T_{\mu\nu} = e u_\mu u_\nu + p[e]\Delta_{\mu\nu} + \pi_{\mu\nu}, \quad \text{where,} \tag{1}$$

$$\Delta_{\mu\nu} = g_{\mu\nu} + u_\mu u_\nu \quad \text{and} \tag{2}$$

$$\pi_{\mu\nu} = -\eta[e]\sigma_{\mu\nu} - \zeta[e]\Delta_{\mu\nu}(\nabla\cdot u) + \mathcal{O}(\partial^2), \quad \text{with} \tag{3}$$

$$\sigma_{\mu\nu} = \Delta_{\mu\alpha}\Delta_{\nu\beta}(\nabla^\mu u^\nu + \nabla^\nu u^\mu) - \frac{2}{3}\Delta_{\mu\nu}\Delta_{\alpha\beta}\nabla^\alpha u^\beta, \tag{4}$$

where e is the proper energy density, u_μ is the fluid velocity, $p[e]$ is the pressure (which is a function of the proper energy via an equation of state), $\pi_{\mu\nu}$ is the shear tensor, η is the shear viscosity and ζ is the bulk viscosity, which vanishes in scale invariant theories such as SYM. Note that the fluid velocity and energy density are defined as the timelike eigenvector and associated eigenvalue of the stress tensor ($T_{\mu\nu}u^\nu = -eu_\mu$) and that $\sigma^{\mu\nu}$ is transverse and traceless: $u^\mu\sigma_{\mu\nu} = \sigma^\mu_\mu = 0$. Alternatively, one can say that the fluid velocity is defined such that when boosting $T_{\mu\nu}$ with velocity u_μ there is no momentum flow, i.e., $T'_{0i} = 0$, which is called the Landau frame.[b] The conservation equation $\nabla^\mu T_{\mu\nu} = 0$ together with the equation of state and the viscosity (computed theoretically or measured) now form a closed system of equations, fully determined by a given initial energy density and fluid velocity.

The equation of state, viscosity and higher-order transport coefficients are not determined by hydrodynamics itself, and are a property of the microscopic theory under consideration. In weakly coupled QCD it is possible to use the Boltzmann equation to perturbatively compute the shear viscosity, which led to[12]

$$\eta_{\text{weak}} = \kappa\frac{T^3}{g^4\log(1/g)} + \mathcal{O}\left(\frac{1}{g^4}\right), \tag{5}$$

with T the temperature, g the gauge coupling and κ a coefficient depending on the number of colors and flavors, which can be computed numerically. Non-perturbatively it is unknown how to compute the viscosity of QCD. However, there

[b]When including a charge density with corresponding current and chemical potential it is also possible to define a frame where there is no charge flow, which is called the Eckart frame.[11] In this review we do not include a charge density.

have been attempts to compute the viscosity using lattice techniques (see Ref. 13 for a review). It was therefore ground-breaking when a few years after the discovery of gauge/gravity duality Policastro, Son and Starinets computed the shear viscosity of a thermal plasma described by strongly coupled SYM theory[14]:

$$\eta = \frac{s}{4\pi} = \frac{\pi}{8} N_c^2 T^3, \tag{6}$$

with s is the entropy density. The factor of $1/4\pi$ is surprisingly universal and valid for any strong coupling and large N_c quantum field theory with a gravitational dual. For instance, adding charge density changes both the shear viscosity and the entropy density, but its ratio is still $1/4\pi$.

Nevertheless, it is possible to have holographic models with different viscosity to entropy ratios, most notably by including higher derivative corrections in the Einstein equations. In this case corrections for finite N_c can actually make the η/s ratio drop below $1/4\pi$, thereby violating the conjectured universal lower bound[15] on the viscosity.[16] Nevertheless, general considerations using the quantum mechanical uncertainty principle still hint at the existence of a lower bound, even though its value is likely smaller than $1/4\pi$.

Holography can also yield finite coupling and finite N_c corrections to the viscosity. For SYM the results read[17–19]:

$$\eta/s = \frac{1}{4\pi}\left(1 + \frac{15}{\lambda^{3/2}}\zeta(3) + \frac{5}{16}\frac{\lambda^{1/2}}{N_c^2}\right), \tag{7}$$

with $\lambda = g^2 N_c$ the 't Hooft coupling. As expected from (5) the viscosity increases as the coupling constant decreases, and the increase can be significant at reasonable values of the coupling constant (for $\lambda \sim 20$ and $N_c \sim 3$ the corrections both are roughly 20%). Current hydrodynamic models of the quark–gluon plasma indeed favor an η/s in the range $1/4\pi$ to $2.5/4\pi$.[20–23]

2.2. *Second-order relativistic hydrodynamics*

Holography and in particular the fluid/gravity duality also has led to a more systematic understanding of (relativistic) hydrodynamics itself.[8,25,26] As alluded to above, via the fluid/gravity correspondence it is possible to systematically solve Einstein's equations. The gradient expansion solution encodes the constitutive relations of hydrodynamics to *all orders* in gradients and associated transport coefficients.[8,25] Let us begin by simply stating the results of Refs. 8 and 25 for the shear tensor $\pi^{\mu\nu}$ up to the second-order in gradients.

Since SYM is conformal, to write $\pi^{\mu\nu}$ at second-order in gradients it is useful to employ derivative operators $\mathcal{D}_\mu$ which transform covariantly under conformal transformations. For tensors $Q^{\mu\cdots}_{\nu\ldots}$ which, under a conformal transformation $g_{\mu\nu} \to e^{2\phi}\tilde{g}_{\mu\nu}$, transform like $Q^{\mu\cdots}_{\nu\ldots} = e^{-w\phi}\widetilde{Q}^{\mu\cdots}_{\nu\ldots}$, $\mathcal{D}$ is defined by[26]

$$\begin{aligned}\mathcal{D}_\sigma Q^{\mu\ldots}_{\nu\ldots} &\equiv \partial_\sigma Q^{\mu\ldots}_{\nu\ldots} + w\mathcal{A}_\sigma Q^{\mu\ldots}_{\nu\ldots} + \gamma^{\mu}_{\sigma\lambda} Q^{\lambda\ldots}_{\nu\ldots} + \cdots - \gamma^{\lambda}_{\sigma\nu} Q^{\mu\ldots}_{\lambda\ldots} - \cdots,\\ \gamma^{\lambda}_{\mu\nu} &\equiv g_{\mu\nu}A^{\lambda} - \delta^{\lambda}_{\mu}A_\nu - \delta^{\lambda}_{\nu}A_\mu,\\ A_\mu &\equiv u^\nu\partial_\nu u_\mu - \frac{\partial_\nu u^\nu}{3}u_\mu.\end{aligned}$$

With this definition derivatives transform covariantly: $\mathcal{D}_\lambda Q^{\mu\cdots}_{\nu\ldots} = e^{-w\phi}\mathcal{D}_\lambda\widetilde{Q}^{\mu\cdots}_{\nu\ldots}$, and for instance Eq. (4) simplifies to $\sigma_{\mu\nu} = \mathcal{D}_\mu u_\nu + \mathcal{D}_\nu u_\mu$. The full second-order conformal shear tensor is now given by[c]:

$$\begin{aligned}\pi_{\mu\nu} = &-\eta\,\sigma_{\mu\nu} + \eta\tau_\pi u^\lambda\mathcal{D}_\lambda\sigma_{\mu\nu} + \lambda_1\left[\sigma_{\mu\lambda}\sigma^{\lambda}_{\nu} - \frac{\Delta_{\mu\nu}}{3}\sigma^{\alpha\beta}\sigma_{\alpha\beta}\right]\\ &+\lambda_2[\omega_{\mu\lambda}\sigma^{\lambda}_{\nu} + \omega_{\nu\lambda}\sigma^{\lambda}_{\mu}] + \lambda_3\left[\omega_{\mu\lambda}\omega^{\lambda}_{\nu} + \frac{\Delta_{\mu\nu}}{3}\omega^{\alpha\beta}\omega_{\alpha\beta}\right],\end{aligned} \tag{8}$$

where $\omega_{\mu\nu} = \mathcal{D}_\mu u_\nu - \mathcal{D}_\nu u_\mu$. In strongly coupled SYM the second-order transport coefficients are given by $\tau_\pi = (2 - \ln(2))/2\pi T$, $\lambda_1 = \eta/2\pi T = 2\lambda_2/\ln(2)$ and $\lambda_3 = 0$.[8] Note that all these coefficients get considerably more complicated when including a chemical potential; in particular λ_3 becomes nonzero.[27]

Second-order hydrodynamics is particularly useful for numerical simulations of viscous hydrodynamics. This is because high momentum modes in first-order relativistic hydrodynamics can propagate faster than the speed of light. Fundamentally this is not a problem, as hydrodynamics is an expansion in long wavelengths, i.e., small momenta, and these acausal modes are outside the regime of its applicability.[28] Nevertheless, these modes make the equations numerically unstable[29] and therefore in numerical simulations a second-order term was added to stabilize the equations, giving Muller–Israel–Stewart hydrodynamics.[30,31]

Muller–Israel–Stewart hydrodynamics can be seen as a special case of second-order hydrodynamics where $\lambda_i = 0$. One insight which holography helped solidify is that, in contrast to the Muller–Israel–Stewart theory, at a given order in the gradient expansion, one must add every possible tensor structure to the constitutive relations which is consistent with the symmetries of the underlying quantum field theory. This had lead to the development of causal viscous relativistic hydrodynamics.[24,32,33] In conformal theories such as SYM, the equations of second-order hydrodynamics give causal propagation for high momenta modes.

While the second-order transport coefficients make simulations much easier to perform, the precise value of these coefficients fortunately does not seem to have a

[c]Most of these computations have been done for fluids living in a curved spacetime,[25] which gives an extra term not considered in this review.

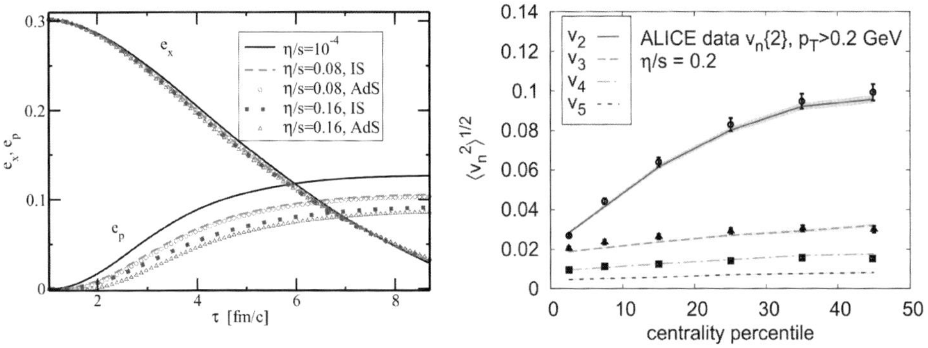

Fig. 1. (Color online) (Left) The second-order hydrodynamics derived using AdS/CFT (AdS) is compared with the older weakly coupled Muller–Israel–Stewart hydrodynamcis (IS), which uses $\tau_\pi = 6\eta/sT$ and does not include the λ_1 or λ_2 in Eq. (8). For these initial conditions the spatial and momentum anisotropy (e_x and e_p) do not depend much on the precise value of the second-order transport coefficients (figure from Ref. 24). (Right) Similar hydrodynamics has been used to obtain recent viscosity over entropy ratios of roughly 0.12 and 0.2 at top RHIC and LHC collisions respectively (LHC estimate shown, figure from Ref. 23).

big influence on the particle spectra in realistic simulations (see Fig. 1 (left)). The latter feature does however crucially depend on using conformal hydrodynamic equations; when using (second-order) non-conformal equations results do seem to depend on the second-order coefficients.[34] One of the biggest successes of these equations are estimates for the shear viscosity of the quark–gluon plasma,[20–23] as shown in Fig. 1 (right).

Another more recent example of a better understanding of hydrodynamics comes from,[35] where in a specific boost-invariant setting all transport coefficients were computed numerically up to order 240 in the gradient expansion. With those coefficients it could be shown that the hydrodynamic gradient expansion is not necessarily convergent, it was found to be an asymptotic series. Moreover, it was understood why hydrodynamics did not converge, since the divergent hydrodynamic expansion contains information about the non-hydrodynamic modes. Within AdS/CFT these modes are included in the so-called quasi-normal modes: vibrations of the black hole horizon. Strikingly, by a Borel resummation the authors of Ref. 35 we able to extract the precise value of the dominant quasi-normal mode from the hydrodynamic expansion, thereby showing how hydrodynamics contains information about its own break-down.

Much work has been devoted to study charged hydrodynamics, extending the previous formulas to a plasma with a conserved current, chemical potential and various new transport coefficients, see for instance. Refs. 27 and 36. More recently, there has been interest in anomalous hydrodynamics, where quantum anomalies lead to modified hydrodynamic equations. A plasma with a nontrivial axial charge density will develop an electric current **J** in the direction of an external magnetic field **B** (chiral magnetic effect or CME)[37,38]:

$$\mathbf{J} = \frac{e^2}{2\pi^2}\mu_5\mathbf{B}, \tag{9}$$

where μ_5 is the chemical potential associated with the axial charge. For averaged heavy ion events μ_5 would be zero, but a full event-by-event anomalous hydrodynamic simulation suggests that this effect is measurable in heavy ion collisions.[39] Similarly to the above analysis, also for the CME holography has proved extremely useful to show how this effect is in fact necessarily included already in first-order charged hydrodynamics, and also to estimate the relevant transport coefficients at strong coupling.[40,41]

Lastly, inspiration from the symmetries present in AdS has led to nontrivial analytic solutions of the hydrodynamic equations. An interesting solution was provided by Gubser,[42] where a solution to viscous conformal hydrodynamics with boost-invariance and a nontrivial expansion in the transverse plane was found. This solution follows from a simple setup in AdS, but then leads to this nontrivial analytic and hence very useful hydrodynamic solution. Recently, this approach has been extended to an exact solution to the relativistic Boltzmann equation.[43]

Interestingly, aside from the computation of transport coefficients, much of the development in hydrodynamics above need not rely on gauge/gravity duality. Nevertheless, inspiration from the duality has led to this more precise understanding of hydrodynamics, by using gauge/gravity as a theoretical arena to get new insights.

3. Fast Thermalization and the Resulting Stress Tensor

3.1. *Simple models*

Perhaps an even more remarkable feature of holography is the straightforward possibility to study far-from-equilibrium dynamics in strongly coupled gauge theories. How do non-equilibrium systems thermalize and relax to local equilibrium? Using holography, answering this question amounts to studying real time dynamics in the gravitational dual, which is not much different from conventional problems in general relativity, such as binary neutron stars or black hole inspirals.[44,45] As is well known, these latter problems can still be very challenging from a numerical point of view, but compared to non-perturbative quantum field theory it is definitely an enormous simplification. For a review of the use of numerical relativity in holography, see for example Ref. 46.

The first studies of holographic thermalization computed how perturbed black holes relax to homogeneous black holes, computing the so-called quasi-normal modes.[47] These are determined by expanding the perturbation of the metric in a sum of modes proportional to $e^{-i\omega_i t}$, with complex frequency ω_i. The Einstein equations and boundary conditions at the horizon and boundary then dictate that there is a discrete spectrum, whereby the mode always decays (i.e., $\Im(\omega_i) < 0$).

The mode with the slowest decay will then determine the late-time decay back to the thermal black hole, and hence can be said to determine the thermalization time in this setting. Dimensional analysis in this scale invariant theory dictates that $\omega_i \sim T$, but the coefficient has to be computed numerically, giving $\Im(\omega_0) \approx 8.6T$.[47]

The estimate from quasi-normal modes could imply a thermalization time $t_{\text{therm}} < 1/T$, which for a temperature of 1000 MeV would be about 0.2 fm/c. Compared to perturbative methods this is very fast, which should be expected for a system at strong coupling. On the other hand, even though the coupling is assumed to be infinite, there is still a finite answer for the thermalization time, leading to the intuition that strong interactions still have some microscopic physics present as the scale $1/T$.

Nevertheless, the quasi-normal mode analysis is only valid near equilibrium: the state is assumed to be a small perturbation of the final thermal state. In heavy ion collisions the initial state is very far-from-equilibrium, so solving the full nonlinear Einstein equations is of great importance. One of the first studies solving this numerically is presented in Refs. 48 and 49, where quark–gluon plasma formation and thermalization was studied in both homogeneous and boost invariant setups (see Fig. 2). The gauge theory dynamics was, by symmetry, effectively $1+0$ dimensional and the dual gravitational dynamics was $1+1$ dimensional. In these papers non-equilibrium states were created by starting in the ground state and turning on background fields. It was found that even in the case where the initial state is the vacuum, a thermal state can be achieved in times of order $1/T$ with T the final state temperature.

An intriguing finding in Ref. 49 was that when the system first starts to behave hydrodynamically, viscous effects were order one: the first-order viscous correction

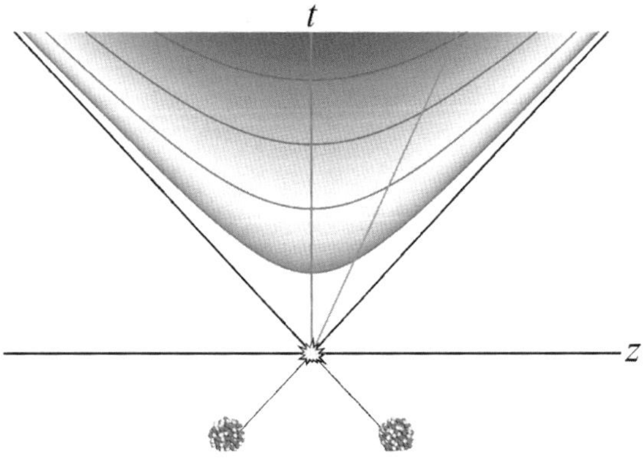

Fig. 2. (Color online) A simple and often used model of a heavy ion collision was proposed by Bjorken,[55] where he assumed that a heavy ion collision is approximately boost-invariant, at least near $z = 0$. This means that all physics only depends on proper time $\tau = \sqrt{t^2 - z^2}$, and hence that all physics experienced by observers, such as a temperature field illustrated here, is independent of its frame. The two red lines would illustrate two such frames, which indeed has equal temperatures at equal τ (figure from Ref. 56).

to the constitutive relations (1) is as big as the leading term. Nevertheless, even with such large viscous corrections the hydrodynamic gradient expansion was seen to be well behaved. Moreover, the applicability of hydrodynamics was seen not to be governed by convergence of the hydrodynamic gradient expansion. Instead, the applicability of hydrodynamics was seen to be governed by the decay of non-hydrodynamic degrees of freedom.

The approach in Ref. 49 was extended in Ref. 50. Here the main feature was the possibility to chose several initial states and to evolve them without turning on any background fields. For all initial states they found a thermalization time less than $1/T$, with T the temperature at the moment of thermalization.

Of particular utility, this paper also introduced a dimensionless quantitiy, $F(w) \equiv \frac{\tau}{w}\partial_\tau w$, where $w \equiv T_{\rm eff}\tau$ can be thought of as dimensionless time and $T_{\rm eff}$ is the temperature associated with the energy density at that time. The dimensionless function $F(w)$ is completely determined within hydrodynamics and for instance equals $2/3$ for ideal hydrodynamics. As $F(w)$ is unique within hydrodynamics, it is clear that differences between $F(w)$ found for the different evolutions in AdS/CFT must be due to genuine non-hydrodynamic effects, as is convincingly shown in Fig. 3 (left).

The right panel of Fig. 3 shows a similar study in a homogeneous setting, with more than 1000 initial states, also leading to thermalization within a time of $1/T$.[51,52] The main new result in this paper was to compare the full nonlinear thermalization process with the linearized quasi-normal mode approach outlined in the second paragraph of this section. Quite surprisingly, it was found that for general initial states the evolution was always well described, within 20%, by the linearized approximation. This then also confirmed that the quasi-normal mode

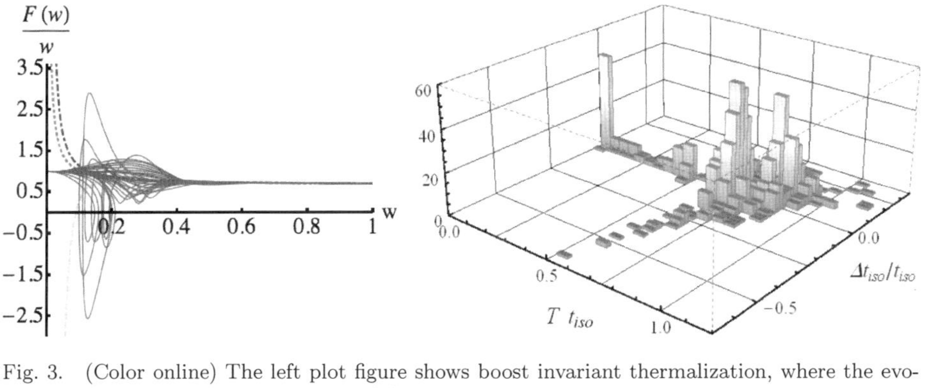

Fig. 3. (Color online) The left plot figure shows boost invariant thermalization, where the evolution of 29 random initial states of $F(w) \equiv \frac{\tau}{w}\partial_\tau w$ as a function of $w \equiv T_{\rm eff}\tau$ is compared with predictions from first-, second- and third-order hydrodynamics (figure from Ref. 50). It can be seen that they all thermalize in a time where $\tau \lesssim 0.5/T_{\rm eff}$, where $T_{\rm eff}$ is the temperature associated with the energy density at that time. The right plot did a similar analysis in a homogeneous setup, using over 1000 initial states. Also there thermalization always happens within $t \lesssim 1/T$. Here it is also shown that the difference between the linearized and fully nonlinear Einstein equations is small, almost always less than 20% (figure from Ref. 51).

estimates of strongly coupled thermalization times were indeed accurate also at the nonlinear level.

Lastly, these studies can be corrected for finite coupling effects, which has recently been done in Refs. 53 and 54. Not surprisingly, finite coupling corrections slow down the thermalization, as in these papers becomes apparent through a smaller imaginary part of the lowest quasi-normal mode.

3.2. *Colliding shock waves*

All models presented above contained enough spacetime symmetry to make the gauge theory dynamics effectively $1+0$ dimensional. Here we shall review results where the gauge theory dynamics are allowed to be $1+1$ and $1+2$ dimensional. In holography the energy–momentum in the gauge theory directly corresponds to the metric in the gravitational theory, so to simulate highly energetic colliding nuclei, it is therefore natural to consider collisions of gravitational shock waves, moving at the speed of light.[57–62] The shock waves can be viewed as analogous to the shock waves constructed in flat space by Dray and 't Hooft,[63] which consist of taking a Schwarzschild black hole and boosting this to the speed of light, keeping its total energy fixed. The shock waves are dual to lumps of energy in the gauge theory, which propagate at the speed of light. When the shock waves collide they can produce a black hole which will then relax to equilibrium. Likewise, in the dual gauge theory, the collision will result in the formation of a non-Abelian plasma which will subsequently thermalize.

One interesting result from the aforementioned studies of shock wave collisions was the entropy production, estimated by the area of the trapped surface of the formed black hole at the time of collision. This trapped surface gives a lower bound in the area of the black hole horizon, which is directly related to the produced entropy in the gauge theory. The authors of Ref. 59 estimated that:

$$S \geq S_{\text{trapped}} \approx 35000 \left(\frac{\sqrt{s_{NN}}}{200\,\text{GeV}} \right)^{2/3}, \tag{10}$$

where S is the entropy produced in a central gold nucleus–nucleus collision with a nucleon–nucleon collision of energy $\sqrt{s_{NN}}$ and S_{trapped} is the entropy of the trapped surface. The entropy increase during the evolution of the quark–gluon plasma is not too big due too to the relatively small viscosity, but can in fact be sizable as highly anisotropic plasmas increase the entropy production. Nevertheless Eq. (10) gives a lower bound on the final entropy produced, which is directly related to the measurable total charged particle multiplicity $N_{\text{tot,ch}} = S/7.5$.[64] In Ref. 56 the estimate Eq. (10) was made somewhat stronger by doing calculations numerically till later times, but two important features remained. Firstly, the numerical value is about right for RHIC collisions, albeit on the high side, which is impressive as Eq. (10) is a direct calculation, without having any input from other experiments. Secondly, the 2/3 power law as a function of $\sqrt{s_{NN}}$ is a robust outcome of a scale invariant theory. Nevertheless, with the LHC results now known,[65] the experimental data favors a power of $1/2$, and Eq. (10) therefore overestimated the multiplicity at LHC collisions. This can be taken as an indication that QCD is not scale invariant at energies probed between RHIC and LHC.

The previous calculations could not say much about the dynamics after the collision, which requires solving the full nonlinear Einstein equations. It was therefore a major breakthrough when a real collision could be simulated within gravity.[66] This study collides planar gravitational shock waves, thereby simplifying the problem by neglecting transverse dynamics. The initial conditions for the shock wave are determined by the stress tensor

$$T_{\pm\pm}(z_\pm) = \kappa h(z_\pm), \tag{11}$$

where $z_\pm = t \pm z$ are null coordinates along the collision axis and $\kappa = N_c^2/2\pi^2$ is a measure of the degrees of freedom of the gauge theory. All other components of the stress tensor are zero, and within pure gravity this fully determines the metric before the collision. For this review we restrict h to be Gaussian:

$$h(z) = \frac{\mu^3}{w\sqrt{2\pi}} \exp\left[-\frac{z^2}{2w^2}\right], \tag{12}$$

where w is the width of the Gaussian and $\kappa\mu^3 = \kappa \int_{-\infty}^{\infty} h(z)dz$ is the total energy per transverse area.

In what follows we make the following choices for parameters. We choose $N_c = 1.8$ such that the ratio $e/T^4 \approx 12$ matches lattice QCD simulations.[59] The choice of $N_c < 3$ compensates for the fact that SYM has approximately three times as many degrees of freedom than QCD. We then fix μ by setting the

$$\kappa\mu^3 \approx \frac{\sqrt{s_{NN}}}{2} \frac{N_{\text{Pb}}}{\pi R^2}, \tag{13}$$

where $N_{\text{Pb}} = 206$ and $R = 6$ fm. The right-hand side of this equation is simply the energy density per unit area near the center of a lead nucleus. We choose the width of the Gaussian to be $w \approx \frac{4}{3} R m_N/\sqrt{s_{NN}}$, with m_N the nucleon mass. Of course real nuclei are not Gaussians (nor is QCD = SYM), so all these numbers have to be taken with a grain of salt. Nevertheless, these choices give a qualitative understanding of the scales in the problem.

After setting up these initial conditions, it is possible to numerically solve the Einstein equations (see Refs. 46 and 56 for an introduction), and thereafter extract the expectation value of the gauge theory stress tensor. Figure 4 shows the energy density obtained in this way for collisions having $\sqrt{s_{NN}} = 19.3\,\text{GeV}$ and $\sqrt{s_{NN}} = 2.76\,\text{TeV}$. For a proper comparison with Refs. 66 and 67, one should notice that the dimensionless product

$$\mu w = \left(\frac{\sqrt{s_{NN}}}{2\kappa} \frac{N_{\text{lead}}}{\pi R^2}\right)^{1/3} \frac{4}{3} R m_N/\sqrt{s_{NN}} \approx 23(\sqrt{s_{NN}}/\text{GeV})^{-2/3}, \tag{14}$$

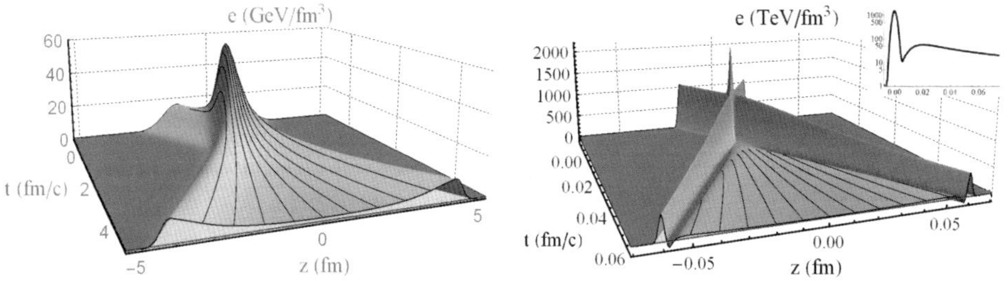

Fig. 4. (Color online) Energy density e for collisions of shock waves at low energy ($\sqrt{s_{NN}}$ = 19.3 GeV, left plot) and high energy ($\sqrt{s_{NN}}$ = 2.76 TeV, right plot) as a function of time and longitudinal coordinate z. The grey planes lie at the origin of the vertical axes. Note that these simulations are planar in the transverse plane and therefore neglect transverse dynamics. For the short time scales presented on the right plot this is probably a good approximation, but the simulation on the left will change when taking transverse expansion into account as well (figures adapted from Ref. 67).

so that 19.3 GeV and 2.76 TeV correspond to a μw of 3.2 and 0.11 respectively, whereby we stress once more that these are ballpark figures, which we think are helpful for a useful comparison of gauge/gravity results to actual heavy ion collisions.

Several curious features are visible in the resulting energy density.[67] Firstly, for the low energy collisions the peak energy density is about 60% higher than the sum of the incoming energies. This is interpreted as a "pile-up" of energy, as the shocks thermalize during the collision. For high energy shocks, there is not enough time for this effect, and the shocks therefore only lose their energy gradually to the plasma, as well as forming a transient region of negative energy density, where it is not even possible to define a local rest frame.[68] Curiously, for the high energy

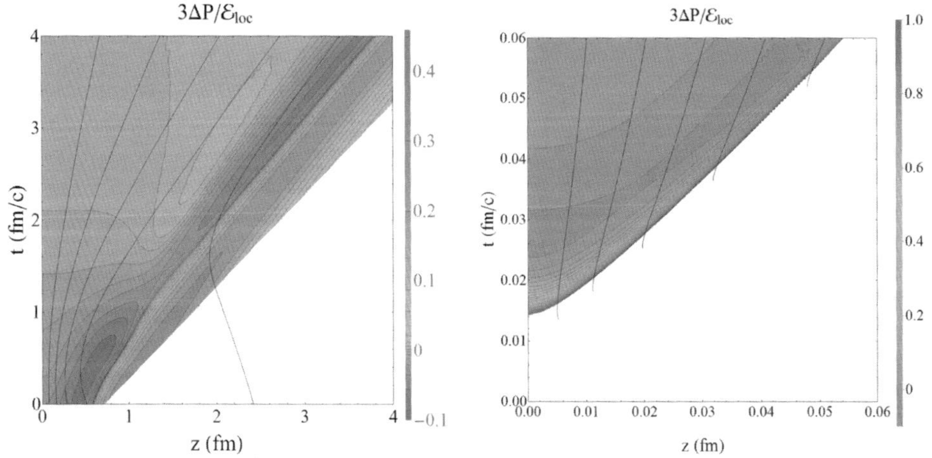

Fig. 5. (Color online) $3\Delta\mathcal{P}_L^{\rm loc}/\mathcal{E}_{\rm loc}$ for low (left plot) and high (right plot) energy shocks, with ΔP the difference between the longitudinal pressure from holography, and the one computed using Eq. (1). The white areas indicate regions outside the light cone or where hydrodynamics deviates by more than 100%. The black lines are again stream lines, which for the high energy shocks stop at the region where no local rest frame exists.[68]

shocks there seems to be a reasonably accurate analytic formulation — at least at small rapidities — by using a complexified boost symmetry, even valid in the far-from-equilibrium regime.[69] At later times the low and high energy collisions behave similarly though; they both thermalize fast (shown in Fig. 5) and deposit all their energy into the plasma.

Interestingly, the formed plasma is quite insensitive to the profile of the initial shocks, provided characteristic length of the profile is not too big.[70] This makes the local energy density of the resulting plasma robust at least for thin and hence high energy shocks. The rapidity profile of these collisions may even be universal at strong coupling, and is therefore of considerable interest (Fig. 6 (right)). The left plot displays the similar profile for low energy collisions, which looks similar in shape, but with a width that becomes smaller as the collision energy decreases.

3.3. *Including transverse dynamics*

Lastly, we would like to mention that it is also possible to include transverse dynamics. Transverse expansion is essential during the late time of the QGPs explosion, but also early time initial velocities in the transverse plane influence the transverse momenta spectra. Nevertheless, a full shock wave collision with transverse expansion requires a $3+1$ dimensional code in AdS spacetime, as there is always the extra AdS direction. As this can be technically complicated it is therefore natural to first try to include transverse dependence while assuming boost invariance in the longitudinal direction.

Assuming boost invariance is a bit subtle, as Fig. 6 convincingly shows a significant dependence on rapidity. Also, using proper time as a variable usually necessitates to start the gravitational simulation a finite time after the collision (see Ref. 71 for a counter example though), which has to be accompanied by an assumption on the dynamics before this time (early time expansion in the case to be presented). Nevertheless, starting the simulation at such a finite time with several simple profiles gave two lessons.[72,73] Firstly, the plasma again thermalized fast, at a similar

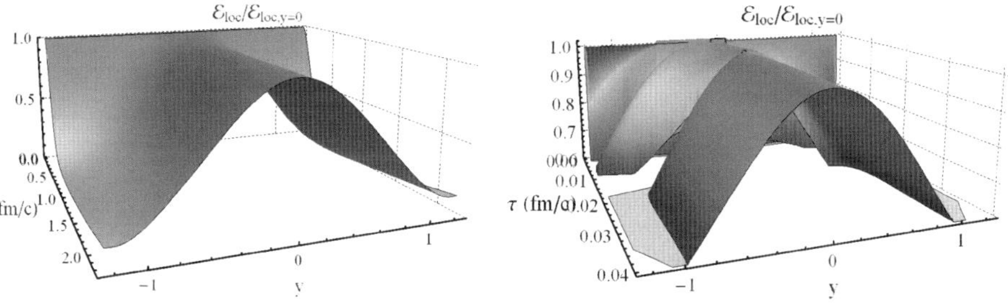

Fig. 6. (Color online) Energy density in the local rest frame around mid-rapidity as a function of spacetime rapidity y and proper time τ for thick (left plot) and thin (right plot) shocks (the same shocks as Fig. 4). On the right plot, we have excluded the region in which the local rest frame is not defined because $2|s| > |e + P_L|$ (figures adapted from Ref. 67).

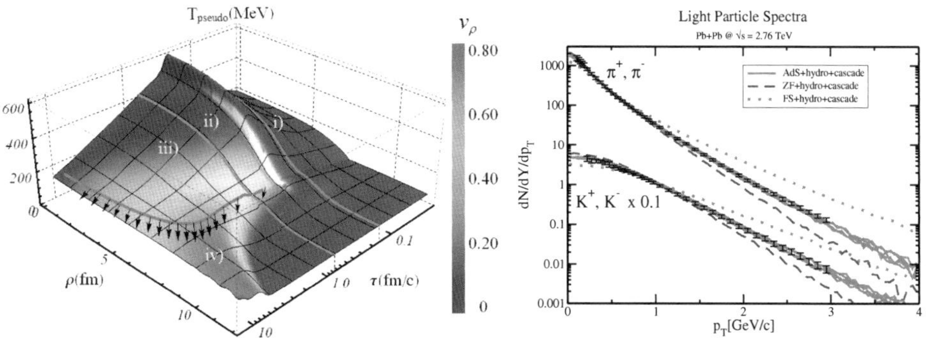

Fig. 7. (Color online) From the local energy density we plot the temperature (assuming equilibrium, therefore labeled T_{pseudo}) and the radial velocity $v_\rho = u_\rho/u_\tau$ for a representative simulation. The plot illustrates four physical tools used: (i) early time expansion, (ii) numerical AdS evolution, (iii) viscous hydrodynamics until $T = 0.17\,\text{GeV}$, (iv) kinetic theory after conversion into particles (indicated by arrows). The right plot shows the resulting particle spectra for pions and kaons, and also includes curves for a zero flow model (ZF) with $v_i = 0$ until $\tau = 1.0$ fm/c, and a free streaming model (FS), with zero longitudinal pressure (figure from Ref. 73, experimental data from ALICE[76]).

time scale as the examples above. Secondly, after the thermalization the transverse velocity was given to a good approximation by the following approximation[73,74]:

$$v_i = -0.33\tau\, \partial_i e/e, \tag{15}$$

with τ the proper time, and e the local energy density in the transverse plane. This formula is in a spirit similar to the universal pre-flow found in Ref. 75, but note that only the flux is universal in the sense of that paper, whereas here we are interested in the local fluid velocity, that depends on the pressures as well.

Using these simulations it was then possible to couple this to relatively standard hydrodynamic and cascade codes, and thereafter obtain the full transverse momenta spectra at mid-rapidity for a central collision, which fits data surprisingly well (Fig. 7). Recently this has also been extended to several other systems, most notably also computing the measurable HBT radii.[74]

4. Jet Energy Loss

4.1. *Are there jets at strong coupling?*

During the early stages of heavy ion collisions — in which the energy density is the highest — energetic partons can be created via hard processes. The resulting excitations — jets — consist of collimated sprays of energy and can traverse the fireball while depositing energy and momentum into the medium. Analysis of the particle correlations in the produced jets can provide useful information about the dynamics of the plasma including the rates of energy loss and momentum broadening.[77–80]

Why study jets and energy loss at strong coupling? One piece of evidence that the quark–gluon plasma produced in heavy ion collision is strongly coupled comes from jet quenching itself.[81–83] Jets traversing the fireball appear to lose their energy very quickly, indicating strong interactions with the medium.

A natural process by which jets can be created is the decay of off-shell particles such as photons. At weak coupling the decay leads to well-collimated sprays of energy with the polarization of the photon imprinting itself in the final state in a so-called "antenna pattern" (see for example Ref. 84 and references therein). However, at strong coupling the situation is very different: the decay of off-shell photons leads to states which are spherically symmetric (in the rest frame of the off-shell photon): all correlation functions at spatial infinity are spherically symmetric and there are no well collimated sprays of energy.[85]

What does this mean for studying jets at strong coupling? Does the observed isotropy mean there are no jets at strong coupling? Where does the isotropy come from? In other words, at what stage in the evolution does the isotropy develop? Does it come from the underlying photon decay or from the subsequent late time evolution as the energy propagates outward to infinity?

In order to understand the above issues, it was useful to study the propagation of radiation in strongly coupled holographic gauge theories. A simple setting to study this problem is that of radiation produced by an accelerated quark in strongly coupled conformal field theories. If one takes a heavy quark and accelerates it along some given trajectory, what does the pattern of emitted radiation look like? This problem is analogous to textbook studies of radiation produced by an accelerated charge in classical electrodynamics. However, instead of radiation carried by a classical $U(1)$ gauge field, the radiation is that of a strongly coupled non-Abelian quantum field theory. Is the emitted radiation isotropic at strong coupling? If not, does it isotropize as it propagates to infinity?

This problem was studied in Refs. 86 and 87. There it was found that the radiation produced by the accelerated quark is not isotropic. Nor does the radiation isotropize as it propagates out to infinity. Remarkably, the angular distribution of power produced by an accelerated quark is the same at weak and strong coupling! This suggests that the isotropy observed in the decay of off-shell photons at strong coupling comes from the production mechanism itself — i.e., the decay — and not the subsequent evolution.

The lesson from this is that one should not treat the jet production mechanism with strong coupling tools. Indeed, in asymptotically free QCD it is natural to expect the production mechanism during the early stages of heavy ion collisions to be governed by weakly coupled physics. Ideally, one should treat the initial hard physics with perturbative QCD and then study soft processes such as energy loss with strongly coupled tools. Simply put, the utility of modeling jets with strongly coupled tools comes from studying their evolution and energy loss *after* they are produced.

4.2. *Holographic models of jets*

While at strong coupling it is impossible to create well-collimated sprays of energy via off-shell photon decays, one can construct states at some initial time $t = t_i$ which share features with jets created via weakly coupled processes. Namely, one can construct states with excitations which are well collimated, have energy $E \gg T$ with T the temperature of the plasma, and which can propagate very far through the plasma before thermalizing. Ideally, one would compute the initial state using perturbative QCD. However, we shall take a different approach and focus on universal features which are insensitive to the precise initial conditions used to construct the state at time t_i. These include the rates of energy loss and momentum broadening and the penetration depth defined as the total distance a jet can propagate through plasma before thermalizing.

Let us focus for simplicity on $q\bar{q}$ jets in strongly coupled conformal field theories (for other holographic models of jets, see for example Refs. 88–91). According to holographic duality, a $q\bar{q}$ pair moving through plasma is dual to a classical string moving through a black hole geometry.[4] The geometry dual to an infinite static plasma at temperature T is the AdS-Schwarzschild geometry, where the metric may be written

$$ds^2 = \frac{L^2}{u^2}\left[-f(u)dt^2 + d\boldsymbol{x}^2 + \frac{du^2}{f(u)}\right], \tag{16}$$

where L is the AdS radius and $f(u) = 1 - u^4/u_h^4$. The horizon of the geometry is at radial coordinate $u = u_h$ with $u_h = 1/\pi T$. The boundary of the geometry, which is where the dual field theory lives, is at radial coordinate $u = 0$.

As it evolves in time the string sweeps out and $1+1$ dimensional *worldsheet*. Events on the worldsheet can be labeled by coordinates (τ, σ) which are related to events in spacetime via the embedding functions $X^M \equiv \{t(\tau,\sigma), \boldsymbol{x}(\tau,\sigma), u(\tau,\sigma)\}$. The dynamics of the embedding functions are governed by the Nambu–Goto action

$$S = -T_0 \int d\tau d\sigma \sqrt{-\gamma}, \tag{17}$$

where the string tension is $T_0 = \frac{\sqrt{\lambda}}{2\pi L^2}$ with λ the 't Hooft coupling and $\gamma = \det \gamma_{ab}$ with

$$\gamma_{ab} = \partial_a X \cdot \partial_b X, \tag{18}$$

the worldsheet metric. Here and below the indices (a, b) run over the worldsheet coordinates τ and σ. The action (17) is simply the area of the worldsheet.

Varying the action we obtain the string equations of motion

$$\partial_\tau \Pi^\tau_\mu + \partial_\sigma \Pi^\sigma_\mu, \tag{19}$$

where the string worldsheet currents Π^a_M are

$$\Pi^a_M = \frac{\delta S}{\delta(\partial_a X^M)} = -T_0\sqrt{-\gamma}\gamma^{ab} g_{MN} \partial_a X^N, \tag{20}$$

where g_{MN} is the AdS-Schwarzschild metric and as usual γ^{ab} is the inverse of γ_{ab}. Note that the energy of the string is

$$E = -\int d\sigma \Pi_0^\tau. \tag{21}$$

The string equations of motion (19) are simply the equations of energy conservation on the worldsheet with Π_0^σ the energy flux down the string.

The endpoints of the string are charged under a bulk $U(1)$ gauge field $\mathcal{A}_M$. The boundary of the AdS geometry behaves as an ideal conductor.[92] Hence the gauge field induces a current J^μ on the boundary, which up to an overall sign has the interpretation as the baryon current of a *dressed* quark.[92,93] This is shown schematically in Fig. 8. Therefore, one can regard the spatial location of the string endpoint as roughly coinciding with the spatial location of the quarks. The boundary conditions at the string endpoints are set by the mass of the quark.[4] For infinitely massive quarks the endpoint is fixed at the boundary. Correspondingly, the current J^μ is localized with delta function support. In contrast, for massless quarks the endpoints are free to fall in the AdS-Schwarzschild geometry. Likewise, the current J^μ spreads out as the endpoints fall. For intermediate mass quarks the endpoint is fixed to be at some radial coordinate u_m.

Via Einstein's equations the presence of the string also perturbs the AdS-Schwarzschild geometry. Just as the near boundary behavior of the gauge field $\mathcal{A}_M$ encodes the quark baryon current J^μ, the near-boundary behavior of the metric perturbation δg_{MN} encodes the stress tensor $T^{\mu\nu}$ of the system. This is also shown schematically in Fig. 8.

With the above preliminaries layed out, the basic strategy then will be to construct initial string data at some time t_i and evolve the string profile forward in time according to the string equations of motion (19). To echo the previous point,

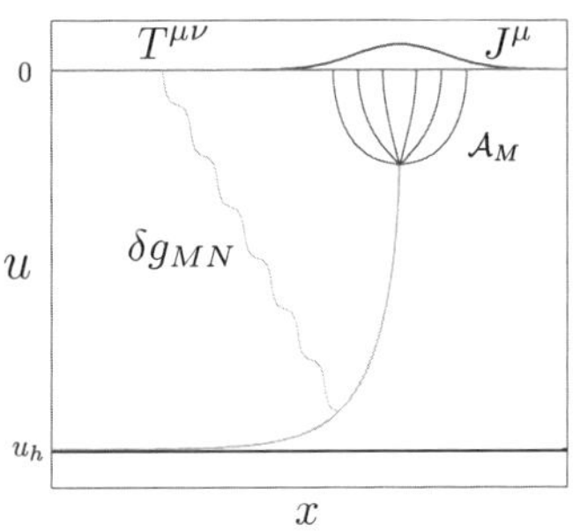

Fig. 8. (Color online) A cartoon showing a string at some fixed time in the AdS-Schwarzschild geometry. The endpoint of the string is charged under a $U(1)$ gauge field $\mathcal{A}_M$. The boundary of the AdS-Schwarzschild geometry — located at radial coordinate $u = 0$ — behaves as an electromagnetic conductor. Up to a minus sign the current J^μ induced on the boundary has the physical interpretation as the baryon current of the quark. Likewise, via Einstein's equations the string perturbs the AdS-Schwarzschild geometry. The near-boundary behavior of the metric perturbation encodes the stress tensor $T^{\mu\nu}$ of the jet.

we seek strings whose boundary energy density T^{00} is that of a well collimated spray of energy which propagates very far before thermalizing. We will then study jet quenching and other observables which are insensitive to the details of the precise form of the initial data. In what follows we will highlight results for infinitely massive quark jets and massless quark jets.

4.3. *Heavy quark energy loss*

One of the first studies[94–96] of energy loss via AdS/CFT came from the study of heavy quarks propagating at constant velocity v through a static box of plasma at temperature T. As the quark moves through the plasma it transfers energy and momentum to the plasma via frictional drag. Hence an external force is required to maintain constant velocity. This can be supplied by an electromagnetic field coupled to the quark's baryon current J^μ.

Under the assumption that the quark has been dragged at constant velocity for an arbitrary duration of time, it is natural to seek a steady-state solution to the string equations of motion. Choosing worldsheet coordinates $\tau = t$ and $\sigma = u$ the steady-state ansatz take the form

$$x(t,u) = vt + vx_0(u), \tag{22}$$

with the endpoint terminating at the boundary $u = 0$. Solving the string equations of motion (19) for $x_0(u)$ yields

$$x_0(u) = \frac{1}{2u_h}\left[\tan^{-1}\frac{u}{u_h} + \frac{1}{2}\log\frac{u_h - u}{u_h + u}\right]. \tag{23}$$

We plot $x_0(u)$ in Fig. 9.

The electromagnetic field dragging the quark (and hence the string endpoint) must be supplying energy at a rate equal to the rate energy is dissipated by drag. In the dual gravitational description the energy lost via drag is encoded by a flux of energy down the string towards the event horizon, $\frac{dE}{dt} = \Pi_0^\sigma$. Plugging the solution

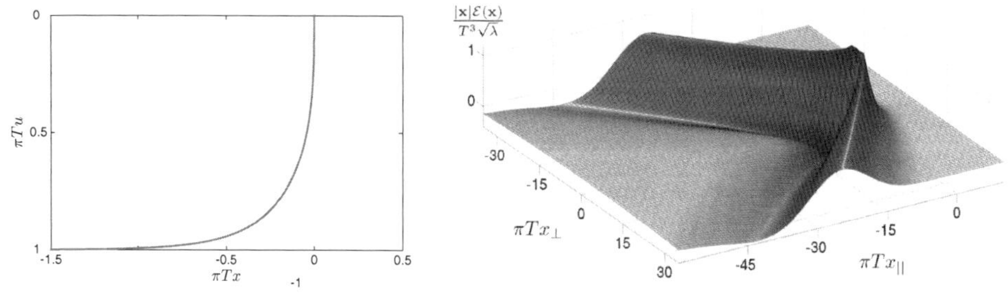

Fig. 9. (Color online) Left: the string profile (23) for a heavy quark. Right: the energy density produced by a heavy quark moving at three quarters the speed of light in the x-direction. A Mach cone is clearly present.

(23) into Eq. (20), the drag reads[94–96]

$$\frac{dE}{dt} = \frac{\pi}{2}\sqrt{\lambda}T^2 \frac{v^2}{\sqrt{1-v^2}}. \tag{24}$$

Hence, as the 't Hooft coupling $\lambda \to \infty$ the quark loses energy faster and faster.

A natural question to ask is where does the energy lost by the quark go? To answer this question it is useful to study the stress tensor $T^{\mu\nu}$. This requires solving Einstein's equations for the metric perturbation δg_{MN} due to the presence of the string and extracting the stress tensor from its near-boundary asymptotics. This calculation was carried out in Refs. 97–100. In the right panel of Fig. 9 we also plot the normalized energy density $\mathcal{E} \equiv \frac{2\pi^2}{N_c^2}T^{00}$ for a heavy quark moving in the x-direction at three quarters the speed of light. The speed of sound in a relativistic conformal field theory is $c_s = 1/\sqrt{3} \approx 0.57$. Hence the quark is moving supersonically. As is clear from the figure, the quark creates a Mach cone. Evidently, the energy lost via drag is carried away by sound waves.

There have since been numerous studies of heavy quark energy loss at strong coupling in more general settings. Some examples include heavy quark energy loss in non-conformal theories,[101] anisotropic plasmas[102] and in far-from-equilibrium states such as the shock wave collisions discussed above.[103] In the latter study it was demonstrated that even in far-from-equilibrium states, such as those produced by colliding shock waves, the equilibrium energy loss formula (24) provides a reasonable estimate of the heavy quark energy loss rate. Additionally, heavy quark energy loss due to deceleration (i.e., non steady-state) was studied in, Ref. 104, and the rates of heavy quark momentum broadening were studied in Refs. 105 and 106.

5. Light Quark Energy Loss

In contrast to strings dual to heavy quarks, where the endpoints are fixed to the AdS boundary, the endpoints of strings dual to massless quarks are allowed to fall unimpeded towards the event horizon of the AdS-Schwarzschild geometry.[4,92] For massless quarks the string equations of motion (19) must be solved subject to the open string boundary conditions

$$\Pi^\sigma_M = 0 \quad \text{at the endpoints.} \tag{25}$$

The fact that string endpoints fall toward the horizon implies there does not exist steady-state solutions to the string equations of motion. Hence studying light quark energy loss is considerably more complicated than studying heavy quark energy loss. Nevertheless, it is a straightforward matter to numerically solve the string equations of motion (19) with the open string boundary conditions (25). For details on how to do this, see for example Ref. 93.

In Fig. 10 we plot a numerically generated string profile at several values of fixed coordinate time t. At time $t = 0$ the initial string profile is simply a point located at the AdS boundary $u = 0$. As time progresses the string expands from a point to an

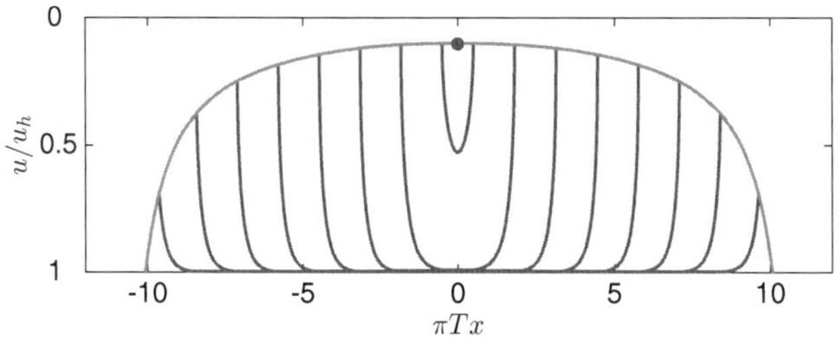

Fig. 10. (Color online) A numerically generated falling string profile dual to a light quark jet. The string is shown at several values of fixed coordinate time t. The string is created at the point $x = u = 0$ and subsequently expands into an extended object. The red lines denote the string endpoint trajectories. The endpoints travel a distance $x_{\text{stop}} \approx 10/\pi T$ before falling into the horizon at $u = u_h$.

extended object and the endpoints fall towards the horizon. The endpoints travel a total distance $x_{\text{stop}} \approx 10/\pi T$ before falling into the horizon. As the endpoint fall into the horizon, the boundary baryon current J^μ (depicted schematically in Fig. 8) becomes delocalized and starts to evolve hydrodynamical according to the diffusion equation.[92] This corresponds to the thermalization of the light quark jet.

The total distance the string endpoints and hence the jet can travel depends on initial conditions. A natural question is for a fixed energy E, what is the maximum distance the string endpoints can travel before falling into the horizon? In the language of the dual field theory, for fixed jet energy E, what is the maximum distance a light quark jet can travel before thermalizing? To answer this question, one can numerically generate an ensemble of solutions to the string equations of motion (19) with different initial conditions. One can then compute the energy (21) and compare the stopping distance x_{stop} to E. This computation was carried out in Ref. 93 where it was found that the maximum distance a jet with energy E can travel before thermalizing is given by

$$x_{\text{stop}} = \frac{\mathcal{C}}{T}\left(\frac{E}{T\sqrt{\lambda}}\right)^{1/3}, \tag{26}$$

where $C \approx 0.526$. The $x_{\text{stop}} \sim E^{1/3}$ scaling was also found in Ref. 107. The constant $\mathcal{C}$ was recently computed analytically and found to be[108]

$$\mathcal{C} = \frac{2^{1/3}\Gamma\left(\frac{5}{4}\right)}{\sqrt{\pi}\Gamma\left(\frac{3}{4}\right)}. \tag{27}$$

We now turn to the light quark energy loss rate. Following Ref. 109, we focus on strings whose x_{stop} is asymptotically large compared to $1/T$. Such string have worldsheets which are approximately null.[93,109] Why? When $x_{\text{stop}} \to \infty$ the scaling (26) requires $E \to \infty$. Since strings have finite tension the $E \to \infty$ limit is generically realized by strings that expand at nearly the speed of light, meaning that the string profile must be approximately that of an expanding filament of null dust. Indeed, null strings have profiles $X^M_{\rm null}$ satisfying $\gamma(X_{\rm null}) = 0$ and from (20) have divergent energy density.

Since null strings satisfy $\gamma(X_{\rm null}) = 0$ they minimize the Nambu–Goto action (17) and are exact albeit singular solutions to the string equations of motion (19). Following Ref. 109, to obtain finite energy solutions to the equations of motion one can expand the string embedding functions about a null string solution

$$X^M = X^M_{\rm null} + \epsilon\delta X^M_{(1)} + \epsilon^2\delta X^M_{(2)} + \dots, \tag{28}$$

where ϵ is a bookkeeping parameter (which can be set equal to 1 at the end of calculations). In what follows it is useful to choose worldsheet coordinates $\tau = t$ and σ such that $\dot{X}_{\rm null} \cdot X'_{\rm null} = 0$ and $\delta X_{(m)} = \{0, \delta\boldsymbol{x}, 0\}$. We denote the location of the string endpoints by $\sigma = \sigma_*$. The string equations of motion can then be solved perturbatively in powers of ϵ.

Focusing on strings propagating in the x-direction, the null string embedding functions can be written

$$X^M_{\rm null} = \{t, x_{\rm geo}(t,\sigma), 0, 0, u_{\rm geo}(t,\sigma)\}, \tag{29}$$

where for each σ $x_{\rm geo}$ and $u_{\rm geo}$ satisfy the null geodesic equations, which read

$$\frac{\partial x_{\rm geo}}{\partial t} = \frac{f}{\xi}, \tag{30}$$

$$\frac{\partial u_{\rm geo}}{\partial t} = \frac{f\sqrt{\xi^2 - f}}{\xi}, \tag{31}$$

where $\xi = \xi(\sigma)$. The parameter ξ determines the initial inclination of the geodesics in the $x - u$ plane and, more fundamentally, specifies the conserved spatial momentum associated with the geodesics, $f(u)^{-1}\partial x_{\rm geo}/\partial t = \xi^{-1}$. The geodesic equations have the solution

$$x_{\rm geo} = -\frac{u_h^2}{u_{\rm geo}}\,{}_2F_1\left(\frac{1}{4}, \frac{1}{2}; \frac{5}{4}; \frac{u_h^4}{\zeta u_{\rm geo}^4}\right) + C(\sigma), \tag{32}$$

where ${}_2F_1$ is the Gauss hypergeometric function, $\zeta \equiv 1/(1-\xi^2)$ and $C(\sigma)$ is an arbitrary function which can be chosen such that $x_{\rm geo}(t=0,\sigma) = 0$.

Henceforth let us focus on strings created at the point $x = u = t = 0$ at the AdS boundary. At asymptotically early times $t \ll u_h$, when the string is close to the AdS boundary, $f = 1$ and the geodesics are given by

$$x_{\rm geo} = t\cos\sigma, \quad u_{\rm geo} = t\sin\sigma. \tag{33}$$

Hence, the worldsheet coordinate σ is simply the initial angle of the geodesic in the $x - u$ plane. Likewise,

$$\xi(\sigma) = \sec\sigma. \tag{34}$$

In Fig. 11 we plot a null string generated by a congruence of geodesics with $\sigma_* = 0.025$. The string profile, denoted by the blue curves, is shown at several

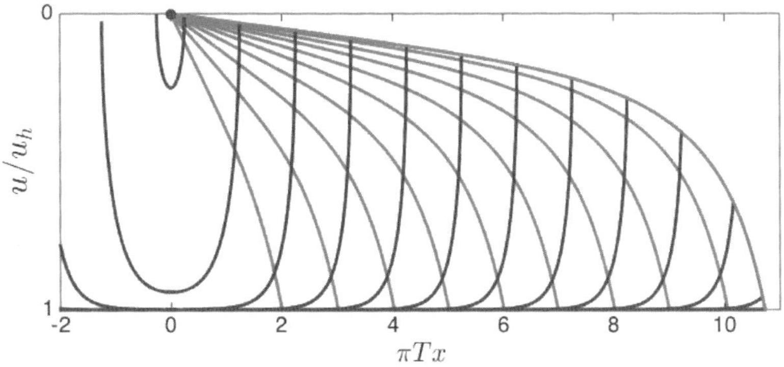

Fig. 11. (Color online) A null string (blue) shown at several different coordinate times t. The string starts off at a point on the boundary and expands at the speed of light while falling towards the horizon. The rust colored curves represent the null geodesics that make up the string and the red curve is the endpoint trajectory.

values of coordinate time t. The string starts off at a point on the boundary and expands at the speed of light while falling towards the horizon. The rust colored curves represent the null geodesics that make up the string and the red curve is the endpoint trajectory. The total distance the string endpoint can travel is entirely controlled by the angle σ_*. As we elaborate on further below, in the limit $\sigma_* \to 0$ the stopping distance diverges.

To study energy loss rate, we must compute the first-order corrections to the string embedding functions. At leading order in the bookkeeping parameter ϵ the worldsheet energy density Π_0^τ and flux Π_0^σ read

$$\Pi_0^\tau = -\frac{T_0 \xi \partial_\sigma u_{\text{geo}}}{u_{\text{geo}}^2} \sqrt{\frac{-\xi}{2\epsilon f \partial_t \delta x_{(1)}}} + O(\sqrt{\epsilon}), \quad \Pi_0^\sigma = O(\sqrt{\epsilon}). \tag{35}$$

Hence at leading order in ϵ the string equations of motion (19) simply read $\partial_t \Pi_0^0 = 0$ so $\Pi_0^\tau = \Pi_0^\tau(\sigma)$. In other words, energy is simply transported on the congruence of geodesics which make up the null string.

We define the energy

$$E(x) \equiv -\int_{\sigma_*}^{\sigma_H(x)} \Pi_0^0(\sigma), \tag{36}$$

where $\sigma_H(x)$ is the σ corresponding the geodesic which impacts the horizon at x. In particular, $\sigma_H(x = x_{\text{stop}}) = \sigma_*$. The energy $E(x)$ is simply the total string energy that makes it past the point x. The energy loss rate per unit length is

$$\frac{dE}{dx} = -\Pi_0^\tau(\sigma_H(x))\sigma_H'(x). \tag{37}$$

We discuss the precise boundary interpretation of dE/dx below. To compute dE/dx we must compute Π_0^τ and $\sigma_H(x)$.

Because Π_0^τ is time independent it is only necessary to compute it at an initial time. Near the AdS boundary, where $f = 1$ and the geodesics are given by (33), the string equation of motion $\partial_t \Pi_0^\tau = 0$ leads to the equation of motion for $\delta x_{(1)}$

$$\partial_t^2 \delta x_{(1)} + \frac{2}{t} \partial_t \delta x_{(1)} = 0, \tag{38}$$

which has the solution

$$\delta x(t, \sigma) = \phi(\sigma) + \frac{1}{t} \psi(\sigma), \tag{39}$$

for arbitrary functions $\phi(\sigma)$ and $\psi(\sigma)$. The open string boundary conditions (25) require $\psi(\sigma_*) = 0$.

Substituting the string solution (39) and the geodesic solution (33) and (34) into (35) we obtain

$$\Pi_0^\tau = -T_0 \csc^2 \sigma \sqrt{\frac{\csc 2\sigma \sin \sigma}{\epsilon \psi(\sigma)}} + O(\sqrt{\epsilon}). \tag{40}$$

With Π_0^τ computed and the congruence of null geodesics specified we can now compute dE/dx. But what is $\psi(\sigma)$! The energy loss rate depends on an arbitrary function. At first sight this seems like a disaster. Not only can we not describe the production mechanism of jets via holography but their energy loss rate, at least for massless jets, seems to be very sensitive to initial conditions. However, this grim diagnosis turns out to be incorrect: the ambiguities in the light quark energy loss rate are largely transient effects which die out over time scales of order $1/T$. These effects are negligible in the high energy limit when $x_{\text{stop}} \to \infty$.

Let us consider $x_{\text{stop}} \to \infty$ limit with x/x_{stop} fixed and first compute $\sigma_H(x)$. $\sigma_H(x)$ can be computed from the geodesics solution (32) by setting $u_{\text{geo}} = u_h$ and solving the resulting transcendental equation for σ. With ξ given by (34) the resulting equation can be solved analytically in the limit $\sigma \to 0$ (i.e., for geodesics which travel very far before falling into the horizon). The solution reads $\sigma_H(x) = \frac{u_h^2 \Gamma(\frac{1}{4})^2}{16\pi x^2}$. Likewise, setting $\sigma_H(x = x_{\text{stop}}) = \sigma_*$, we see that the stopping distance is related to σ_* via

$$x_{\text{stop}} = \frac{u_h \Gamma\left(\frac{1}{4}\right)^2}{4\sqrt{\pi \sigma_*}}. \tag{41}$$

As previously advertised, $x_{\text{stop}} \to \infty$ when $\sigma_* \to 0$. We therefore may write

$$\sigma_H(x) = \sigma_* \left(\frac{x_{\text{stop}}}{x}\right)^2. \tag{42}$$

Therefore, when we take $x_{\text{stop}} \to \infty$ with x/x_{stop} fixed, we are forced to take $\sigma_H \to 0$. Simply put, when $x_{\text{stop}} \to \infty$, energy is transported along geodesics which originate asymptotically close to the string endpoint. Only these geodesics propagate to distances $x = O(x_{\text{stop}})$. This behavior can clearly be seen in Fig. 10.

Therefore, to get the energy loss rate we may expand Eq. (37) about $\sigma_H \sim \sigma_* \to 0$. The function $\psi(\sigma)$ may be approximated by $\psi(\sigma) \approx \psi'(\sigma_*)(\sigma - \sigma_*)$. Hence, near the string endpoint

$$\Pi_0^\tau \approx -\frac{T_0}{\sigma^2\sqrt{2\epsilon\psi'(\sigma_*)(\sigma-\sigma_*)}}. \tag{43}$$

The energy loss will depend on the unknown parameter $\psi'(\sigma_*)$ which can be repackaged in terms of the initial energy

$$E(x=0) = \frac{\pi T_0}{2\sigma_*^{3/2}\sqrt{2\epsilon\psi'(\sigma_*)}} + O(1/\sqrt{\sigma_*}). \tag{44}$$

Substituting (43) and (41) in (37) and dividing by (44) we then secure[109]

$$\frac{1}{E(x=0)}\frac{dE}{dx} = -\frac{4x^2}{\pi x_{\text{stop}}^2\sqrt{x_{\text{stop}}^2 - x^2}}. \tag{45}$$

Equation (45) together with the stopping distance (26) provides a complete description of energy loss in terms of the initial energy, 't Hooft coupling λ and temperature T.

We now return to the boundary interpretation of dE/dx. It turns out that it is possible to solve the gravitational backreaction problem analytically in the long wavelength limit.[d] In doing so the hydrodynamic limit of the stress tensor $T^{\mu\nu}_{\text{hydro}}$ can be extracted from the gravitational field perturbation in closed form. How is the string energy loss rate (45) encoded in $T^{\mu\nu}_{\text{hydro}}$? The answer is

$$\partial_\mu T^{\mu\nu}_{\text{hydro}} = f^\nu(x)\delta^3(\boldsymbol{x} - \hat{x}t), \tag{46}$$

with $f^0 = f^x = \frac{dE}{dx}\theta(x_{\text{stop}} - x)$. Simply put, the energy loss rate (45) describes the rate in which energy is transferred from the jet to hydrodynamic modes.

The light quark energy loss rate (45) stands in stark contrast to the heavy quark counterpart (24). Heavy quarks lose most of their energy in the early stages of their trajectory when they are moving fastest. In contrast, from (45) we see that the light quark energy loss rate actually increases as x increases! This behavior, which was first pointed out in Ref. 93, is reminiscent of a Bragg peak where the energy loss rate near thermalization is highest. Indeed, the energy loss rate (45) diverges as $x \to x_{\text{stop}}$.[e]

Recently the light quark energy loss rate (45) was employed in a phenomenological hybrid model of jet quenching which included both strong and weak coupling physics.[78,79] In the hybrid model the jet consists of a shower of partons which can split as the jet evolves. The splitting events were treated using perturbative QCD.

[d] This will be expanded upon in a coming paper.

[e] The rate that energy flows into hydrodynamic modes is only defined over time and length scales $\gg 1/T$. As such (45) should be regarded as the gradient expansion of the instantaneous jet energy loss rate. The gradient expansion of a function $f(x)$ can diverge even if $f(x)$ is itself finite. Hence, the instantaneous energy loss rate need not diverge as $x \to x_{\text{stop}}$.

Via soft interactions, in between splitings the partons can interact with the thermal medium and lose energy. The energy loss rate of the individual partons was treated using the light quark energy loss rate (45). The coefficient of the $E^{1/3}$ scaling in the stopping distance (26) was allowed to vary providing a one parameter fit to data. This gave a good fit to the data provided that the resulting stopping distances are a factor 3 to 4 times larger in QCD than that in SYM at the same temperature and energy. This is natural since SYM has more degrees of freedom for energetic partons to interact with than QCD.

6. Discussion with Future Perspectives

We hope to have shown some remarkable applications of holography to heavy ion collisions, while emphasizing that these results were obtained in a theory quite different from QCD itself. Most notably almost all results are obtained in the limit of infinitely strong coupling. Nevertheless, many of the works presented promise to have a degree of universality: they may apply to a large range of strongly coupled gauge theories. One particular example of such a hope is Fig. 6, which could display a universal rapidity profile of the plasma formed in high energy collisions in a strongly coupled gauge theories. Indications are that this profile can match experimental data for RHIC energies, while being perhaps too narrow at LHC energies.

The latter may also be expected; it was already known that LHC collisions did not agree well with the holographic prediction of the total multiplicity of charged particles,[59,61] as expanded upon in Subsection 3.2. One natural explanation would be that at such high energies the coupling constant is intermediate, thereby invalidating the assumption of an infinitely strong coupling. Indeed models with weak coupling tend to give wider rapidity distributions, so from RHIC to LHC we may be watching a cross-over. On the other hand, it is necessary to make the AdS/CFT models more realistic before drawing a firm conclusion, which can for instance be done by having less symmetry, preferably containing event-by-event fluctuations in the initial profile.

While this review naturally restricts to heavy ion collisions, we also wish to mention that holography is used for a much broader range of strongly coupled systems. These include non-fermi liquids[110] and holographic superconductors,[111] whereby the latter are believed to possibly have some relation with high temperature superconductors.

Lastly, we wish to express our excitement for the coming time, with several new experimental runs with different energies, different nuclei and higher statistics. Including features such as event-by-event distributions, nontrivial correlations, rare photon jets will give a very constraining data set, which will make it possible to distinguish all scenarios presented. This will hopefully give major lessons for both QCD, quark–gluon plasma physics, and applications of the gauge/gravity duality in a broad sense.

Acknowledgments

We thank Krishna Rajagopal for interesting discussions and comments on the manuscript. WS wishes to thank Michal Heller, David Mateos, Jorge Casalderrey, Paul Romatschke and Scott Pratt for collaborations on much of the work presented here. PC is supported by the Fundamental Laws Initiative of the Center for the Fundamental Laws of Nature at Harvard University. WS is supported by the U.S. Department of Energy under Grant Contract Number DE-SC0011090.

References

1. J. M. Maldacena, *AIP Conf. Proc.* **484** (1999) 51–63.
2. K. Aamodt *et al.*, *Phys. Rev. Lett.* **105** (2010) 252302.
3. E. Witten, *Adv. Theor. Math. Phys.* **2** (1998) 505–532.
4. A. Karch and E. Katz, *JHEP* **0206** (2002) 43.
5. E. Witten, *Adv. Theor. Math. Phys.* **2** (1998) 253–291.
6. O. Aharony, S. S. Gubser, J. M. Maldacena, H. Ooguri and Y. Oz, *Phys. Rept.* **323** (2000) 183–386.
7. S. Hawking, *Commun. Math. Phys.* **43** (1975) 199–220.
8. S. Bhattacharyya, V. E. Hubeny, S. Minwalla and M. Rangamani, *JHEP* **0802** (2008) 45.
9. V. E. Hubeny, S. Minwalla and M. Rangamani, The fluid/gravity correspondence, in *String Theory and Its Applications: From MeV to the Planck Scale (TASI 2010)*, eds. M. Dine, T. Banks and T. Sachdev (World Scientific, 2012), pp. 817–860, Conference C10-06-01.1.
10. L. D. Landau and E. M. Lifshits, *Fluid Mechanics* (Pergamon Press, 1959).
11. C. Eckart, *Phys. Rev.* **58** (1940) 919–924.
12. P. B. Arnold, G. D. Moore and L. G. Yaffe, *JHEP* **0011** (2000) 1.
13. H. B. Meyer, *Eur. Phys. J. A* **47** (2011) 86.
14. G. Policastro, D. Son and A. Starinets, *Phys. Rev. Lett.* **87** (2001) 081601.
15. P. Kovtun, D. T. Son and A. O. Starinets, *Phys. Rev. Lett.* **94** (2005) 111601.
16. Y. Kats and P. Petrov, *JHEP* **0901** (2009) 44.
17. A. Buchel, J. T. Liu and A. O. Starinets, *Nucl. Phys. B* **707** (2005) 56–68.
18. A. Buchel, *Nucl. Phys. B* **803** (2008) 166–170.
19. R. C. Myers, M. F. Paulos and A. Sinha, *Phys. Rev. D* **79** (2009) 041901.
20. P. Romatschke and U. Romatschke, *Phys. Rev. Lett.* **99** (2007) 172301.
21. U. Heinz, C. Shen and H. Song, *AIP Conf. Proc.* **1441** (2012) 766–770.
22. B. Schenke, P. Tribedy and R. Venugopalan, *Phys. Rev. C* **86** (2012) 034908.
23. C. Gale, S. Jeon and B. Schenke, *Phys. Rev. Lett.* **110** (2013) 012302.
24. M. Luzum and P. Romatschke, *Phys. Rev. C* **78** (2008) 034915.
25. R. Baier, P. Romatschke, D. T. Son, A. O. Starinets and M. A. Stephanov, *JHEP* **0804** (2008) 100.
26. R. Loganayagam, *JHEP* **0805** (2008) 87.
27. J. Erdmenger, M. Haack, M. Kaminski and A. Yarom, *JHEP* **0901** (2009) 055.
28. R. P. Geroch, *J. Math. Phys.* **36** (1995) 4226.
29. W. A. Hiscock and L. Lindblom, *Phys. Rev. D* **31** (1985) 725–733.
30. I. Muller, *Z. Phys.* **198** (1967) 329–344.
31. W. Israel and J. Stewart, *Annals Phys.* **118** (1979) 341–372.
32. R. Baier and P. Romatschke, *Eur. Phys. J. C* **51** (2007) 677–687.
33. P. Romatschke, *Int. J. Mod. Phys. E* **19** (2010) 1–53.

34. H. Song and U. W. Heinz, *Phys. Lett. B* **658** (2008) 279–283.
35. M. P. Heller, R. A. Janik and P. Witaszczyk, *Phys. Rev. Lett.* **110** (2013) 211602.
36. N. Banerjee, J. Bhattacharya, S. Bhattacharyya, S. Dutta, R. Loganayagam *et al.*, *JHEP* **1101** (2011) 94.
37. D. Kharzeev and A. Zhitnitsky, *Nucl. Phys. A* **797** (2007) 67–79.
38. D. E. Kharzeev, *Prog. Part. Nucl. Phys.* **75** (2014) 133–151.
39. Y. Hirono, T. Hirano and D. E. Kharzeev, The chiral magnetic effect in heavy-ion collisions from event-by-event anomalous hydrodynamics (2014), arXiv:1412.0311.
40. D. T. Son and P. Surowka, *Phys. Rev. Lett.* **103** (2009) 191601.
41. A. Gynther, K. Landsteiner, F. Pena-Benitez and A. Rebhan, *JHEP* **1102** (2011) 110.
42. S. S. Gubser, *Phys. Rev. D* **82** (2010) 085027.
43. G. S. Denicol, U. W. Heinz, M. Martinez, J. Noronha and M. Strickland, *Phys. Rev. Lett.* **113**(20) (2014) 202301.
44. L. Lehner, *Class. Quant. Grav.* **18** (2001) R25–R86.
45. F. Pretorius, *Phys. Rev. Lett.* **95** (2005) 121101.
46. P. M. Chesler and L. G. Yaffe, *JHEP* **1407** (2014) 086.
47. G. T. Horowitz and V. E. Hubeny, *Phys. Rev. D* **62** (2000) 024027.
48. P. M. Chesler and L. G. Yaffe, *Phys. Rev. Lett.* **102** (2009) 211601.
49. P. M. Chesler and L. G. Yaffe, *Phys. Rev. D* **82** (2010) 026006.
50. M. P. Heller, R. A. Janik and P. Witaszczyk, *Phys. Rev. Lett.* **108** (2012) 201602.
51. M. P. Heller, D. Mateos, W. van der Schee and D. Trancanelli, *Phys. Rev. Lett.* **108** (2012) 191601.
52. M. P. Heller, D. Mateos, W. van der Schee and M. Triana, *JHEP* **1309** (2013) 26.
53. D. Steineder, S. A. Stricker and A. Vuorinen, *Phys. Rev. Lett.* **110**(10) (2013) 101601.
54. S. A. Stricker, *Eur. Phys. J. C* **74** (2014) 2727.
55. J. Bjorken, *Phys. Rev. D* **27** (1983) 140–151.
56. W. van der Schee, *Gravitational Collisions and the Quark–Gluon Plasma*, PhD thesis, Utrecht University (2014).
57. R. A. Janik and R. B. Peschanski, *Phys. Rev. D* **73** (2006) 045013.
58. J. L. Albacete, Y. V. Kovchegov and A. Taliotis, *JHEP* **0807** (2008) 100.
59. S. S. Gubser, S. S. Pufu and A. Yarom, *Phys. Rev. D* **78** (2008) 066014.
60. D. Grumiller and P. Romatschke, *JHEP* **0808** (2008) 27.
61. S. Lin and E. Shuryak, *Phys. Rev. D* **79** (2009) 124015.
62. I. Y. Aref'eva, A. Bagrov and E. Guseva, *JHEP* **0912** (2009) 9.
63. T. Dray and G. 't Hooft, *Nucl. Phys. B* **253** (1985) 173.
64. B. Muller and K. Rajagopal, *Eur. Phys. J. C* **43** (2005) 15–21.
65. A. Toia, *J. Phys. G* **38** (2011) 124007.
66. P. M. Chesler and L. G. Yaffe, *Phys. Rev. Lett.* **106** (2011) 021601.
67. J. Casalderrey-Solana, M. P. Heller, D. Mateos and W. van der Schee, *Phys. Rev. Lett.* **111** (2013) 181601.
68. P. Arnold, P. Romatschke and W. van der Schee, *JHEP* **1410** (2014) 110.
69. S. S. Gubser and W. van der Schee, *JHEP* **01** (2014) 28.
70. J. Casalderrey-Solana, M. P. Heller, D. Mateos and W. van der Schee, *Phys. Rev. Lett.* **112** (2014) 22.
71. M. P. Heller, R. A. Janik and P. Witaszczyk, *Phys. Rev. D* **85** (2012) 126002.
72. W. van der Schee, *Phys. Rev. D* **87** (2013) 061901.
73. W. van der Schee, P. Romatschke and S. Pratt, *Phys. Rev. Lett.* **111** (2013) 222302.
74. M. Habich, J. Nagle and P. Romatschke, *Eur. Phys. J. C* **75** (2015) 15.
75. J. Vredevoogd and S. Pratt, *Phys. Rev. C* **79** (2009) 044915.
76. B. Abelev *et al.*, *Phys. Rev. Lett.* **109** (2012) 252301.

77. J. Casalderrey-Solana, E. Shuryak and D. Teaney, Hydrodynamic flow from fast particles (2006), arXiv:hep-ph/0602183.
78. J. Casalderrey-Solana, D. C. Gulhan, J. G. Milhano, D. Pablos and K. Rajagopal, *JHEP* **1410** (2014) 19.
79. J. Casalderrey-Solana, D. C. Gulhan, J. G. Milhano, D. Pablos and K. Rajagopal, *Nucl. Phys. A* **931** (2014) 487–492.
80. A. Adare *et al.*, *Phys. Rev. Lett.* **98** (2007) 172301.
81. S. Adler *et al.*, *Phys. Rev. Lett.* **97** (2006) 052301.
82. S. Chatrchyan *et al.*, *Phys. Rev. C* **84** (2011) 024906.
83. G. Aad *et al.*, *Phys. Rev. Lett.* **105** (2010) 252303.
84. W. T. Giele, D. A. Kosower and P. Z. Skands, *Phys. Rev. D* **78** (2008) 014026.
85. D. M. Hofman and J. Maldacena, *JHEP* **0805** (2008) 12.
86. C. Athanasiou, P. M. Chesler, H. Liu, D. Nickel and K. Rajagopal, *Phys. Rev. D* **81** (2010) 126001.
87. Y. Hatta, E. Iancu, A. Mueller and D. Triantafyllopoulos, *Nucl. Phys. B* **850** (2011) 31–52.
88. P. M. Chesler, Y.-Y. Ho and K. Rajagopal, *Phys. Rev. D* **85** (2012) 126006.
89. P. Arnold and D. Vaman, *JHEP* **1010** (2010) 99.
90. P. Arnold and D. Vaman, *JHEP* **1104** (2011) 27.
91. P. Arnold and D. Vaman, *J. Phys. G* **38** (2011) 124175.
92. P. M. Chesler, K. Jensen and A. Karch, *Phys. Rev. D* **79** (2009) 025021.
93. P. M. Chesler, K. Jensen, A. Karch and L. G. Yaffe, *Phys. Rev. D* **79** (2009) 125015.
94. C. Herzog, A. Karch, P. Kovtun, C. Kozcaz and L. Yaffe, *JHEP* **0607** (2006) 13.
95. S. S. Gubser, *Phys. Rev. D* **74** (2006) 126005.
96. J. Casalderrey-Solana and D. Teaney, *Phys. Rev. D* **74** (2006) 085012.
97. P. M. Chesler and L. G. Yaffe, *Phys. Rev. Lett.* **99** (2007) 152001.
98. P. M. Chesler and L. G. Yaffe, *Phys. Rev. D* **78** (2008) 045013.
99. S. S. Gubser, S. S. Pufu and A. Yarom, *Phys. Rev. Lett.* **100** (2008) 012301.
100. J. J. Friess, S. S. Gubser, G. Michalogiorgakis and S. S. Pufu, *Phys. Rev. D* **75** (2007) 106003.
101. A. Ficnar, J. Noronha and M. Gyulassy, *J. Phys. G* **38** (2011) 124176.
102. M. Chernicoff, D. Fernandez, D. Mateos and D. Trancanelli, *JHEP* **1208** (2012) 41.
103. P. M. Chesler, M. Lekaveckas and K. Rajagopal, *JHEP* **1310** (2013) 13.
104. M. Chernicoff and A. Guijosa, *JHEP* **0806** (2008) 5.
105. J. Casalderrey-Solana and D. Teaney, *JHEP* **0704** (2007) 39.
106. S. S. Gubser, *Nucl. Phys. B* **790** (2008) 175–199.
107. S. S. Gubser, D. R. Gulotta, S. S. Pufu and F. D. Rocha, *JHEP* **0810** (2008) 52.
108. A. Ficnar and S. S. Gubser, *Phys. Rev. D* **89** (2014) 026002.
109. P. M. Chesler and K. Rajagopal, *Phys. Rev. D* **90** (2014) 025033.
110. T. Faulkner, H. Liu, J. McGreevy and D. Vegh, *Phys. Rev. D* **83** (2011) 125002.
111. S. A. Hartnoll, C. P. Herzog and G. T. Horowitz, *Phys. Rev. Lett.* **101** (2008) 031601.

Jet Structure in Heavy Ion Collisions

J.-P. Blaizot
Institut de Physique Théorique de Saclay,
CNRS/URA2306, CEA Saclay,
F-91191 Gif-sur-Yvette, France

Y. Mehtar-Tani
Institute for Nuclear Theory,
University of Washington,
Seattle, WA 98195-1550, USA

We review recent theoretical developments in the study of the structure of jets that are produced in ultra relativistic heavy ion collisions. The core of the review focusses on the dynamics of the parton cascade that is induced by the interactions of a fast parton crossing a quark–gluon plasma. We recall the basic mechanisms responsible for medium induced radiation, underline the rapid disappearance of coherence effects, and the ensuing probabilistic nature of the medium induced cascade. We discuss how large radiative corrections modify the classical picture of the gluon cascade, and how these can be absorbed in a renormalization of the jet quenching parameter $\hat{q}$. Then, we analyze the (wave)-turbulent transport of energy along the medium induced cascade, and point out the main characteristics of the angular structure of such a cascade. Finally, color decoherence of the in-cone jet structure is discussed. Modest contact with phenomenology is presented towards the end of the review.

1. Introduction

A *jet* is a cone-shaped beam of energetic particles which originate from a hard parton produced in a high energy collision. Jets have played an important role in the development of Qauntum Chromodynamics (QCD), with three jet events in e^+e^- annihilation providing a direct experimental evidence for the existence of the gluon, and the quark–gluon vertex.[1,2] Moreover, perturbative QCD correctly describes the soft hadronic activity surrounding jets. For instance, the inclusive intrajet hadron distribution, the so-called fragmentation function, exhibits a humpbacked shape that reflects the suppression of soft gluon radiation at large angle as a result of destructive interferences, as predicted by QCD.[3–5] Today, jet technology has become a precision tool embodied in powerful Monte Carlo event generators,[6–8] and involve elaborate perturbative QCD calculations (culminating at the NNLO level), that are not only relevant for further testing strong and weak interactions at high energy, but also in searching for physics beyond the Standard Model at the LHC (see e.g. Refs. 9 and 10 recent reviews).

In the context of heavy ion collisions, jets occupy a very special place. They are produced in hard collisions over tiny space time regions, and the partons that originate form such hard collisions propagate through the hot matter produced in the bulk of the heavy ion collisions. As such they constitute ideal "test particles" whose study could reveal the properties of the medium they traverse before hadronizing. It was recognized long ago,[11] for instance, that quark and gluon jets may lose energy while crossing a quark–gluon plasma produced in heavy ion collisions, resulting in a strong suppression of high-p_t hadrons. This phenomenon, commonly referred to as *jet-quenching*, was in fact observed at RHIC[12–15] and at the LHC,[16–18] providing direct evidence for the creation of dense QCD mater in heavy ion collisions. The high energy of the LHC make jets a very powerful diagnostic tool: they are produced in large number, have large energies (with p_t ranging from 100 to 300 GeV, an order of magnitude larger than at RHIC). Furthermore, the detector capabilities of ALICE, ATLAS and CMS, allow for the exploration of a variety of new and more detailed observables, among which, the jet nuclear modification factor, the dijet asymmetry, the jet profiles.[16,18–27] All this is opening new perspectives in the field and invites new theoretical efforts.

The theory of jet quenching in heavy ion collisions started developing in the early '90s with the first studies of medium induced gluon bremsstrahlung by Gyulassy and Wang,[28,29] accompanied by those of Baier, Dokshitzer, Peigné, Mueller, Schiff[30–32] and Zakharov,[33,34] where it was realized that radiative parton energy loss is the dominant mechanism for jet quenching in large QCD media. This pioneering works were followed by many others emphasizing various aspects of the phenomena and using different formalisms. For instance, assuming the medium to be dilute, Gyulassy, Levai and Vitev[35,36] (see also Ref. 37) developed a formalism in which the interactions with the plasma can be calculated order by order in opacity. In a similar approach, due to Guo and Wang, the so-called High-Twist (HT),[38,39] the first medium correction to the the leading order splitting functions is computed and embedded in a medium-modified version of the DGLAP equation. Considering the plasma to be dense, an approximation of multiple soft collisions was applied by Armesto, Salgado and Wiedemann[40,41] where multiple gluon radiations are treated as independent and described with a Poisson distribution.[42] An analogous approach, accounting for thermal effects, is due to Arnold, Moore and Yaffe.[43–45]

While based on similar assumptions (see Refs. 46–48 for reviews), the various approaches fail to provide a consensual quantitative description of the RHIC data.[49] There are perhaps two reasons for that. First, the relatively low momentum range accessible at RHIC, of the order of 10 GeV, may question the applicability of weak coupling techniques. Second, each of the models has limitations related to approximations done regarding the nature of the medium and the kinematics. For instance, none of the models addresses from first principles the issue of in-medium multiple parton branchings and resulting interferences, and thus rely on strong assumptions on the space–time structure of the in-medium parton shower. Moreover, at RHIC, scenarios based on radiative energy loss proved to be inconsistent with the strong suppression of the single non-photonic electrons resulting from semi-leptonic decays of hadrons carrying heavy quarks,[50] reviving the interest for collisional energy loss.[51–54] At the same time, evidence for almost perfect fluid behavior of the quark–gluon plasma,[55–57] diverted the attention towards strong coupling techniques, providing an alternative perspective on jet quenching based on the AdS/CFT correspondence (see Ref. 58 for a recent review, and references therein).

In parallel to the aforementioned developments, the theory of jet-quenching based on essentially weak techniques of QCD, has progressed within the last few years. The set goals are the understanding of the structure of jets in matter, and how this structure is modified with respect to that of well understood jets in vacuum[59–72] (see Ref. 73 for a recent review), and the calculation of higher-order corrections,[74–80] addressing some of the limitations of the standard approaches of energy loss. Among the major results of these efforts, it was shown that in the limit of asymptotically large or dense media successive medium-induce branchings are quasi-local and independent from one another,[65,78] providing a theoretical basis for the approximations made in some models of jet quenching.[40,42,45] Color coherence, that characterizes jet structure in vacuum, has been shown to be significantly altered in a QCD medium.[59–61] As a result, medium-induced color decoherence suppresses the interferences between successive gluon radiations. There is also an effort to systematize the calculations of multiple parton branching in the framework of Soft-Collinear Effective Theory (SCET).[76,81] Last but not least, radiative corrections to some jet-quenching observables were recently carried out,[74,75,77–79,82] in addition to calculations of the jet quenching parameter on the lattice.[83,84] The present review will present a summary of some of these developments.

We should also mention that the theoretical developments mentioned above are accompanied by ongoing efforts to develop Monte Carlo event generators, that encompass some of the recent advances in the microscopic physics, as well as a realistic modeling of the geometry of the plasma and its hydrodynamical evolution, in order to carry out systematic studies of jet observables with all sorts of experimental biases, such as, p_t cuts, triggers, etc. The most notable ones are: JEWEL,[85,86] Q-PYTHIA[87] and MARTINI,[88,89] that account for the BDMPS-Z radiation mechanism, in addition to PYQUEN/HYDJET,[90] YAJEM[91] and MATTER.[92] These developments will not be covered in the review. Also, the phenomenology of jet the quenching is beyond the scope of this review and will be covered in another contribution to the present volume.[93]

Before we embark on the theoretical discussion, a word of warning. For the sake of simplicity we shall most of the time deal with the ideal scenario in which the parton cascade is purely gluonic and interacts with a uniform and static plasma. We also assume the medium to be dense enough so that we can apply the multiple soft scattering approximation. Corrections to this idealized picture can be calculated, and this will be mentioned at proper places, but those corrections do not change the qualitative picture. The review is organized as follows. In Sec. 2, the physics of parton branching in vacuum and the BDMPS-Z mechanism for medium-induced gluon radiation are introduced. In Sec. 3, we generalize the BDMPS-Z spectrum to multiple branchings and develop a probabilistic picture of in-medium gluon cascades. Then, in Sec. 4, we analyze radiative corrections to momentum broadening and parton energy loss and show that they can be reabsorbed in a renormalization of the quenching parameter. A complementary discussion on NLO corrections to energy loss and $p_\perp$-broadening in high temperature plasma is presented in this volume Ref. 94: note however that these corrections are related to soft plasma modes and are of a different nature than the radiative corrections involved in the renormalization of $\hat{q}$. In Sec. 5, we examine the time evolution of the angular and energy distribution along the medium-induced cascade, away from the jet core. In Sec. 6, we discuss the physics of in-medium jet decoherence that affects the inner jet structure. Finally, Section 7 presents a brief summary and outlook.

2. Jets and Parton Cascades of Two Kinds

Perturbative QCD successfully predicts the emergence of jets in the final state of high energy collisions such as e^+e^- annihilation and hadronic collisions. As a matter of fact, the collimation of jet particles relates directly to the collinear nature of QCD parton splittings. The underlying mechanism is akin to QED bremsstrahlung: an electric charge that suffers an acceleration subsequently radiates photons collinearly to its direction of motion. Likewise, quarks (and gluons) carry a color charge and thus can act as sources of collinear gluon bremsstrahlung. Besides the confining forces of strong interactions that eventually turn patrons into observable hadrons, the difference with QED (at the partonic level) lies in the fact that, contrary to photons, gluons may also radiate affecting the general pattern of the collimated cascade. The nascent parton possesses typically a large virtuality Q^2, and therefore radiates gluon, which themselves radiate other gluons, forming a so-called *parton cascade*. Because this cascade generically propagates in vacuum, we shall refer to it as a *vacuum cascade*. Such a cascade is accompanied by a degradation of the virtuality Q, which can be used to define an evolution "time", $t \equiv \ln Q^2/Q_0^2$, where Q_0 is the scale that marks the end point of the parton cascade where hadronization takes place. Typically, $Q_0 \sim \Lambda_{\rm QCD}$. Hadronization is a non perturbative phenomenon and one may fear that it erases all memory of the detailed structure of the preceding partonic phase. However, this is not so. It turns out that inclusive observables are to a large extent determined by the parton dynamics prior to hadronization. This remarkable property, known as the local parton–hadron duality,[3,4] is largely responsible for the success of perturbative QCD in describing jet observables to a high degree of precision.

When the energetic parton propagates in a medium, another mechanism may generate radiation, beyond that due to the possible virtuality of the parton: during its propagation, the parton collides with the constituents of the medium, and each of these collisions may in principle trigger radiation. In fact, because of this mechanism, even an on-shell parton entering a medium will be able to radiate and generate a cascade. We shall refer to such a cascade as to a *medium-induced cascade.*

We therefore encounter two types of parton cascades:

(1) The vacuum cascade that forms after a hard process that creates a parton with a large virtuality.
(2) The medium-induced parton cascade that forms due to multiple interactions of the jet with a QCD medium.

This review will be mostly devoted to an analysis of the medium induced cascade. But in our discussion we shall often refer to the vacuum cascade in order to emphasize some of the distinctive properties of the medium induced cascade.

In both cascades, coherence and interferences play an important role by suppressing radiation in particular regions of the spectrum. For the vacuum cascade, this is the case of gluons with large transverse wavelengths which do not resolve the individual color charges carried by the jet, and thus are radiated coherently off the total color charge of the jet. In the case of medium induced radiation, we have gluons with long formation times that do not "see" the individual constituents on which they scatter, and appear to be radiated coherently off many scattering centers. In both cases the gluon bremsstrahlung spectrum is suppressed with respect to the incoherent case. We now discuss these phenomena in some more detail.

2.1. *Jet structure in vacuum: Transverse color coherence*

Experimentally, a jet is defined by a collection of particles that are contained in a cone of given opening angle $\Theta_{\rm jet}$. The characteristic hard scale for the perturbative evolution is then $Q \equiv E\Theta_{\rm jet} \gg \Lambda_{\rm QCD}$. It corresponds to the largest transverse scale (with respect to the jet axis) that can be carried by one of the jet constituents and serves as an upper bound for the jet invariant transverse mass. The phase space between the hard scale Q and the non-perturbative scale $Q_0 \sim \Lambda_{\rm QCD}$ is filled by the vacuum cascade.

The microscopic mechanism generating the cascade is bremssthalung. It is common to both the vacuum and in-medium shower. In the collinear limit, the branching probability into two offsprings, with energy fractions z and $1 - z$ respectively, reads

$$dP \equiv \frac{\alpha_s}{4\pi} \frac{dk_\perp^2}{k_\perp^2} P(z)dz, \tag{1}$$

where α_s is the strong coupling constant, and $P(z)$ the Altarelli–Parisi splitting function.[95] The expression (1) of the branching probability exhibits a collinear divergence that corresponds to a mass singularity, and a soft divergence, when $k_\perp \to 0$ and $z \to 0$ ($z \to 1$), respectively. The latter is cured in inclusive quantities by real-virtual cancellations, while the former is regularized by cutting off the transverse momentum integration at the scale Q_0. The collinear divergence provides a large logarithm $\ln Q^2/Q_0^2$ that may compensate the small coupling constant in the expression (1) of the emission probability. When $\bar{\alpha} \ln Q^2/Q_0^2 \sim 1$, multiple branchings need to be resummed. In this resummation, successive branchings are strongly ordered in their transverse momenta, with the largest momentum emitted first.[95–97]

It was soon realized that this picture is not complete and fails to describe the soft sector of jet observables. This is because it does not fully take into account the effects of color coherence: The radiation pattern of a collection of color charges that form a jet, as they originate from a single parton, does not simply reduce to the incoherent sum of the radiation spectra of the individual charges. If the transverse wave length $\lambda_\perp \sim 1/k_\perp$ of the radiated gluon is larger than the transverse size of the jet $r_\perp \sim \Theta_{\text{jet}} t_f$, at the time $t_f = \omega/k_\perp^2$ when the gluon forms, the latter does not resolve the fine structure of the jet. The radiation is then triggered instead by the total charge of the jet, i.e., the color charge of the original parton. This leads to a suppression of the radiation as compared to that expected form the various color charges contained in the jet.

Such considerations, including the fact that gluons carry a color charge and thus can radiate as well,[98–101] and accounting also for for energy conservation, constitute the basis of the jet algorithms in vacuum that were first introduced in the Marchesini–Webber model.[102] Along a QCD shower a strict ordering in angles accounts for the color coherence and the evolution equation for the inclusive parton distribution reads,

$$\frac{\partial D(x,Q)}{\partial \ln(Q^2/Q_0^2)} = \int_0^1 \frac{dz}{z} \frac{\alpha_s(k_\perp)}{2\pi} \hat{P}(z) D(x/z, zQ), \tag{2}$$

where $k_\perp \equiv z(1-z)Q$ and $\hat{P}(z)$ stands for the regularized Altarelli–Parisi splitting function[95] (the dependence on different parton species is omitted). Equation (2) is referred to as the Modified-Leading-Log-Approximation (MLLA)[3,100,101] as it differs from DGLAP[95–97] by the fact that on the r.h.s. the parton distribution scales with zQ rather than Q to account for angular ordering. Equation (2) describes in reality the evolution of the parton shower as a function of the jet-opening angle Θ_{jet} with $Q \equiv \Theta_{\text{jet}} E$ and E the energy of the parent parton: The infinitesimal variation of the parton distribution is given by an initial splitting of the parent parton generating a subjet characterized by the energy $E' = zE$ and the corresponding scale is $Q' = E'\Theta_{\text{jet}} \equiv zQ$.

2.2. *Medium-induced gluon radiation: Longitudinal coherence*

In a medium of size L, an energetic parton experiences multiple collisions with the medium constituents. In principle each of these collisions puts the gluon off-shell, so that it may subsequently emit radiation, after a typical time, commonly referred to as the *formation time*, $t_f \sim \omega/k_\perp^2$, where ω and $k_\perp$ are the energy and the transverse momentum of the radiated gluon. However, this simple picture may be strongly affected by coherence effects in multiple scattering.

Such effects were studied long ago in the context of QED, by Landau, Pomeranchuk and Migdal (LPM),[103,104] and are generically referred to as the *Landau–Pomeranchuk–Migdal (LPM) effect*. The QCD analog of the LPM effect[a] was investigated by Gyulassy and Wang (GW)[28,29] first, then by Baier, Dokshitzer, Mueller, Peign, Schiff and Zakharov (BDMPS-Z).[30–34] The main focus of these early studies was the energy loss caused by induced gluon radiation.[42] For large media, radiative energy loss scales as L^2 where L is the length of the matter traversed, as opposed to collisional energy loss that scales as L.[105]

We shall return extensively in this review to the BDMPS-Z mechanism, but it is useful at this stage to have an heuristic understanding of the basic physics involved. Consider then a hard quark or gluon of energy E propagating in a dense QCD medium. The collisions of the parton with the medium constituents induce radiation, with the typical formation time of a radiated gluon given by $t_f \sim \omega/k_\perp^2$. Now, during that time scale, the gluon fluctuation experiences multiple kicks, which may cause its transverse momentum to increase by an amount $k_\perp^2 \sim \hat{q}t_f$ where $\hat{q}$ is a diffusion coefficient to be specified shortly. Assuming that these multiple kicks are the only source of transverse momentum, and combining the two relations above one finds that the typical time for induced emission, to which we shall refer here as the *branching time* is given by

$$\tau_{\rm br} = \sqrt{\frac{\omega}{\hat{q}}}. \tag{3}$$

Multiple collisions cause the broadening of the transverse momentum. In the regime where the collisions are soft, and where a large number of collisions are needed to change the transverse momentum by a significant amount, a diffusion approximation becomes appropriate. In this case, it is customary to define a transport coefficient $\hat{q}$ that describes the diffusion in momentum space. Assuming that the dominant color field fluctuations of the medium are characterized by a scale m_D, to which we refer as the Debye screening mass, the typical transverse momentum exchanged with the medium in a collision is $\sim m_D$. Then the transverse momentum squared acquired by the hard parton traveling a distance L is, by definition of $\hat{q}$, $\hat{q}L$. Equivalently, if $\lambda_{\rm el}$ denotes the elastic mean free path, we have $m_D^2 \simeq \hat{q}\lambda_{\rm el}$. The coefficient $\hat{q}$ is referred to as the *jet quenching parameter*.

[a]See Ref. 106 for a recent review on the LPM effect in QED and QCD.

Now, when $\tau_{\rm br} \lesssim \lambda_{\rm el}$, each of the $N = L/\lambda_{\rm el}$ collisions act incoherently and produce a maximal radiation intensity $\propto N$. This is the so-called Bethe–Heitler regime which yields a spectrum

$$\omega \frac{dI}{d\omega} \simeq \bar{\alpha} \frac{L}{\lambda_{el}}, \quad \left(\bar{\alpha} \equiv \frac{\alpha_s N_c}{\pi} \right). \tag{4}$$

On the other hand, when $\tau_{\rm br} \gg \lambda_{\rm el}$, a number $N_{\rm coh} = \tau_{\rm br}/\lambda_{\rm el}$ of the collisions act coherently as a single collision to radiate one gluon. As a result the radiation spectrum scales as $N/N_{\rm coh}$ and reads.[31–34,b,c]

$$\omega \frac{dI}{d\omega} \simeq \bar{\alpha} \sqrt{\frac{\omega_c}{\omega}}, \quad \omega_c \equiv \hat{q} L^2, \tag{5}$$

where ω_c is such that $\tau_{\rm br}(\omega_c) = L$. When $\omega = \omega_c$, $\omega dI/d\omega \sim \bar{\alpha}$, as if we had a single collision. The full Bethe–Heitler result $\omega dI/d\omega \sim \bar{\alpha} N$ is recovered when $\omega = \hat{q}\lambda_{\rm el}^2 \equiv \omega_{BH}$.

Note that for $\omega \gtrsim \omega_c$ the spectrum is strongly suppressed. In this case, the radiated gluon cannot distinguish the hard event that produced its radiator from the multiple scattering of the latter with the medium. As a result, the suppression $1/\sqrt{\omega}$ turns into a stronger one, that is $1/\omega^2$. A complete calculation, accounting for finite size effects, with the two limiting cases above, yields[31]

$$\omega \frac{dI}{d\omega} \simeq 2\bar{\alpha} \ln |\cos(\Omega L)|, \tag{6}$$

where $\bar{\alpha} \equiv \alpha_s C_R/\pi$, with $C_R \equiv C_{F,A}$ is the color charge of the source of radiation, and

$$\Omega = \frac{1+i}{2} \sqrt{\frac{\hat{q}}{\omega}} \sim \frac{1}{\tau_{\rm br}(\omega)}. \tag{7}$$

By radiating gluons, an energetic parton, with energy $E \gg \omega_c$, that traverses a dense QCD medium, loses a mean energy,[d]

$$\Delta E \equiv \int_0^\infty d\omega \omega \frac{dI}{d\omega} = \frac{\pi}{4} \bar{\alpha} \hat{q} L^2. \tag{8}$$

Indeed, since the spectrum Eq. (6) falls as $1/\sqrt{\omega}$ when $\omega \ll \omega_c$ (cf. Eq. (5)), the ω integral in (8) is dominated by gluon frequencies of the order of ω_c. This formula relates the mean-energy loss to the typical transverse momentum squared accumulated by the gluons during the maximal formation time $t_f \sim L$, $\langle k_\perp^2 \rangle \sim \hat{q} L$, that is, $\Delta E \sim \langle k_\perp^2 \rangle L$.

[b] I stands for "Intensity", a dimensionless quantity proportional to the number of gluons.
[c] In QED, the spectrum scales as $\alpha_s \sqrt{\omega \omega_c}/E$. It depends explicitly on the energy of the projectile. At asymptotically high energies, the photon radiation dies out, while according to Eq. (5) gluon radiation does not.
[d] For parton energies $E \ll \omega_c$ we find that $\Delta E \sim \sqrt{E}$.

Consider now the number of radiated gluons with frequency larger than ω,

$$N(\omega) = \int_{\omega}^{\omega_c} d\omega' \frac{dI}{d\omega}. \tag{9}$$

When $\omega \ll \omega_c$, the number of radiated gluons is dominated by the lower bound, ω, of the integral in Eq. (9), for the spectrum rises as $1/\omega^{3/2}$. Thus, we obtain

$$N(\omega) \simeq \sqrt{\frac{\omega_s(L)}{\omega}}, \tag{10}$$

where

$$\omega_s(L) \equiv \bar{\alpha}^2 \omega_c(L). \tag{11}$$

As long as $N(\omega)$ is smaller than one it can be identified with a radiation probability, i.e., $N(\omega) \sim P(\omega) < 1$. When the average number of emitted gluons exceeds unity, higher orders in perturbation theory, that involve multiple gluon radiations, become relevant processes that need to be dealt with. According to Eq. (10), this regime is achieved when $\omega \lesssim \omega_s(L)$. This criterion can be looked at from a different perspective. As we shall see later, the branching rate for a gluon with energy ω is given by

$$\frac{\bar{\alpha}}{\tau_{\rm br}(\omega)} \equiv \frac{1}{t_*(\omega)}, \tag{12}$$

where, accordingly (see Eq. (3))

$$t_*(\omega) = \frac{1}{\bar{\alpha}} \sqrt{\frac{\omega}{\hat{q}}} \tag{13}$$

is the time it takes a gluon with energy ω to branch, a quantity that one could also refer to as an "inelastic mean-free path". Then, the condition $\omega \lesssim \omega_s(L)$ is equivalent to the condition $t_*(\omega) \lesssim L$, that is, during the time $t_*(\omega)$ a gluon with energy ω will branch at least once during the time it spends in the matter. When applied to the leading particle, the quantity $t_*(E)$ will be referred to as the stopping time in Sec. 5.

3. Probabilistic Picture of Medium-Induced Gluon Cascade

We start now examining more closely the properties of the medium-induced cascade and begin by putting in place various elements of the formalism that will be used throughout this review.

3.1. *Non-relativistic dynamics in the transverse plane*

A high energy jet originates from an ultrarelativistic parton, whose direction of propagation is chosen as the z axis. We choose light-cone variables[e] such that the

[e]With this choice, the 4-coordinate vector is $x \equiv (x^+, x^-, \boldsymbol{x})$, where $x^+ = (t+z)/2$, $x^- = t - z$, and $\boldsymbol{x} \equiv (x, y)$.

4-momentum p of a parton reads

$$p^+ = \frac{1}{2}(E + p_z), \quad p^- = E - p_z \quad \text{and} \quad \boldsymbol{p} \equiv (p_x, p_y), \tag{14}$$

where E is the energy, p_z the longitudinal momentum, and $\boldsymbol{p}$ the transverse momentum. The angle θ formed by the 3-momentum $\boldsymbol{p}$ and the jet axis is given by $p_\perp \equiv |\boldsymbol{p}| \sin\theta$. Considering that jets are collimated objects, and that their angular broadening is small, we assume in all our calculations that $\theta \ll 1$, and accordingly,

$$p^+ \simeq E \quad \text{and} \quad |\boldsymbol{p}| \equiv p_\perp \simeq E\theta, \tag{15}$$

while the light-cone time can be identified with the real time,

$$x^+ \simeq t \simeq z. \tag{16}$$

We work in the light-cone gauge $A^+ = 0$, a convenient choice in the present context. This is because the current J^-, which describes the target through which the jet propagates, is coupled to the $+$ component of the gauge field, which vanishes in this gauge. In turn, gluon radiation off particles propagating in the $-z$ direction is suppressed, and only the jet constituents, propagating in the $+z$ direction, radiate. Furthermore, the light-cone gauge allows for a partonic picture of the radiation dynamics.

The hard gluons that compose a collimated jet travel along nearly straight-line trajectories, and interact with the color field of the medium. Only the $-$ component of this field couples to the high energy gluon with the helicity conserving vertex, $V^{ij+}A^- \sim p^+ A^- \delta^{ij}$, where $i, j = 1, 2$ depict the two transverse components of the physical polarization vector ϵ^i_λ. The transverse component of the field can be removed by making use of the additional degree of freedom of the light-cone gauge.[107] Moreover, the gluon with very large longitudinal momentum p^+, propagates near the light-cone branch, $x^- = 0$, so that, finally, the only component of the medium field that matters is $A(t, \boldsymbol{x}) \equiv A^-(x)|_{x^-=0}$.

The choice of the light-cone gauge along with the eikonal approximation for the vertex imply that the gluon propagator in the medium background field is diagonal in helicity space, i.e., $G^{ij} \sim \delta^{ij} G$, with G obeying the following equation,

$$[\Box - ig\partial^+ A^-(x) \cdot T] G(x, x') = \delta(x - x'), \tag{17}$$

where g is the gauge coupling constant. We have already mentioned that the jet probes the medium gauge field only near the light cone, i.e. $x^- \approx 0$. We may then assume that over such small distance, the medium gauge field is constant, and perform a Fourier transform with respect to x^-.[f] We define

$$(\boldsymbol{x}|\mathcal{G}(t, t')|\boldsymbol{x}') = \frac{i}{2E} \int \frac{dx^-}{2\pi} \mathrm{e}^{-iE(x-x')^-} G(x - x'), \tag{18}$$

[f] Since x^- is the conjugate variable to $p^+ \sim E$, assuming that the field A does not depend on x^- leads to the conservation of p^+ at each vertex coupling the field to the parton. Alternatively, this conservation law could be obtained by observing that the transfer of $+$ momentum from the plasma to the parton is negligible compared to the large energy E of the parton.

and from Eq. (17) above, one finds that $\mathcal{G}(t,t')$ obeys the following 2-dimensional Schrödinger equation

$$\left[i\frac{\partial}{\partial t}+\frac{\partial_\perp^2}{2E}+gA(t,\boldsymbol{x})\right](\boldsymbol{x}|\mathcal{G}(t-t')|\boldsymbol{x}')=i\delta(t-t')\delta(\boldsymbol{x}-\boldsymbol{x}'), \tag{19}$$

where the background field, $A \equiv A_a T^a$, with T^a the generators of $SU(3)$ in the adjoint representation, is a matrix in color space, and so is the propagator $\mathcal{G}$. In the high energy limit, the interesting dynamics is therefore localized in the transverse plane. It is the 2-dimensional equivalent of a non-relativistic particle in a background potential, where the invariant $p^+ = E$ component of the momentum plays the role of the mass.

The propagation of the gluon from $(t_0, \boldsymbol{x}_0)$ to $(t, \boldsymbol{x})$ is described by the Green's function $(\boldsymbol{x}|\mathcal{G}(t,t_0)|\boldsymbol{x}_0)$ that solves the 2-dimensional Schrödinger equation (19) (see e.g. Ref. 108). Its solution can be formally represented by a path integral in transverse coordinate space:

$$(\boldsymbol{x}|\mathcal{G}(t,t_0)|\boldsymbol{x}_0)=\Theta(t-t_0)\int_{\boldsymbol{x}_0}^{\boldsymbol{x}}\mathcal{D}\boldsymbol{r}\exp\left[i\frac{E}{2}\int_{t_0}^{t}dt'\dot{\boldsymbol{r}}^2(t')\right]\mathcal{U}_{\boldsymbol{r}}(t,t_0), \tag{20}$$

where retarded conditions are used and

$$\mathcal{U}_{\boldsymbol{r}}(t,t_0)=\mathcal{T}\exp\left[ig\int_{t_0}^{t}dt'A(t',\boldsymbol{r}(t'))\cdot T\right] \tag{21}$$

is a Wilson-line along the path $\boldsymbol{r}(t')$, with $\boldsymbol{r}(t_0)\equiv\boldsymbol{x}_0$ and $\boldsymbol{r}(t)\equiv\boldsymbol{x}$. Between two scatterings, the gluon propagation is free,

$$(\boldsymbol{x}|\mathcal{G}_0(t,t_0)|\boldsymbol{x}_0)=\Theta(t-t_0)\frac{E}{2\pi i(t-t_0)}\exp\left[\frac{iE(\boldsymbol{x}-\boldsymbol{x}_0)^2}{2(t-t_0)}\right]. \tag{22}$$

The Green's function $\mathcal{G}(t,t')$ plays the role of a unitary time evolution operator for a free particle evolving from the position state $|\boldsymbol{x})$ to the state $|\boldsymbol{x}')$. In momentum representation,

$$|\boldsymbol{p})=\int d^2\boldsymbol{x}\mathrm{e}^{-i\boldsymbol{p}\cdot\boldsymbol{x}}|\boldsymbol{x}),\quad \int\frac{d^2\boldsymbol{p}}{(2\pi)^2}|\boldsymbol{p})(\boldsymbol{p}|=1, \tag{23}$$

we define the amplitude for a gluon of (large) momentum $p_0\equiv(p_0^+,\boldsymbol{p}_0=0)$, present in the system at time t_0, to evolve in the medium into a gluon with momentum $p_1=(p_1^+,\boldsymbol{p}_1)$, where $p_0^+=p_1^+\equiv E$ (to within an irrelevant phase factor),

$$\mathcal{M}(p_1\,|\,p_0)=(\boldsymbol{p}_1|\mathcal{G}(t_1,t_0)|\boldsymbol{p}_0). \tag{24}$$

The amplitude $\mathcal{M}\equiv\mathcal{M}_{ab}$ is a matrix in color space, propagating a gluon with color b to color a. It is also a (diagonal) matrix in the helicity space. We do not write explicitly color and helicity indices to alleviate the notation.

The free Hamiltonian in 2-dimensional space, reads

$$\mathcal{H}_0 = \frac{\boldsymbol{p}^2}{2E}. \tag{25}$$

The interaction part of the hamiltonian is the sum of two components, $\delta\mathcal{H} = \delta\mathcal{H}_1 + \delta\mathcal{H}_2$,

$$(\boldsymbol{q}|\delta\mathcal{H}_1|\boldsymbol{p}) = gA(\boldsymbol{q}-\boldsymbol{p},t) \quad \text{and} \quad (\boldsymbol{q};\boldsymbol{q}'|\delta\mathcal{H}_2|\boldsymbol{p}) = g(\boldsymbol{q};\boldsymbol{q}'|V|\boldsymbol{p}), \tag{26}$$

where $\delta\mathcal{H}_1$ represents the elastic scattering of a particle off the potential $A(\boldsymbol{q},t)$, while $\delta\mathcal{H}_2$ represents the splitting of a particle into two. Note that the latter process does not conserve the non-relativistic energy $\boldsymbol{p}^2/2E$, but it conserves the "mass" E. When using the propagator (20) we effectively treat $\delta\mathcal{H}_1$ to all orders. Later, we shall calculate the branching probability of a gluon to first order in $\delta\mathcal{H}_2$. Multiple branchings will be treated via probabilistic cascades.

3.2. *Independent multiple scatterings approximation*

The color field of the plasma, through which the jet propagates, fluctuates from time to time and place to place. The typical scale characterizing such fluctuations is the Debye screening mass m_D, which is much smaller than the energy E of the jet. As we have seen earlier, the energy E plays the role of the mass in the 2-dimensional non-relativistic dynamics in the transverse plane. Because this mass is large, the motion in the transverse plane is a slow motion, compared to which the dynamics of the plasma appears to be very fast. It follows that the interactions mediated by exchange of plasma fluctuations are thus nearly instantaneous.[g] In performing the average over the plasma fluctuations, we may then use the following correlator

$$\langle A^a(t,\boldsymbol{x})A^{*b}(t',\boldsymbol{y})\rangle = n\delta^{ab}\delta(t'-t)\gamma(\boldsymbol{x}-\boldsymbol{y}), \tag{27}$$

where n is the density of color charges, and $\gamma(\boldsymbol{x}-\boldsymbol{y})$ will be specified shortly. This form of correlator is somewhat analogous to that used in the McLerran–Venugopalan model[109,110] that deals with cold nuclear matter of large nuclei. It is also in line with the Gyulassy–Wang model, in which the medium is modeled as a collection of static scattering centers.[28,29,111]

Assuming that it can be reliably calculated from perturbative QCD, the correlator $\gamma(\boldsymbol{x})$ is given in leading order by

$$\gamma(\boldsymbol{x}) = g^2\int_{\boldsymbol{q}} \frac{\mathrm{e}^{i\boldsymbol{q}\cdot\boldsymbol{x}}}{\boldsymbol{q}^4}, \tag{28}$$

which is proportional to the Coulomb elastic cross-section, with $\boldsymbol{q}$ the transverse momentum transfer. Here we introduced a shorthand notation for transverse

[g]Said differently, because of Lorentz time dilation, the high energy parton has a very poor resolution in x^+, so that the medium correlations appear to him as effectively local in time.

momentum integrations, to be used throughout this review: $\int_{\boldsymbol{q}} \equiv d^2\boldsymbol{q}/(2\pi)^2$. A somewhat more elaborate version of γ for a QCD plasma in thermal equilibrium is obtained from the substitution[112] $\boldsymbol{q}^4 \to \boldsymbol{q}^2(\boldsymbol{q}^2 + m_D^2)$.

The picture that emerges form this discussion is one where the jet propagates through a random background field, to be averaged over with the correlation function given by Eq. (28). Using another language, the high energy jet collides with the plasma constituents at a rate determined by the Coulomb cross-section. Note that this depends on the medium solely through the density of color charges, and the Debye mass. The motion of the jet in the transverse plane is a slow motion, and the interactions with the plasma constituents are effectively instantaneous.

The treatment of collisions in terms of the propagation in a random field with an instantaneous correlation function implicitly assumes that the collisions are independent. This requirement is satisfied in a weakly coupled quark–gluon plasma: The Debye mass is $m_D \sim gT$, with T the temperature. The elastic mean free path for soft collisions (with momentum transfer of order m_D) can be estimated as

$$\lambda_{\text{el}} \sim \frac{1}{n\sigma} \sim \frac{1}{g^2 T}, \tag{29}$$

where we have used $\sigma \sim g^4/m_D^2$ and $n \sim T^3$. This is parametrically larger (by one power of the coupling constant) than the duration of the interactions, that is $\lambda_{\text{el}} \gg m_D^{-1}$.

3.3. *Momentum broadening*

The multitude of kicks that it receives from its collisions with the medium constituents affects the propagation of a high energy parton. In particular, it modifies its transverse momentum. This leads to the so-called *momentum broadening*, which we now discuss.

The amplitude (24) introduced above allows us to calculate the transverse momentum spectrum of a gluon at time t, given the spectrum at time t_0:

$$\frac{dN}{d\Omega_p} = \int_{\boldsymbol{p}_0} \mathcal{P}(\boldsymbol{p}, t \,|\, \boldsymbol{p}_0, t_0) \frac{dN}{d\Omega_{p_0}}, \quad d\Omega_{\boldsymbol{p}_0} \equiv \frac{dE d^2\boldsymbol{p}_0}{2E(2\pi)^3}, \tag{30}$$

where $dN/d\Omega_{p_0}$ is the spectrum at time t_0, and $\mathcal{P}(\boldsymbol{p}, t \,|\, \boldsymbol{p}_0, t_0)$ will be referred to as the $p_\perp$-broadening probability. This is calculated by squaring the amplitude, averaging over the initial color and polarization, summing over final polarization and color, and finally averaging over the background field. In doing this calculation, it is convenient to keep the momentum $\boldsymbol{p}_0$ in the amplitude distinct from that, $\bar{\boldsymbol{p}}_0$, in the complex conjugate amplitude. We get then

$$\begin{aligned}\mathcal{M}(\boldsymbol{p} \,|\, \boldsymbol{p}_0)\mathcal{M}^*(\boldsymbol{p} \,|\, \bar{\boldsymbol{p}}_0) &= \frac{1}{N_c^2 - 1} \langle \text{Tr}(\boldsymbol{p}|\mathcal{G}(t_1, t_0)|\boldsymbol{p}_0)(\bar{\boldsymbol{p}}_0|\mathcal{G}^\dagger(t_0, t_1)|\boldsymbol{p}) \rangle \\ &= (2\pi)^2 \delta(\bar{\boldsymbol{p}}_0 - \bar{\boldsymbol{p}}) \mathcal{P}(\boldsymbol{p}, t_1 \,|\, \boldsymbol{p}_0, t_0),\end{aligned} \tag{31}$$

where the trace is over color indices and in the last line we have factored out the momentum conserving delta function that reflects translational invariance. The angular brackets in the first line denote medium average.

In Eq. (31), appears the Fourier transform of a fundamental quantity, which we refer to as a 2-point function, $(X_1|S^{(2)}(t,t_0)|X_0)$ (see Eq. (32)). This is one of a collection of n-point functions that play an important role in the discussion. This 2-point function, is not a usual field theory 2-point function. It is built from the product of a propagator $\mathcal{G}$ of an amplitude and the corresponding propagator $\mathcal{G}^\dagger$ in the complex conjugate amplitude. This is given, quite generally, by

$$(X_1|S^{(2)}(t,t_0)|X_0) = \frac{1}{N_c^2-1}\langle \mathrm{Tr}\,(\boldsymbol{x}_1|\mathcal{G}(t_1,t_0)|\boldsymbol{x}_0)(\bar{\boldsymbol{x}}_0|\mathcal{G}^\dagger(t_0,t_1)|\bar{\boldsymbol{x}}_1)\rangle, \tag{32}$$

where $X_i = (\boldsymbol{x}_i;\bar{\boldsymbol{x}}_i)$. This two point function describes the evolution from coordinates $|X_0) \equiv |\boldsymbol{x}_0,\bar{\boldsymbol{x}}_0)$ to $|X_1) \equiv |\boldsymbol{x}_1,\bar{\boldsymbol{x}}_1)$,[h] of a (fictitious) dipole made of two gluons in a color singlet state, one in the amplitude and the other in the complex conjugate amplitude. From the equation satisfied by the propagator $\mathcal{G}$, Eq. (19), one easily deduces the equation satisfied by $S^{(2)}$, or equivalently that satisfied by the broadening probability $(\mathcal{P}(\boldsymbol{p},t\,|\,\boldsymbol{p}_0,t_0) \equiv \mathcal{P}(\boldsymbol{p},t))$[65]

$$\frac{\partial}{\partial t}\mathcal{P}(\boldsymbol{p},t) = -\frac{N_c}{2}n\int_{\boldsymbol{q}}\sigma(\boldsymbol{q})\mathcal{P}(\boldsymbol{p}-\boldsymbol{q},t), \tag{33}$$

where $\sigma(\boldsymbol{q})$ is the Fourier transform of the so-called dipole cross-section

$$\sigma(\boldsymbol{x}) = 2g^2[\gamma(\boldsymbol{0}) - \gamma(\boldsymbol{x})], \tag{34}$$

with $\boldsymbol{x}$ corresponding to the transverse size of the dipole. Eq. (33) can be solved by Fourier transforming to coordinate space. The r.h.s. of Eq. (33) becomes local in $\boldsymbol{x}$, and the solution exponentiate. Moving back to Fourier space, it is straightforward to get (for $t \equiv L$)

$$\mathcal{P}(\boldsymbol{p},L) = \int d^2\boldsymbol{x}\,\exp\left[-\frac{N_c}{2}n\sigma(\boldsymbol{x})L - i\boldsymbol{p}\cdot\boldsymbol{x}\right], \tag{35}$$

which relates the gluon scattering cross-section at high energy with the S-matrix for a dipole passing trough the medium, in the independent multiple scattering approximation.[113]

In the diffusion limit, namely when the final momentum $\boldsymbol{p}$ is acquired by a large number of small momentum transfers, i.e., $\boldsymbol{q} \ll \boldsymbol{p}$, Eq. (33) takes the form of a

[h] Some explanation about the notation is needed here. We regard $S^{(2)}(t_1,t_0)$ as a matrix in transverse space, $(Y|S^{(2)}(t_1,t_0)|X)$, with $X = (\boldsymbol{x};\bar{\boldsymbol{x}})$, $Y = (\boldsymbol{y};\bar{\boldsymbol{y}})$ with the coordinates $\boldsymbol{x},\bar{\boldsymbol{x}},\boldsymbol{y},\bar{\boldsymbol{y}}$ attached to $\mathcal{G}$ and $\mathcal{G}^\dagger$ as indicated in Eq. (32) (the time variables are not considered as matrix indices, and $S^{(2)}(t_1,t_0)$ is proportional to $\theta(t_1-t_0)$).

Fokker–Planck equation

$$\frac{\partial}{\partial t}\mathcal{P}(\boldsymbol{p},t) = \frac{1}{4}\left(\frac{\partial}{\partial \boldsymbol{p}}\right)^2 [\hat{q}(\boldsymbol{p}^2)\mathcal{P}(\boldsymbol{p},t)], \tag{36}$$

where the (momentum dependent) diffusion coefficient

$$\hat{q}(\boldsymbol{p}^2) = -\frac{N_c}{2}n\int_{\boldsymbol{q}} \boldsymbol{q}^2\sigma(\boldsymbol{q}) \simeq \frac{g^4 N_c n}{4\pi}\ln\frac{\boldsymbol{p}^2}{m_D^2}, \tag{37}$$

stands for the jet-quenching parameter,[32] estimated here to logarithmic accuracy: the lower cut-off of the logarithmic transverse momentum integration is given by the Debye screening mass m_D. We shall see later that this coefficient receives potentially large radiative corrections. Another article in this volume discusses an independent set of corrections to the leading-order expression (37) (see also Ref. 114).

Finally, assuming $\hat{q}$ to be roughly constant, one obtains from the diffusion equation (36) that the transverse momentum squared acquired by a high energy gluon traversing a length L of the medium is

$$\langle p_\perp^2\rangle_{\rm typ} = \hat{q}L. \tag{38}$$

It can also be shown that, in the harmonic approximation, the dipole cross-section (34) relates to the quenching parameter as follows

$$N_c n\sigma(\boldsymbol{x}) \simeq \frac{1}{2}\boldsymbol{x}^2\hat{q}. \tag{39}$$

3.4. *Quasi-local gluon branching*

We turn now to inelastic processes that are responsible for the formation of the gluon cascade. This will be handled in first order in the Hamiltonian $\delta\mathcal{H}_2$. We follow here the derivation presented in Ref. 65.

We aim to extract the inelastic transition probability $\mathcal{P}_2(\boldsymbol{k}_a,\boldsymbol{k}_b,L\,|\,\boldsymbol{p}_0,t_0)$, that relates the 2-gluon spectrum to the single gluon spectrum

$$\frac{dN}{d\Omega_{k_a}d\Omega_{k_b}} = \int_{\boldsymbol{p}_0}\mathcal{P}_2(\boldsymbol{k}_a,\boldsymbol{k}_b,L\,|\,\boldsymbol{p}_0,t_0)\frac{dN}{d\Omega_{p_0}}. \tag{40}$$

This is obtained from the amplitude for a gluon of momentum $p_0 \equiv (E,\boldsymbol{p}_0)$ at time t_0 to propagate until the time t_1 where it has acquired the transverse momentum $\boldsymbol{p}_1$ and branches into two gluons with momenta $q_1 \equiv (zE,\boldsymbol{q}_1)$ and $q_1' \equiv ((1-z)E,\boldsymbol{q}_1')$ respectively. The offspring gluons propagate then until they escape the medium at L with momenta $k_a \equiv (zE,\boldsymbol{k}_a)$ and $k_b \equiv ((1-z)E,\boldsymbol{k}_b)$. This amplitude is given by (we omit the color indices to simplify the notations),

$$\frac{g}{2E}\int dt_1\int_{\boldsymbol{q}_1\boldsymbol{q}_1'\boldsymbol{p}_1}(\boldsymbol{k}_a|\mathcal{G}(L,t_1)|\boldsymbol{q}_1)\,(\boldsymbol{k}_b|\mathcal{G}(L,t_1)|\boldsymbol{q}_1')\,(\boldsymbol{q}_1;\boldsymbol{q}_1'|V|\boldsymbol{p}_1)(\boldsymbol{p}_1|\mathcal{G}(t_1,t_0)|\boldsymbol{p}_0). \tag{41}$$

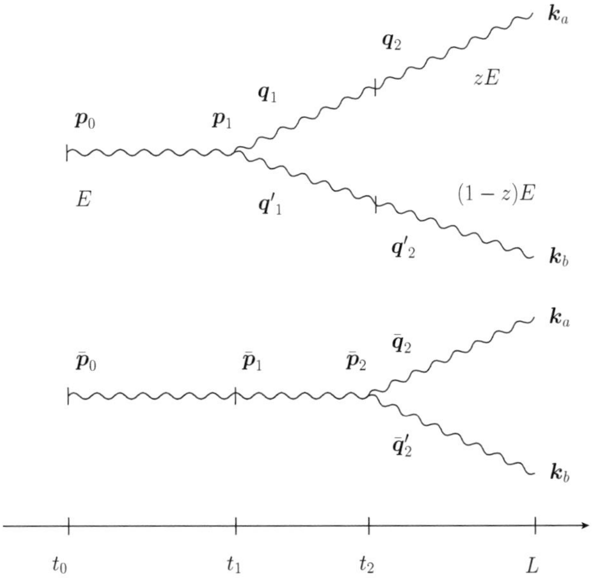

Fig. 1. A diagrammatic representation of the time evolution of the amplitude for a gluon branching given by Eq. (41) (upper diagram) and the complex conjugate amplitude (lower diagram) involved in the calculation of the branching probability Eq. (45), form t_0 to L. The branching takes place at t_1 in the amplitude and t_2 in the complex conjugate amplitude (see text for further explanations).

It is represented in the upper diagram of Fig. 1. The 3-gluon vertex is given by

$$(\boldsymbol{q}a;\boldsymbol{q}'b|V|\boldsymbol{p}c) = \delta(\boldsymbol{q}-\boldsymbol{q}'-\boldsymbol{p})2f^{\mathrm{abc}}\epsilon^i(\boldsymbol{q})\epsilon^j(\boldsymbol{q}')\epsilon^k(\boldsymbol{p})\Gamma^{ijk}(\boldsymbol{q}-z\boldsymbol{p},z), \tag{42}$$

with

$$\Gamma^{ijk}(\boldsymbol{Q},z) = \frac{1}{1-z}Q^k\delta^{ij} - Q^i\delta^{jk} + \frac{1}{z}Q^j\delta^{ik} \tag{43}$$

and $z=\omega_a/E$ is the energy fraction of the parent gluon carried by gluon a. The particular combination of momenta $\boldsymbol{Q}=\boldsymbol{q}-z\boldsymbol{p}$, reflects a Gallilean invariance of the effective 2-dimensional dynamics. A detailed derivation of this formalism from Feynman diagrams in light-cone gauge, $A^+=0$, is given in Appendix A of Ref. 65. Eq. (41) provides the order g correction to the evolution operator $\mathcal{G}$.

By squaring the amplitude (41) in order to obtain the 2-gluon production distribution, we are left with the average over the gauge field. Because the correlations of two A fields is instantaneous, this operation is conveniently visualized by plotting the amplitude and its complex conjugate as in Fig. 1). By performing the average over the field configurations with the help of Eq. (27), we identify then three factors corresponding to the three different time intervals, involving two, three and four propagators respectively. The 2-point function, $S^{(2)}$ of the first interval $[t_0,t_1]$ is related to the broadening probability, cf. Eq. (30). The 3-point function, $S^{(3)}$ of the second interval, $[t_1, t_2]$, will determine the branching rate as will shall see shortly. Finally, the last interval, involves a 4-point function, $S^{(4)}$ which factorizes into two 2-point function in the limit of a short formation time,[65]

$$(\boldsymbol{k}_a\boldsymbol{k}_b;\boldsymbol{k}_a\boldsymbol{k}_b|S^{(4)}(L,t_2)|\bar{\boldsymbol{q}}_2\bar{\boldsymbol{q}}_2';\boldsymbol{q}_2\boldsymbol{q}_2')$$
$$\simeq \delta(\boldsymbol{q}_2-\boldsymbol{q}_2')\delta(\bar{\boldsymbol{q}}_2-\bar{\boldsymbol{q}}_2')\mathcal{P}(\boldsymbol{k}_a,L\,|\,\boldsymbol{q}_2,t_2)\mathcal{P}(\boldsymbol{k}_b,t\,|\,\boldsymbol{q}_2',t_2), \tag{44}$$

up to terms of the order of $\tau_{\rm br}/L$ that we neglect. One can indeed show that the color correlations between the offspring gluons are exponentially suppressed after a time of the order of the branching time $\tau_{\rm br}$, as they move apart. This property is closely related to that of color decoherence that will be discussed in more detail in Sec. 6.

By integrating over the intermediate states, summing over the final polarization vectors and averaging over the initial ones using the completeness relation $\sum_{\lambda=1,2}\epsilon_\lambda^i(\boldsymbol{k})\epsilon_\lambda^{*j}(\boldsymbol{k}) = \delta^{ij}$, one finds that the amplitude squared depends on two quantities, the broadening probability and the reduced 3-point function as follows

$$\mathcal{P}_2(\boldsymbol{k}_a,\boldsymbol{k}_b,L\,|\,\boldsymbol{p}_0,t_0) = \frac{g^2}{2E^2}2\mathrm{Re}\int_0^L dt_1\int_0^{t_1}dt_2\int_{\boldsymbol{Q}_1,\boldsymbol{Q}_2,\boldsymbol{l}}\Gamma^{ijk}(\boldsymbol{Q}_1)\Gamma^{ijk}(\boldsymbol{Q}_2)$$
$$\times\,\mathcal{P}(\boldsymbol{k}_a,t|\boldsymbol{q}_2,t_2)\mathcal{P}(\boldsymbol{k}_b,t\,|\,\boldsymbol{q}_2',t_2)\tilde{S}^{(3)}$$
$$\times\,(\boldsymbol{Q}_2,\boldsymbol{Q}_1\boldsymbol{l};t_1,t_2)\mathcal{P}(\boldsymbol{p}_1,t\,|\,\boldsymbol{p}_0,t_2), \tag{45}$$

where the 3-point function $\tilde{S}^{(3)}$ reads (cf. Ref. 78)

$$\tilde{S}^{(3)}(\boldsymbol{Q}_1,\boldsymbol{Q}_2,\boldsymbol{l};t_1,t_2) = \int d\boldsymbol{u}_1 d\boldsymbol{u}_1 d\boldsymbol{v}\, e^{i\boldsymbol{u}_1\cdot\boldsymbol{Q}_1 - i\boldsymbol{u}_2\cdot\boldsymbol{Q}_2 - i\boldsymbol{v}\cdot\boldsymbol{l}}\int_{\boldsymbol{u}_1}^{\boldsymbol{u}_2}\mathcal{D}\boldsymbol{u}$$
$$\times\exp\left\{\frac{iz(1-z)E}{2}\int_{t_1}^{t_2}dt\dot{\boldsymbol{u}}^2 - \frac{N_c n}{4}\right.$$
$$\left.\times\int_{t_1}^{t_2}dt[\sigma(\boldsymbol{u})+\sigma(\boldsymbol{v}-z\boldsymbol{u})+\sigma(\boldsymbol{v}+(1-z)\boldsymbol{u})]\right\}, \tag{46}$$

with $\boldsymbol{Q}_1 = \boldsymbol{q}_1 - z\boldsymbol{p}_1$, $\boldsymbol{Q}_2 = \boldsymbol{q}_2 - z\boldsymbol{p}_2$ and $\boldsymbol{l} = \bar{\boldsymbol{p}}_2 - \bar{\boldsymbol{p}}_1$. Momentum conservation at each of the two vertices implies that $\boldsymbol{p}_1 = \boldsymbol{q}_1 + \boldsymbol{q}_1'$ and $\bar{\boldsymbol{p}}_2 = \bar{\boldsymbol{q}}_2 + \bar{\boldsymbol{q}}_2'$. The function $\tilde{S}^{(3)}$ is an important quantity that encodes the evolution, from t_1 to t_2, of the the system of two gluons with energies zE and $(1-z)E$ respectively, and one gluon with energy E. The transverse vector $\boldsymbol{u}$ corresponds to the transverse distance between the offspring gluons a and b, while $\boldsymbol{v}$ corresponds to the coordinate of the center of mass of the system.

Furthermore, the sum of gluon momenta in the amplitude and the complex conjugate amplitude vanishes at any given time, $\boldsymbol{q}_2+\boldsymbol{q}_2' = \bar{\boldsymbol{q}}_2+\bar{\boldsymbol{q}}_2'$, $\boldsymbol{p}_1 = \bar{\boldsymbol{p}}_2$ and at each vertex we have $\boldsymbol{q}_1' = \boldsymbol{p}_1 - \boldsymbol{q}_1$ and $\bar{\boldsymbol{q}}_2' = \bar{\boldsymbol{p}}_2 - \bar{\boldsymbol{q}}_2$. The contraction of the three gluon vertices in the amplitude and complex conjugate amplitude yields

$$\frac{1}{2}\Gamma^{ijk}(\boldsymbol{Q}_1)\Gamma^{ijk}(\boldsymbol{Q}_2) = \frac{2}{z(1-z)}P(z)(\boldsymbol{Q}_1 \cdot \boldsymbol{Q}_2), \tag{47}$$

where

$$P(z) = N_c\left[\frac{z}{1-z} + \frac{1-z}{z} + z(1-z)\right], \tag{48}$$

is the gluon–gluon Altarelli–Parisi splitting function.[95] Equation (45) can be further simplified in the short formation time approximation, i.e, $\tau_{\rm br} \ll L$.

The difference between t_1 in the amplitude and t_2 in the complex conjugate amplitude, $\tau = \Delta t = t_2 - t_1$ is at most of the order of $\tau_{\rm br}$. The 3-point function decays exponentially for $\tau > \tau_{\rm br}$. It follows that the upper bound of the integral over τ can be send to infinity,

$$\int_0^L dt_2 \int_0^{t_2} dt_1 = \int_0^L dt_2 \int_0^{t_2} d\tau \approx \int_0^L dt_2 \int_0^{\infty} d\tau. \tag{49}$$

Similarly, the relative transverse momenta generated in the branching, $\boldsymbol{Q}_1$, $\boldsymbol{Q}_2$, $\boldsymbol{l}$ are typically of the order of $k_{\rm br} \sim \sqrt{\hat{q}\tau_{\rm br}}$ and thus, can be neglected compared to the external momenta. This is the collinear branchings approximation. Hence, one can integrate over these internal transverse momenta and τ, and this allows us to define the transverse momentum integrated kernel,[33,115,i]

$$\mathcal{K}(z,E) = \frac{P(z)}{[z(1-z)E]^2}{\rm Re}\int_0^\infty d\tau \int_{\boldsymbol{Q}_1\boldsymbol{Q}_2\boldsymbol{l}} (\boldsymbol{Q}_1 \cdot \boldsymbol{Q}_2)\tilde{S}^{(3)}(\boldsymbol{Q}_1, \boldsymbol{Q}_2, \boldsymbol{l}, t_1, \tau). \tag{50}$$

In the harmonic approximation, we can ignore the logarithmic dependence of $\hat{q}$ that appears in the dipole cross-section (39), and the reduced 3-point function (46) can be explicitly calculated. This yields for the kernel (50),

$$\mathcal{K}(z,E) = \frac{[1-z+z^2]^{5/2}}{[z(1-z)]^{3/2}}\sqrt{\frac{\hat{q}}{E}}. \tag{51}$$

Finally, the transition probability (45) takes the compact form[65] (letting $t_1 \equiv t$)

$$\mathcal{P}_2(\boldsymbol{k}_a, \boldsymbol{k}_b, z; L\,|\,\boldsymbol{p}_0, t_0) = 2g^2 z(1-z)\int_0^L dt \mathcal{P}(\boldsymbol{k}_a, L\,|\,z\boldsymbol{p}, t)\mathcal{P}(\boldsymbol{k}_b, L\,|\,(1-z)\boldsymbol{p}, t)$$
$$\times\, \mathcal{K}(z,E)\mathcal{P}(\boldsymbol{p}, t\,|\,\boldsymbol{p}_0, t_0). \tag{52}$$

When the branching time is comparable to the size of the system, $\tau_{\rm br} \sim L$, finite size effects are no longer negligible but in this regime only a single branching occurs.[65,69] The corresponding branching probability integrated over transverse momentum generalizes Eq. (6) including finite energy corrections, that is $z \sim 1-z$, can be found in Refs. 33, 115 and 116. Also, the generalization to arbitrary parton species is straightforward and can be recovered from the aforementioned references.

[i] The vacuum part of the 3-point point function, $\tilde{S}_0^{(3)}$, is implicitly subtracted.

This is a leading order result that describes the instantaneous branching of a hard gluon inside a medium of size L. The probability distributions (35) and (52) can serve as building blocks for the in-medium gluon cascade that we shall present in Subsec. 3.5.

3.5. *Independent multiple branchings approximation*

The probability distribution (52) is parametrically of the form

$$\mathcal{P}_2 \sim \frac{\bar{\alpha}}{\tau_{\rm br}} L, \tag{53}$$

where $\bar{\alpha}/\tau_{\rm br}$ is the rate of (quasi-instantaneous) branching, and the factor L reflects the fact that the branching can occur anywhere along the medium, i.e., $0 < t < L$. Consider now the branching probability $\mathcal{P}_3$, where two successive branchings produce three gluons in the final state. The two branchings occur successively at times t_1 and t_2 in the amplitude and at times t'_1 and t'_2 in the complex conjugate amplitude. When the time intervals $[t_1, t'_1]$ and $[t_2, t'_2]$ do not overlap, that is, $t_1 < t'_1 < t_2 < t'_2$ the resulting successive integrations over the different times yields $L^2\,\tau_{\rm br}^2$, provided $\tau_{\rm br} \sim \tau_{1\rm br} \sim \tau_{2\rm br} \ll L$. In this case, the branchings are independent from one another and the 3-gluon probability factorizes as

$$\mathcal{P}_3^{\rm indep} \sim (\mathcal{P}_2)^2 \sim \left(\bar{\alpha}\frac{L}{\tau_{\rm br}}\right)^2. \tag{54}$$

On the other hand, interferences between the subsequent branchings may occur when the time intervals $[t_1, t'_1]$ and $[t_2, t'_2]$ overlap. For example, this may happen when $t_1 < t_2 < t'_1 < t'_2$ with $\tau_1, \tau_2 < \tau_{\rm br} \ll L$. The corresponding time integrals yield a factor $L\,\tau_{\rm br}^3$. In that case, the contribution to $\mathcal{P}_3$ from the interferences, $\mathcal{P}_3^{\rm inter}$, is suppressed by a factor $\tau_{\rm br}/L$ with respect to the factorized form (54), and it is obvious to generalize the argument to all possible time configurations.

For consistency, and to complete the picture of independent multiple branching approximation, it is of course necessary that the time between successive branching be large compare to the branching time. But this is guaranteed for inclusive quantities at weak coupling since $t_*(\omega) = \tau_{\rm br}(\omega)/\bar{\alpha} \gg \tau_{\rm br}(\omega)$.

It follows that, as long as the length L of the system is much larger that the branching time $\tau_{\rm br}$, successive branchings of a gluons in a large medium can be treated in first approximation as a probabilistic cascade. The properties of this cascade can be deduced from a set of n-point functions for which a generating functional was constructed in Ref. 78.

The two building blocks needed in this construction are the elastic and collinear branching rates, or equivalently the probabilities P_1 and P_2 given by

$$\begin{aligned} P_1(k\,|\,p) &= 2(2\pi)E\delta(\omega - E)\mathcal{P}(\boldsymbol{k}, t\,|\,\boldsymbol{p}, t), \\ P_2(k_1, k_2\,|\,p) &= 2(2\pi)E\delta(\omega_1 + \omega_2 - E)\delta(\boldsymbol{p} - \boldsymbol{k}_1 - \boldsymbol{k}_2) \\ &\quad \times \delta(\boldsymbol{k}_1 - z\boldsymbol{p})\mathcal{P}_2(\boldsymbol{k}_1, \boldsymbol{k}_2, z; t\,|\,\boldsymbol{p}, t_0). \end{aligned} \tag{55}$$

In this review, we shall focus solely on the lowest of the aforementioned n-point functions, namely the single-inclusive gluon distribution

$$D(\omega, \boldsymbol{k}) = \omega \frac{dN}{d\omega d^2\boldsymbol{k}}$$
$$= \frac{1}{2(2\pi)^3} \sum_{n=1}^{\infty} \frac{1}{n!} \int \prod_{i=1}^{n} d\Omega_i n P_n(k, k_1, \ldots, k_{n-1}). \tag{56}$$

It obeys the evolution equation

$$\frac{\partial}{\partial t} D(x, \boldsymbol{k}, t) = \frac{N_c n}{2} \int_{\boldsymbol{q}} \sigma(\boldsymbol{l}, t) D(x, \boldsymbol{k} - \boldsymbol{l}, t)$$
$$+ \alpha_s \int_0^1 dz \left[\frac{2}{z^2} \mathcal{K}\left(z, \frac{x}{z} E\right) D\left(\frac{x}{z}, \frac{\boldsymbol{k}}{z}, t\right) - \mathcal{K}(z, xE) D(x, \boldsymbol{k}, t) \right]. \tag{57}$$

The first and second terms on the r.h.s. correspond, respectively, to the elastic and the branching rates represented in Fig. 2, with in the latter case $\boldsymbol{l} = 0$.

3.6. *The gluon distribution beyond the collinear approximation*

In constructing the probabilistic picture in the previous section, we have explicitly assumed that the dominant contribution to the transverse momentum broadening along the cascade comes from elastic scatterings. Some increase in transverse momentum can also arise from the branching processes.

Let us then estimate the contribution of multiple radiations to transverse momentum broadening. Recall that the typical transverse momentum generated by one radiation of frequency ω is $k_{\rm br}(\omega) \sim (\omega \hat{q})^{1/4}$. Combining this with the radiative spectrum, $\omega dI/d\omega \sim L/t_*(\omega)$ (cf. Eq. (5)), we obtain the following parametric estimate for the total squared transverse momentum generated in splittings

$$\langle k_\perp^2 \rangle_{\rm typ,rad} \sim \int_{\omega_{\rm BH}}^{\omega_c} \frac{d\omega}{\omega} k_{\rm br}^2(\omega) \frac{L}{t_*(\omega)} \sim \bar{\alpha} \hat{q} L \ln \frac{L}{\lambda_{\rm el}}, \tag{58}$$

where we have integrated over the relevant range of gluon energies from $\omega_{\rm BH}$ to ω_c, and we have used the relation $\omega_{\rm BH} = \hat{q}\lambda_{\rm el}$. Compared to the typical transverse

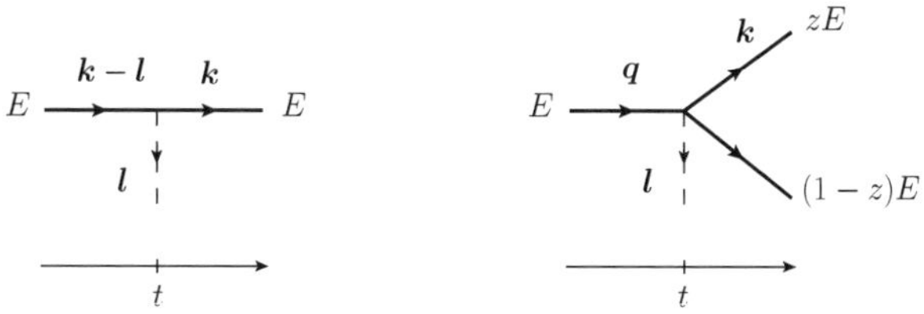

Fig. 2. Graphical representation of the elastic (left panel) and inelastic (right panel) rates. For collinear branching, $\boldsymbol{q} = \boldsymbol{k}/z$ and $\boldsymbol{l} = \boldsymbol{0}$.

momentum acquired by elastic scattering, the latter estimate is suppressed by a factor $\bar{\alpha}$, yet enhanced by a potentially large logarithm $\ln L/\lambda_{\rm el}$. We will see that this not the dominant contribution. A more detailed analysis reveals a double logarithmic enhancement[75] that is missed by the estimate we just did. We shall return in the next section to a detailed discussion of the origin of this correction. Here, we just indicate how it can be simply obtained by taking into account the momentum broadening taking place during the branching process. This forces us to go beyond the collinear approximation, and hence perform a calculation whose validity is not immediately obvious, but will be justified later.

We note that the 3-point function defined earlier contains the information about the transverse momentum broadening accompanying a single branching. Hence, instead of integrating over $\boldsymbol{l}$ and $\boldsymbol{Q}$, as we did for Eq. (50), we define the unintegrated kernel

$$\mathcal{K}(\boldsymbol{Q},\boldsymbol{l},z,E) = \frac{P(z)}{[z(1-z)E]^2}\mathrm{Re}\int_0^\infty d\tau\int_{\boldsymbol{Q}}(\boldsymbol{Q}_1\cdot\boldsymbol{Q})\tilde{S}^{(3)}(\boldsymbol{Q}_1,\boldsymbol{Q},\boldsymbol{l},t_1,\tau). \tag{59}$$

One can thus generalize Eq. (60) as

$$\frac{\partial}{\partial t}D(x,\boldsymbol{k},t) = \alpha_s\int_0^1 dz\int_{\boldsymbol{Q},\boldsymbol{l}}\left[\frac{2}{z^2}\mathcal{K}\left(\boldsymbol{Q},\boldsymbol{l},z,\frac{x}{z}E\right)D\left(\frac{x}{z},\boldsymbol{q},t\right)\right.$$
$$\left. -\,\mathcal{K}(\boldsymbol{Q},\boldsymbol{l},z,xE)D(x,\boldsymbol{k}-\boldsymbol{l},t)\right] - \frac{N_c n}{2}\int_{\boldsymbol{l}}\sigma(\boldsymbol{l})D(x,\boldsymbol{k}-\boldsymbol{l},t), \tag{60}$$

where $\boldsymbol{Q}\equiv\boldsymbol{k}-z(\boldsymbol{q}+\boldsymbol{l})$. We shall now follow the strategy that we adopted in order to reduce Eq. (33) to the diffusion equation (36). This involves an expansion around the large momentum $\boldsymbol{k}$ of the followed gluon. The momenta $\boldsymbol{Q}$ and $\boldsymbol{l}$, which are at most of the order of $k_{\rm br}\equiv\sqrt{z(1-z)\hat{q}E}$, are small compared to $\boldsymbol{k}$ which is typically of the order of $\hat{q}L$. The expansion of the gluon distributions around $\boldsymbol{k}$ yields for the first term of Eq. (60)

$$D\left(\frac{x}{z},\frac{\boldsymbol{k}-\delta\boldsymbol{k}}{z}\right) = D\left(\frac{x}{z},\frac{\boldsymbol{k}}{z}\right) - \delta\boldsymbol{k}\cdot\frac{\partial}{\partial\boldsymbol{k}}D\left(\frac{x}{z},\frac{\boldsymbol{k}}{z}\right)$$
$$+\frac{1}{2!}\,\delta k^i\delta k^j\frac{\partial}{\partial k_i}\frac{\partial}{\partial k_j}D\left(\frac{x}{z},\frac{\boldsymbol{k}}{z}\right)+\cdots \tag{61}$$

where we have set $\delta\boldsymbol{k}\equiv\boldsymbol{Q}+z\boldsymbol{l}$. One expands similarly $D(x,\boldsymbol{k}-\boldsymbol{l})$. It is easy to see that the leading terms reproduce Eq. (57). The linear terms vanish upon angular integration. Remain the quadratic terms, whose contribution can be cast in the form of the diffusion term, thereby exhibiting a correction $\Delta\hat{q}$ to the jet quenching parameter. For consistency, we shall also simplify the collision term by using the diffusion approximation. At this point, we anticipate that evaluation of the correction $\Delta\hat{q}$ meets with logarithmic divergences. These arise from the region $z\sim 1$. To the leading-logarithmic accuracy, we can set $z=1$ everywhere, except in the dominant singularity. The dominant contribution to $\Delta\hat{q}$ can be then written as

$$\Delta\hat{q}(\boldsymbol{k}^2) = 2\alpha_s \int_x^1 dz \int_{\boldsymbol{Q},\boldsymbol{l}} [(\boldsymbol{Q}+\boldsymbol{l})^2 - \boldsymbol{l}^2]\mathcal{K}(\boldsymbol{Q},\boldsymbol{l},z,xE), \tag{62}$$

where the $\boldsymbol{k}^2$ dependence arises from the integration boundary $\boldsymbol{Q}^2, \boldsymbol{l}^2 \ll \boldsymbol{k}^2$. The complete calculation of the integral (62) is presented in the next section. One gets[78]

$$\Delta\hat{q}(\boldsymbol{k}^2) = \frac{\alpha_s N_c}{2\pi}\hat{q}\ln^2\frac{\boldsymbol{k}^2}{\hat{q}\tau_0}, \tag{63}$$

where τ_0^{-1} is the maximum energy that can be extracted from the medium in a single scattering (e.g. $\tau_0 = 1/T$ for a weakly coupled plasma with temperature T). We return to this result in the next section. Finally, Eq. (60) can be recast in a form that is similar to that of Eq. (57) (after performing the diffusion approximation),[78]

$$\begin{aligned}\frac{\partial}{\partial t}D(x,\boldsymbol{k},t) &= \frac{1}{4}\left(\frac{\partial}{\partial \boldsymbol{k}}\right)^2 [(\hat{q}(\boldsymbol{k}^2) + \Delta\hat{q}(\boldsymbol{k}^2))D(x,\boldsymbol{k}-\boldsymbol{q},t)] \\ &\quad + \alpha_s \int_0^1 dz \left[\frac{2}{z^2}\mathcal{K}\left(z,\frac{x}{z}E\right)D\left(\frac{x}{z},\frac{\boldsymbol{k}}{z},t\right) - \mathcal{K}(z,xE)D(x,\boldsymbol{k},t)\right],\end{aligned} \tag{64}$$

where $p_\perp$-broadening due to soft gluon radiation is now taken into account effectively via a redefinition of the quenching parameter $\hat{q} \to \hat{q} + \Delta\hat{q}$.

We are now ready to revisit the standard estimate of the typical momentum broadening Eq. (38), making use of the redefinition of the quenching parameter above, substituting $\boldsymbol{k}^2 \sim \hat{q}L$ in Eq. (63), we obtain

$$\langle k_\perp^2\rangle_{\text{typ}} \simeq \hat{q}L\left[1 + \frac{\alpha_s N_c}{\pi}\ln^2\frac{L}{\tau_0}\right]. \tag{65}$$

This result agrees with that obtained in Ref. 75 using a different approach.[j] Note that we obtain a double logarithm of the medium length, as opposed to the heuristic estimate (58), which involve a single logarithm. Although, the soft (short time) logarithmic enhancement is correctly captured in this estimate, the single scattering that contributes to the branching probability that in turn leads to the second logarithm, is missing in the BDMPS-Z spectrum that is used Eq. (58).

4. Renormalization of the Jet Quenching Parameter

The correction to the jet quenching parameter discussed in the previous section points to the existence of potentially important radiative corrections. The calculation that we just presented focussed on a correction that is singular, and this is what allowed us to retain in the branching kernel contributions that should not be

[j]Note that NLO corrections to $p_\perp$-broadening have been also investigated in Deep Inelastic Scattering in the High-Twist approach.[77]

kept *a priori* since, if they were not divergent, they would be of the same order of magnitude as terms that have been consistently neglected in arriving at Eq. (57). In this section we discuss a more systematic approach to the calculation of radiative corrections. Still, at present, the program of calculating the radiative corrections is incomplete.[74,75,78,79,117,118] As we shall see, one is able so far to complete the calculation only for the leading singular part. Furthermore, it is only this singular part that can be interpreted as a correction to the jet quenching parameter; less singular corrections would presumably be non local, thereby spoiling their interpretation in terms of a correction to the jet quenching parameter.

4.1. *Radiative corrections to momentum broadening*

We start by revisiting the calculation of the momentum broadening. Recall that this is described by the 2-point function $S^{(2)}(t_1, t_0)$, Eq. (32), for which the following equation is easily established

$$S^{(2)}(t_1, t_0) = S_0^{(2)}(t_1, t_0) + \int dt_3 \int dt_2 S_0^{(2)}(t_1, t_3)\Sigma^{(2)}(t_3, t_2)S^{(2)}(t_2, t_0). \quad (66)$$

In this equation, the instantaneous interaction kernel $\Sigma^{(2)}$ is related to the dipole cross-section (34),

$$(\boldsymbol{x}|\Sigma^{(2)}(t_3, t_2)|\boldsymbol{y}) = \frac{N_c n}{2}\delta(\boldsymbol{x} - \boldsymbol{y})\delta(t_3 - t_2)\sigma(\boldsymbol{x}). \quad (67)$$

We are interested in the radiative correction to $\Sigma^{(2)}$. The corresponding diagrams are displayed in Fig. 3. The important par is that involving the three gluons evolving between the times t_2 and t_3, and to which we refer as a 3-point function. The dominant, double logarithmic contribution corresponds to the case where the radiated gluon (the wavy lines in Fig. 3) is soft, i.e., its energy ω is $\omega \ll E$.

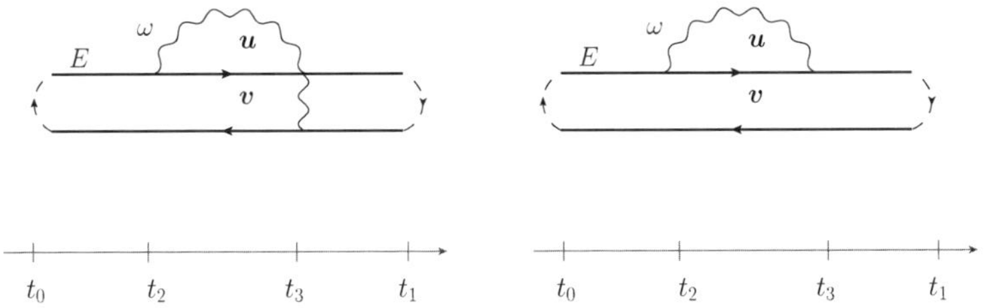

Fig. 3. Real (left) and virtual (right) contributions to the radiative correction to the 2-point function. There are other diagrams that are not shown, corresponding to a different time ordering of the emission and absorption vertices in the real term, and a diagram where the virtual correction is attached to the lower line. The transverse vector $\boldsymbol{v}$ stands for the transverse size of the frozen hard dipole (represented by the straight lines), whose constituents carry an energy E, and $\boldsymbol{u}$ corresponds to the time-dependent transverse distance between the radiated gluon (represented by a wavy line) and its radiator whose dynamics is encoded in the 3-point function $\tilde{S}^{(3)}$. The integration over $\boldsymbol{u} \sim \boldsymbol{q}^{-1}$ in the region $\boldsymbol{u} \gg \boldsymbol{v}$ yields the second logarithmic enhancement of the radiative correction to $\hat{q}$ (cf. Eq. (74)).

A simplification then occurs in the calculation of the diagrams since we may assume that the transverse coordinates of the hard gluons (in the amplitude and its complex conjugate) are frozen during the lifetime of the fluctuation, i.e., between t_2 (in the amplitude) and t_3 (in the complex conjugate amplitude). The calculation leads then, for the real correction, and using a symbolic matrix notation,

$$\Delta S^{(2)}(t_1,t_0) = -\int_{t_0}^{t_1} dt_3 \int_{t_0}^{t_3} dt_2 S^{(2)}(t_1,t)\Delta\Sigma^{(2)}(t_3,t_2)S^{(2)}(t_2,t_0), \tag{68}$$

with

$$(X_3|\Delta\Sigma^{(2)}(t_3,t_2)|X_2) = \delta(X_3 - X_2)\frac{\alpha_s N_c}{2}\, 2\mathrm{Re} \times \int \frac{d\omega}{\omega^3}\, \boldsymbol{\partial}_{\boldsymbol{u}_2}\cdot\boldsymbol{\partial}_{\boldsymbol{u}_3}\tilde{S}^{(3)}(\boldsymbol{u}_2,\boldsymbol{u}_3,\boldsymbol{v};\tau)|_{\boldsymbol{u}_2=0,\boldsymbol{u}_3=-\boldsymbol{v}}, \tag{69}$$

where $\tau \equiv t_3 - t_2$. The 3-point function $\tilde{S}^{(3)}(\boldsymbol{u}_2,\boldsymbol{u}_3,\boldsymbol{v};\tau)$ encodes the dynamics in the transverse plane of the system made of a frozen dipole and a soft gluon. It corresponds to the soft limit, i.e. $z = \omega/E \ll 1$, of Eq. (46). Here, $\boldsymbol{v}$ stands for frozen transverse size of the hard dipole during the time interval τ, while $\boldsymbol{u}_2 = \boldsymbol{0}$ and $\boldsymbol{u}_3 = -\boldsymbol{v}$ stand for the transverse separation between the radiated gluon and its radiator (the hard gluon in the amplitude) at the times t_2 and t_3 respectively (see left panel of Fig. 3 for an illustration). The virtual correction is given by the same formula in which $\boldsymbol{u}_3 = \boldsymbol{0}$ since the gluon is emitted and reabsorbed by the same line. Anticipating on an argument that will be provided later, we treat the correction $\Sigma^{(2)}(t_3,t_2)$ effectively as a local correction, i.e., set formally $\Sigma^{(2)}(t_3,t_2) \propto \delta(t_3-t_2)$. After replacing in Eq. (68) the factors $S^{(2)}$ external to the radiative correction by free 2-point functions (which are independent of time), and performing the t_2 integration, we get

$$(X_1|\Delta S^{(2)}(t_1,t_0)|X_0) = -(t_1-t_0)\delta(X_1-X_0)\frac{\alpha_s N_c}{2}\int\frac{d\omega}{\omega^3}\int d\tau K(\boldsymbol{v},\tau), \tag{70}$$

with

$$K(\boldsymbol{v},\tau) \equiv 2\mathrm{Re}\int_{\boldsymbol{q}_2,\boldsymbol{q}_3,\boldsymbol{l}} (\mathrm{e}^{i(\boldsymbol{l}-\boldsymbol{q}_3)\cdot\boldsymbol{v}} - \mathrm{e}^{i\boldsymbol{l}\cdot\boldsymbol{v}})(\boldsymbol{q}_2\cdot\boldsymbol{q}_3)\tilde{S}^{(3)}(\boldsymbol{q}_2,\boldsymbol{q}_3,\boldsymbol{l};\tau), \tag{71}$$

where we have performed a Fourier transform of the 3-point function. Besides an integration over τ, we recognize in Eq. (71) the soft limit, i.e. $z \ll 1$ of the kernel Eq. (59). Note the vanishing of this expression when $\boldsymbol{v} \to 0$. It results from a cancellation between real and virtual terms, reflecting the property of color transparency of a dipole cross-section.

The identification of the correction to the dipole cross-section proceeds by comparison with the integral equation (66) (more properly, its generalization, whose validity relies on the locality assumption) for fixing the relation between $\Delta\sigma$ and

$\Delta\Sigma^{(2)}$. One then gets, leaving the bounds on the τ integration unspecified for the moment,

$$\frac{N_c n}{2}\Delta\sigma(\boldsymbol{v}) = \alpha_s \frac{N_c}{2}\int \frac{d\omega}{\omega^3}\int d\tau K(\boldsymbol{v},\tau) \approx \frac{1}{4}\boldsymbol{v}^2\Delta\hat{q}, \tag{72}$$

where the last equality stems from Eq. (39) expressing $\Delta\sigma(\boldsymbol{v})$ at small $\boldsymbol{v}$ in the harmonic approximation, by expanding the phases in Eq. (71). We use this relation to interpret the correction $\Delta\sigma(\boldsymbol{v})$ as a correction to the parameter $\hat{q}$. Note that in this approximation the resulting integrations over the transverse momenta $\boldsymbol{q}_2$ and $\boldsymbol{l}$ are bounded from above by $\boldsymbol{v}^{-2}$.

The explicit calculation of $\Delta\hat{q}$, using the expression (46) of the reduced 3-point function, yields[78,79]

$$\begin{aligned}\Delta\hat{q} &= \frac{\alpha_s N_c}{\pi} 2\mathrm{Re}\int d\omega \int d\tau \frac{i\Omega^3}{\sinh(\Omega\tau)}\left[1+\frac{4}{\sinh^2(\Omega\tau)}\right],\\ &\simeq \frac{\alpha_s N_c}{\pi}\int\frac{d\omega}{\omega}\int^{\Omega^{-1}}\frac{d\tau}{\tau}\hat{q},\end{aligned} \tag{73}$$

which as anticipated exhibits a double logarithmic divergence, when $\tau \to 0$ and $\omega \to 0$. As clear from Eq. (73) the τ integral is bounded at the upper end by $|\Omega|^{-1} \sim \tau_{\rm br}(\omega) = \sqrt{\omega/\hat{q}}$ (cf. Eq. (7)) corresponding to the onset of the multiple scattering regime and the LPM effect: the relevant gluon fluctuation experiences a single scattering with the medium constituents. In order to systematically account for the boundaries of the double integral,[k] it is in fact more convenient to change variables, from (ω,τ) to $(\boldsymbol{q},\tau)$, with $\tau \equiv \omega/\boldsymbol{q}^2$ the formation time of the radiated gluon, and $\boldsymbol{q}$ its transverse momentum which can run up to $\boldsymbol{p}^2 \equiv \boldsymbol{v}^{-2}$ (the logarithmic transverse momentum integration can be rephrased in terms of the transverse coordinate of the radiated gluon $\boldsymbol{q}^{-1} \sim \boldsymbol{u} \gg \boldsymbol{v}$, see Fig. 3). We obtain then

$$\Delta\hat{q}(\tau_{\rm max},\boldsymbol{p}^2) \equiv \frac{\alpha_s N_c}{\pi}\int_{\tau_0}^{\tau_{\rm max}}\frac{d\tau}{\tau}\int_{\hat{q}\tau}^{\boldsymbol{p}^2}\frac{d\boldsymbol{q}^2}{\boldsymbol{q}^2}\hat{q}(\boldsymbol{q}^2), \tag{74}$$

where we have explicitly indicated the scale dependence of $\hat{q}$. The boundary corresponding to the region of multiple scattering now appears as the lower bound of the $\boldsymbol{q}$ integration, $\boldsymbol{q}^2 \gg k_{\rm br}^2 \equiv \hat{q}\tau$. Since medium-induced gluons forms inside the medium, the largest value for τ is the length of the medium $L \sim t - t_0$. As for the lowest value τ_0, it can be interpreted as the inverse of the largest energy that can be extracted from the medium through a single scattering.

For a constant $\hat{q}$ in the integral, one can easily perform the integrations, and by keeping the leading contributions, we recover the result first derived in Refs. 75

[k]For a detailed discussion on the boundaries of the logarithmic integrals see Refs. 75 and 118.

and 78,

$$\Delta\hat{q} \simeq \frac{\alpha_s C_A}{2\pi}\hat{q}\ln^2\left(\frac{L}{\tau_0}\right), \tag{75}$$

where we have used the fact that $\boldsymbol{p}^2 \sim \hat{q}L$.

4.2. *Locality of radiative corrections*

So far we have discussed a single radiative correction. The logarithmic phase space for radiative corrections extends from τ_0 to L which makes them explicitly non-local. One may therefore question why we have argued that these corrections could be absorbed into a redefinition of $\hat{q}$, considered to be a local transport coefficient. Indeed, the quenching parameter makes sense strictly speaking only in the approximation of independent multiple scatterings. This requires the duration of each interactions to be much smaller than the mean-free path.

The key lies in the nature of the logarithmic integral whose boundaries do not affect the overall multiplicative constant. Thus, to the extent that one restricts oneself to a leading order calculation, one can proceed as if the lifetimes of the fluctuations involved in the radiative corrections were small, and treat them as local. The coefficient of the leading double logarithm is calculated correctly, and corrections coming from overlapping contributions will be subleading. In short, the leading corrections can be treated as being effectively local in time and thus independent.[79] This argument is what allows us to generalize Eq. (66) to a $\Sigma(t,t')$ that includes the radiative corrections, to treat the double logarithmic correction to $\Sigma(t,t')$ as effectively local, and multiple radiative corrections as independent. As we have seen, such corrections can be interpreted, in the coordinate space description, as a modification of the dipole cross-section $\sigma(\boldsymbol{v}) \rightarrow \sigma(\boldsymbol{v}) + \Delta\sigma(\boldsymbol{v},\tau_{\max})$, where $\tau_{\max} \equiv t_1 - t_0$, or in momentum space as a modification of the equation for the momentum broadening probability[79]

$$\frac{\partial}{\partial t}\mathcal{P}(\boldsymbol{p},t) = \frac{1}{4}\left(\frac{\partial}{\partial \boldsymbol{p}}\right)^2 [\hat{q}(\boldsymbol{p}^2) + \Delta\hat{q}(\boldsymbol{p}^2)]\mathcal{P}(\boldsymbol{p},t), \tag{76}$$

which generalizes Eq. (36). In both cases, the radiative corrections are accounted for by a correction to the jet quenching parameter.

4.3. *Radiative corrections to energy loss and universality*

We now address the question of whether the absorption of the dominant logarithmic contributions to the radiative corrections in a redefinition of $\hat{q}$ is universal, that is, whether it holds for other observables than the broadening probability. This has important implications as it would affect the branching rate in Eq. (64), that controls the number of final state gluons. In order to address this issue, we consider radiative corrections to the BDMPS-Z spectrum (5) that can be recovered in limit $\omega = zE \ll E$ of the branching rate (50).

The BDMPS-Z spectrum (5) of radiated gluons with frequency $\omega \ll E$, can be calculated from the reduced 3-point function (46),[33] according to[l]

$$\frac{dI}{d\omega dt} \equiv \frac{\alpha_s N_c}{\omega^2} 2\text{Re} \int_0^\infty d\tau \partial_{\boldsymbol{u}} \cdot \partial_{\boldsymbol{u}'} \tilde{S}^{(3)}(\boldsymbol{u}, \boldsymbol{u}', \boldsymbol{v}; \tau)|_{\boldsymbol{v}=\boldsymbol{u}=\boldsymbol{u}'=\boldsymbol{0}}. \tag{77}$$

Here the variable t runs up to $\sim L$. This spectrum is valid in the large medium length limit where the gluon branching time $\tau_{\text{br}} = \sqrt{\omega/\hat{q}} \ll L$ and the integration over the gluon formation time $\tau < L$ is suppressed exponentially beyond τ_{br}. The transverse coordinates $\boldsymbol{u}$ and $\boldsymbol{u}'$ correspond to initial and final coordinates of the radiated gluon, and $\boldsymbol{v}$ corresponds to the size of the energetic dipole that radiates the gluon (see Fig. 4). This is why we can integrate τ up to infinity in Eq. (77).

The 3-point function $\tilde{S}^{(3)}$ is given explicitly by Eq. (46) in the harmonic approximation, and $\tau = t_1 - t_0$, with t_0 and t_1 denoting the times of the emission in the amplitude and its complex conjugate, respectively (see Fig. 4). The BDMPS-Z spectrum of Eq. (5) is easily recovered from this expression by performing the integrations over the transverse momenta and over τ.[65,73] As is clear from Eq. (46) the reduced 3-point function depends explicitly on $\hat{q}$. One can show that the leading radiative corrections do not modify $\tilde{S}^{(3)}$, except for a change in the value of the parameter $\hat{q}$, the correction to $\hat{q}$ being, besides, the same as that calculated in the previous section for momentum broadening.

Quite generally, we are concerned with the radiative corrections of the 3-point function $S^{(3)}$, whose graphical interpretation is given in Fig. 4. The diagram displayed there corresponds typically to a branching process where a gluon with initial energy E (represented by the lower two thick lines in the amplitude and the conjugate amplitude, respectively), emits a gluon at t_0 in the amplitude and reabsorbs it in the complex conjugate amplitude at $t_1 = t_0 + \tau$, with energy $\omega = zE$. The correction to the 3-point function (in the limit $\boldsymbol{v} \to \boldsymbol{0}$ as required by the integration over $\boldsymbol{l}$ in Eq. (46)) can now be written by analogy with Eq. (68) as (see Fig. 5 for

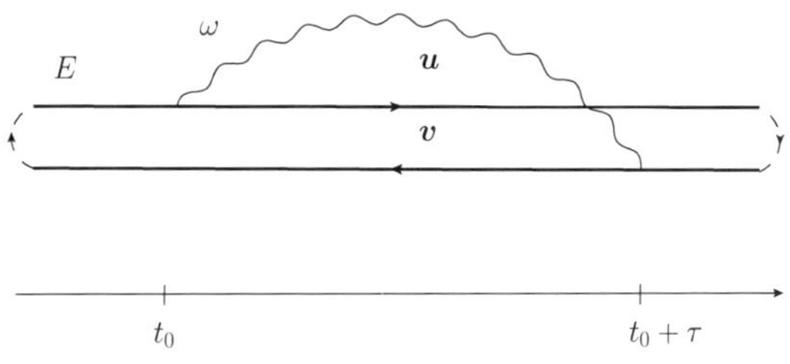

Fig. 4. Diagrammatic illustration of the BDMPS-Z spectrum (77) that corresponds to the radiation of a soft gluon with energy ω. We denote by $\boldsymbol{u}$ the transverse position of the radiated gluon with respect its radiator, the energetic dipole whose size is denoted by $\boldsymbol{v}$.

[l]The vacuum part of the 3-point point function, $\tilde{S}_0^{(3)}$, is implicitly subtracted.

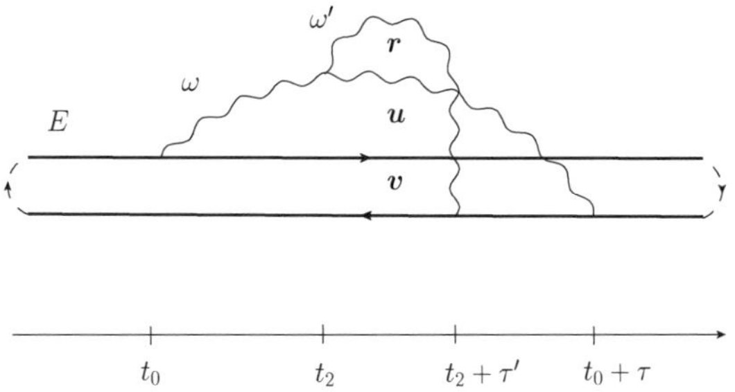

Fig. 5. Diagrammatic illustration of the radiative correction to the BDMPS-Z spectrum (77). The gluon ω' is integrated over and account for real and virtual contributions. Here one topology among nine is displayed corresponding to the radiation of the gluon ω' from gluon ω and its absorption by the gluon E in the complex conjugate amplitude. The labeling is as in the previous figure, with here, in addition, $\boldsymbol{r}$ denoting the transverse coordinate of the gluon fluctuation ω'. The integration over $\boldsymbol{r} \sim \boldsymbol{q}^{-1}$ in the region $\boldsymbol{r} \gg \boldsymbol{u}$ yields the second logarithmic enhancement of the radiative correction to $\hat{q}$ (cf. Eq. (73)).

an illustration)

$$\Delta S^{(3)}(t_1, t_0) \simeq - \int_{t_0}^{t_1} dt_3 \int_{t_0}^{t_3} dt_2 S^{(3)}(t_1, t_2) \Delta\Sigma^{(3)}(t_3, t_2) S^{(3)}(t_2, t_0), \tag{78}$$

where

$$(X_3 | \Delta\Sigma^{(3)}(t_3, t_2) | X_2) = \delta(X_3 - X_2) \frac{\alpha_s N_c}{2} \int \frac{d\omega'}{\omega'^3} K(\boldsymbol{u}, \tau'). \tag{79}$$

At this point we proceed as for the 2-point function, and treat the correction $\Delta\Sigma^{(3)}(t_3, t_2)$ as a local correction. By comparing with Eq. (46) in the limits $z \to 0$ and $\boldsymbol{v} \to 0$, one then gets

$$\frac{N_c n}{2} \Delta\sigma(\boldsymbol{u}) = \frac{\alpha_s N_c}{2} \int^{\tau_{\max}} d\tau' \int \frac{d\omega'}{\omega'^3} K(\boldsymbol{u}, \tau'). \tag{80}$$

Thus, to double logarithmic accuracy, the radiative corrections to the reduced 3-point function are accounted for by correcting the dipole cross-section and thus the jet quenching parameter. This result applies to the particular 3-point function involved in the calculation of the gluon spectrum (77). In order to evaluate the corrected spectrum, we need to perform the integration over τ in Eq. (77), replacing $\hat{q}$ by $\hat{q} + \Delta\hat{q}(\tau_{\max})$. However, since the correction to $\hat{q}$ is computed to double logarithmic accuracy one can simply replace the variable $\tau_{\max}$ in Eq. (80) (see also Eq. (74)) by its typical value in the radiation process, i.e., $\tau_{\max} \simeq \tau_{\rm br} \equiv \sqrt{\omega/\hat{q}}$. The integral over τ in the BDMPS-Z spectrum (77) can then be performed as for the case with no radiative correction. Doing so, we obtain[79]

$$\frac{dI}{d\omega dt} \equiv \frac{\alpha_s N_c}{\pi} \sqrt{\frac{\hat{q} + \Delta\hat{q}}{\omega}}, \tag{81}$$

where for a constant $\hat{q}$ one gets, from Eq. (74) letting $\boldsymbol{p}^2 \simeq k_{\rm br}^2(\omega) \equiv \sqrt{\omega\hat{q}}$,

$$\hat{q} + \Delta\hat{q} \approx \hat{q} \left[1 + \frac{\alpha_s N_c}{2\pi} \ln^2 \sqrt{\frac{\omega}{\hat{q}\tau_0^2}} \right]. \tag{82}$$

It can be shown that this result extends to the full kernel (50) in the large N_c limit.[79] These results were recently confirmed by an alternative calculation of the radiative corrections to the mean-energy loss to double logarithmic accuracy.[117]

4.4. *Renormalization of the quenching parameter*

For large media, as soon as $\bar{\alpha}\ln^2(L/\tau_0) \sim 1$, one has to resum the double logarithmic power corrections. Unlike the previous resummation of independent multiple radiative corrections, this now involves radiative corrections that are correlated to each other. To understand how this resummation proceeds, we denote the standard leading order definition of the jet-quenching parameter by $\hat{q}_0$ and we note that the first correction to the jet-quenching parameter, $\hat{q}_1(\tau, \boldsymbol{k}^2) \equiv \Delta\hat{q}(\tau, \boldsymbol{k}^2)$ is proportional to the 3-point function, $S^{(3)}[\hat{q}_0]$ which is itself a function of the leading order $\hat{q}_0$. As we have shown, under radiative corrections, the 3-point function gets renormalized by a simple modification of $\hat{q}$, that is, $S^{(3)}[\hat{q}_0] \to S^{(3)}[\hat{q}_0 + \hat{q}_1]$. This allows us to compute the second correction from Eq. (74),

$$\hat{q}_2(\tau, \boldsymbol{k}^2) = \bar{\alpha} \int_{\tau_0}^{\tau} \frac{d\tau'}{\tau'} \int_{\hat{q}_0\tau'}^{1/\boldsymbol{x}^2} \frac{d\boldsymbol{q}^2}{\boldsymbol{q}^2} \hat{q}_1(\tau', \boldsymbol{q}^2). \tag{83}$$

One sees emerging a self-similarity that results from the separation of time scales involved in the computation of the leading logarithms. The structure of the first double logarithmic corrections being set, the next corrections that yield double logarithms will follow the same systematics, with successive gluonic fluctuations ordered in formation time $\tau_0 \ll \tau_1 \ll \cdots \ll \tau_n \equiv \tau_{\max}$, or in transverse size $\boldsymbol{r}_0 \gg \boldsymbol{r}_1 \gg \cdots \gg \boldsymbol{r}_n \equiv \boldsymbol{r}_{\max}$, or in transverse momentum $m_D \ll \boldsymbol{q}_1 \ll \cdots \ll \boldsymbol{q}_n \equiv \boldsymbol{k}$. A diagrammatic illustration is given in Fig. 6. The difference with the standard Double-Logarithmic Approximation (DLA) is the limits of logarithmic phase-space set by the LPM effect since, i.e., multiple-scatterings since in the DLA

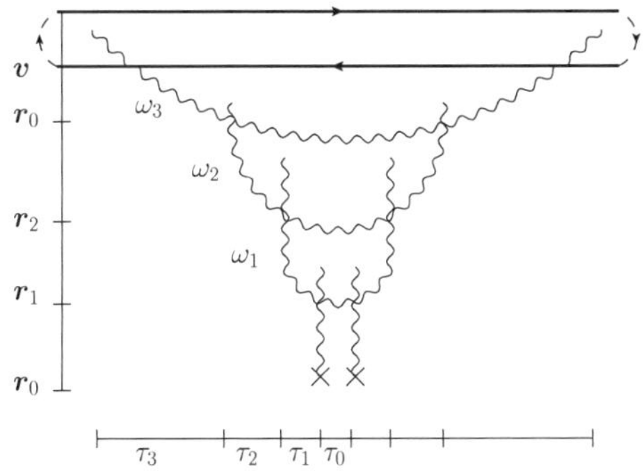

Fig. 6. A diagrammatic representation of the resummation of double logarithmic contributions for a single scattering. Here three gluon emissions are depicted. In the double logarithmic approximation successive radiations are strongly ordered in formation time, i.e. $\tau_0 \ll \tau_1 \ll \tau_2 \ll \tau_3 \ll L$, and in transverse sizes, i.e., $1/m_D \sim \boldsymbol{r}_0 \gg \boldsymbol{r}_1 \gg \boldsymbol{r}_2 \gg \boldsymbol{r}_3 \gg \boldsymbol{v}$. For the original dipole the subsequent radiations look *as if* they were instantaneous.

only a single scattering contributes, which imposes that the formation time of a fluctuation to be smaller than the BDMPS-Z formation time, or in terms of our transverse momentum variables, $\boldsymbol{q}^2 \gg \hat{q}_0\tau$.[m] The following equation resums the double logarithmic corrections to all orders

$$\frac{\partial \hat{q}(\tau, \boldsymbol{k}^2)}{\partial \ln \tau} = \int_{\hat{q}\tau}^{\boldsymbol{k}^2} \frac{d\boldsymbol{q}^2}{\boldsymbol{q}^2} \bar{\alpha}(\boldsymbol{q}) \hat{q}(\tau, \boldsymbol{q}^2). \tag{84}$$

with some initial condition $\hat{q}(\tau_0, \boldsymbol{k})$. We have let the coupling running at the transverse scale $\boldsymbol{q}$. The important feature of this equation is that it predicts the evolution of the jet-quenching parameter from an initial condition $\hat{q}_0$ (which can be computed, e.g. on the lattice, or to leading order in α_s, which implies, $\hat{q}(\tau_0) \equiv \hat{q}_0$ as given by the leading order result (37). The τ_0 cut-off that was introduced to cut the logarithmic divergence in the radiative corrections, can be seen as a factorization scale. The solution to this equation was derived in Ref. 75 for the p_t-broadening in the case where $\hat{q}_0 = \hat{q}(\tau_0)$ is constant and for a final $\tau = L$ and $\boldsymbol{k}^2 = \hat{q}_0 L$, merging the two independent variables at the end of the evolution. The solution reads

$$\hat{q}(L) = \frac{1}{\sqrt{\bar{\alpha}}} I_1 \left(2\sqrt{\bar{\alpha}} \ln \frac{L}{\tau_0} \right) \hat{q}(\tau_0). \tag{85}$$

A semi-analytic analysis of this equation including the running of the coupling has been performed in Ref. 120. For large L, the quenching parameter scales like $\hat{q}(L) \sim L^{\gamma}$, with the anomalous

$$\gamma = 2\sqrt{\bar{\alpha}}. \tag{86}$$

Interestingly, the resummation of large double logarithms modifies the scaling of the energy loss with L,

$$\Delta E \sim L^{2+\gamma}, \tag{87}$$

a correction that seems to fall between the standard small coupling result, $\Delta E \sim L^2$ and the strong coupling result obtained with the help of the AdS-CFT correspondence in $\mathcal{N} = 4$ SYM theory,[121,122] $\Delta E \sim L^3$.

Finally, let us make now a rough estimate of the renormalized quenching parameter. The standard perturbative estimate yields a value of about $\hat{q}_0 \sim 1\,\mathrm{GeV}^2/\mathrm{fm}$.[32] For $\alpha_s \sim 0.5$, $L \sim 5$ fm and $T = \tau_0^{-1} \sim 500$ MeV, Eq. (85) yields a sizable increase by a factor 2, $\hat{q} \sim 2\,\mathrm{GeV}^2/\mathrm{fm}$. This result is in the ballpark of the jet quenching parameter values extracted from the data.[123]

5. Energy Flow and Angular Structure of the In-Medium Cascade

We return now to the medium induced cascade, and investigate some of its characteristic features as revealed by the analysis of the inclusive distribution function

[m] The resummation of double logarithms in the quenching parameter was postulated earlier in Ref. 119 where the LPM suppression was not taken into account.

$D(x, \boldsymbol{k})$. We shall discuss in particular the energy and angular distributions of the emitted gluons. As we shall see, these properties make the BDMPS-Z cascade very different from the more common parton cascades in vacuum that are described by the DGLAP equation.

Our starting point is Eq. (57) for the inclusive distribution. Since the transverse momentum $\boldsymbol{k}$ remains small compared to the longitudinal momentum, it is actually convenient to use angle variables rather than transverse momenta. Accordingly, we set

$$D(x, \boldsymbol{\theta}) \equiv (2\pi)^2 x \frac{dN}{dx d^2\boldsymbol{\theta}}, \tag{88}$$

where

$$x = \frac{\omega}{E} \quad \text{and} \quad \boldsymbol{\theta} \equiv \frac{\boldsymbol{k}}{\omega} = \frac{\boldsymbol{k}}{xE}. \tag{89}$$

Note that $\boldsymbol{\theta}$ is a 2-dimensional vector collinear to $\boldsymbol{k}$, whose (small) magnitude equals the polar angle of the emitted gluon with respect to the initial direction of the leading particle. In this new variable, Eq. (60) reads

$$\begin{aligned} \frac{\partial}{\partial t} D(x, \boldsymbol{\theta}, t) &= \int dz \mathcal{K}(z) \left[\frac{D(x/z, \boldsymbol{\theta}, t)}{t_*(x/z)} - \frac{D(x, \boldsymbol{\theta}, t)}{t_*(x)} \right] \\ &\quad + \frac{N_c n}{2} \int \frac{d^2\boldsymbol{\theta}'}{(2\pi)^2} \sigma(\boldsymbol{\theta}', x) D(x, \boldsymbol{\theta} - \boldsymbol{\theta}', t), \end{aligned} \tag{90}$$

with (see Eq. (34))

$$\sigma(x, \boldsymbol{\theta}) = (xE)^2 \sigma(\boldsymbol{q}) = \frac{32\pi^2 \alpha_s^2}{(xE)^2} \left[\frac{1}{\boldsymbol{\theta}^4} - \delta(\boldsymbol{\theta}) \int \frac{d^2\boldsymbol{\theta}'}{\boldsymbol{\theta}'^4} \right]. \tag{91}$$

An illustration of the gluon cascade described by Eq. (90) is given Fig. 7. Note that the locality (in angle) of the splitting term reflects the effective collinearity of the splitting.

In addition to changing to angular variables, in Eq. (90) we have explicitly factorized the kernel (50) into

$$\alpha_s \mathcal{K}(z, xE) = \frac{\mathcal{K}(z)}{t_*(x)}, \tag{92}$$

where now

$$\mathcal{K}(z) = \frac{[1 - z(1-z)]^{5/2}}{z^{3/2}(1-z)^{3/2}} \quad \text{and} \quad t_*(x) = \frac{1}{\bar{\alpha}} \sqrt{\frac{xE}{\hat{q}}}. \tag{93}$$

We shall refer to $\mathcal{K}(z)$ as the *reduced kernel.* The quantity

$$t_*(E) \equiv \frac{1}{\bar{\alpha}} \sqrt{\frac{E}{\hat{q}}} \tag{94}$$

is a characteristic time scale of the BDMPS-Z cascade. After a time of order $t_*(E)$, most of the initial energy has been radiated into soft gluons (provided $L > t_*(E)$, of

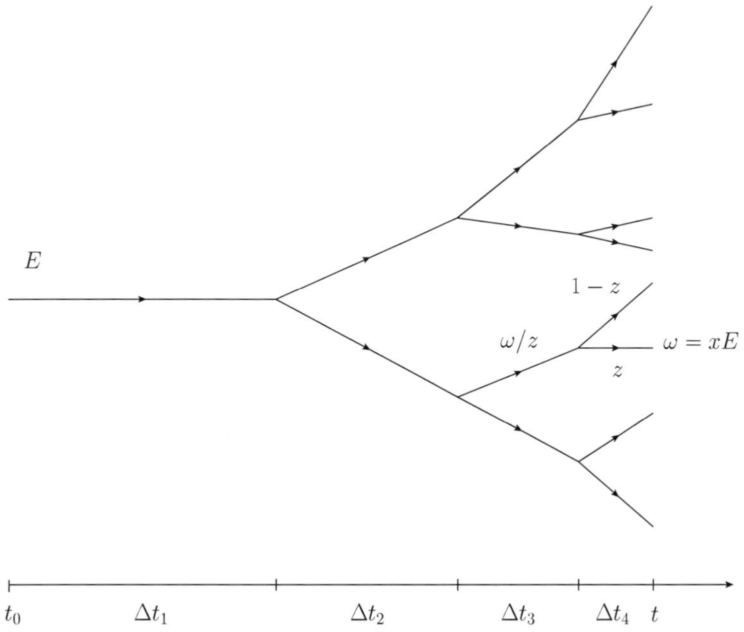

Fig. 7. Illustration of a gluon cascade that is initiated by a gluon with energy E. Four generations are displayed. The branching time $\Delta t_i \sim t_*(x_i)$, that corresponds to the lifetime of generation i, decreases after each branching as in a BDMPS-Z cascade. The inclusive distribution $D(x, \boldsymbol{\theta})$ measures the probability to find in the cascade, at time t, a gluon with energy xE. The rate equation (90) describes how this distribution evolves with time t.

course), hence the often used denomination of *stopping time*, or *stopping distance*.[124] In the following, we often denote $t_*(E)$ simply by t_*, except when confusion may arise. Note that $t_*(x) = t_*\sqrt{x}$.

As such the kernel (93) does not account for finite size effects that become important when $t_*(x) > L$, corresponding to $x_c > 1$ ($\omega_c > E$) (cf. Eq. (6)). In the following discussion we neglect these corrections, we thus implicitly restrict jet energies to be $E \sim \omega_c$ or smaller (see Refs. 70 and 71 for discussions on this matter).

In the next subsection, we analyze the energy distribution, obtained by integrating $D(x, \boldsymbol{\theta}, t)$ over the angle. That is, we define

$$D(x, t) = \int \frac{d^2\boldsymbol{\theta}}{(2\pi)^2} D(x, \boldsymbol{\theta}, t), \tag{95}$$

which obeys the following equation[44,45,66,125]

$$\frac{\partial}{\partial t} D(x, t) = \int dz \mathcal{K}(z) \left[\frac{D(x/z, t)}{t_*(x/z)} - \frac{D(x, t)}{t_*(x)} \right] \equiv \mathcal{I}[D]. \tag{96}$$

The angular distribution of the radiated gluons will be discussed next.

5.1. Energy distribution and wave turbulence

Since initially the total energy is carried by a single gluon, the initial condition for Eq. (96) is simply $D_0(x) = \delta(1-x)$. In the regime where the length of the medium is so short that at most a single branching can be induced, $L \lesssim t_*(x(1-x))$, Eq. (96) can be solved by iterations, with the first one given by

$$D_1(x, L) = \frac{L}{t_*} x \mathcal{K}(x), \tag{97}$$

which coincides with the BDMPS-Z spectrum (5) at small x. In the opposite regime where $L \gg t_*(x)$, which corresponds to gluon energies $\omega \ll \omega_s \equiv \bar{\alpha}^2 \, \hat{q} \, L^2$, multiple branchings are important. A non-perturbative solution becomes mandatory.

Equation (96) can be solved exactly for a simplified version of the reduced kernel (93) in which one neglects the z dependence of the numerator, that is

$$\mathcal{K}(z) \approx \frac{1}{z^{3/2}(1-z)^{3/2}}. \tag{98}$$

This turns out to be an excellent approximation. In fact, we shall argue later that the exact form of the kernel plays a minor role in the determination of the general features of the cascade. The solution of Eq. (96) for the simplified kernel reads[66,72]

$$D(x, \tau) = \frac{\tau}{\sqrt{x}(1-x)^{3/2}} \exp\left(-\pi \frac{\tau^2}{1-x}\right), \quad \text{with } \tau = \frac{L}{t_*}. \tag{99}$$

This solution exhibits two remarkable features: a peak near $x = 1$ associated with the leading particle, and a scaling behavior in $1/\sqrt{x}$ at small x where the x dependence factorizes from the time dependence, i.e.

$$D(x, \tau) \approx \frac{\tau}{\sqrt{x}} \mathrm{e}^{-\pi \tau^2}. \tag{100}$$

An illustration of this solution is given in Fig. 8, left panel. The energy of the leading particle, initially concentrated in the peak at $x \lesssim 1$, gradually disappears into radiated soft gluons, and after a time $t \sim t_*$ (i.e. $\tau \sim 1/\sqrt{\pi} \approx 0.5$) most of the energy is to be found in the form of radiated soft ($x \lesssim 0.1$) gluons. This is also the time at which the peak corresponding to the leading particle disappears (see Fig. 8). At the same time the occupation of the small x modes increases (linearly) with time, keeping the characteristic form of the scaling spectrum. When the peak has disappeared, the cascade continues to lower x, causing a uniform, shape conserving, decrease of the occupations of the modes, and a flow of energy towards small x.

We define

$$\mathcal{F}(x_0, \tau) = -\frac{\partial \mathcal{E}(x_0, \tau)}{\partial \tau}, \quad \mathcal{E}(x_0, \tau) \equiv \int_{x_0}^{1} dx D(x, \tau), \tag{101}$$

where $\mathcal{E}(x_0, \tau)$ is the amount of energy contained in the modes with $x > x_0$, and $\mathcal{F}(x_0, \tau)$ is the corresponding flux of energy, counted positively for energy moving

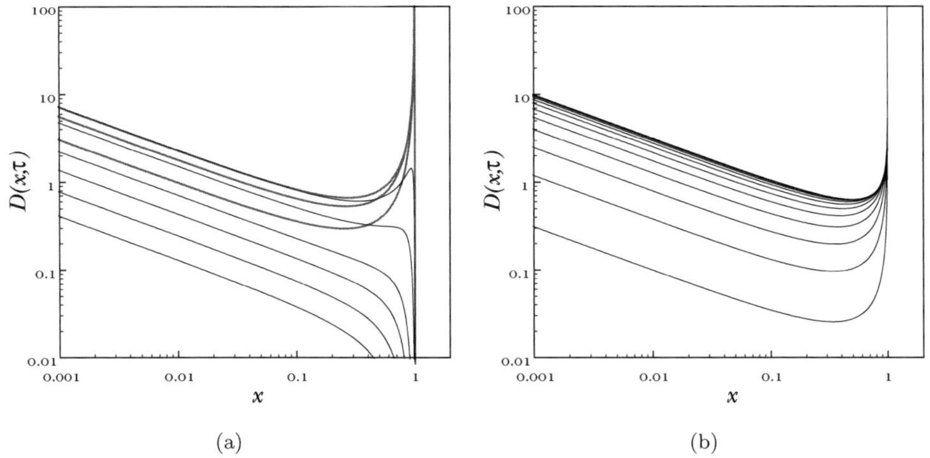

Fig. 8. (Color online) The function $D(x, \tau)$ (Eq. (99)) at various times. Left panel: the filling of the modes, which proceeds till the disappearance of the leading particle peak. The values of τ are, for the thick (blue) curves, from bottom to top: 0.1, 0.2, 0.3 (during this stage the leading particle acts as a source for soft gluon radiation), and for the thin (black) curves, from top to bottom: 0.5, 0.7, 0.9,1.0,1.1,1.2 (the leading parton has exhausted its energy and the peak has disappeared, while energy continues to flow to small x, the amount of energy in each mode decreasing exponentially fast). Right panel: energy is constantly injected into the system by a source located at $x = 1$ (see Eq. (106)). After a transitory regime, characterized by a uniform increase with time of the scaling spectrum, the system reaches a steady state. The values of τ are, from bottom to top: 0.1,0.2,0.3, 0.4, 0.5, 0.6,0.7,0.8,0.9,1.0.

to values of x smaller than x_0. These quantities can be calculated explicitly. We have for instance

$$\mathcal{E}(x_0, \tau) = \int_{x_0}^{1} dx D(x, \tau) = \mathrm{e}^{-\pi\tau^2} \mathrm{erfc}\left(\sqrt{\frac{\pi x_0}{1 - x_0}}\tau\right), \tag{102}$$

with erfc(x) the complementary error function. We note that the fraction of the total energy "stored in the spectrum", namely

$$\lim_{x_0 \to 0} \mathcal{E}(x_0, \tau) = \mathrm{e}^{-\pi\tau^2}, \tag{103}$$

decreases with time, and accordingly there is a non-vanishing flux of energy reaching $x = 0$

$$\mathcal{F}(0, \tau) = 2\pi\tau \mathrm{e}^{-\pi\tau^2}. \tag{104}$$

It follows that the complete, energy conserving, solution involves a contribution

$$D_c(x) = n_c(\tau)\delta(x) \quad \text{with } n_c(\tau) = 1 - \mathrm{e}^{-\pi\tau^2}, \tag{105}$$

somewhat analogous to a condensate where the radiated energy accumulates. Note that when $\tau \sim 1/\sqrt{\pi}$, corresponding to the disappearance of the leading particle into soft radiation, about 60% of the initial energy has flown into the condensate.

It is interesting to consider also the situation where the leading particle is replaced by a source that injects energy at a constant rate $\mathcal{A}$ at $x = 1$. In this case, we are led to look for the solution of the following equation

$$\frac{\partial}{\partial t} D(x,\tau) = \mathcal{A}\delta(1-x) + \mathcal{I}[D], \tag{106}$$

where $\mathcal{I}[D]$ denotes the r.h.s. of Eq. (96). The exact solution of Eq. (106) with initial condition $D(x, \tau = 0) = 0$ reads

$$D_{\rm st}(x,\tau) = \frac{\mathcal{A}}{2\pi\sqrt{x(1-x)}}\left(1 - {\rm e}^{-\pi\frac{\tau^2}{1-x}}\right), \tag{107}$$

and is plotted in Fig. 8, right panel. As time goes on, this solution converges to the stationary solution $(\mathcal{A}/2\pi)/\sqrt{x(1-x)}$, keeping the shape of the small x spectrum, with just an overall time-dependent scaling. Remarkably, the small x scaling form $D(x,t) = f(t)/\sqrt{x}$ is reached well before the stationary state is achieved. When the steady state is reached, all the energy provided by the source flows throughout the entire system towards the condensate at $x = 0$, that plays the role of a sink, while the population of the various modes stays unchanged.

Let us contrast these properties with those of the DGLAP cascade.[95] A simplified version of the corresponding evolution equation for the inclusive one particle distribution reads

$$\frac{\partial}{\partial t} D(x,t) = \bar{\alpha}\int_x^1 \frac{dz}{z(1-z)} D(x/z,t) - \bar{\alpha}\int_0^1 \frac{dz}{1-z} D(x,t), \tag{108}$$

where here the time variable is related to the virtuality Q^2 of the emitting parton, $t \equiv \ln Q^2/Q_0^2$. This equation has the form of Eq. (96) if one identifies

$$\frac{1}{t_*(x)} = \bar{\alpha} \quad \text{and} \quad \mathcal{K}(z) = \frac{1}{z(1-z)}. \tag{109}$$

Thus the DGLAP equation differs from the BDMPS-Z equation in two major aspects. First, the kernel $\mathcal{K}(z)$ is less singular near $z = 0$ and $z = 1$. Second, the rate of successive branchings is independent of the parent energy, i.e., it is constant along the cascade. We shall see that the latter property is what makes the major difference between the BDMPS-Z and the DGLAP cascades.

The solution of Eq. (108) is obtained using a Mellin transform.[72] The initial condition reads $\tilde{D}(\nu, 0) = 1$, with $\tilde{D}(\nu, 0)$ the Mellin transform of $D(x, 0) = \delta(1-x)$. The solution can be expressed as the following integral

$$D(x,t) = \int_{c-i\infty}^{c+i\infty} \frac{d\nu}{2\pi i} \exp\left[-(\psi(\nu)+\gamma)t + \nu\ln\frac{1}{x}\right], \tag{110}$$

where $\phi(\nu)$ stands for the digamma function and $\gamma \simeq 0.58$ stands for the Euler constant.

This integral representation allows us to verify a few properties that are relevant for our discussion. First, it is easy to check that the energy is conserved by the evolution: from Eq. (110) we have $\mathcal{E}(0,t) = \int_0^1 dx D(x,t) = \tilde{D}(1,t) = 1$. Furthermore, Eq. (110) has following asymptotic behavior when $\ln 1/x \gg t$:

$$D(x,t) \approx \left(\frac{1}{x}\right)^{2\sqrt{\frac{t}{\ln 1/x}}}. \tag{111}$$

Here, in contrast to what happens with the BDMPS-Z cascade, the time dependence of the energy distribution does not factorize, and no obvious scaling behavior emerges at small x. The growth of the spectrum at small x is in fact tamed by the exponent in Eq. (111), making the spectrum integrable when $\ln 1/x > t$: all the energy remains in the spectrum, and no energy flows to $x = 0$, unlike in the BDMPS-Z cascade.

These features are illustrated in Fig. 9 where a numerical calculation of the DGLAP cascade is presented. One recognizes from the left panel features analogous to that of the BDMPS-Z cascade, in particular the disaperence of the leading particle peak, but no simple scaling pattern can be identified. On the right panel, the the solution in the presence of a source is displayed. The solution in that case approaches the stationary solution $D(x) \sim \mathcal{A}/x$ at not too small x, but it takes a long time to populate the very small x modes according to this $1/x$ law. In fact, for any time t, the true solution at very small x deviates from the stationary solution in the same fashion as in Eq. (111).

The properties of the two cascades that we have discussed in this section can be understood in a broader context.[72] The two cascades are governed by the same

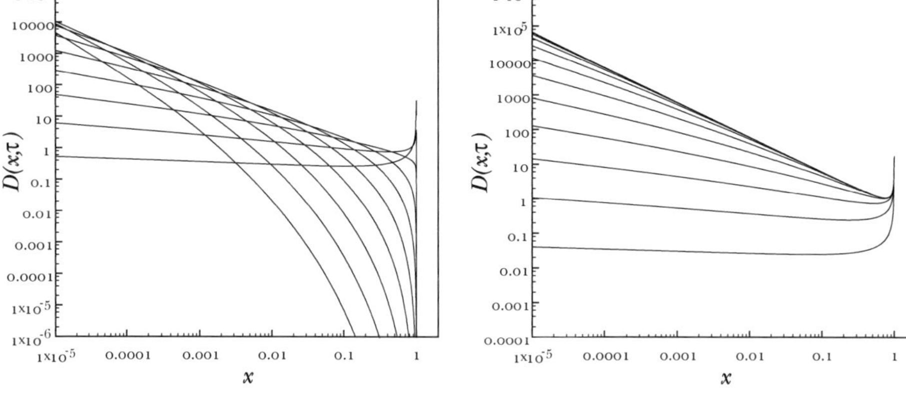

(a) (b)

Fig. 9. Model of a DGLAP cascade according to Eq. (108) (left panel), and with a source added at $x = 1$ (right panel). The curves corresponds to $\tau = 0.2, 0.4, 0.6, 0.8, 1.0, 1.2, 1.4, 1.6, 1.8, 2.0$ from bottom to top. On the right panel, the emergence of the scaling solution $D(x) \sim 1/x$ is clearly visible, as well as the persistent deviation from it of the true solution at very small x.

general equation, Eq. (96) with, in each case, a specific kernel $\mathcal{K}(z)$ and a specific time scale $t_*(x)$. The kernel $\mathcal{K}(z)$ controls how, in a given splitting, the energy is shared between the two offsprings. The time scale $t_*(x)$ controls the rate at which successive splittings occur. The characteristic features of the cascades result from two properties: energy conservation and the existence of approximate scaling solutions. The latter involves crucially $t_*(x)$. The role of the kernel $\mathcal{K}(z)$ is less important.

To better understand the role of $t_*(x)$, let us observe that Eq. (96) admits an approximate fixed point solution, which we refer to as a scaling solution, of the form

$$D_{\rm sc}(x,t) = \frac{t_*(x)}{x}. \tag{112}$$

Indeed, in the region where $D(x,\tau) \sim D_{\rm sc}(x,\tau)$, there is complete cancellation between gain and loss terms. Note that this compensation occurs only for $z \geq x$ in the integrals of Eq. (96), so that the scaling solution cannot be an exact fixed point of the equation. However, the general solution is driven to this approximate fixed point.

Now, since energy is conserved, and since it is continuously moving towards the lower values of x, the distribution will generically develop a divergent behavior at small x. Depending on the behavior of the scaling spectrum (112) when $x \to 0$, the approximate fixed point solution may lead to a divergent expression for the total energy $\mathcal{E}(0,t)$ contained in the modes between 0 and 1. In order for this not to happen, either the fixed point solution is never reached at very small x, or $t_*(x)$ is an increasing function of x. The former occurs in the DGLAP cascade for which $t_*(x) = \text{cst}$, while the latter is what occurs in the BDMPS-Z cascade for which $t_*(x) \sim \sqrt{x}$. Another way to see the role of the x dependence of $t_*(x)$ is to observe that in the DGLAP cascade, where t_* is independent of x, it takes an infinite time to propagate a finite amount of energy from $x = 1$ to $x = 0$. In contrast, in the BDMPS-Z cascade, because the branching rate accelerates along the cascade (since $t_*(x)$ decreases with decreasing x (see Fig. 7 for an illustration)), the corresponding time is finite, and in fact of the order of t_*. This is the reason why there is a finite flow of energy all the way down to $x = 0$ in the BDMPS-Z cascade while the energy remains stored in the spectrum in the DGLAP cascade. For a more detail discussion see Ref. 72.

At this point, it is useful to comment briefly on a peculiarity of the BDMPS-Z cascade, that makes the transition between the dilute, single-branching regime, and the the multiple branching regime completely smooth, with no sign of a change of regime as x crosses the value x_s. Indeed, the perturbative solution, valid at $\tau \ll 1$), is proportional to the BDMPS-Z spectrum

$$D(x,\tau) \simeq \frac{\tau}{\sqrt{x}}, \tag{113}$$

and this is already in the scaling form. We shall see that the scale x_s becomes visible in the angular distribution, to which we now turn.

The properties of the cascades depend also on the splitting kernel, that is, on the way the energy is distributed between the offsprings during a splitting. However, this turns out to have a minor effect on the main characteristics of the cascade, as compared to that of the transport time scale just mentioned, at least as long as the splitting kernel is not too singular. In that case the cascades develop as if the branching were completely democratic, with the two offsprings taking each half the energy of the parent gluon. The interactions responsible for the splittings can then be considered as local (in energy space). As already mentioned, these properties of the BDMPS-Z cascade that we have briefly listed, are typical of wave turbulence.[126]

5.2. *Angular distribution*

We turn now the angular distribution. Our starting point is Eq. (90), that we write in Fourier space as

$$\frac{\partial}{\partial t}D(x,\boldsymbol{u},t) = \frac{1}{t_*}\int dz\mathcal{K}(z)\left[\sqrt{\frac{z}{x}}D\left(\frac{x}{z},\boldsymbol{u},t\right) - \frac{z}{\sqrt{x}}D(x,\boldsymbol{u},t)\right] + \frac{N_c n}{2}\sigma(x,\boldsymbol{u})D(x,\boldsymbol{u},t), \tag{114}$$

with the Fourier transform given by

$$D(x,\boldsymbol{u},t) = \int \frac{d^2\boldsymbol{\theta}}{(2\pi)^2}D(x,\boldsymbol{\theta},t)\mathrm{e}^{-i\boldsymbol{\theta}\cdot\boldsymbol{u}} \tag{115}$$

and similarly for $\sigma(x,\boldsymbol{u})$. We shall also consider the diffusion approximation ($\boldsymbol{\theta}' \ll \boldsymbol{\theta}$) of Eq. (90)

$$\frac{\partial}{\partial t}D(x,\boldsymbol{\theta},t) = \int dz\mathcal{K}(z)\left[\frac{D(x/z,\boldsymbol{\theta},t)}{t_*(x/z)} - \frac{D(x,\boldsymbol{\theta},t)}{t_*(x)}\right] + \frac{1}{4(xE)^2}\left(\frac{\partial}{\partial\boldsymbol{\theta}}\right)^2[\hat{q}D(x,\boldsymbol{\theta},t)], \tag{116}$$

with $\hat{q} \simeq 4\pi\alpha_s^2 N_c n \ln(\boldsymbol{\theta}^2/\boldsymbol{\theta}_D^2)$ and $\boldsymbol{\theta}_D \equiv m_D/\omega = m_D/(xE)$.

Solving Eq. (114) exactly is difficult. Part of the difficulty comes from the fact that the angular distribution has two distinct components: a hard component, corresponding to large angles produced by single hard scatterings, and a soft component that can be obtained as the solution of the diffusion equation (116). Each of these two components is strongly modified by gluon branching. The soft component admits moments, which is not the case for the hard component. The characteristic angle that marks the boundary between the soft and the hard components depends on x, i.e., on the amount of branching. It can be estimated by calculating the mean squared angle of the soft component. This provides a first orientation into the various regimes that characterize the general solution of Eq. (90), and that are illustrated in Fig. 10.

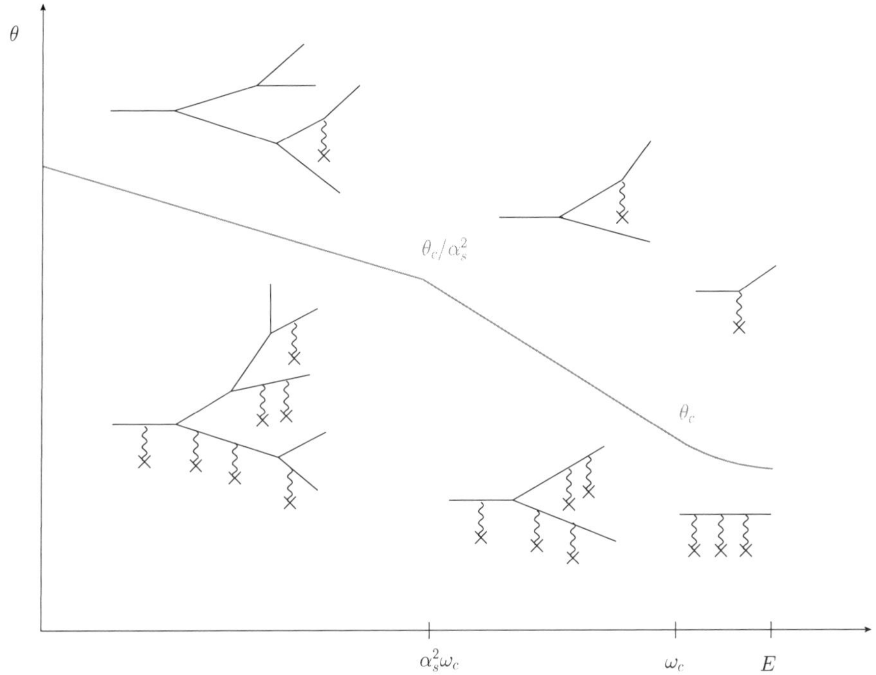

Fig. 10. (Color online.) The various regimes of the physical processes (branching and scattering) that accompany the propagation of a fast parton with energy $E \gtrsim \omega_c$ in a dense medium. The red curve stands for the root of $\langle\theta^2\rangle$. It separates the regime of single hard, large angle, scattering, from that of soft, small angle, multiple scattering. Accordingly, the soft component of the angular distribution lies below the red line, the hard component above. The line $\alpha_s^2\omega_c \sim \omega_s$ separates the region dominated by single branching from that of multiple branchings. $\omega \sim E$ indicates the region where the leading particle propagates without splitting, and merely suffers momentum broadening due to its collisions with the constituents of the medium.

We define the typical squared angle as follows

$$\langle\theta^2\rangle = \frac{1}{D(x,L)}\int \frac{d^2\boldsymbol{\theta}}{(2\pi)^2}\boldsymbol{\theta}^2 D(x,\boldsymbol{\theta},L), \tag{117}$$

where $D(x,\boldsymbol{\theta},L)$ is here the solution of Eq. (116). This can be determined explicitly in the three regimes characterizing the following ranges of x values: (i) $x \lesssim 1$, corresponding to the leading particle; (ii) $x_s \ll x \ll 1$ where single emission dominates; (iii) $x \ll x_s$, the regime of multiple emissions. We refer to Ref. 70 for the details of the calculation, and discuss now the main results.

When $x \lesssim 1$,

$$\langle\theta^2\rangle = \frac{\hat{q}L}{E^2} = \theta_s^2(1,L), \quad \theta_s^2(x,L) \equiv \frac{\hat{q}L}{\omega^2} = \frac{\hat{q}L}{x^2E^2}. \tag{118}$$

This angle reflects directly the momentum broadening of the leading particle propagating in the medium over a distance L.

In the regime $x_s \ll x \ll 1$, where typically one emission occurs, the dominant contribution comes from momentum broadening of the radiated gluon from the time of its emission. The other contribution, corresponding to the momentum broadening of the leading particle before the splitting, is suppressed by a factor x^2. Keeping the dominant contribution, we get

$$\langle \theta^2 \rangle = \frac{1}{2}\theta_s^2(x, L). \tag{119}$$

The factor 1/2 originates from the average over the emission time.

Note that for realistic situations $\omega_c < E$, the maximal frequency is thus $x_c \equiv \omega_c/E < 1$. In this case, $x_s \ll x \ll x_c$, and hence, these rare radiations are on average confined to the angular corona,

$$\theta_c \ll \theta \ll \frac{1}{\alpha_s^2}\theta_c, \tag{120}$$

where the critical angle

$$\theta_c = (\hat{q}L^3)^{-1/4}, \tag{121}$$

corresponds to the smallest BDMPS-Z radiation angle. The third regime, $x \ll x_s$, is dominated by multiple branchings at parametrically large angles,

$$\theta \gg \frac{1}{\alpha_s^2}\theta_c. \tag{122}$$

It requires a non perturbative treatment and the result is (see Refs. 67 and 70 for details and Ref. 68 for a similar discussion)

$$\langle \theta^2 \rangle = \frac{1}{4\bar{\alpha}}\left[\frac{\hat{q}}{(xE)^3}\right]^{1/2} \equiv \theta_*^2(x). \tag{123}$$

As was the case for the previous regime, this angle can be identified with that corresponding to the momentum broadening of the observed gluon since the time of its emission. The major difference with the previous regime is that now $\theta_*(x)$ does not depend on the size of the medium. This is because, in the multiple branching regime, only gluons that are emitted off the leading particle at a distance, that is shorter than their characteristic stopping time $t_*(x)$, from the medium boundary L, escape the medium. Accordingly, their measured squared angle is typically $\theta_*^2(x) \sim \hat{q}t_*(x)/(xE)^2$, which recalling Eq. (94) agrees with Eq. (123).

The complete distribution in the three regimes can also be determined explicitly. Again, we refer to Ref. 70 for details.

When $x \lesssim 1$, the distribution takes the factorized form

$$D(x, \boldsymbol{\theta}, L) \simeq \mathcal{P}(\boldsymbol{\theta}, L)D(x, L), \tag{124}$$

where $D(x, L)$ is the energy distribution given by Eq. (99) (for $x \lesssim 1$). In the regime of multiple scatterings, i.e., for $\theta^2 \lesssim \theta_s^2(L, x)$

$$\mathcal{P}(\boldsymbol{\theta}, L) = \frac{4\pi}{\theta_s^2(x, L)} \exp\left[-\frac{\boldsymbol{\theta}^2}{\theta_s^2(x, L)}\right]. \tag{125}$$

In the opposite case, $\theta \gg \theta_s(x, L)$, a single hard scatterings dominates the distribution and we have

$$\mathcal{P}(\boldsymbol{\theta}, L) \approx \frac{nN_c g^4 L}{(xE)^2 \boldsymbol{\theta}^4}. \tag{126}$$

In the second regime the energy distribution is given by the leading order BDMPS-Z distribution,

$$D(x, L) \simeq \frac{L}{t_* \sqrt{x}} \tag{127}$$

and we expect also the angular distribution to be given by a single radiation that undergoes multiple scatterings. For sufficiently small x, as already emphasized, the angular deviation of the radiated gluon is larger than that of the leading parton. Accordingly one can neglect the momentum broadening before the emission. We get,[37]

$$D(x, \boldsymbol{\theta}, L) \simeq \int_0^L \frac{dt}{t_*} \mathcal{P}(x, \boldsymbol{\theta}, L-t) x \mathcal{K}(x). \tag{128}$$

Finally, in the fully non-perturbative regime, i.e., $x \ll x_s$, and in the soft region, both multiple scatterings and multiple branchings must be resummed. A solution can be obtained as a power series in the number of scatterings.[70] This is achieved by using an approximation that is equivalent to the diffusion approximation, hence not accurate for the very the tail of the distribution (see below), but it allows us to determine the distortion of the main peak. The result reads

$$D(x, \boldsymbol{\theta}, L) = \frac{4\pi}{\theta_*^2(x)} \eta\left(\frac{\boldsymbol{\theta}^2}{\theta_*^2(x)}\right) D(x, L), \tag{129}$$

where the entire dependence on $\boldsymbol{\theta}$ is in the scaling function η, given by

$$\eta(z) = \int_0^\infty d\alpha J_0(2\sqrt{z\alpha}) \sum_{n=0}^{\infty} c_n (-\alpha^2)^n, \tag{130}$$

where J_0 is a Bessel function, and we have the normalization property.

$$\int_0^\infty dz \eta(z) = 1. \tag{131}$$

In Fig. 11, we have plotted the angular distribution $\eta(z)$ in the multiple branching regime. For the numerical evaluation we have computed the first 500 terms in the multiple scattering series (130).[70]

Consider finally the single scattering limit achieved when $\theta \gg \theta_*(x)$. This is easily found to be

$$D(x, \boldsymbol{\theta}, L) \approx \frac{N_c n g^4 t_*(x)}{(2xE)^2 \boldsymbol{\theta}^4} D(x, L), \tag{132}$$

where $N_c n g^4 t_*(x)/(2xE)^2 \sim \theta_*^2(x)$ is the typical angle squared of gluons in the multiple branching regime.

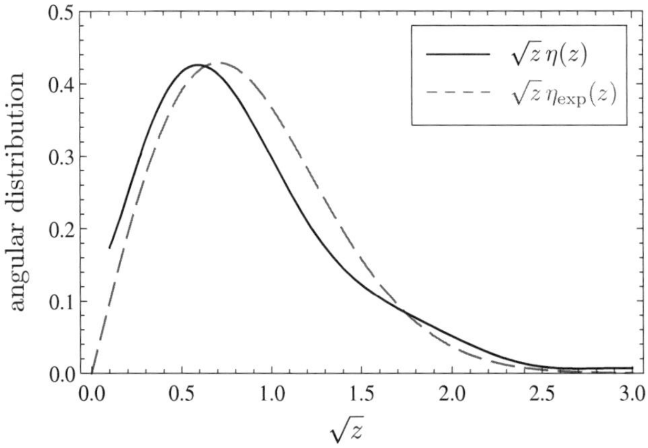

Fig. 11. (Color online) The angular distribution (130) of a gluon in the multiple branching regime (solid black line) compared to an exponential distribution with identical first two moments (dashed purple line). Here $z \equiv \theta^2/\theta_*^2(x)$.

5.3. *Application: Energy imbalance in dijet events*

As we have argued in this section, a striking feature of the BDMPS cascade is that it provides an efficient mechanism to transport energy down to the lowest accessible gluon frequencies. Since soft gluons are typically emitted at large angles, the cascade naturally populates soft gluon modes at large angles with respect to the jet axis. We end this section with a brief analysis of the phenomenological implications of this mechanism.

Recently, the CMS collaborations investigated the energy imbalance in strongly asymmetric dijet events.[18,26] As they are born back-to-back jets have roughly equal energy and momenta. The observed asymmetry is due to the fact that the leading jet (the most energetic of the two) traverses a shorter distance in the medium than the subleading one, and hence loses less energy. Surprisingly, the energy balance is recovered at quite large angles, $R > 0.8$, and most of the "missing" energy is carried by very soft particles.[18]

As a very rough model for an asymmetric dijet event, we consider a leading jet and a subheading one that have traversed, respectively, the path lengths $L = 5$ fm and $L = 1$ fm in opposite directions. We use the following set of parameters: $\hat{q} = 1\,\text{GeV}^2/\text{fm}$, $\bar{\alpha} = 0.3$, and $E \equiv E_1 = E_2 = 100\,\text{GeV}$. First, we find that the broadening of the subleading jet, $\theta_s \sim 0.02$ (cf. Eq. (118)) is negligible compared to the opening angle of jet $\Theta_{\rm jet} = 0.3$, as observed in the data.

We evaluate the energy imbalance by integrating the difference $D_2(x,\theta) - D_1(x,\theta)$, where D_1 and D_2 are the medium-induced gluon distributions of the leading and sub-leading jet respectively, in the different bins in energy and angle. The minus sign accounts for the fact that the constituents of the subleading jet propagate in opposite direction to those of the leading jet. The angular distribution is assumed to be Gaussian with a width determined by an interpolation between

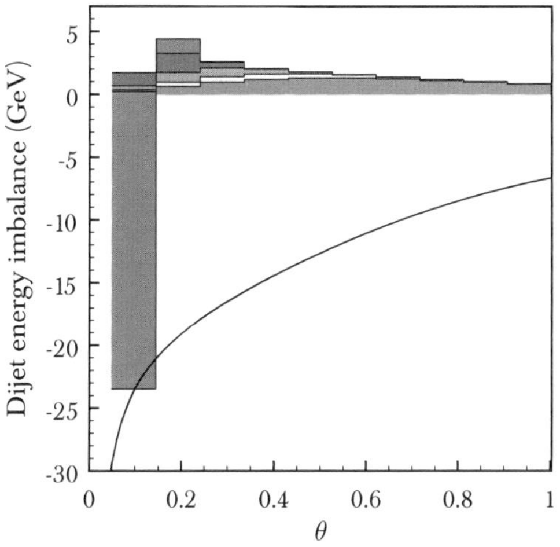

Fig. 12. (Color online) The difference of the angular and energy distributions of sub-leading and leading jets for $\omega_{\rm BH} = 0.5\,\rm GeV$. The histograms account for the four binnings of energies: [0–1] GeV (gray), [1–2] GeV (yellow), [2–4] GeV (orange), [4–8] GeV (green), [8–100] GeV (red). In the first angular bin we observe a large imbalance of energy in the hard particles. This energy is partly recovered at large angles by very soft particles. The cumulative energy is given by the full line.

the mean squared angles in the three regimes discussed above. Finally, the evolution equation Eq. (96) is solved numerically where soft emissions with frequencies $\omega < \omega_{\rm BH}$ are cut-off to regulate the unphysical soft singularity in the branching kernel (51).[67]

In Fig. 12, is plotted, in the form of histograms, the energy imbalance integrated on various angular and energy intervals.[67] At small angle $\theta \lesssim 0.2$ there is a deficit of hard particles with energies $\omega > 8\,\rm GeV$ (red negative histogram). As we increase the angle the energy imbalance turns in favor of the subleading jet and it is dominated by very soft particles with energies $\omega < 1\,\rm GeV$ (positive gray histograms). The full line corresponds to the integrated energy imbalance for angles from 0 to θ. The imbalance is expected to be recovered at large θ, when $\theta \to \pi/2$. The energy transported at large angles is clearly dominated by very soft gluons.[n] The jet structure that emerges from this plot is in semi-quantitative agreement with the recently released data by CMS.[26]

6. Decoherence of the Jet Core

In this section, we discuss briefly how the two cascades, vacuum and medium induced, can interfere in a jet produced in heavy ion collisions. The issue we want to address is to what extent color coherence, that constrains the pattern of successive

[n] These features are absent in the angular distribution of primary gluon radiations, as shown in the early work,[127] confirming the important of multiple branchings to transport energy at large angles.

parton branchings in vacuum, is altered when the jet propagates in a medium. As a paradigmatic example, we consider the radiation (in vacuum or in a QCD medium) of a quark–antiquark pair in the singlet state (the generalization to different color configurations is straightforward). This pair, which we refer to as an antenna, can be thought of as being produced by the decay of the highly virtual primary parton.[59–64]

In vacuum, the radiated spectrum off the quark reads[3] (see discussion in Subsec. 2.1)

$$dP_{gq} = \frac{\alpha_s C_F}{\pi} \Theta(\theta_{q\bar{q}} - \theta) \frac{d\theta^2}{\theta^2} \frac{d\omega}{\omega}, \tag{133}$$

and similarly for the antiquark, i.e., $dP \equiv dP_{gq} + dP_{g\bar{q}}$. Here, the theta function accounts on average for the suppression of large angle radiation ($\theta \gg \theta_{q\bar{q}}$) caused by color coherence. The total radiation spectrum exhibits two radiation cones aligned on the quark and the antiquark momenta, respectively. Eq. (133) is at the basis of Eq. (2).

The probability for the quark of the antenna to radiate soft gluons after traversing a QCD medium is given by,[59,60]

$$dP_{gq} = \frac{\alpha_s C_F}{\pi} [\Theta(\theta_{q\bar{q}} - \theta) + \Delta_{\rm med} \Theta(\theta - \theta_{q\bar{q}})] \frac{d\theta^2}{\theta^2} \frac{d\omega}{\omega}, \tag{134}$$

where

$$\Delta_{\rm med} \equiv \frac{1}{N_c^2 - 1} \langle {\rm Tr}\, \mathcal{U}_{\bar{p}}^\dagger \mathcal{U}_{\boldsymbol{p}} \rangle \approx 1 - \exp\left[-\frac{r_\perp^2}{l_\perp^2} \right]. \tag{135}$$

and $\mathcal{U}$ is a Wilson line in the adjoint representation, evaluated along the trajectory of the quark, and the harmonic approximation has been used to evaluate the correlator of the Wilson lines. Here $r_\perp \equiv \theta_{q\bar{q}} L$ stands for the antenna transverse size and

$$l_\perp \equiv \sqrt{\frac{12}{\hat{q} L}} \tag{136}$$

is a length characterizing the range of coherence in the transverse plane. The quantity $\Delta_{\rm med}$ may indeed be interpreted as a decoherence parameter: when $r_\perp \ll l_\perp$, or equivalently $\theta_c \gg \Theta_{\rm jet}$, where the BDMPS-Z critical angle θ_c is defined in Eq. (121), we have $\Delta_{\rm med} \sim r_\perp^2 / l_\perp^2 \ll 1$ and we recover Eq. (133). In this regime the inner structure of the antenna is not resolved by the medium and the antenna radiates coherently as in vacuum.

On the other hand, the case $r_\perp \gg l_\perp$ ($\theta_c \ll \Theta_{\rm jet}$), corresponds to full decoherence of the antenna, and we have $\Delta_{\rm med} \simeq 1$. Here, the angular constraint disappears, and the quark and the antiquark radiate independently from one another,[60,o]

$$dP_{gq} \approx \frac{\alpha_s C_F}{\pi} \frac{d\theta^2}{\theta^2} \frac{d\omega}{\omega}. \tag{137}$$

[o] A analogous phenomena occurs between initial and final state radiation in a dense medium.[128,129]

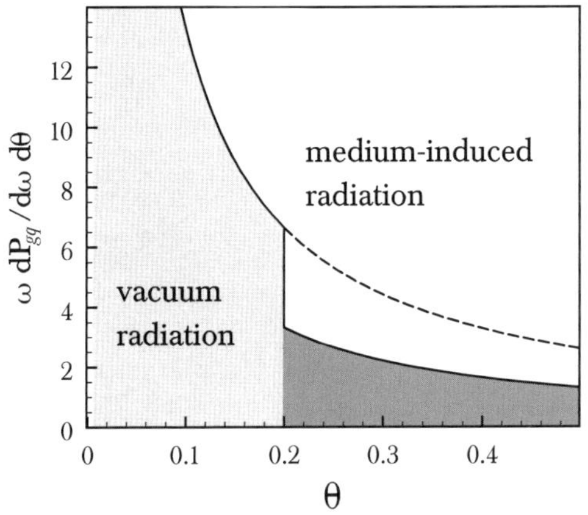

Fig. 13. The soft gluon emission spectrum off the quark constituent of a singlet antenna with opening angle $\theta_{q\bar{q}} = 0.2$, according to Eq. (134), in the presence of a medium with $\Delta_{\rm med} = 0.5$ (solid line). Here $\equiv \alpha_s C_F/\pi = 1$. On average vacuum radiation is confined within $\theta < \theta_{q\bar{q}}$, while the medium-induced radiation is radiated at $\theta > \theta_{q\bar{q}}$. The limit of opaque medium, given by $\Delta_{\rm med} = 1$, is marked by the dashed line. Figure taken from Ref. 60.

This discussion applies also to gluons radiated inside the medium.[61–64] The decoherence parameter given by (135) can be rewritten as $\Delta_{\rm med} \sim 1 - \exp(L/t_{\rm decoh})$, from which we extract the decoherence time,

$$t_{\rm decoh} \equiv \left(\frac{1}{\hat{q}\theta_{q\bar{q}}^2}\right)^{1/3}. \tag{138}$$

The antenna spectrum (134) is plotted in Fig. 13 for $\Delta_{\rm med} = 0.5$.

The antenna set-up serves as guideline to discuss the decoherence that a jet experiences when passing through a dense medium. Consider a jet with opening angle $\Theta_{\rm jet}$. While inclusive jet observables in vacuum are characterized by two scales, $Q \equiv \Theta_{\rm jet}E$, that correspond to the jet transverse mass typically of the order of 100 GeV and $Q_0 \sim \Lambda_{\rm QCD}$, in the presence of a dense medium, an additional (hard) scale comes into play, $Q_{\rm med} = \max(r_\perp^{-1}, \sqrt{\hat{q}L})$, where here $r_\perp \equiv \Theta_{\rm jet}L$. This medium-generated scale is of the order of a few GeV's, therefore, we typically have $Q_{\rm med} \ll Q$. The medium affects the collimated parton shower, and thus causes its decoherence, only at scales of the order or smaller than $Q_{\rm med}$. Hence, the medium alters the property of angular ordering by allowing for additional soft gluon radiations up to the limiting angle $Q_{\rm med}/\omega$.[63,64] As a result of this aforementioned separation of scales a large portion of the jet is barely affected by its interactions with the medium.

Comparing the two medium scales one is left with two regimes:

(1) $r_\perp \ll l_\perp$ (or $\Theta_{\rm jet} \ll \theta_c$): in this situation the jet is on average not resolved by the medium and remains color coherent. Multiple interactions with the medium will induced gluon cascades off the total charge, typically at large angles (cf.

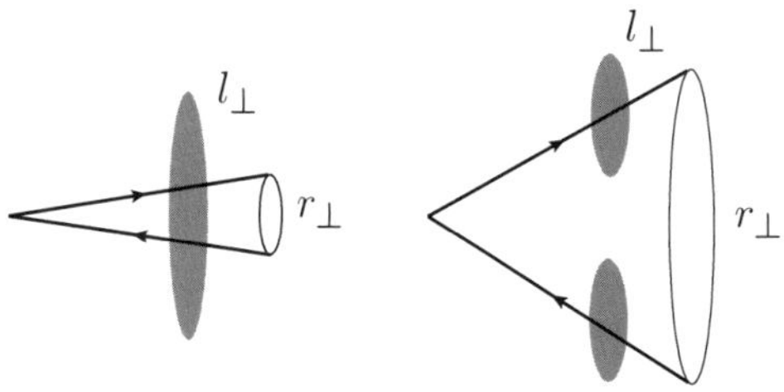

Fig. 14. An illustration of a quark–antiquark antenna of typical transverse size $r_\perp = \Theta_{q\bar{q}} L$ in a medium of length L, in the coherent regime (left) where the typical in-medium resolution length $l_\perp \equiv (\hat{q}L)^{-1/2} \gg r_\perp$. In this case, the medium doest not resolve the inner structure of the antenna, and thus interacts with its total charge. In the opposite case (right panel), $l_\perp \equiv\ll r_\perp$, the medium resolves the quark and antiquark color charges, causing their decoherence. Then each of the antenna constituents radiates independently.

Eq. (122)),

$$\theta \gtrsim \frac{1}{\alpha_s^2}\theta_c \gg \Theta_{\rm jet}. \tag{139}$$

The unresolved vacuum cascade looks as if it developed outside the medium.

(2) In the opposite case, $r_\perp \gg l_\perp$ (or $\Theta_{\rm jet} \gg \theta_c$), the medium resolves more charges inside the jet, which results in an enhancement of the medium-induced radiation rate. In other words, the medium resolves a certain number of sub-structures characterized by the opening angle θ_c, that evolve independently of each other. On the other hand, the inner structure of these sub-jets remain unresolved by the medium.[130] Consequently, the unresolved system loses energy *coherently* with a rate proportional to the total charge contained within θ_c.

For realistic situations it was noted[130] that a core containing most of the jet energy remains unresolved by the medium. Secondary substructures carry typically smaller energy fractions and are expected to be influenced significantly by medium effects. Under these circumstances, an unresolved jet lose energy as a single parton whose final distribution is then given by the rate equation (96). In this approximation, we can write the jet spectrum in nucleus–nucleus collision as a convolution of the hard parton spectrum due to incoherent binary collisions, and the probability distribution to find a parton with energy $p_\perp$ that originates from the nascent parton with momentum $p'_\perp = p_\perp/x$,

$$\frac{dN^{AA}_{\rm jet}}{d^2p_\perp} \equiv N_{\rm coll}\int_0^1 \frac{dx}{x} D(x, p_\perp/x)\frac{dN^{pp}_{\rm jet}}{d^2p_\perp}(p_\perp/x), \tag{140}$$

where the distribution $D(x, p'_\perp)$, was discussed in Subsec. 5.1, identifying $p'_\perp \equiv E$ and $N_{\rm coll}$ stands for the number of binary collisions. It is obtained by solving numerically Eq. (96), where finite size effects are taken into account (cf. Eq. (6)). This procedure is analogous to that used to computed quenching of hadron spectra,[40,42,131] but proves to be best justified for high-p_t jet spectra. The resulting nuclear modification factor, $R_{AA} \equiv N^{-1}_{\rm coll} dN^{AA}_{\rm jet}/dN^{pp}_{\rm jet}$, is plotted in Fig. 15(a),[132]

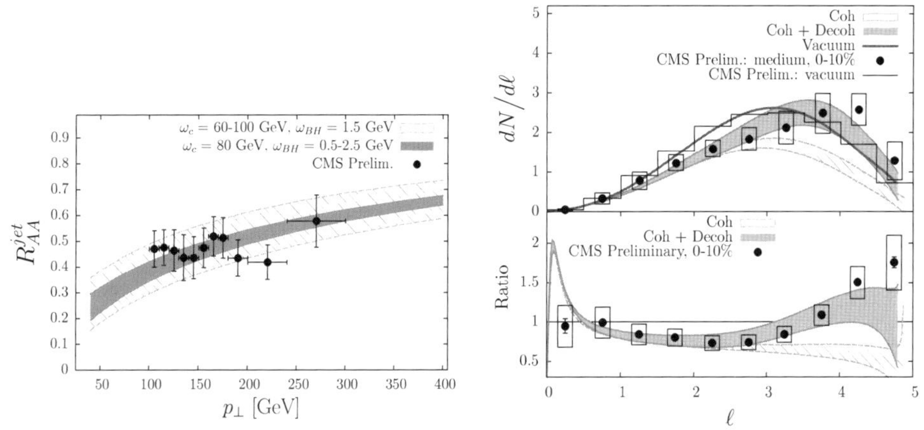

(a) (b)

Fig. 15. (Color online) (a) Calculation of the nuclear modification factor with $\omega_c = 80\,\text{GeV}$ as a function of jet $p_\perp$ for central Pb–Pb collisions. The solid (red) band includes the variation of ω_{BH} around a central value of 1.5 GeV. The dashed (gray) band includes, in addition, a variation of $\omega_c \in [60, 100]$ GeV. The experimental data are taken from.[24] (b) Upper panel: the longitudinal fragmentation function plotted as a function of $\ell = \ln 1/x$. Lower panel: the ratio of medium-modified and vacuum fragmentation functions. The experimental data are taken from Ref. 133. See text for further details.

where the uncertainty bands shows the sensitivity of this calculation to variations of the medium length L and the in-medium diffusion coefficient $\hat{q}$, via the variable $\omega_c = 60-100\,\text{GeV}$. In addition the sensitivity to the soft scale $\omega_{\text{BH}} = 0.5-1.5\,\text{GeV}$ is also shown, is implemented as a cut-off of the unphysical singularity in the branching kernel (51). For $\hat{q} = 2\,\text{GeV}^2/\text{fm}$ and $L = 4$ fm, we get $\omega_c = 80\,\text{GeV}$ which corresponds to the average value. These values are consistent with the usual expectations.[123]

In the CMS analysis the jet opening angle is chosen to be $\Theta_{\text{jet}} = 0.3$, it follows that $r_\perp \simeq 1.5$ fm and $l_\perp \simeq 1.1$ fm. Obviously, the jet is neither in the fully coherent regime nor in the decoherent one. Hence, decoherent radiation off secondary substructures may be important. It was suggested in Ref. 132 that a signature of partial decoherence may be found in intrajet structures, e.g., in the fragmentation functions, or equivalently the energy distribution of particles inside the jet-cone, $D(x) = xdN/dx$.

The strategy to identify such effects consists in looking for departures from the fully coherent limit. In the coherence regime, the unresolved vacuum shower loses energy out of the jet-cone as a single parton. Hence, we can effectively factor out the in-medium distribution $D_{\text{med}}(z, p_\perp)$ described by Eq. (96) from the vacuum evolution (see Ref. 132 for details):

$$D^{\text{coh}}(x) \equiv \int_x^1 \frac{dz}{z} D_{\text{vac}}(x/z, Q) D_{\text{med}}(z, p_\perp), \tag{141}$$

where $D_{\rm vac}(x/z, Q)$ is solution to the MLLA evolution equation (2) with $Q \equiv \Theta_{\rm jet} p_\perp$. The result is plotted in Fig. 15(b) and is given by the gray band. Equation (141) properly matches the pp baseline with $D_{\rm med}(z, p_\perp) = \delta(1-z)$ (see the blue curve in Fig. 15(b)). The semi hard sector is well described and is consistent with the nuclear modification factor above. However, there is an excess of few soft particles that is not accounted for. We include now the decoherent contribution, $\Delta D^{\rm decoh}$ which is computed to double logarithmic accuracy by convoluting the probability to form gluon–quark antenna in the medium which then subsequently radiates decoherent soft gluons according to Eq. (134) (see Ref. 132 for details). The total distribution, $D^{\rm tot} \simeq D^{\rm coh} + \Delta D^{\rm decoh}$ is displayed as red band in Fig. 15. This in-cone decoherent radiation proves to be sufficient to account for the excess of soft gluons seen in the data.

7. Conclusions and Outlook

In this review, we have analyzed the dominant features of the medium-induced parton cascade, relying on the BDMPS-Z mechanism for medium induced gluon radiation. The medium induced cascade exhibits remarkable properties that originate on the one hand from the specific mechanism of medium induced radiation, on the other hand from the rapid loss of coherence among the radiated gluons caused by their multiple interactions with the medium constituents. These properties manifest themselves in a specific mechanism for the transport of energy along the cascade, with a characteristic angular structure. The transport of energy involves a constant flow down to the lowest accessible frequencies. The angular structure is such that the softest quanta are emitted at large angles. We emphasize that these features, reminiscent of those revealed by the analysis of recent CMS data on the angular distribution of the missing energy in inbalanced dijet events, are intrinsic properties of the BDMPS-Z cascade. They do not result from a specific coupling to the surrounding medium, nor are they influenced by the collective dynamics of this medium. In fact, the only dependence on the medium is hidden in the value of the jet quenching parameter $\hat{q}$.

Much remains to be understood about the BDMPS cascade. In particular the treatment of what happens at the lower end of the cascade, where gluon frequencies become smaller than $\omega_{\rm BH}$ needs to be improved. In view also of the potential role of this cascade in the thermalization of the quark–gluon plasma,[125,138] understanding how energy carried by the soft radiation is dissipated in the medium is an important task. An important ingredient entering the description of the medium induced cascade is the jet quenching parameters, $\hat{q}$. We have seen that this coefficient receives sizable radiative corrections. The calculations of these corrections is at present limited to the most singular contributions (double logarithmic). We have argued that it is the singular nature of the correction that allows us to absorb it in a redefinition of $\hat{q}$. It is unclear whether the simple description of energy loss and momentum broadening as local transport phenomena would hold in a more accurate treatment of these radiative corrections.

Aside from the medium-induced cascade that we have just discussed, a jet in a medium is accompanied by another, vacuum-like, cascade. The two cascades are geometrically separated. The vacuum-like cascade is a collimated and coherent cascade, constituting the jet core, characterized by collinear and angular ordered splittings triggered by the initial hard collision. The medium-induced cascade develops at parametrically large angles by successive branchings of the primary radiated gluons off the unresolved total color charge of the jet core. Much remains to be understood about the interaction between the vacuum cascade and the vacuum induced cascade. An interesting phenomenon takes place when the medium charges resolve substructures in the inner jet core. The coherence of the resolved charges is broken, allowing them to radiate independently in the soft sector. This mechanism for medium-induced soft radiation is of a different nature than the BDMPS-Z mechanism. It causes the decoherence of vacuum radiation which is not broadened by the medium and hence is expected to remain collimated. This mechanism may provide a natural explanation for the excess of soft particles measured in the jet fragmentation region. Finally, in addition to the effects on the intrajet structure, color decoherence is expected to be significant for processes at NLO, e.g., observables involving multi-jet events.

A complete description of parton cascades in a QCD medium goes along with a refined understanding and modeling of the medium dynamics.[92,134–136] This, together with the details of the experimental setup, requires the development of event generators,[92,137] an effort undertaken by many groups.[85–89,92] Clearly, the microscopic physics that has been discussed in this review needs to be implemented in these generators. While this is the case, to some extent, of the BDMPS-Z cascade,[85,88,89] as far as we know, the decoherence mechanism as well as the renormalization of the quenching parameter are not yet accounted for.

Acknowledgments

We would like to thank N. Armesto, J. Casalderrey-Solana, F. Dominguez, L. Fister, E. Iancu, M. Martinez, J. G. Milhano, A. H. Mueller, C. A. Salgado, K. Tywoniuk, B. Wu for many discussions on the issues discussed in this paper, as well as for direct collaboration on some of the works reported here. Y. M.-T. is specially grateful to C. A. Salgado and K. Tywoniuk for collaborating on Refs. 59 and 60. This research is supported by the European Research Council under the Advanced Investigator Grant ERC-AD-267258.

References

1. J. R. Ellis, M. K. Gaillard and G. G. Ross, *Nucl. Phys. B* **111** (1976) 253, [Erratum-ibid. **130** (1977) 516].

2. B. H. Wiik, *Conf. Proc. C* **7906181** (1979) 113.
3. Y. L. Dokshitzer, V. A. Khoze, A. H. Mueller and S. I. Troian, "Basics of perturbative QCD," Gif-sur-Yvette, France: Ed. Frontieres (1991).
4. V. A. Khoze and W. Ochs, *Int. J. Mod. Phys. A* **12** (1997) 2949, [hep-ph/9701421].
5. Y. I. Azimov, Y. L. Dokshitzer, V. A. Khoze and S. I. Troyan, *Z. Phys. C* **31** (1986) 213.
6. T. Sjostrand, S. Mrenna and P. Z. Skands, *JHEP* **0605** (2006) 026, [hep-ph/0603175].
7. M. Bahr, S. Gieseke, M. A. Gigg, D. Grellscheid, K. Hamilton, O. Latunde-Dada, S. Platzer and P. Richardson *et al.*, arXiv:0809.2002 [hep-ph].
8. T. Gleisberg, S. Hoeche, F. Krauss, M. Schonherr, S. Schumann, F. Siegert and J. Winter, *JHEP* **0902** (2009) 007, [arXiv:0811.4622 [hep-ph]].
9. J. Currie, A. Gehrmann-De Ridder, E. W. N. Glover and J. Pires, *JHEP* **1401** (2014) 110, [arXiv:1310.3993 [hep-ph]].
10. J. Pires, arXiv:1412.3427 [hep-ph].
11. J. Bjorken, FERMILAB-PUB-82-059-THY, FERMILAB-PUB-82-059-T (1982).
12. I. Arsene *et al.* [BRAHMS Collaboration], *Nucl. Phys. A* **757** (2005) 1, [nucl-ex/0410020].
13. B. B. Back, M. D. Baker, M. Ballintijn, D. S. Barton, B. Becker, R. R. Betts, A. A. Bickley and R. Bindel *et al.*, *Nucl. Phys. A* **757** (2005) 28, [nucl-ex/0410022].
14. J. Adams *et al.* [STAR Collaboration], *Nucl. Phys. A* **757** (2005) 102, [nucl-ex/0501009].
15. K. Adcox *et al.* [PHENIX Collaboration], *Nucl. Phys. A* **757** (2005) 184, [nucl-ex/0410003].
16. G. Aad *et al.* [ATLAS Collaboration], *Phys. Rev. Lett.* **105** (2010) 252303, [arXiv:1011.6182 [hep-ex]].
17. K. Aamodt *et al.* [ALICE Collaboration], *Phys. Lett. B* **696** (2011) 30, [arXiv:1012.1004 [nucl-ex]].
18. S. Chatrchyan *et al.* [CMS Collaboration], *Phys. Rev. C* **84** (2011) 024906, [arXiv:1102.1957 [nucl-ex]].
19. G. Aad *et al.* [ATLAS Collaboration], *Eur. Phys. J. C* **71** (2011) 1795, [arXiv:1109.5816 [hep-ex]].
20. G. Aad *et al.* [ATLAS Collaboration], *Phys. Lett. B* **719** (2013) 220, [arXiv:1208.1967 [hep-ex]].
21. G. Aad *et al.* [ATLAS Collaboration], *Phys. Lett. B* **739** (2014) 320, [arXiv:1406.2979 [hep-ex]].
22. G. Aad *et al.* [ATLAS Collaboration], *Phys. Rev. Lett.* **114**(7) (2015) 072302, [arXiv:1411.2357 [hep-ex]].
23. S. Chatrchyan *et al.* [CMS Collaboration], *JHEP* **1210** (2012) 087, [arXiv:1205.5872 [nucl-ex]].
24. CMS Collaboration [CMS Collaboration], CMS-PAS-HIN-12-004.
25. S. Chatrchyan *et al.* [CMS Collaboration], arXiv:1310.0878 [nucl-ex].
26. CMS Collaboration [CMS Collaboration], CMS-PAS-HIN-14-010.
27. J. Adam *et al.* [ALICE Collaboration], arXiv:1502.01689 [nucl-ex].
28. M. Gyulassy and X. N. Wang, *Nucl. Phys. B* **420** (1994) 583, [nucl-th/9306003].
29. X. N. Wang, M. Gyulassy and M. Plumer, *Phys. Rev. D* **51** (1995) 3436, [hep-ph/9408344].
30. R. Baier, Y. L. Dokshitzer, S. Peigne and D. Schiff, *Phys. Lett. B* **345** (1995) 277, [hep-ph/9411409].
31. R. Baier, Y. L. Dokshitzer, A. H. Mueller, S. Peigne and D. Schiff, *Nucl. Phys. B* **483** (1997) 291, [hep-ph/9607355].

32. R. Baier, Y. L. Dokshitzer, A. H. Mueller, S. Peigne and D. Schiff, *Nucl. Phys. B* **484** (1997) 265, [hep-ph/9608322].
33. B. G. Zakharov, *JETP Lett.* **63** (1996) 952, [hep-ph/9607440].
34. B. G. Zakharov, *JETP Lett.* **65** (1997) 615, [hep-ph/9704255].
35. M. Gyulassy, P. Levai and I. Vitev, *Nucl. Phys. B* **571** (2000) 197, [hep-ph/9907461].
36. M. Gyulassy, P. Levai and I. Vitev, *Nucl. Phys. B* **594** (2001) 371, [nucl-th/0006010].
37. U. A. Wiedemann, *Nucl. Phys. B* **588** (2000) 303, [hep-ph/0005129].
38. X. F. Guo and X. N. Wang, *Phys. Rev. Lett.* **85** (2000) 3591, [hep-ph/0005044].
39. X. N. Wang and X. f. Guo, *Nucl. Phys. A* **696** (2001) 788, [hep-ph/0102230].
40. C. A. Salgado and U. A. Wiedemann, *Phys. Rev. D* **68** (2003) 014008, [hep-ph/0302184].
41. N. Armesto, C. A. Salgado and U. A. Wiedemann, *Phys. Rev. D* **69** (2004) 114003, [hep-ph/0312106].
42. R. Baier, Y. L. Dokshitzer, A. H. Mueller and D. Schiff, *JHEP* **0109** (2001) 033, [hep-ph/0106347].
43. P. B. Arnold, G. D. Moore and L. G. Yaffe, *JHEP* **0206** (2002) 030, [hep-ph/0204343].
44. P. B. Arnold, G. D. Moore and L. G. Yaffe, *JHEP* **0301** (2003) 030, [hep-ph/0209353].
45. S. Jeon and G. D. Moore, *Phys. Rev. C* **71** (2005) 034901, [hep-ph/0309332].
46. J. Casalderrey-Solana and C. A. Salgado, *Acta Phys. Polon. B* **38** (2007) 3731, [arXiv:0712.3443 [hep-ph]].
47. D. d'Enterria, Landolt-Bornstein **23** (2010) 471, [arXiv:0902.2011 [nucl-ex]].
48. A. Majumder and M. Van Leeuwen, *Prog. Part. Nucl. Phys. A* **66** (2011) 41, [arXiv:1002.2206 [hep-ph]].
49. N. Armesto, B. Cole, C. Gale, W. A. Horowitz, P. Jacobs, S. Jeon, M. van Leeuwen and A. Majumder *et al.*, *Phys. Rev. C* **86** (2012) 064904, [arXiv:1106.1106 [hep-ph]].
50. S. S. Adler *et al.* [PHENIX Collaboration], *Phys. Rev. Lett.* **96** (2006) 032301, [nucl-ex/0510047].
51. M. G. Mustafa, *Phys. Rev. C* **72** (2005) 014905, [hep-ph/0412402].
52. S. Peigne, P. B. Gossiaux and T. Gousset, *JHEP* **0604** (2006) 011, [hep-ph/0509185].
53. S. Wicks, W. Horowitz, M. Djordjevic and M. Gyulassy, *Nucl. Phys. A* **784** (2007) 426, [nucl-th/0512076].
54. M. Djordjevic, *Phys. Rev. C* **74** (2006) 064907, [nucl-th/0603066].
55. P. F. Kolb and U. W. Heinz, In *Hwa, R.C. (ed.) et al.: Quark gluon plasma* 634–714, [nucl-th/0305084].
56. P. Romatschke and U. Romatschke, *Phys. Rev. Lett.* **99** (2007) 172301, [arXiv:0706.1522 [nucl-th]].
57. U. Heinz and R. Snellings, *Ann. Rev. Nucl. Part. Sci.* **63** (2013) 123, [arXiv:1301.2826 [nucl-th]].
58. J. Casalderrey-Solana, H. Liu, D. Mateos, K. Rajagopal and U. A. Wiedemann, arXiv:1101.0618 [hep-th].
59. Y. Mehtar-Tani, C. A. Salgado and K. Tywoniuk, *Phys. Rev. Lett.* **106** (2011) 122002, [arXiv:1009.2965 [hep-ph]].
60. Y. Mehtar-Tani, C. A. Salgado and K. Tywoniuk, *Phys. Lett. B* **707** (2012) 156, [arXiv:1102.4317 [hep-ph]].
61. Y. Mehtar-Tani and K. Tywoniuk, *JHEP* **1301** (2013) 031, [arXiv:1105.1346 [hep-ph]].
62. J. Casalderrey-Solana and E. Iancu, *JHEP* **1108** (2011) 015, [arXiv:1105.1760 [hep-ph]].

63. Y. Mehtar-Tani, C. A. Salgado and K. Tywoniuk, *JHEP* **1204** (2012) 064, [arXiv:1112.5031 [hep-ph]].
64. Y. Mehtar-Tani, C. A. Salgado and K. Tywoniuk, *JHEP* **1210** (2012) 197, [arXiv:1205.5739 [hep-ph]].
65. J. P. Blaizot, F. Dominguez, E. Iancu and Y. Mehtar-Tani, *JHEP* **1301** (2013) 143, [arXiv:1209.4585 [hep-ph]].
66. J. P. Blaizot, E. Iancu and Y. Mehtar-Tani, *Phys. Rev. Lett.* **111** (2013) 052001, [arXiv:1301.6102 [hep-ph]].
67. J. P. Blaizot, Y. Mehtar-Tani and M. A. C. Torres, arXiv:1407.0326 [hep-ph].
68. A. Kurkela and U. A. Wiedemann, *Phys. Lett. B* **740** (2014) 172, [arXiv:1407.0293 [hep-ph]].
69. L. Apolinrio, N. Armesto, J. G. Milhano and C. A. Salgado, *JHEP* **1502** (2015) 119, [arXiv:1407.0599 [hep-ph]].
70. J. P. Blaizot, L. Fister and Y. Mehtar-Tani, arXiv:1409.6202 [hep-ph].
71. L. Fister and E. Iancu, arXiv:1409.2010 [hep-ph].
72. J. P. Blaizot and Y. Mehtar-Tani, arXiv:1501.03443 [hep-ph].
73. Y. Mehtar-Tani, J. G. Milhano and K. Tywoniuk, *Int. J. Mod. Phys. A* **28** (2013) 1340013, [arXiv:1302.2579 [hep-ph]].
74. B. Wu, *JHEP* **1110** (2011) 029, [arXiv:1102.0388 [hep-ph]].
75. T. Liou, A. H. Mueller and B. Wu, *Nucl. Phys. A* **916** (2013) 102, [arXiv:1304.7677 [hep-ph]].
76. M. Fickinger, G. Ovanesyan and I. Vitev, *JHEP* **1307** (2013) 059, [arXiv:1304.3497 [hep-ph]].
77. Z. -B. Kang, E. Wang, X. -N. Wang and H. Xing, *Phys. Rev. Lett.* **112** (2014) 102001, [arXiv:1310.6759 [hep-ph]].
78. J. P. Blaizot, F. Dominguez, E. Iancu and Y. Mehtar-Tani, *JHEP* **1406** (2014) 075, [arXiv:1311.5823 [hep-ph]].
79. J. P. Blaizot and Y. Mehtar-Tani, *Nucl. Phys. A* **929** (2014) 202, [arXiv:1403.2323 [hep-ph]].
80. P. Arnold and S. Iqbal, arXiv:1501.04964 [hep-ph].
81. G. Ovanesyan and I. Vitev, *Phys. Lett. B* **706** (2012) 371, [arXiv:1109.5619 [hep-ph]].
82. J. Ghiglieri, J. Hong, A. Kurkela, E. Lu, G. D. Moore and D. Teaney, *JHEP* **1305** (2013) 010, [arXiv:1302.5970 [hep-ph]].
83. A. Majumder, *Phys. Rev. C* **87**(3) (2013) 034905, [arXiv:1202.5295 [nucl-th]].
84. M. Panero, K. Rummukainen and A. Schfer, *Phys. Rev. Lett.* **112**(16) (2014) 162001, [arXiv:1307.5850 [hep-ph]].
85. K. Zapp, J. Stachel and U. A. Wiedemann, *Phys. Rev. Lett.* **103** (2009) 152302, [arXiv:0812.3888 [hep-ph]].
86. K. C. Zapp, J. Stachel and U. A. Wiedemann, *JHEP* **1107** (2011) 118, [arXiv:1103.6252 [hep-ph]].
87. N. Armesto, L. Cunqueiro and C. A. Salgado, *Eur. Phys. J. C* **63** (2009) 679, [arXiv:0907.1014 [hep-ph]].
88. B. Schenke, C. Gale and S. Jeon, *Phys. Rev. C* **80** (2009) 054913, [arXiv:0909.2037 [hep-ph]].
89. C. Young, B. Schenke, S. Jeon and C. Gale, *Phys. Rev. C* **86** (2012) 034905, [arXiv:1111.0647 [nucl-th]].
90. I. P. Lokhtin, L. V. Malinina, S. V. Petrushanko, A. M. Snigirev, I. Arsene and K. Tywoniuk, *Comput. Phys. Commun.* **180** (2009) 779, [arXiv:0809.2708 [hep-ph]].
91. T. Renk, *Int. J. Mod. Phys. E* **20** (2011) 1594, [arXiv:1009.3740 [hep-ph]].
92. A. Majumder, *Phys. Rev. C* **88** (2013) 014909, [arXiv:1301.5323 [nucl-th]].

93. G. Qin and X. -N. Wang," Jet quenching in heavy-ion collisions, this volume." (2015).
94. J. Ghiglieri and D. Teaney, "Parton energy loss and momentum broadening at NLO in high temperature QCD plasmas", (2015).
95. G. Altarelli and G. Parisi, *Nucl. Phys. B* **126** (1977) 298.
96. V. N. Gribov and L. N. Lipatov, *Sov. J. Nucl. Phys.* **15** (1972) 438, [*Yad. Fiz.* **15** (1972) 781].
97. Y. L. Dokshitzer, *Sov. Phys. JETP* **46** (1977) 641, [*Zh. Eksp. Teor. Fiz.* **73** (1977) 1216].
98. A. H. Mueller, *Phys. Lett. B* **104** (1981) 161.
99. B. I. Ermolaev and V. S. Fadin, *JETP Lett.* **33** (1981) 269, [*Pisma Zh. Eksp. Teor. Fiz.* **33** (1981) 285].
100. A. Bassetto, M. Ciafaloni, G. Marchesini and A. H. Mueller, *Nucl. Phys.* **B207** (1982) 189.
101. A. Bassetto, M. Ciafaloni and G. Marchesini, *Phys. Rept.* **100** (1983) 201.
102. G. Marchesini and B. R. Webber, *Nucl. Phys. B* **310** (1988) 461.
103. L. D. Landau and I. Pomeranchuk, *Dokl. Akad. Nauk Ser. Fiz.* **92** (1953) 735.
104. A. B. Migdal, *Phys. Rev.* **103** (1956) 1811.
105. E. Braaten and M. H. Thoma, *Phys. Rev. D* **44** (1991) 2625.
106. S. Peigne and A. V. Smilga, *Phys. Usp.* **52** (2009) 659, [*Usp. Fiz. Nauk* **179** (2009) 697] [arXiv:0810.5702 [hep-ph]].
107. J. P. Blaizot and Y. Mehtar-Tani, *Nucl. Phys. A* **818** (2009) 97, [arXiv:0806.1422 [hep-ph]].
108. A. Kovner and U. A. Wiedemann, in *Quark Gluon Plasma 3*, eds. R. C. Hwa and X.-N. Wang (World Scientific, 2004), pp. 192–248, [hep-ph/0304151].
109. L. D. McLerran and R. Venugopalan, *Phys. Rev. D* **49** (1994) 2233, [hep-ph/9309289].
110. L. D. McLerran and R. Venugopalan, *Phys. Rev. D* **49** (1994) 3352, [hep-ph/9311205].
111. X. N. Wang and M. Gyulassy, *Phys. Rev. Lett.* **68** (1992) 1480.
112. P. Aurenche, F. Gelis and H. Zaraket, *JHEP* **0205** (2002) 043, [hep-ph/0204146].
113. A. H. Mueller and S. Munier, *Nucl. Phys. A* **893** (2012) 43, [arXiv:1206.1333 [hep-ph]].
114. S. Caron-Huot, *Phys. Rev. D* **79** (2009) 065039, [arXiv:0811.1603 [hep-ph]].
115. R. Baier, Y. L. Dokshitzer, A. H. Mueller and D. Schiff, *Nucl. Phys. B* **531** (1998) 403, [hep-ph/9804212].
116. P. B. Arnold, *Phys. Rev. D* **79** (2009) 065025, [arXiv:0808.2767 [hep-ph]].
117. B. Wu, *JHEP* **1412** (2014) 081, [arXiv:1408.5459 [hep-ph]].
118. E. Iancu, *JHEP* **1410** (2014) 95, [arXiv:1403.1996 [hep-ph]].
119. J. Casalderrey-Solana and X. N. Wang, *Phys. Rev. C* **77** (2008) 024902, [arXiv:0705.1352 [hep-ph]].
120. E. Iancu and D. N. Triantafyllopoulos, *Phys. Rev. D* **90**(7) (2014) 074002, [arXiv:1405.3525 [hep-ph]].
121. Y. Hatta, E. Iancu and A. H. Mueller, *JHEP* **0801** (2008) 063, [arXiv:0710.5297 [hep-th]].
122. P. M. Chesler, K. Jensen, A. Karch and L. G. Yaffe, *Phys. Rev. D* **79** (2009) 125015, [arXiv:0810.1985 [hep-th]].
123. K. M. Burke *et al.* [JET Collaboration], *Phys. Rev. C* **90**(1) (2014) 014909, [arXiv:1312.5003 [nucl-th]].
124. P. B. Arnold, S. Cantrell and W. Xiao, *Phys. Rev. D* **81** (2010) 045017, [arXiv:0912.3862 [hep-ph]].

125. R. Baier, A. H. Mueller, D. Schiff and D. T. Son, *Phys. Lett. B* **502** (2001) 51, [hep-ph/0009237].
126. S. Nazarenko, Wave Turbulence, Lecture Notes in Physics, Vol. 825 (Springer, 2011).
127. C. A. Salgado and U. A. Wiedemann, *Phys. Rev. Lett.* **93** (2004) 042301, [hep-ph/0310079].
128. N. Armesto, H. Ma, M. Martinez, Y. Mehtar-Tani and C. A. Salgado, *JHEP* **1312** (2013) 052, [arXiv:1308.2186 [hep-ph]].
129. N. Armesto, H. Ma, M. Martinez, Y. Mehtar-Tani and C. A. Salgado, *Phys. Lett. B* **717** (2012) 280, [arXiv:1207.0984 [hep-ph]].
130. J. Casalderrey-Solana, Y. Mehtar-Tani, C. A. Salgado and K. Tywoniuk, *Phys. Lett. B* **725** (2013) 357, [arXiv:1210.7765 [hep-ph]].
131. K. J. Eskola, H. Honkanen, C. A. Salgado and U. A. Wiedemann, *Nucl. Phys. A* **747** (2005) 511, [hep-ph/0406319].
132. Y. Mehtar-Tani and K. Tywoniuk, *Nucl. Phys. A* **931** (2014) 475, [arXiv:1401.8293 [hep-ph]].
133. CMS Collaboration [CMS Collaboration], CMS-PAS-HIN-12-013.
134. A. D. Polosa and C. A. Salgado, *Phys. Rev. C* **75** (2007) 041901.
135. A. Dainese, C. Loizides and G. Paic, *Eur. Phys. J. C* **38** (2005) 461.
136. A. Majumder, arXiv:1405.2019 [nucl-th].
137. T. Renk, *Phys. Rev. C* **88**(5) (2013) 054902, [arXiv:1212.0646 [hep-ph]].
138. A. Kurkela and E. Lu, *Phys. Rev. Lett.* **113**(18) (2014) 182301, [arXiv:1405.6318 [hep-ph]].

Parton Energy Loss and Momentum Broadening at NLO in High Temperature QCD Plasmas

Jacopo Ghiglieri
Institute for Theoretical Physics,
Albert Einstein Center,
University of Bern,
Sidlerstrasse 5, 3012 Bern, Switzerland
jacopo.ghiglieri@itp.unibe.ch

Derek Teaney
Department of Physics and Astronomy,
Stony Brook University,
Stony Brook, New York 11794-3800, USA
derek.teaney@stonybrook.edu

We present an overview of a perturbative-kinetic approach to jet propagation, energy loss, and momentum broadening in a high temperature quark–gluon plasma. The leading-order kinetic equations describe the interactions between energetic jet-particles and a non-abelian plasma, consisting of on-shell thermal excitations and soft gluonic fields. These interactions include $2 \leftrightarrow 2$ scatterings, collinear bremsstrahlung, and drag and momentum diffusion. We show how the contribution from the soft gluonic fields can be factorized into a set of Wilson line correlators on the light-cone. We review recent field-theoretical developments, rooted in the causal properties of these correlators, which simplify the calculation of the appropriate Wilson lines in thermal field theory. With these simplifications lattice measurements of transverse momentum broadening have become possible, and the kinetic equations describing parton transport have been extended to next-to-leading order in the coupling g.

1. Introduction

The suppression of highly energetic jets (or jet quenching) is one of the most striking findings of the experimental program of heavy-ion collisions.[1–5] A comprehensive review[6] can be found in this volume, which also contains another contribution reviewing a specific approach in greater detail.[7]

In this review, we concentrate on a weakly-coupled kinetic approach describing the propagation of high momentum jet-like particles through a Quark–Gluon Pasma (QGP). A detailed perturbative description of the QGP and jet-quenching is available when the temperature is high $T \gg \Lambda_{\mathrm{QCD}}$, and the momentum of the jet-particles is much larger than the temperature, $p \gg T$.

The kinetic picture of the high temperature QGP, which has emerged through a combination of physical intuition, direct diagrammatic analysis, and a gradient expansion, is characterized by hard particles and random classical non-abelian fields.[8–11] We will describe a Boltzmann equation for the propagation of jet-like particles interacting with the hard particles and random fields which comprise this idealized weakly coupled plasma. Here the jet-particles have momentum $p \gg T$, while the hard particles have momentum $\sim T$, and the random classical fields are soft with momenta of order gT, $p \gg T \gg gT$, where g is the strong coupling constant. For simplicity, we will limit the discussion to pure gauge theory. As discussed in Sec. 2, to leading order in the coupling constant there are three processes relevant to the transport of jet-particles in the high temperature plasma: (i) $2 \leftrightarrow 2$ elastic scatterings with hard particles, (ii) collinear radiation, and (iii) drag and momentum diffusion driven by the soft classical background.

The interactions between the jet-particles and the classical non-abelian fields, requires a resummation scheme known as the Hard Thermal Loop (HTL) effective theory,[8,9] which is the QCD analog of the Vlasov equations.[10] This effective theory is necessary to compute the drag coefficients and collinear bremsstrahlung rates which describe the propagation of jets. The computational complexity of the Vlaslov equations would, at first sight, make any extension beyond leading order in the coupling extremely challenging. However, HTLs correlators (and statistical correlators more generally) simplify greatly when evaluated at lightlike separations.[12] It is precisely such light-like correlators which must be evaluated to determine the drag, diffusion, and collinear bremmstrahlung rates of jet-particles propagating in plasma.

These simplifications are a consequence of the following physical picture: Since the hard and jet-particles are propagating almost exactly along the light-cone, they are probing an essentially undisturbed plasma, at least as far as the soft classical background is concerned. Informally, we say that the soft classical background "can't keep up" with the hard or jet-particle traversing the plasma. Thus, the soft correlations that are probed by the hard and jet-particles are statistical in nature rather than dynamical.

In Sec. 3 we will illustrate the basics of HTLs and then proceed to introduce the field-theoretical arguments for the light-like simplifications, originally due to Caron-Huot.[12] We then use these simplifications at leading order to determine the transverse momentum diffusion and drag coefficients for high momentum probes.

An important consequence of these lightlike simplifications and their physical origin (see above) is that lattice techniques can be used to compute several lightlike correlators, which arise in jet-quenching physics. Due to the Minkowskian nature of the problem and the large energies at play, a direct *ab initio* lattice calculation of jet propagation in medium is not feasible. However, as we show in detail, the soft contribution to quantities such as $\hat{q}$ (the jet quenching parameter) becomes Euclidean and three-dimensional, which makes it amenable to lattice measurements.

In Sec. 4 we illustrate the basic principles of the dimensionally reduced (three-dimensional) effective theory, and review the first lattice calculation[13] of the soft contribution to $\hat{q}$.

Of course, the special properties of HTLs on the lightcone also simplify perturbative calculations. Indeed, Caron-Huot employed these results to compute $\hat{q}$ and the transverse scattering kernel to NLO[12] in the strong coupling constant g. Building on this and related developments, the photon and low-mass dilepton emission rates were also computed to NLO.[14,15] These electromagnetic rates at NLO correct the strictly collinear approximations made at leading order. Finally, this review anticipates an upcoming publication by the current authors where the LO Boltzmann equation for jet-transport described above is extended to NLO.[16] In Sec. 5 we sketch the Boltzmann equation to this order, which includes the drag coefficient at NLO and semi-collinear emission rates describing bremsstrhalung at wider angles. Many of the parameters of the effective kinetic theory such as drag coefficients, thermal masses, and the transverse collision kernel, can be computed using the three-dimensional lattice described above, providing a tantalizing semi-perturbative description of energy loss. These results and future directions are summarized in Sec. 6.

2. The Kinetic Picture at Leading Order

In this section we will summarize the kinetics of high temperature weakly coupled non-abelian plasmas. As we mentioned in the introduction, this kinetic picture of the QGP is characterized by hard particles and random classical non-abelian fields.[8–11] First we will describe the interactions between these constituents qualitatively, and subsequently give a more quantitative description. We are particularly interested in formulating a Boltzmann equation for the transport of high momentum gluons $p \gg T$ traversing the QGP, i.e. a Boltzmann equation for jet-particles interacting with the hard particles and random classical fields which comprise the high temperature plasma. The phase space distribution of the jet-particles is notated with $f(t, \boldsymbol{x}, \boldsymbol{p})$ (or $f_{\boldsymbol{p}}$ when clear from context) while the phase space distribution of the equilibrium hard particles is notated with $n(t, \boldsymbol{x}, \boldsymbol{p})$ (or $n_{\boldsymbol{p}}$).

The interactions between the particles and the classical fields of the QGP are characterized by three processes:

(1) First, there are $2 \leftrightarrow 2$ collisions between the hard and the jet-particles. These scattering events are strongly localized in space–time.
(2) Second, the random classical fields in the plasma induce drag and momentum diffusion and change the momenta of the jet-particles in characteristic ways. The drag and diffusion coefficients are calculated by examining the response of the classical fields to the non-equilibrium jet-particle.
(3) Finally, the momentum diffusion produced by the random classical field induces collinear bremsstrahlung, causing the jet-particles to split collinearly.

At leading order in the coupling g, the Boltzmann equation for the jet-particles takes the form

$$\left(\frac{\partial}{\partial t} + \boldsymbol{v_p} \cdot \frac{\partial}{\partial \boldsymbol{x}}\right) f(t, \boldsymbol{x}, \boldsymbol{p}) = C_{2\leftrightarrow 2}[\mu] + C_{\rm diff}[\mu] + C_{\rm coll}, \tag{1}$$

where the three rates on the right-hand side reflect these three processes. We have anticipated that the $2 \leftrightarrow 2$ and diffusion rates depend on a separation scale μ, but the dependence on μ cancels when both processes are included. In the remainder of this section we will discuss each of these rates in greater detail.

The hard particles are approximately on-shell and carry the majority of the energy and momentum of the plasma. For this reason these hard constituents are the most important for the thermodynamics of the QGP, and the soft fields are only important insofar as they influence the kinetics of the jet and hard particles. The typical momentum of these excitations is of order the temperature, but the virtuality is of order $\sim g^2T^2$. Specifically, for a hard particle moving in the positive z direction, the scaling of momentum in lightcone coordinates is[a]

$$P^\mu = (p^+, p^-, \boldsymbol{p}_\perp) \tag{3}$$

$$= (T, g^2T, gT). \tag{4}$$

Squaring this momentum shows that the virtuality of these on shell constituents is of order $P^2 \sim g^2T^2$. To estimate the contribution of these on-shell constituents to the plasma energy density we use the equilibrium distribution function, $n_{\boldsymbol{p}} = 1/(e^{E_p/T} - 1)$, and write

$$e(t, \boldsymbol{x}) = 2d_A \int_{\sim gT}^{\infty} \frac{d^3p}{(2\pi)^3} E_{\boldsymbol{p}} n_{\boldsymbol{p}}. \tag{5}$$

First, we note that the integral is cut off when the three-momentum of the constituents is soft, $p \sim gT$. At this point the power counting implicit in Eq. (3) no longer applies. Taking the free gas dispersion relation $\epsilon_{\boldsymbol{p}} = p$, we see that the contribution of the soft modes ($p \lesssim gT$) is of order g^3T^4, while the leading contribution is of order T^4.

The first correction to the free gas result for thermodynamic properties the QGP can be found by determining the correction to the dispersion curve for the on-shell particles. Specifically, the dispersion curve (the relation between $E_{\boldsymbol{p}}$ and $\boldsymbol{p}$) is found by evaluating the one-loop self energy for an approximately on shell

[a] We use a notation where $p^+ = \frac{p^0+p^z}{2}$, $p^- = p^0 - p_z$, four-vectors are denoted by uppercase letters and the metric is the "mostly-plus" one, so that $P^2 = -2p^+p^- + p_\perp^2$. The integration measure is

$$\int \frac{d^4P}{(2\pi)^4} = \int \frac{dp_+ dp_- d^2p_\perp}{(2\pi)^4}. \tag{2}$$

gluon. The self-energy diagram involves internal lines of momentum

$$Q^\mu = (q^+, q^-, q_\perp) \sim (T, T, T), \tag{6}$$

and are thus highly virtual compared to the on-shell particles. The effect of these virtual modes (which are integrated out in any kinetic description) is to change the dispersion relation of the on-shell modes

$$E_{\boldsymbol{p}} = p + \frac{m_\infty^2}{2p} + \cdots, \tag{7}$$

where the ellipses indicate higher order corrections. Here we have defined the leading order gluon asymptotic mass in pure gauge theory at high temperature[17–20]

$$m_\infty^2 = 2g^2 C_A \int \frac{d^3p}{(2\pi)^3} \frac{n_B}{p} = \frac{1}{6} g^2 C_A T^2. \tag{8}$$

As discussed in Sec. 5, the asymptotic mass can be expressed as a correlator of field strengths along the light-cone, which is closely related to an analogous correlator for $\hat{q}$, see Eq. (80).

These hard modes with $p \sim T$ occasionally collide with the jet partons, transferring momenta of order the temperature. At leading order these collisional processes are $2 \leftrightarrow 2$ processes, which can be localized in space–time to within a distance of $\sim 1/T$. A kinetic equation for the phase space distribution function of the jet partons including such scattering processes takes the form

$$\left(\frac{\partial}{\partial t} + \boldsymbol{v}_{\boldsymbol{p}} \cdot \frac{\partial}{\partial \boldsymbol{x}} \right) f_{\boldsymbol{p}}(t, \boldsymbol{x}) = C_{2\leftrightarrow 2}[\mu] + \cdots, \tag{9}$$

where the ellipses will be described below. The velocity in this equation is the group velocity, $\boldsymbol{v}_{\boldsymbol{p}} = \partial E_{\boldsymbol{p}} / \partial \boldsymbol{p}$. The collision rate between the hard modes and the jet particles takes the form[b]

$$\begin{aligned} -C_{2\leftrightarrow 2}[\mu] = \frac{1}{4d_A} \int_{\boldsymbol{p}'\boldsymbol{k}\boldsymbol{k}'} & \frac{|\mathcal{M}|^2}{(2p)(2p')(2k)(2k')} (2\pi)^4 \delta^4(P + P' - K - K') \\ & \times [f_{\boldsymbol{p}} n_{\boldsymbol{p}'}(1+n_{\boldsymbol{k}})(1+n_{\boldsymbol{k}'}) - f_{\boldsymbol{k}} n_{\boldsymbol{k}'}(1+n_{\boldsymbol{p}'}) - n_{\boldsymbol{k}} f_{\boldsymbol{k}'}(1+n_{\boldsymbol{p}'})], \end{aligned} \tag{10}$$

where in the second line we have dropped terms that are exponentially suppressed for $p \gg T$.[c] The kinematics of the collisional process with momentum exchange of order $\sim T$ is shown in Fig. 1. The squared matrix element (summed over initial and final state colors and helicities) may be obtained from vacuum perturbation theory:

$$|\mathcal{M}|^2 = 16 d_A g^4 C_A^2 \left(3 - \frac{su}{t^2} - \frac{st}{u^2} - \frac{tu}{s^2} \right). \tag{11}$$

[b]We use a familiar shorthand notation, $\int_{\boldsymbol{p}\boldsymbol{k}\boldsymbol{k}'} \equiv \int \frac{d^3p}{(2\pi)^3} \frac{d^3k}{(2\pi)^3} \frac{d^3k'}{(2\pi)^3}$, and we are neglecting the thermal mass when writing Eq. (10). The leading factor of $1/(4d_A)$ reflects an average over spins and colors of the external gluon and a symmetry factor 1/2 for the sum over final states.

[c]We consider $\exp(-p/T) \ll 1$. In the remainder of this review we sometimes treat T/P as an explicit expansion parameter for illustration purposes. The detailed exposition and the calculations in[16] are performed without treating T/P as an expansion parameter.

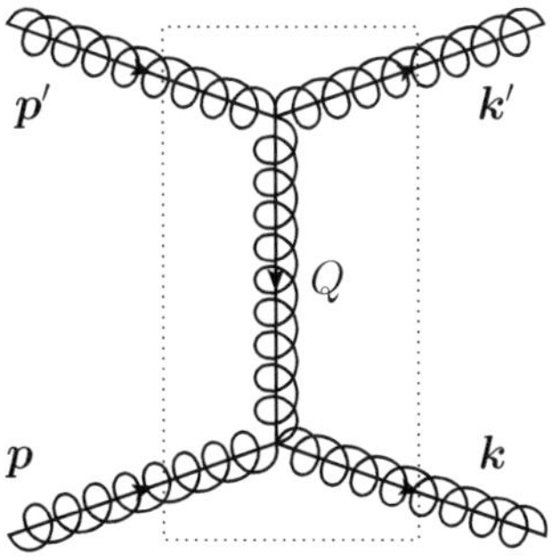

Fig. 1. Hard $2 \leftrightarrow 2$ collision contributing the collision rate $C_{2\leftrightarrow 2}[\mu]$. Only hard lines which enter or exit the boxed region are included in an effective Boltzmann description.

The $2 \leftrightarrow 2$ scattering rate is divergent in the infrared, and therefore we have notated the dependence on the separation scale μ in Eq. (10). There are many ways this divergence can be regulated. At leading and next-to-leading we find it convenient[16] to simply cutoff the transverse momentum exchange at small $q_\perp$, $q_\perp > \mu$. It is not difficult to extract the logarithmic dependence on μ for $\mu \ll T$. Indeed, let us consider for illustration a leading-log approximation to $C_{2\leftrightarrow 2}[\mu]$: we expand the distribution function and matrix elements to second-order in the exchange momentum Q and arrive at a Fokker–Planck equation[21–23] for $f_{\boldsymbol{p}}$

$$C_{2\leftrightarrow 2}[\mu] = \hat{e}_{UV}(\mu) v^i \frac{\partial f_{\boldsymbol{p}}}{\partial p^i} + \frac{1}{2}\hat{q}^{ij}_{UV}(\mu)\frac{\partial^2 f_{\boldsymbol{p}}}{\partial p^i \partial p^j} + \mathcal{O}\left(\frac{T}{p}\right) + \mu\text{-independent}. \tag{12}$$

In writing this equation we have dropped terms suppressed by T/p. Here $\hat{v}$ is a unit vector in the direction of $\boldsymbol{p}$, and the diffusion tensor $q^{ij}_{UV}(\mu)$ controls the longitudinal and transverse momentum diffusion,

$$\hat{q}^{ij}_{UV}(\mu) \equiv \hat{q}_{L,UV}(\mu)\hat{v}^i\hat{v}^j + \frac{1}{2}\hat{q}_{UV}(\mu)(\delta^{ij} - \hat{v}^i\hat{v}^j). \tag{13}$$

The values of these coefficients are found from the expansion of Eq. (10), and for pure gauge are at leading log

$$\hat{q}_{UV}(\mu) = g^2 C_A T \frac{m_D^2}{2\pi} \log\left(\frac{T}{\mu}\right), \tag{14}$$

$$\hat{q}_{L,UV}(\mu) = g^2 C_A T \frac{m_\infty^2}{2\pi} \log\left(\frac{T}{\mu}\right). \tag{15}$$

Here the Debye mass is given by the integral over distribution functions

$$m_D^2 = 2g^2 C_A \int \frac{d^3p}{(2\pi)^3} \frac{n_p(1+n_p)}{T} = \frac{1}{3} g^2 C_A T^2, \tag{16}$$

and the asymptotic mass is given by a similar integral in Eq. (8). At this point the interpretation of these thermodynamic integrals as the Debye and asymptotic masses is premature. This interpretation will be clear from Sec. 3, which explains physically why the transverse diffusion coefficient involves the Debye mass, while the longitudinal diffusion involves the asymptotic mass.

The fluctuation-dissipation theorem relates the coefficient of $\hat{q}_{L,UV}(\mu)$ to $\hat{e}_{UV}(\mu)$

$$q_{L,UV}(\mu) = 2T\hat{e}_{UV}(\mu) + \mathcal{O}\left(\frac{T}{p}\right). \tag{17}$$

This Einstein relation relation is necessary in order for $f_{\boldsymbol{p}}$ to reach its equilibrium form $e^{-p/T}$ under the Fokker–Planck evolution in Eq. (12). Since the equilibrium distribution is a stationary solution to the full $2 \leftrightarrow 2$ Boltzmann equation (Eq. (10)) this relation between the drag and the longitudinal diffusion coefficient automatically arises from a direct expansion of the Boltzmann equation.[16,24,25]

As we will see, the cutoff dependence in the $2 \leftrightarrow 2$ contribution to the Boltzmann equation cancels when the interactions with the soft background fields are included. The description of these soft modes is our next task. As discussed above, the virtuality of the hard modes is of order $P^2 \sim (gT)^2$ with the momentum scaling given in Eq. (3). The soft gluonic excitations, on the other hand, have a momentum scaling

$$P^\mu = (gT, gT, gT), \tag{18}$$

but also have virtuality of order $P^2 \sim (gT)^2$. Since these modes are soft, they are highly occupied due to the Bose–Einstein distribution function

$$n_B(p^0) \simeq \frac{T}{p^0} \sim \frac{1}{g}. \tag{19}$$

Thus, at leading order these soft modes can be treated as a classical gluonic field interacting with the hard on-shell modes.

It is important to emphasize that the energy in these soft modes constitutes a small fraction of the total energy density. Nevertheless, they significantly influence the kinetics of the on-shell particles. Specifically, these soft modes collide frequently with the hard and jet particles and exchange soft momenta of order gT. It seems intuitive that the effect of these soft scatterings can be incorporated into a Fokker–Planck equation describing the drag and momentum diffusion of the jet (and hard) particles

$$\left(\partial_t + v_{\boldsymbol{p}} \cdot \frac{\partial}{\partial \boldsymbol{x}}\right) f_{\boldsymbol{p}} = C_{\text{diff}}[\mu] + \cdots. \tag{20}$$

The collision (or Fokker–Planck) operator can be written as

$$C_{\text{diff}}[\mu] = \hat{e}(\mu) v^i \frac{\partial f}{\partial p^i} + \frac{1}{2}\hat{q}^{ij}(\mu)\frac{\partial^2 f}{\partial p^i \partial p^j} + \mathcal{O}\left(\frac{T}{P}\right), \tag{21}$$

where the momentum diffusion parallel and perpendicular to the direction of motion $\boldsymbol{p}$ are quantified by $\hat{q}^{ij}(\mu)$ and the fluctuation-dissipation theorem is satisfied as described above.[d] Terms of higher order in T/p are discussed in detail elsewhere.[16]

[d] In the remainder of this review $\hat{q}$ describes only the contribution of soft gT to transverse momentum broadening, and perhaps should be written as $\hat{q}_{\text{soft}}$. In most of the literature on jet energy loss

We have emphasized that the parameters of the Fokker–Planck evolution depend on the scale μ separating the scattering from soft modes, and the scattering from hard $2 \leftrightarrow 2$ collisions. As the separation scale is changed, the parameters of the Fokker–Planck equation change and the drag and diffusion rates change accordingly. Indeed, Sec. 3 shows that at leading order the drag and diffusion coefficients are

$$\hat{q}(\mu) = g^2 C_A T \frac{m_D^2}{2\pi} \log\left(\frac{\mu}{m_D}\right), \tag{22a}$$

$$\hat{q}_L(\mu) = g^2 C_A T \frac{m_\infty^2}{2\pi} \log\left(\frac{\mu}{m_\infty}\right). \tag{22b}$$

Thus, a change in these coefficients due to a change in μ is compensated by a corresponding change in the hard $2 \leftrightarrow 2$ collision operator in Eq. (12).

As is well known the Fokker–Planck evolution specified by Eq. (21) is equivalent to a Langevin process,[27] where each particle in the phase space follows an equation of motion specified by a drag and a stochastic force ξ^i, whose variance is specified by the diffusion coefficients

$$\frac{dp^i}{dt} = -\hat{e}(\mu)v^i + \xi^i, \quad \langle \xi^i(t)\xi^j(t')\rangle = \hat{q}^{ij}(\mu)\delta(t-t'). \tag{23}$$

The Langevin description is valid on timescales which are long compared to the underlying correlation times between the microscopic forces in the medium. For a classical particle satisfying Newton's Law, $dp^i/dt = \mathcal{F}^i$, one would adjust the diffusion coefficient so that the integrated squared variance between the microscopic and stochastic forces agree

$$\hat{q}^{ij}\mathcal{T} \equiv \int dt \int dt' \,\langle \xi^i(t)\xi^j(t')\rangle = \int dt \int dt' \,\langle \mathcal{F}^i(t)\mathcal{F}^j(t')\rangle \,. \tag{24}$$

Here $\mathcal{T}$ is the total time, and the forces are to be evaluated along the trajectory of the particle, i.e. for a particle traveling at the speed of light along the z-axis the coordinates are $x^\mu = (x^+, x^-, \boldsymbol{x}_\perp) = (t, 0, \boldsymbol{0}_\perp)$. For a hard particle in representation R interacting with a soft classical background QCD field, the forces are $gT_R^a F_a^{i\mu}(x^+)v_\mu$ and are dressed with Wilson lines following the trajectory of the particle along the light-cone from past infinity[e]

$$\mathcal{F}^i(x^+) \equiv U_R^\dagger(x^+, -\infty) g F^{i\mu}(x^+) v_\mu U_R(x^+, -\infty). \tag{25}$$

(see for example Ref. 26) $\hat{q}$ refers to the sum of the soft and hard contribution, with a logarithmic sensitivity on the jet energy p.

[e]Here and below we use a matrix notation $F^{\mu\nu} = F_a^{\mu\nu} T_R^a$. $\boldsymbol{v}^\mu$ is a lightlike four vector $v^\mu = (1, \boldsymbol{v})$ in the direction of motion of the particle, which is conventionally taken to be along the z-axis. Thus, $F^{i\mu}v_\mu = F^{i-}$ and note that $F^{z-} = F^{+-}$. More explicitly, the Wilson lines are $U(x^+, 0) = Pe^{-i\int_0^{x^+} dx^+ A^-}$.

The appropriate formulas for $\hat{q}$ and q_L are then

$$\hat{q}(\mu) = \frac{1}{d_R}\int_{-\infty}^{\infty} dx^+ \mathrm{Tr}\left\langle \boldsymbol{F}_\perp(x^+)\boldsymbol{F}_\perp(0)\right\rangle, \tag{26}$$

$$\hat{q}_L(\mu) = \frac{1}{d_R}\int_{-\infty}^{\infty} dx^+ \mathrm{Tr}\left\langle \boldsymbol{F}^z(x^+)\boldsymbol{F}^z(0)\right\rangle, \tag{27}$$

where the two transverse directions are summed over in Eq. (26), and the trace averages over the colors of the incoming particle. For a classical background field these correlators can also be written as[f]

$$\hat{q}(\mu) = \frac{g^2 C_R}{d_A}\int_{-\infty}^{\infty} dx^+ \langle v_\mu F_a^{\mu\nu}(x^+) U_A^{ab}(x^+,0) v_\rho F^{\rho}{}_{\nu,b}(0)\rangle, \tag{29}$$

$$\hat{q}_L(\mu) = \frac{g^2 C_R}{d_A}\int_{-\infty}^{\infty} dx^+ \left\langle v_\mu F_a^{\mu +}(x^+) U_A^{ab}(x^+,0) v_\rho F_b^{\rho +}(0)\right\rangle, \tag{30}$$

where $U_A^{ab}(x^+,0)$ is the adjoint Wilson line.

The evaluation of the transport parameters $\hat{q}_L(\mu)$ and $\hat{q}(\mu)$ using these correlators involves understanding how the hard particles interact with the classical gluon fields at leading and next-to-leading order. This had been carefully examined by Blaizot and Iancu who worked in a background field gauge and systematically employed a gradient expansion to determine the appropriate kinetic equations which are the non-abelian generalization of the Vlaslov equations.[10] We will write the non-abelian Vlaslov equations in the next section and exhibit the appropriate HTL diagrammatic rules for evaluating these soft classical gluon correlators. Then in Sec. 3 we will show how these rules can be used to arrive at Eq. (22).

The random classical gluonic fields cause the hard and jet particles to diffuse in momentum. This random walk induces collinear bremsstrahlung in medium. We will follow the AMY formalism for computing the collinear bremsstrahlung rate,[11] and refer to Ref. 28 that compares the AMY formalism to other approaches, which are described in the reviews.[3,5] Collinear bremsstrahlung can be included into the Boltzmann equation and takes the form of a local rate $C_{\rm coll}$ for a hard (approximately massless) gluon to branch into two hard gluons moving in approximately the same direction as their parent[29–31]

$$(\partial_t + v_{\boldsymbol{p}} \cdot \partial_{\boldsymbol{x}}) f_{\boldsymbol{p}} = C_{\rm coll} + \cdots. \tag{31}$$

To describe the gluon bremsstrahlung rate we refer to Fig. 2 which schematically illustrates the collinear radiation induced by the soft scatterings. The kinematics are shown for a gluon with momentum $p = |\boldsymbol{p}|$ in the longitudinal direction to

[f] To see this, take $dx^+ = \Delta x^+$ small. Then, evaluate the following trace to first order in Δx^+ to see the structure of the adjoint Wilson line emerge

$$\mathrm{Tr}[\mathcal{F}^z(x^+)\mathcal{F}^z(0)] \simeq v_\mu F_a^{z\mu}(x^+) v_\rho F_b^{z\rho}(0)\mathrm{Tr}[(1 + ig\Delta x^+ A^-)T^a(1 - ig\Delta x^+ A^-)T^b]. \tag{28}$$

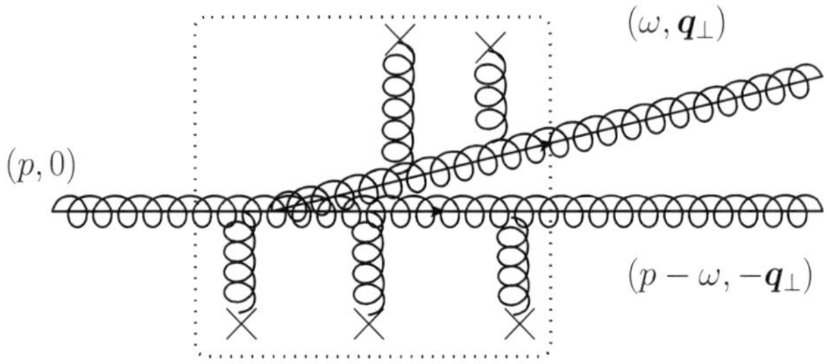

Fig. 2. Schematic Feynman diagram contributing to the leading order collinear bremsstrahlung rate. Hard gluon lines are labeled by their three momentum $(p_z, \boldsymbol{p}_\perp)$. The interactions with the random classical background bath are illustrated by the gluon lines with crosses. Only hard lines which enter or exit the boxed region are included in an effective Boltzmann description.

branch into two gluons moving in approximately the same direction with longitudinal momentum ω and $p-\omega$. Although the bremsstrahlung rate enters the Boltzmann equation as a local rate, it must be understood that the emission process can only be localized to within a time scale set by the formation time of the radiation. The inverse formation time will be defined as the energy difference between the initial and final states

$$(\tau_{\rm form})^{-1} \equiv \delta E(h, p, \omega) = (E_\omega + E_{p-\omega}) - E_p. \tag{32}$$

Using the dispersion relation for the hard particles this reads

$$\delta E(h, p, \omega) \simeq \frac{h^2}{2p\omega(p-\omega)} + \frac{m^2_{\infty\omega}}{2\omega} + \frac{m^2_{\infty p-\omega}}{2(p-\omega)} - \frac{m^2_{\infty p}}{2p}, \tag{33}$$

where $m^2_{\infty,p}$ is the asymptotic mass of the particle with momentum p, as summarized in Eq. (8). We have further defined

$$\boldsymbol{h} \equiv p\boldsymbol{q}_\perp. \tag{34}$$

As seen from the figure and described below, $\boldsymbol{h}/p$ is a transverse momentum vector which is conjugate to the (transverse) coordinate separation $\mathbf{x}_\perp$ between the initial and final states.

The bremsstrahlung rate $C_{\rm coll}$ is determined by the rate of transverse momentum kicks (of magnitude $q_\perp$) which a hard particle experiences traversing the soft classical fields:

$$\mathcal{C}_R(q_\perp) \equiv \lim_{p\to\infty} (2\pi)^2 \frac{d\Gamma_R(\boldsymbol{p}, \boldsymbol{p}+q_\perp)}{d^2 q_\perp}. \tag{35}$$

Here $\boldsymbol{p}$ is the momentum of the hard particle, which is large ($\boldsymbol{p} \to \infty$) relative to the typical momentum, $\sim gT$, of the background fields. The collision kernel $\mathcal{C}_R$ can be expressed as a Wilson loop in the $(x^+, x_\perp)$ plane evaluated in the classical background,[12,32,33] as sketched in Fig. 3. To motivate the appropriate Wilson loop, we note that the average squared momentum transfer per unit time (i.e. $\hat{q}$) is determined by $\mathcal{C}_R(q_\perp)$,

$$\hat{q}(\mu) = \int^{\mu} \frac{d^2 q_\perp}{(2\pi)^2} q_\perp^2 \mathcal{C}_R(q_\perp). \tag{36}$$

As described above $\hat{q}$ can be expressed as a correlation of field strengths (see Eq. (26)). Numerically at least, this correlation function should be understood as the limit of a Wilson loop (see Fig. 3) as the transverse separation approaches zero and the length approaches infinity

$$\lim_{x_\perp \to 0} \lim_{L \to \infty} \langle W_R(x_\perp, L) \rangle = \exp\left(-\frac{1}{4} \hat{q}(\mu) x_\perp^2 L \right). \tag{37}$$

It is not difficult to show that this Wilson loop expression agrees with Eq. (26) in the limit of small separation. Although it is not entirely obvious, it is certainly not surprising that the Wilson loop is sufficient to completely determine $\mathcal{C}_R(\boldsymbol{q}_\perp)$. Specifically, the impact parameter representation of the collision kernel

$$\mathcal{C}_R(x_\perp) = \int \frac{d^2 q_\perp}{(2\pi)^2} (1 - e^{i\boldsymbol{q}\cdot \mathbf{x}_\perp}) \mathcal{C}_R(\boldsymbol{q}_\perp), \tag{38}$$

is determined by the asymptotic behavior of the same Wilson loop at large L

$$\lim_{L \to \infty} \langle W_R(x_\perp, L) \rangle = \exp(-\mathcal{C}_R(\mathbf{x}_\perp) L). \tag{39}$$

Expressions of this form have been used to compute $\hat{q}(\mu)$ at leading and next-to-leading order.

We have described the collision kernel in detail, since it is a key parameter in determining the collinear bremsstrahlung rate. A derivation of this rate is beyond the scope of this review, and here we will simply quote the relevant formulas and describe the relevant physics. The radiation rate for a gluon of momentum p to radiate a gluon of longitudinal momentum ω in approximately the same direction

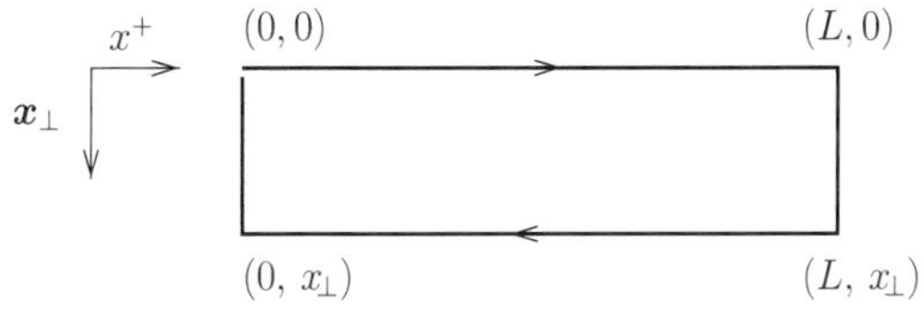

Fig. 3. Wilson loop leading to $\hat{q}(\mu)$. Here the side rails, $Pe^{-i\int_0^L dx^+ A^-(x^+, 0)}$ and $Pe^{-i\int_L^0 dx^+ A^-(x^+, x_\perp)}$, are separated by transverse coordinates $\mathbf{x}_\perp$ and extend in $x^+ = (t+z)/2$ from $x^+ = 0$ up to $x^+ = L$. x^- is zero in this figure. The Wilson loop is traced over colors, $\frac{1}{d_R}\mathrm{Tr}\ldots$, as indicated by the gap at the top left corner.

is[5,31]

$$\left.\frac{d\Gamma(p,\omega)}{d\omega}\right|_{\rm coll} = \frac{g^2 C_A}{16\pi p^7}(1+n(\omega))(1+n(p-\omega))\frac{1+x^4+(1-x)^4}{x^3(1-x)^3} \times \int \frac{d^2h}{(2\pi)^2} 2\mathbf{h}\cdot \mathrm{Re}\,\mathbf{F}(\mathbf{h},p,\omega), \tag{40}$$

where $x \equiv \omega/p$ is the longitudinal momentum fraction of one of the outgoing gluons. The leading factor $(1+x^4+(1-x)^4)/(x^3(1-x)^3)$ records the collinear emission probability accompanying a scattering process. Similarly $\mathbf{F}(\mathbf{h},p,\omega)$ records the time evolution of the current over the formation time. In momentum space this time evolution results in an integral equation for $\mathbf{F}(\mathbf{h},p,\omega)$ which resums the influence of multiple soft scatterings[g] on the emission process[5,31]

$$2\mathbf{h} = i\delta E(\mathbf{h},p,\omega)\mathbf{F}(\mathbf{h}) + \int \frac{d^2k_\perp}{(2\pi)^2}\frac{\mathcal{C}_A(k_\perp)}{2}\{[\mathbf{F}(\mathbf{h})-\mathbf{F}(\mathbf{h}-\omega\mathbf{k}_\perp)] + [\mathbf{F}(\mathbf{h})-\mathbf{F}(\mathbf{h}+p\mathbf{k}_\perp)] + [\mathbf{F}(\mathbf{h})-\mathbf{F}(\mathbf{h}-(p-\omega)\mathbf{k}_\perp)]\}. \tag{41}$$

Finally, the full contribution to the Boltzmann equation takes the form of a gain term (reflecting the radiation from particles with momentum $p+\omega$) and a loss term (reflecting the radiation from the particle with longitudinal momentum p):[29,34]

$$C_{\rm coll} = \int_{-\infty}^{\infty} d\omega\, f_{p+\omega}\left.\frac{d\Gamma(p+\omega,\omega)}{d\omega}\right|_{\rm coll} - \theta(p-2\omega) f_p \left.\frac{d\Gamma(p,\omega)}{d\omega}\right|_{\rm coll}. \tag{42}$$

The region with $\omega < 0$ represents the collinear merging of two gluons to make a final state gluon. We also remark that Eq. (42) is finite in the infrared, as the power-law divergences for $\omega \to 0$ and $\omega \to p$ cancel between the gain and loss terms.

In practice, the integral equation for the bremsstrahlung rate (Eq. (41)) is most easily solved by working in coordinate space, which also illustrates the role of the coordinate space Wilson loop in shown in Fig. 3. To facilitate this rewriting, we recall $\boldsymbol{h}/p$ is conjugate to $\boldsymbol{x}_\perp$, which we relabel as $\mathbf{b}$. Dividing Eq. (41) by p and Fourier transforming with respect to $\boldsymbol{h}/p$ we arrive at a coordinate space representation of the integral equation:

$$-2i\boldsymbol{\nabla}\delta^2(\mathbf{b}) = \frac{i}{2\,x\,(1-x)}(m_\infty^2(1-x+x^2)-\nabla_{\mathbf{b}}^2)\boldsymbol{f}(\mathbf{b}) + \left(\frac{\mathcal{C}_A(x\mathbf{b})}{2}+\frac{\mathcal{C}_A(\mathbf{b})}{2}+\frac{\mathcal{C}_A((1-x)\mathbf{b})}{2}\right)\boldsymbol{f}(\mathbf{b}), \tag{43}$$

where we have defined

$$\boldsymbol{f}(\mathbf{b}) = \int \frac{d^2h}{(2\pi)^2 p^3} e^{i\frac{\mathbf{b}\cdot\mathbf{h}}{p}}\mathbf{F}(\mathbf{h}), \tag{44}$$

[g] Their effect on the emission process is usually called the Landau–Pomeranchuk–Migdal (LPM) effect.

and note that the integral entering the rate (40) becomes

$$\mathrm{Re}\int \frac{d^2h}{(2\pi)^2 p^4} 2\mathbf{h}\cdot\mathbf{F}(\mathbf{h}) = 2\,\mathrm{Im}(\nabla_b \cdot \boldsymbol{f}(0)). \tag{45}$$

This differential equation (Eq. (43)) can be solved to determine the bremsstrahlung rate.

The coordinate representation of the integral equation (Eq. (43)) naturally involves the Fourier-transformed collision kernel $\mathcal{C}_R(\mathbf{b})$ stemming from the Wilson line discussed above (see Fig. 3 and Eq. (39)). Let us remark that the second line in Eq. (43) contains a sum of three two-body contributions, with the transverse distance b weighted by the ratio of two of the three energies p, ω and $p-\omega$. Indeed, it is easy to show[31,35–38] that, at leading order, this is the only possibility. Beyond leading order, one could expect a more complicated three-pole structure, with three Wilson lines on the light-cone, properly linked by transverse lines at their endpoints. However, at least up to NLO the symmetries of the system yield the simple planar dependence in the transverse plane reflected in Eq. (43).[12]

To summarize, we have identified three processes at leading order which are relevant to the transport of high momentum jet particles in a weakly coupled QGP. The first process is straightforward $2 \leftrightarrow 2$ collisions. This rate depends on the separation scale μ between the $\sim T$ scale, characterized by hard particles, and the $\sim gT$ scale, characterized by classical fields. The second process is momentum diffusion, which describes the interactions between the classical fields and the jet-particles. This processes is described by a Fokker–Planck equation, whose parameters $\hat{e}(\mu)$ and $\hat{q}(\mu)$ also depend on the separation scale. However, the sum of the $2 \leftrightarrow 2$ collisions and the Fokker–Planck evolution is independent of μ. Finally, the third process is collinear bremsstrahlung, which is independent of the separation scale to this order. The parameters entering the collinear bremsstrahlung rate are the asymptotic masses m_∞^2 and the scattering kernel $\mathcal{C}(q_\perp)$. Both of these quantities can be expressed as (closely related) correlators of field strengths along the light-cone, Eqs. (29), (39) and (80).

Anticipating Sec. 5, at next-to-leading order we will find corrections to the parameters of the Fokker–Planck equation, $\hat{e}(\mu)$ and $\hat{q}(\mu)$, and corrections to the parameters of the collinear bremsstrahlung rate, m_∞^2 and $\mathcal{C}(q_\perp)$. As we will see the NLO corrections to $\hat{e}(\mu)$ depend logarithmically on the separation scale μ. However, at NLO a correction to the collinear bremsstrahlung rate must also be included, which accounts for not-so-collinear emissions. This *semi-collinear* emission rate also depends on the separation scale μ, and the complete NLO rate is independent of μ. Thus, the NLO correction to the collision term takes the schematic form

$$\delta C = \delta C_{\mathrm{coll}} + \delta C_{\mathrm{diff}}[\mu] + \delta C_{\text{semi-coll}}[\mu]. \tag{46}$$

Before we describe these developments in detail, we will show how the computation of lightlike correlators in thermal field theory (such as $\hat{e}(\mu)$ and $\hat{q}(\mu)$) can be dramatically simplified using the causal properties of these correlators. This is the task of the next section.

3. Hard Thermal Loops on the Light-Cone: A Modern Perspective

In the previous section we described a kinetic equation for jet-like particles based on $2 \leftrightarrow 2$ collisions, diffusion and drag, and collinear bremsstrahlung. The diffusion and drag coefficients are determined by classical statistical correlators of the soft background field on the light-cone, Eqs. (26) and (27). These correlation functions are to be evaluated using the HTL effective theory, which describes the response of the coupled Maxwell–Boltzmann system. Similarly, the collision kernel $\mathcal{C}_R(\boldsymbol{q}_\perp)$ is also determined through Eq. (39) by the interaction between the classical particles and the the background field on the light-cone.

HTLs involve a complex set of medium-modified propagators and vertices. Already at leading order the evaluation of the necessary correlators is seemingly complex. However, as we will describe in this section the evaluation of *lightcone* HTL correlators turns out to be surprisingly simple. For certain operators, such as those associated with transverse momentum diffusion Eq. (26), and soft corrections to the asymptotic mass, a Euclidean formalism can be used.[12] On the other hand, other operators such as those associated with longitudinal diffusion and drag are surprisingly insensitive to the $\sim gT$ sector, and lead to contact terms which correct the collinear emission.

These simplifications are a consequence of the following physical picture. Since the hard and jet-particles are propagating almost exactly along the light-cone, they are probing an essentially undisturbed plasma, at least as far as the soft classical background is concerned. Thus, the soft correlations that are probed by are the hard and jet-particles are statistical rather than dynamical, i.e. these are correlations that arise from the partition function, rather than a retarded response averaged over the correlated initial conditions determined by the partition function. This heuristic explanation underlies the dimensionally-reduced lattice calculation of $\hat{q}$ which is described in Sec. 4.

At a technical level, these simplifications were first realized using light-cone causality of retarded propagators,[12] which says that

$$G_R(p^+,p^-,\boldsymbol{p}_\perp) = \int dx^+dx^-d^2x_\perp e^{i(p^+x^-+p^-x^+-\boldsymbol{p}_\perp\cdot\boldsymbol{x}_\perp)}G_R(x^+,x^-,\boldsymbol{x}_\perp) \quad (47)$$

is an analytic function of p^+ in the upper half-plane at fixed p^- and $p_\perp$. This is because the retarded response function is only non-zero in the forward light-cone $2x^+x^- \geq x_\perp^2$. Thus the integral in Eq. (47) has support only for $x^- > 0$, and the Fourier integral provides an analytic continuation in the upper half p^+ plane, due to the decreasing exponential $e^{ip^+x^-}$. To illustrate these simplifications in a concrete context, we will evaluate the drag and transverse momentum diffusion coefficients, Eqs. (26) and (27), at leading order. To do this we will first briefly review Hard Thermal Loops in Subsec. 3.1. Then we will use the HTL theory to evaluate Eqs. (26) and (27) in Subsecs. 3.2 and 3.3, using light-cone causality to simplify the calculation.

3.1. *Hard thermal loops and light-cone correlation functions*

To briefly describe the HTLs consider a scalar field ϕ_a transforming in the representation R (where a is a color index). There is a single particle density matrix $N_{ba}(t, \boldsymbol{p}, \boldsymbol{x})$ transforming in the $R \times \bar{R}$ representation which records the statistics of the color orientations of a quasi-particle excitation. Heuristically,[39] $N_{ba}(t, \boldsymbol{p}, \boldsymbol{x})$ is a quasi-local expectation value of $\langle a^{\dagger}_{a,\boldsymbol{p}}(t, \boldsymbol{x}) a_{b,\boldsymbol{p}}(t, \boldsymbol{x})\rangle$ at the space time point $(t, \boldsymbol{x})$. More formally, $N_{ba}(t, \boldsymbol{p}, \boldsymbol{x})$ can be defined as the Wigner transform of the gauge-covariant Wightman function.[10] The gluon singlet distribution function (denoted $n_{\boldsymbol{p}}$ above) is proportional to the color trace of a single particle density matrix,

$$n_{\boldsymbol{p}} \equiv \frac{1}{d_R} \mathrm{Tr}[N(t, \boldsymbol{x}, \boldsymbol{p})] \tag{48}$$

while the color current is associated with the gluon distribution is[h]

$$J^{A,\mu}(t, \boldsymbol{x}) = 2g \int \frac{d^3\boldsymbol{p}}{(2\pi)^3} \mathrm{Tr}[T^A N(t, \boldsymbol{x}, \boldsymbol{p})] v^{\mu}_{\boldsymbol{p}}, \tag{49}$$

where $v^{\mu}_{\boldsymbol{p}} = (1, \hat{\boldsymbol{p}})$. The single particle density matrices N_{ba} obey a Vlaslov equation

$$(D_t + \boldsymbol{v} \cdot \boldsymbol{D}_{\boldsymbol{x}}) N_{\boldsymbol{p}} + \frac{1}{2} g \left\{ (\boldsymbol{E} + \boldsymbol{v} \times \boldsymbol{B})_i, \frac{\partial N_{\boldsymbol{p}}}{\partial \boldsymbol{p}_i} \right\} = 0, \tag{50}$$

where $D_\mu = \partial_\mu - ig[A_\mu, \cdot]$ denotes the appropriated gauge covariant derivative for the $R \times \bar{R}$ representation, and the curly brackets denote an anti-commutator. The particles source the classical chromo-electric and magnetic fields

$$-D_\mu F^{\mu\nu}(t, \boldsymbol{x}) = J^{\nu}(t, \boldsymbol{x}). \tag{51}$$

The HTL retarded Green function is derived from this Maxwell–Boltzmann system of equations, Eqs. (50) and (51), by expanding around the equilibrium state, $n_{\boldsymbol{p}} = 1/(e^{E_{\boldsymbol{p}}/T} - 1)$

$$N_{ba} = n_{\boldsymbol{p}} \delta_{ba} + \delta N_{ba}, \tag{52}$$

$$A_\mu = 0 + \delta A_\mu, \tag{53}$$

and solving for the response of the classical field to an external current, $A^{\mu}(\omega, \boldsymbol{q}) = iG^{\mu\nu}_R(\omega, \boldsymbol{q}) J_{\nu,\mathrm{ext}}(\omega, \boldsymbol{q})$. We will not go through this set of steps but refer to the

[h]The leading factor of 2 is the spin of the gluon, and for a scalar field this factor would not appear.

literature.[10,40] The resulting retarded Green functions in the Coulomb gauge are

$$G_R^{00}(\omega, \boldsymbol{q}) = \left. \frac{-i\eta^{00}}{q^2 + \Pi_L(q^0/q)} \right|_{q^0=\omega+i\epsilon}, \tag{54a}$$

$$G_R^{ij}(\omega, \boldsymbol{q}) = \left. \frac{-i(\delta^{ij} - \hat{q}^i\hat{q}^j)}{-(q^0)^2 + q^2 + \Pi_T(q^0/q)} \right|_{q^0=\omega+i\epsilon}, \tag{54b}$$

where $\eta^{00} = -1$ and the self-energies are

$$\Pi_L(x) = m_D^2 \left(1 - \frac{x}{2} \log\left(\frac{x+1}{x-1}\right)\right), \tag{55a}$$

$$\Pi_T(x) = \frac{m_D^2}{2} \left(x^2 - \frac{(x^2-1)x}{2} \log\left(\frac{x+1}{x-1}\right)\right), \tag{55b}$$

with $x = q^0/q$. The spectral density is the difference between the retarded and advanced green functions

$$\rho^{\mu\nu}(\omega) = G_R^{\mu\nu}(\omega, \boldsymbol{q}) - G_A^{\mu\nu}(\omega, \boldsymbol{q}), \tag{56}$$

where $G_A(\omega, \boldsymbol{q})$ is found by setting $q^0 = \omega - i\epsilon$ in Eq. (54). The HTL spectral density and its properties are discussed in Ref. 10.

As we described above, $\hat{q}$ and $\hat{q}_L$ are determined by the propagation of the hard and jet particles in a random classical background field. In equilibrium, the fluctuation-dissipation theorem relates the symmetrized two-point function $G_{rr}(\omega, \boldsymbol{q})$ to the imaginary part of the retarded response[i]

$$G_{rr}^{\mu\nu}(\omega, \boldsymbol{q}) \equiv \frac{1}{2} \langle \{A^\mu(\omega, \boldsymbol{q}), A^\nu\} \rangle = \left(n_B(\omega) + \frac{1}{2}\right) \rho^{\mu\nu}(\omega, \boldsymbol{q}). \tag{57}$$

For a classical background field A_μ, the symmetrization of the field operators is irrelevant, and the symmetrized two point function simply records the two-point statistics of the classical field. Approximating Eq. (57) for $\omega \ll T$ yields

$$G_{rr}^{\mu\nu}(\omega, \boldsymbol{q}) = \langle A^\mu(\omega, \boldsymbol{q}) A^\nu \rangle = \frac{T}{\omega} \rho_{HTL}^{\mu\nu}(\omega, \boldsymbol{q}). \tag{58}$$

We will use these statistics of the classical soft modes when evaluating the longitudinal drag and diffusion coefficient in the next section.

In addition to these rather complex propagators, the Maxwell–Boltzmann system leads to additional HTL vertices which describe how the soft classical background interacts with itself. The resulting Feynman rules describing this response are rather complex as well. To date, this elaborate set of Feynman vertices has been used to compute the dilepton emission rate at small invariant mass[42] and the diffusion coefficient of a heavy quark at NLO.[43] As discussed above and in the next sections, these intricate HTL rules simplify dramatically when evaluating the light-cone correlators involved in energy loss. Indeed, the NLO computations of energy loss described in Sec. 5 would not have been possible without these simplifications.

[i]Our notation G_{rr} follows the one of the so-called r/a formalism of Thermal Field Theory. We refer to Ref. 41 for a comprehensive set of Feynman rules for the HTL theory in that formalism.

3.2. *Transverse diffusion and Euclidean operators*

In this section we will compute $\hat{q}$ and $\mathcal{C}(q_\perp)$ at LO using Eqs. (26), (35) and (39) as our starting point. Recall that these Wilson line correlators describe the propagation of a hard or jet particle traversing the classical field provided by the soft modes. The statistics of these classical fluctuations are given by Eqs. (57) and (58). As we will see, light-cone causality Eq. (47) dramatically simplifies the calculation, leading to a Euclidean formulation.

For $\mathcal{C}$, applying the definition (39), one finds (in any non-singular gauge) that the diagrams in Fig. 4 contribute. Clearly, the diagram on the right and its symmetric give rise to the $x_\perp$-independent part in Eq. (38), whereas the one on the left gives rise to the $x_\perp$-dependent exponential, so that, at leading order, for an adjoint source (a gluon) one has

$$\mathcal{C}(q_\perp) = g^2 C_A \int \frac{dq^0 dq_z}{(2\pi)^2} 2\pi\delta(q^0 - q_z) G_{rr}^{--}(Q), \tag{59}$$

$$\mathcal{C}(x_\perp) = \int \frac{d^2q_\perp}{(2\pi)^2}(1 - e^{i\mathbf{x}_\perp\cdot\mathbf{q}_\perp})\mathcal{C}(q_\perp), \tag{60}$$

where $G_{rr}^{--}(Q) = v_\mu v_\nu G_{rr}^{\mu\nu}(Q)$ is given for soft momenta by Eq. (58). Since $q^0 = q^z$ enforces space-like kinematics, it would then seem that one is left with an integral of the non-trivial HTL spectral function Eq. (56) over the entire Landau cut, $|q^0| < q$. Such integral poses no particular challenge numerically,[44] but any extension beyond leading order, with loops composed of HTL propagators and vertices generated by the rules discussed in the previous Subsec. 3.1, would be extremely challenging from the computational viewpoint. We refer to the computation to NLO[43] of κ, the analogue of $\hat{q}$ for a very massive quark, for an example of such an intricate calculation. In the present case, however, there is a considerable simplification. As we have remarked in Sec. 2, $\mathcal{C}(q_\perp)$ can be obtained from a Wilson loop supported on a light front (a null plane), as show in Fig. 3. The special analiticity properties of objects of this kind are the key to their simplification. Indeed, Alves, Das and Perez,[45] and Weldon,[46,47] introduced the technique of light-front quantization into Euclidean Thermal Field Theory, extending the usual imaginary time formalism from the space-like domain to light fronts. Subsequently, and independently, Caron-Huot[12] introduced a similar quantization technique and showed how it can reduce

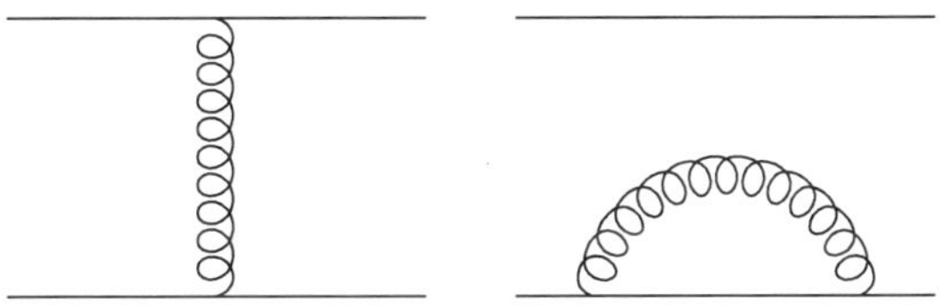

Fig. 4. Diagrams for the leading-order evaluation of $\mathcal{C}(x_\perp)$ in a non-singular gauge. The plain lines represent the Wilson lines along the light-cone, at transverse separation $x_\perp$. The diagram with the self-energy on the upper Wilson line is not shown.

the calculation of (the soft contribution to) this Wilson loop to a much simpler Euclidean calculation in dimensionally-reduced EQCD.

The first step in this endeavor consists in showing that the calculation of n-point correlation functions, where all fields lie on a spacelike hypersurface, can be carried out by Euclidean techniques.[12,45,46] The null correlators we need can also be computed provided that they are free of collinear singularities, which they are up to NLO included.[12] The proofs presented in Refs. 45, 46 and 12 are rather complex and technical. Here we will present a much simpler derivation, also due to Caron-Huot[j], already collected in Ref. 14 , which we follow closely. It is an illustration of the two-point function case, which, as one can see from Eq. (59), will nicely yield the LO result for $\mathcal{C}(q_\perp)$.

Consider the symmetrized correlator of some operator (such as G^{--} in our case), $G_{rr}(x^0, \boldsymbol{x})$ with $|x^z| > |x^0|$. Since the separation is spacelike, operators commute, and therefore G_{rr} equals $G^<$, or $G^>$, as shown in Eq. (58). We are interested in the lightlike limit $x^z = x^0$. There are additional subtleties in taking this limit, but these are irrelevant for the soft and leading order contributions to such correlators. Let us write G_{rr} in terms of its Fourier representation, i.e. the inverse of Eq. (47),

$$G_{rr}(x^0, \boldsymbol{x}) = (2\pi)^{-4} \int d\omega \int dp^z d^2p_\perp e^{i(x^z p^z + \mathbf{x}_\perp \cdot \mathbf{p}_\perp - \omega x^0)}\, G_{rr}(\omega, p^z, p_\perp) \quad (61)$$

and use the exact relation between the spectral function and the symmetric propagator, Eq. (57). We then shift one integration variable to $\tilde{p}^z = p^z - (x^0/x^z)\omega$:

$$G_{rr}(x^0, \boldsymbol{x}) = \int \frac{d\omega d\tilde{p}^z d^2p_\perp}{(2\pi)^4} e^{i(x^z \tilde{p}^z + \mathbf{x}_\perp \cdot \mathbf{p}_\perp)} \left(n_{\rm B}(\omega) + \frac{1}{2}\right) \times \rho(\omega, \tilde{p}^z + \omega(x^0/x^z), p_\perp), \quad (62)$$

so that the Fourier exponent is now frequency-independent and we can perform the ω integration by contour methods. The spectral weight is written as

$$\rho(\omega, \boldsymbol{p}) = G_R(\omega, \boldsymbol{p}) - G_A(\omega, \boldsymbol{p}), \quad (63)$$

and the integral over $G_R(\omega, p)$ is performed by closing the contour in the uppper half plane, while the integration of $G_A(\omega, p)$ is performed by closing the contour in the lower half plane. Since $|x^0/x^z| < 1$, $G_R(\omega, \tilde{p}^z + \omega(x^0/x^z), p_\perp)$ is an analytic function of ω in the upper complex ω plane. This is a consequence of (slightly generalized) light-cone causality[k] as described in and after Eq. (47). The advanced function is similarly an analytic function in the lower complex ω plane. Therefore the only singularities encountered in continuing the frequency integration are those in the statistical function $(n_{\rm B}(\omega) + 1/2)$, which has poles at $\omega_n = 2\pi i n T$ with

[j] S. Caron-Huot, oral presentation at the Institute for Nuclear Theory (Seattle), 29 March 2012.
[k] This is precisely a statement of lightcone causality as described in Eq. (47) when $x^0/x^z = 1$. In this case, $\tilde{p}^z = -p^-$ and ω parametrizes the analytic continuation in p^+ at fixed p^- and $p_\perp$.

$n = (\ldots - 1, 0, 1, \ldots)$ and residues equal to T. Closing the contour around these poles, and renaming $\tilde{p}^z$ to p^z, we find[1]

$$G_{rr}(x^0, \mathbf{x}) = T \sum_n \int \frac{d^3p}{(2\pi)^3} e^{i\boldsymbol{x}\cdot\boldsymbol{p}} G_E \left(\omega_n, p^z + i\omega_n \frac{x^0}{x^z}, p_\perp \right), \quad \omega_n = 2\pi n T, \tag{64}$$

where we have recalled that the retarded Green function is determined by an analytic continuation of the Euclidean function, $G_R(\omega, p) = -iG_E(\omega_E = -i(\omega + i\epsilon), p)$.

When we need to compute the soft gT contribution to such a correlator, one may drop the nonzero Matsubara frequency contributions; that is, we keep only the $n = 0$ term in the sum. This term, $G_E(\omega_n, p^z + i\omega_n(x^0/x^z), p_\perp) \to G_E(0, p^z, p_\perp)$, is the Euclidean correlation function of the three-dimensional dimensionally reduced EQCD theory, which we shall describe in more detail in Sec. 4. Thus, the soft contribution to Euclidean operators at spacelike and lightlike separations is time-independent.

Let us go back to the computation of $\mathcal{C}(x_\perp)$. As all points in that Wilson loop are either at a space-like or light-like separation, the above introduced EQCD reduction is applicable. Hence, at leading order, Eq. (59) becomes

$$\begin{aligned} \mathcal{C}(q_\perp) &= g^2 C_A T \int \frac{dq_z}{(2\pi)} 2\pi\delta(q_z) G_E^{--}(0, q_z, q_\perp) \\ &= g^2 C_A T \left(\frac{1}{q_\perp^2} - \frac{1}{q_\perp^2 + m_D^2} \right). \end{aligned} \tag{65}$$

The result is easily understood as the difference between the massless, transverse and massive, longitudinal EQCD propagators, which also arise as the $q^0 = q^z = 0$ limit of Eq. (54). For (the soft contribution to) $\hat{q}$ we then have the well-known logarithmic dependence on the cutoff, which we anticipated in Eq. (22) in Sec. 2, i.e.

$$\hat{q}(\mu) = g^2 C_A T \int^\mu \frac{d^2q_\perp}{(2\pi)^2} \frac{m_D^2}{q_\perp^2 (q_\perp^2 + m_D^2)} = \frac{g^2 C_A T m_D^2}{2\pi} \ln \frac{\mu}{m_D}. \tag{66}$$

Furthermore, we remark that Eq. (65) was already obtained[48] using a sum rule. Caron-Huot has shown[12] that the two approaches are equivalent, and that the sum rule is possible because of the aforementioned analyticity of retarded and advanced functions. These very same properties will be of great importance for the evaluation of non-Euclidean operators.

Finally, let us consider the effect of the non-zero Matsubara modes we have neglected in Eq. (65). If $q_\perp \sim gT$ is kept soft, the subtleties in putting $x^0/x^z = 1$ will need to be reexamined when including the non-zero Matsubara modes in the sum. The evaluation of these contributions is related to the collinear singularities that should show up at NNLO. It would be interesting to clarify these collinear

[1]The pole at $n = 0$ is an artifact of the separation of ρ, which vanishes for $\omega = 0$, into G_R and G_A. The individual poles there can then be dealt with in a principal value prescription, for instance.

singularities with direct calculations. If one considers shorter transverse separations (with $q_\perp \sim T$) then all Matsubara modes will contribute straightforwardly, and the Euclidean light-front computation in Eq. (64) may be more cumbersome than a direct computation in the real-time formalism. The leading-order contribution to $\hat{q}$ from the scale T has been computed[26] using real time methods.

3.3. *Longitudinal diffusion and non-Euclidean operators*

As we mentioned at the beginning of Sec. 3, not all lightcone or light-front supported operators admit a three-dimensional, Euclidean description for the soft modes. A prime example is the longitudinal momentum diffusion coefficent $\hat{q}_L$, as given by Eq. (27). At leading order it is given by the diagram shown in Fig. 5. In any non-singular gauge, it reads

$$\hat{q}_L = g^2 C_A \int_{-\infty}^{+\infty} dx^+ \int \frac{d^4Q}{(2\pi)^4} e^{-iq^- x^+} (q^+)^2 G_{rr}^{--}(Q), \tag{67}$$

where again $G_{rr}^{--}(Q)$ is given by Eq. (58). The x^+ integration sets q^- to zero. We clearly see that, although originating from a lightcone operator, q^+ cannot be evaluated in EQCD: indeed, the zero-mode contribution exactly vanishes when the previous techniques are applied.

We can however evaluate Eq. (67) by employing sum rules that are rooted in the same analyticity properties that were used in the derivation of Eq. (64). In detail, we plug the result of Eq. (58) in Eq. (67). Up to $\mathcal{O}(g^2)$ corrections[m] we then have

$$\hat{q}_L = g^2 C_A \int \frac{dq^+ d^2 q_\perp}{(2\pi)^3} T q^+ (G_R^{--}(q^+, q_\perp) - G_A^{--}(q^+, q_\perp)). \tag{68}$$

This too would be a simple enough numerical integral[49] over the HTL spectral function in the Landau cut, of difficult extension to higher orders. However, as we have previously remarked, retarded (advanced) two-point functions are analytic in the upper (lower) half-plane in any timelike or light-like variable. We can thus deform the integration contours[16] away from the real axis onto $\mathcal{C}_R$ ($|q^+| = \mu^+ \gg gT$,

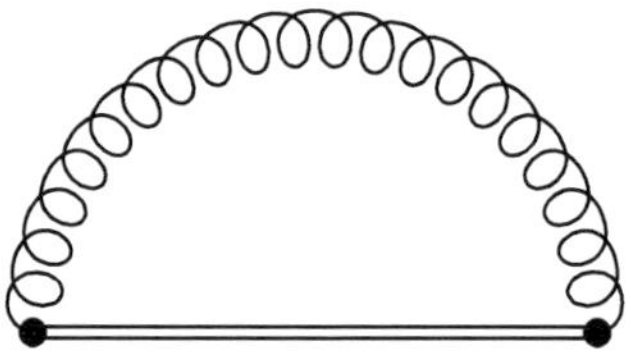

Fig. 5. The leading-order soft contribution to $\hat{q}_L$. The two dots are the two field strengths and the double line is the adjoint Wilson line connecting them. The curly line is a soft HTL gluon.

[m]When expanding the statistical factor in the soft region in Eq. (58), one has $n_\mathrm{B}(\omega) + 1/2 = T/\omega(1 + \mathcal{O}(g^2))$.

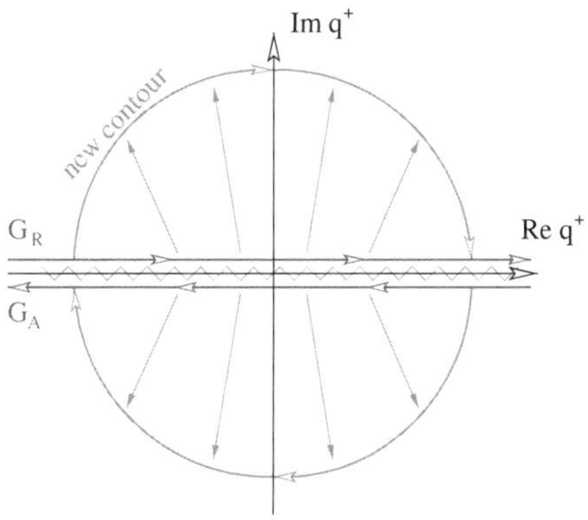

Fig. 6. (Color online) Integration contour in the complex q^+ integration, and the deformation we use to render $q^+ \gg gT$. G_R runs above the real axis and G_A below.

$\mathrm{Im}\, q^+ > 0$) and $\mathcal{C}_A$ ($|q^+| = \mu^+ \gg gT$, $\mathrm{Im}\, q^+ < 0$), as depicted in Fig. 6.[n] μ^+ is a UV cutoff on q^+, to make sure that we consider only the soft region. As we shall see, it will have no effect.

Along the arcs q^+ is then large, albeit complex. The longitudinal and transverse propagators simplify greatly, i.e.

$$G_R^{--}(q^+, q_\perp) = G_R^L(q^+, q_\perp) + \frac{q_\perp^2}{q^2} G_R^T(q^+, q_\perp) \tag{69}$$

$$= \frac{i}{(q^+)^2} + \frac{q_\perp^2}{(q^+)^2} \frac{-i}{q_\perp^2 + m_\infty^2} + \mathcal{O}\left(\frac{1}{(q^+)^3}\right), \tag{70}$$

where we have used the Coulomb gauge propagators given in Eq. (54) and $m_\infty^2 = m_D^2/2$ is the LO gluon asymptotic thermal mass as given in Eq. (8). What this equation is telling us is that, for large (complex) q^+, the longitudinal part is unchanged with respect to its unresummed version, whereas the transverse part acquires the asymptotic mass, which is typical of collinear, transverse excitations.

Upon plugging Eq. (70) in Eq. (68), the q^+ integration along the arcs is trivial and the end result is[16]

$$\hat{q}_L(\mu) = g^2 C_A T \int^\mu \frac{d^2 q_\perp}{(2\pi)^2} \frac{m_\infty^2}{q_\perp^2 + m_\infty^2} = \frac{g^2 C_A T}{2\pi} m_\infty^2 \ln \frac{\mu}{m_\infty}, \tag{71}$$

where contributions smaller than $1/(q^+)^2$ in Eq. (70) are not needed, as they would only give rise to power-law terms in the cutoff on q^+, which would then cancel against contributions from larger scales. The transverse integration has been regulated in the UV with $\mu \gg gT$, as in Eq. (66). Again, this result had been anticipated

[n]The longitudinal and transverse contributions to $G_R^{--}(Q)$ contain poles at $q^+ = q^-/2 \pm iq_\perp$ ($q^2 = 0$), which, being on both sides of the complex plane, appear to violate analyticity. However their residue cancels in the sum of longitudinal and transverse components. As observed in Ref. 12 , they are artifacts of the decomposition into Lorentz-variant longitudinal and transverse modes and their contribution has to vanish in all gauge-invariant quantities.

in Eq. (22). As we commented after Eq. (16), this calculation, together with the one in the previous subsection, clearly shows the physical reasons causing the Debye mass m_D to appear in $\hat{q}(\mu)$ and the asymptotic mass m_∞ to appear in $\hat{q}_L(\mu)$.

The sum rule we have just illustrated has been derived in this fashion in Ref. 16. It can also be seen as the bosonic equivalent of the one presented in the calculation of the thermal photon rate to NLO[14] and of the thermal ultrarelativistic right-handed neutrino rate at LO.[50] We also remark that the same result (71) was previously obtained in a different way[51] for the drag coefficient $\hat{e}$, which is related by the Einstein relation Eq. (17). As it is shown there,[51] once the difference in regularization between $q_\perp < \mu_\perp$ and $q < \mu_q$ is taken into account, Eq. (71) agrees with the numerical results of Braaten and Thoma[49] for $v \to 1$.

In summary, in this section we have introduced HTLs and their simplification on the light-cone. In Subsec. 3.2 we have analyzed the subset of operators, including transverse momentum diffusion, that admit a three-dimensional, Euclidean description in the soft region, whereas in the current subsection we have analyzed those that do not, showing how they still considerably simplify. In the following section we will show how Euclidean operators can be calculated on the lattice, whereas in Sec. 5 we will use the results of this section to generalize the perturbative kinetic approach to NLO.

4. Non-Perturbative Determinations of $\mathcal{C}(x_\perp)$ and $\hat{q}$

As we mentioned previously, the field-theoretical developments introduced by Caron-Huot[12] make it possible to map certain operators supported on light fronts and light-cones to the Euclidean domain, where the soft contribution is dominated by the dimensionally reduced zero modes of 3D EQCD. Recently, this has been exploited to compute the soft contribution to $\mathcal{C}(x_\perp)$ and $\hat{q}$ on the lattice. Laine and Rothkopf[52] have performed a first study of the Wilson loop defined in Fig. 3 in classical gauge theory, which allows real-time lattice studies and catches the qualitative nature of the soft physics, which, has we have argued in Sec. 2, is of classical origin. In particular, they also examined the dependence on v of the Wilson loop (i.e. whether the approach to the light-cone from above or below is different for classical fields), finding it modest.

Panero, Rummukainen and Schäfer[13] have instead computed the soft contribution to $\mathcal{C}(x_\perp)$ and to $\hat{q}$ by means of lattice simulations of EQCD.[o] In the remainder of this section, we will illustrate the basic principles of this treatment and its potential application to other light-cone observables that are suited to it, such as $\hat{q}(\delta E)$, which will be defined in Sec. 5, or the asymptotic masses. To this end, we start by reviewing Electrostatic QCD (EQCD),[54–58] which is a dimensionally-reduced EFT of QCD describing the time-independent, long-wavelength $\lambda \sim 1/(gT)$ modes of

[o] A different method, not based on the properties of Sec. 3, for estimating $\hat{q}$ on the lattice has been proposed.[53]

the latter theory. If treated perturbatively, it breaks down at very long wavelengths (very small momenta) $\lambda \sim 1/(g^2T)$ ($k \sim g^2T$) where magnetic non-perturbative contributions[54,59] appear. If, on the other hand, it is treated non-perturbatively, both the electric and magnetic scales are treated correctly. Hence, a lattice calculation in EQCD, such as the one of $\hat{q}$ we are considering, determines the soft (gT) and ultrasoft (g^2T) contributions. Indeed, non-perturbative EQCD has been employed for many studies of the thermodynamics of QCD, as in this example Ref. 60, and of the gauge sector of the standard model, as in this review Ref. 61 and this recent work Ref. 62.

The Lagrangian of EQCD reads

$$\mathcal{L} = \frac{1}{4}F^a_{ij}F^a_{ij} + \mathrm{Tr}((D_iA_0)^2) + m^2_{\scriptscriptstyle E}\mathrm{Tr}(A^2_0) + \lambda_3(\mathrm{Tr}(A^2_0))^2, \tag{72}$$

where only operators of dimension up to three have been kept. $D_i = \partial_i - ig_E[A_i, \cdot]$ is the standard covariant derivative, so that the Lagrangian describes a 3D SU(3) Yang-Mills field minimally coupled to an adjoint scalar field A_0, which has a mass and a quartic coupling (adjoint Higgs). Fermions, having only odd Matsubara frequencies proportional to πT, are completely integrated out in the dimensional reduction.

At leading order the matching between QCD and EQCD yields[54,56,58]

$$g_E^2 = g^2(\mu)T, \quad m^2_{\scriptscriptstyle E} = \left(\frac{N_c}{3} + \frac{n_f}{6}\right) g^2(\mu)T^2, \quad \lambda_3 = \frac{9-n_f}{24\pi^2}g^4(\mu)T, \tag{73}$$

where g_E is the dimensionful coupling of EQCD and $g(\mu)$ is the scale-dependent coupling of QCD. The mass parameter is equal to the leading-order Debye mass. The positive dimensionality of g_E makes the theory super-renormalizeable.

The EQCD action can be readily discretized for lattice studies We refer to Refs. 13 and 60 for details on this procedure. The implementation of the Wilson loop[12,32,33] depicted in Fig. 3 is on the other hand somewhat non-trivial in the dimensionally-reduced theory, where the time dimension is integrated out. In the continuum, the Wilson line along the light-cone direction changes from

$$U(x_f^+; x_i^+) = \mathcal{P}\exp\left[-ig\int_{x_i^+}^{x_f^+} dx^+(A^0(x^+) - A^z(x^+))\right] \text{ in Minkowskian QCD,} \tag{74}$$

to

$$U(x_{z,f}; x_{z,i}) = \mathcal{P}\exp\left[-ig\int_{x_{z,i}}^{x_{z,f}} dx_z\,(iA_0(x_z) - A_z(x_z))\right] \text{ in EQCD,} \tag{75}$$

where the A^0 in Eq. (74) is the temporal gauge field of Minkowskian 4D QCD, whereas the one in Eq. (75) is the adjoint scalar of Euclidean, 3D EQCD, the extra i coming from the Wick rotation. The other coordinates ($\mathbf{x}_\perp$ and, in Eq. (74), x^-) are understood to be constant and are not shown.

The lattice version of Eq. (75) is also non-trivial, because in the discretized action the 3D gauge fields are naturally re-expressed as gauge links between the lattice sites, while the A_0 scalar field is defined on the lattice sites. This leads to, in the notation of[13]

$$L_3(x, an_\ell) = \prod_{n=0}^{n_\ell - 1} U_3(x + an\hat{3}) H(x + a(n+1)\hat{3}), \tag{76}$$

where L_3 is the lattice equivalent of Eq. (75), with $n_\ell = \ell/a$ the length of the Wilson line in units of the lattice spacing a. U_3 is a gauge link in the z direction and $H(x)$ is a Hermitian, rather than unitary, matrix obtained by exponentiation of $A_0(x)$, i.e.

$$H(x) = \exp[-ag_E^2 A_0(x)]. \tag{77}$$

Note that $H(x)$ represents a parallel transporter along a *real-time* interval of length a. The transverse Wilson lines L_1 are instead the usual product of gauge links, so that, schematically, the Wilson loop takes the form depicted in Fig. 7.

The lattice measurements[13] of $\mathcal{C}(x_\perp)$ through the above definition are summarized in Fig. 8. As the authors remarked, once the value the Debye mass which includes the non-perturbative relative $\mathcal{O}(g)$ correction[63] is used to set the scale and as input for the perturbative calculation of Caron-Huot, the agreement between lattice and perturbation theory becomes surprisingly good. If, on the other hand, the perturbative Debye mass is used, perturbation theory yields a significantly smaller $\mathcal{C}(x_\perp)$ at intermediate distances.

The soft contribution to $\hat{q}$ can then be extracted from the short-distance behavior of $\mathcal{C}(x_\perp)$ along the line of Eq. (37), i.e.[12,14,64]

$$\mathcal{C}(x_\perp) \xrightarrow{x_\perp m_D \ll 1} c_1 x_\perp + \hat{q}_{\rm soft}\frac{x_\perp^2}{4} + \mathcal{O}(x_\perp^3). \tag{78}$$

The $\hat{q}_{\rm soft}$ extracted in this way also contains a residual logarithmical dependence on $x_\perp$, reflecting the leading-order UV divergence of $\hat{q}_{\rm soft}$ (see Eq. (66)) that is

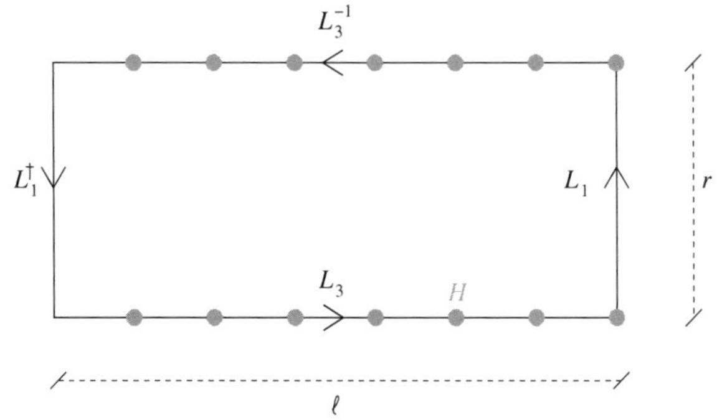

Fig. 7. The Wilson loop yielding $\mathcal{C}(x_\perp = r)$ in EQCD. The dots on the lines in the horizontal direction represent the insertion of the Hermitian operator H at the lattice sites, whereas the lines connecting them are the gauge links in the z direction. Figure taken from the original reference.[13]

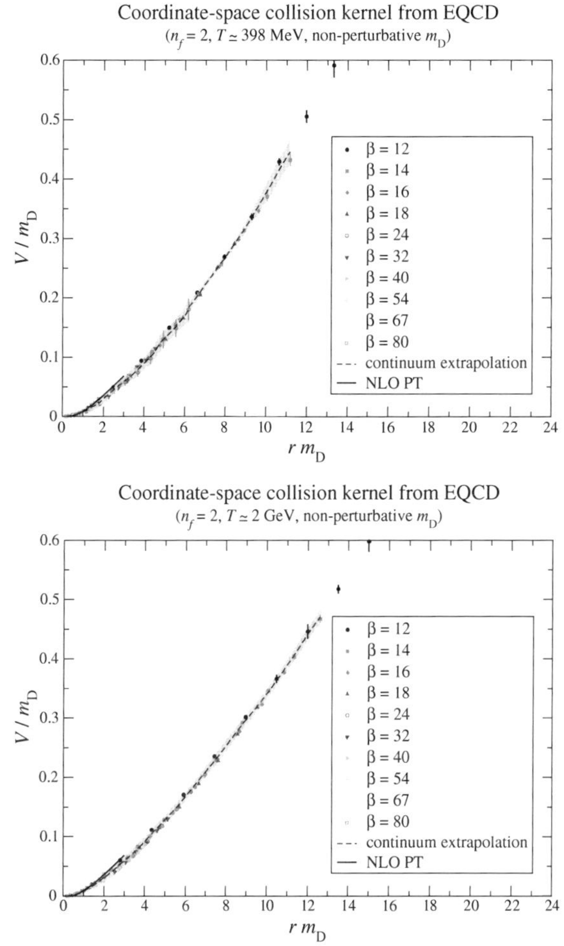

Fig. 8. (Color online) Results for $\mathcal{C}(x_\perp)$ at two different values of the temperature. The Debye mass used as a scale for the axes and plugged in the NLO perturbative calculation[12] is the non-perturbative one.[63] Figure taken from the original reference.[13]

absorbed by the hard contribution. We refer to Refs. 64 and 13 for further details on the matching procedure and the addition of the leading-order hard contribution.[26] Once this has been performed, Panero *et al.* report an estimate for $\hat{q}$ of 6 GeV^2/fm at their lower temperature of 398 MeV, with an estimated uncertainty of 15 to 20%. We remark that higher-order contributions from the hard scale, as well as the possible collinear contribution mentioned in Subsec. 3.3, are absent from this determination.

A possible limitation to the approach we have outlined is that it relies on a separation of scale between the hard particles, with momenta of order πT, and the soft and ultrasoft fields, characterized by gT and g^2T. However, as the authors remark, the literature suggests (see for instance Refs. 60, 65 and 66) that analytical computations relying on this separation of scales may be sufficiently accurate down to low temperatures, perhaps, suprisingly, down to[67] $T \sim 2T_c$.

It is worth remarking that the NLO perturbative calculation predicts at the origin a negative linear slope,[p] i.e. $c_i < 0$ in Eq. (78), which is not observed in the lattice calculation. This can be attributed to discretization errors, which are more severe at short distances, corresponding to the UV region $p_\perp \gg m_D$ in momentum space. At leading order in PT, the dominant UV behavior $(1/p_\perp^2)$ cancels between the longitudinal and transverse one-gluon exchanges, as shown in Eq. (65), leaving a $m_D^2/p_\perp^4$ correction. This cancellation, while exact in the continuum, is only approximate on the lattice. D'Onofrio, Kurkela and Moore[68] have estimated the associated error to be of order $a/x_\perp$ and hence especially relevant at short distances. Their computation of the renormalization properties of Eq. (76) to order a can help alleviating these discretization effects.

We also note that, by comparing their results with the estimate of the magnetic contribution only to $\hat{q}$,[33,64] which can be performed similarly by simulating a Wilson loop in the pure 3D gauge theory, Panero *et al.* have been able to establish that the magnetic contribution (g^2T only) is subleading compared to the full non-perturbative EQCD determination, which includes the gT and g^2T contributions.

Future directions for this very promising new direction can include more refined measurements of $\hat{q}$, for instance improving the approach to the continuum at short distances along the lines discussed before or extending the range of explored temperatures. An altogether similar approach can be used to determine non-perturbatively the soft contribution to other operators that are amenable to a Euclidean treatment, such as $\hat{q}(\delta E)$ or the condensates appearing in the definition of the asymptotic masses (see Subsec. 5.1). One could then envisage for the future the establishment of a factorization program for the computation of jet-related observables: the soft physics, where pQCD struggles, is encoded into these effective operators measured on the lattice, whereas the harder scales are treated perturbatively.[q] EFT techniques could be used to make the factorization well defined and deal with possible large logarithms. A first step in this direction has been taken in this study.[70] It compares the screening masses of a specific correlator, extracted from lattice measurements, with their determination through a differential equation that is very similar to Eq. (43), up to a complex phase for $\mathcal{C}(x_\perp)$. By using the lattice determination of the latter the authors find a very good agreement with the non-perturbative screening masses.

[p] Due to the super-renormalizability of EQCD, each loop order causes a different power-law behavior for $\mathcal{C}$, so that higher orders cannot contaminate this effect.

[q] A factorized approach sharing the same spirit, but using holographic methods for the description of the medium, has recently been presented in Ref. 69.

5. Perturbative Next-to-Leading Order

As we have anticipated, the recent technological developments outlined in Sec. 3 make the extension to next-to-leading order of the kinetic approach sketched in Sec. 2 possible. In that section we have introduced the three fundamental processes that make up the leading-order collision operator, i.e. diffusion, large-angle scatterings and collinear splittings/joinings. Next-to-leading order contributions represent a relative $\mathcal{O}(g)$ correction and come about because, as we have mentioned, soft gluons are highly occupied, with $n_{\rm B}(q^0 \sim gT) \approx T/q^0 \sim 1/g$. For this reason one can show that loops composed of soft propagators only are suppressed by a factor of g rather than g^2, giving rise to *soft loop corrections*. Another source of $\mathcal{O}(g)$ corrections comes from regions of the LO calculation where a gluon becomes soft, without being properly treated, i.e. with HTL resummations. For instance, in large-angle scatterings, the integrations over the energies of the incoming and outgoing gluons stretch down to zero, including an $\mathcal{O}(g)$ region of phase space where they are soft, but treated as massless, rather than HTL, excitations. At NLO one then needs to replace the improper LO evaluation of these *mistreated regions* with the proper one.

In the following we will proceed to sketch the kinematical regions affected by NLO corrections and their evaluation. We will not dwell on mistreated regions, which, in a nutshell, require the identification and subtraction of the relevant counterterms in the corresponding points of the NLO calculations.

Large-angle scatterings are not sensitive to soft loop corrections, as they would require internal and/or external soft lines, which are either excluded by construction or suppressed. Hard loops represent an $\mathcal{O}(g^2)$ correction, so that this region is not affected at NLO.

Collinear processes are sensitive to soft loop corrections: they give rise to the $\mathcal{O}(g)$ correction[12] to $\mathcal{C}(x_\perp)$ which we have mentioned in the previous sections, as well as to an $\mathcal{O}(g)$ correction[20] to the dispersion relations of the hard, collinear particles.

Diffusion processes are also naturally sensitive to $\mathcal{O}(g)$ corrections, since the typical momenta there are by construction soft. The soft loop corrections to $\hat{q}$[12] have also been mentioned before, while those to $\hat{q}_L$[16] will be introduced in the following.

Finally, a new process appears at NLO. We label it *semi-collinear*. It is related to the example we made earlier on in this section of a large-angle scattering with an incoming (or outgoing) soft gluon, which causes the angle between the two outgoing (or incoming) hard ones to be smaller than in the large-angle case but larger than in the pure collinear case. Indeed, as we shall see, this process will represent a bridge between the other three.

Schematically, the collision operator then takes this form

$$\delta C = \delta C_{\rm coll} + \delta C_{\rm diff}[\mu] + \delta C_{\rm semi\text{-}coll}[\mu], \tag{79}$$

where we anticipate, as noted by the μ dependence, that the last two processes will require regulation, but that the dependence on the regulator will vanish in the sum.

5.1. *The collinear region*

In the collinear regime the $\mathcal{O}(g)$ corrections enter then in two places, as we anticipated: both the effective thermal masses squared m_∞^2 and the collision kernel $\mathcal{C}(x_\perp)$ get $\mathcal{O}(g)$ corrections.[r]

For what concerns the thermal masses, an expression in terms of gauge-invariant fermionic and bosonic condensates supported on the light-cone is known.[20] It reads[s]

$$m_\infty^2 = g^2 C_A Z_g \quad Z_g \equiv \frac{1}{d_A}\left\langle v_\mu F^{\mu\nu a}\left(\frac{1}{(v\cdot D)^2}\right)_{ab} v_\rho F^{\rho\, b}_{\nu}\right\rangle, \tag{80}$$

where $v = (1, \boldsymbol{v})$ is a null vector ($|\boldsymbol{v}| = 1$). Intuitively, Z_g describes how thermal fluctuations of the gauge fields affect the propagation of a fast, light-like particle. It can be rewritten as

$$Z_g = \frac{-1}{d_A}\int_0^\infty dx^+ x^+ \langle v_\mu F_a^{\mu\nu}(x^+) U_A^{ab}(x^+,0) v_\rho F^\rho_{b\,\nu}(0)\rangle, \tag{81}$$

which makes the connection with $\hat{q}$ clearer: by comparing with Eq. (29), it is easy to see that they share the same Lorentz structure. At leading order Z_g is dominated by hard ($p \sim T$) momenta and one easily has

$$Z_g^{\rm LO} = 2\int \frac{d^3p}{(2\pi)^3 p} n_{\rm B}(p) = \frac{T^2}{6}. \tag{82}$$

The $\mathcal{O}(g)$ contribution arises from the proper treatment of the IR limit of the above integration, requiring in principle HTL resummation. However, Z_g is supported on a lightcone, so that the methods of Sec. 3 become applicable. As shown by Caron-Huot[20] (further details can also be found here[14]) the $\mathcal{O}(g)$ contribution is easily obtained from the Euclidean zero-modes, yielding

$$\delta Z_g = T\int \frac{d^3q}{(2\pi)^3}\frac{q_\perp^2}{(q_z+i\epsilon)^2}\left(\frac{-1}{q^2+m_D^2}+\frac{1}{q^2}\right) = -\frac{Tm_D}{2\pi}, \tag{83}$$

so that the NLO correction to the asymptotic mass δm_∞^2 reads

$$\delta m_\infty^2 = g^2 C_A\left(-\frac{Tm_D}{2\pi}\right). \tag{84}$$

For what concerns the scattering kernel, Caron-Huot[12] has shown that three-body contributions vanish in the three-pole operator sketched below Eq. (43). Hence, also at NLO one just needs the sum of three two-body contributions at

[r]For simplicity we do not illustate the two further mistreated regions[16] in the soft and semi-collinear limits respectively.

[s]In QCD with quarks there is also a fermionic condensate. Both condensates contribute to the asymptotic masses of gluons and quarks.

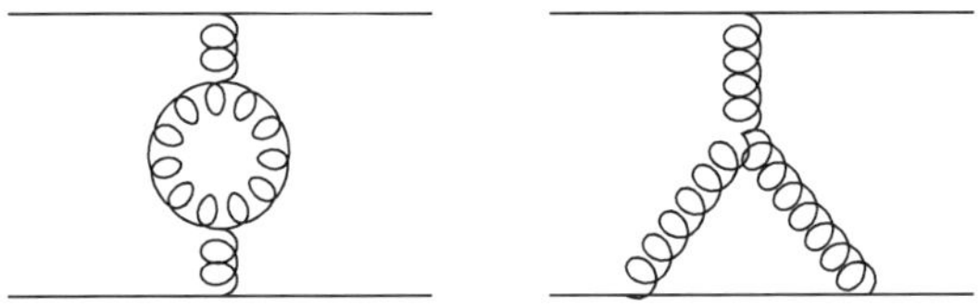

Fig. 9. Two sample diagrams necessary for the NLO evaluation of $\mathcal{C}(q_\perp)$. The conventions are as in Fig. 4.

different transverse separations. The computation of these is nothing but the NLO contribution to $\mathcal{C}(x_\perp)$. We have shown in detail in Subsec. 3.3 how to "Euclideanize" the leading-order calculation. At NLO one then needs to compute one-loop diagrams in three-dimensional EQCD, such as those shown in Fig. 9. The result for $\delta\mathcal{C}(q_\perp)$[12] and the corresponding $\delta\mathcal{C}(x_\perp)$[14] are not shown here for brevity. We point to the references for their expressions. We would like to stress again how the the reduction to the Euclidean, three-dimensional theory makes this calculation tremendously simpler. If we were to compute the diagrams of Fig. 9 in the HTL-resummed theory in four Minkowskian dimensions, we would have to deal with complicated denominators of (up to 4) HTL propagators (see Eq. (54)) and the intricate HTL vertices. Indeed, in both our example diagrams we would need to consider them along the standard bare vertices. In the Euclideanized, three-dimensional theory, on the other hand, the propagators become simple massive or massless ones and the effective vertices just disappear, as they depend linearly on the external frequencies, which vanish in dimensional regularization.[t]

In order to obtain the NLO correction to the collinear rate (40), Eq. (43) is then to be perturbed by δm_∞^2 and $\delta\mathcal{C}$. The first order in this perturbation $\delta \boldsymbol{f}(\mathbf{b})$, is plugged in Eqs. (45) and (40). Details on this procedure may be found in these references.[14,16] Algorithms for the solution of the differential equation with the correct boundary conditions are discussed in these other works Refs. 14, 15, 71 and 72.

We furthermore stress that, should a factorized approach based on lattice measurements of soft observables, as outlined at the end of Sec. 4, be applied here, one would need a non-perturbative measurement of the aforementioned three-pole operator. It would be interesting to analyze the non-perturbative appearance of three-body forces in such an object.

Finally, we remark that recently it has been pointed out[73,74] that collinear processes are sensitive to a large, double-logarithmic correction, which also shows up as a double-logarithmic correction to $\hat{q}$. This term represent a relative $\mathcal{O}(g^2)$ correction in our framework and would hence contribute to NNLO. It would arise from considering the first correction to the eikonal approximation in deriving Eq. (42).

[t]Indeed, there are no resummed vertices of the same order of the bare ones in EQCD, as we have shown in Sec. 4.

5.2. *The diffusion sector at NLO*

As we anticipated, the Fokker–Planck diffusion equation (21) still applies. We then need the $\mathcal{O}(g)$ corrections to $\hat{q}$ and $\hat{q}_L$, which unambiguously determine $\hat{e}$ at NLO through the procedure sketched in Eq. (17). Both can be calculated from their Wilson line definitions (26) and (27). The former is computed with the Euclidean technology of Subsec. 3.2 and the latter with the sum rules of Subsec. 3.3. However, before going in the details of the two separate calculations, one can proceed to a simplification,[16,43] which is attained by using the equation of motion of the Wilson line $D^-_{x^+}U(x^+;0)=0$ to simplify $F^{-\mu}$ to $\partial^\mu A^-$ (in all gauges but the $A^-=0$ one) and by noting again that operator ordering is not relevant at NLO: classical soft gluons commute, as we have used in obtaining the simplified forms (29) and (30). The Wilson line structure then simplifies to

$$\hat{q}=\frac{g^2C_R}{d_A}\int_{-\infty}^{+\infty}dx^+\langle\partial^\perp A^{-a}(x^+)U_A^{ab}(x^+,0)\partial^\perp A^{-b}(0)\rangle, \tag{85}$$

$$\hat{q}_L=\frac{g^2C_R}{d_A}\int_{-\infty}^{+\infty}dx^+\langle\partial^+ A^{-a}(x^+)U_A^{ab}(x^+,0)\partial^+ A^{-b}(0)\rangle. \tag{86}$$

Their evaluation requires the computation of the diagrams shown in Fig. 10.

For what concerns $\hat{q}$, the reduction to the Euclidean, three-dimensional theory introduced before makes this calculation too tremendously simpler. One then obtains for $\delta\hat{q}$, the NLO contribution to $\hat{q}$[12,u]

$$\delta\hat{q}=\frac{g^4C_RC_AT^3}{32\pi^2}\frac{m_D}{T}(3\pi^2+10-4\ln 2). \tag{87}$$

For what concerns $\delta\hat{q}_L$, the sum-rule technology we have introduced in Subsec. 3.3 can be applied and leads to another tremendous computational simplification.[16] As a general strategy, we remark that the Wilson line propagators depend only on the minus components of the momenta, so that the plus component, which we call q^+, can be deformed again. This corresponds to expanding those diagrams for large, complex q^+. The leading contribution can be of order $(q^+)^0$ and the subleading one of order $(q^+)^{-1}$. Higher-order terms are suppressed and can be

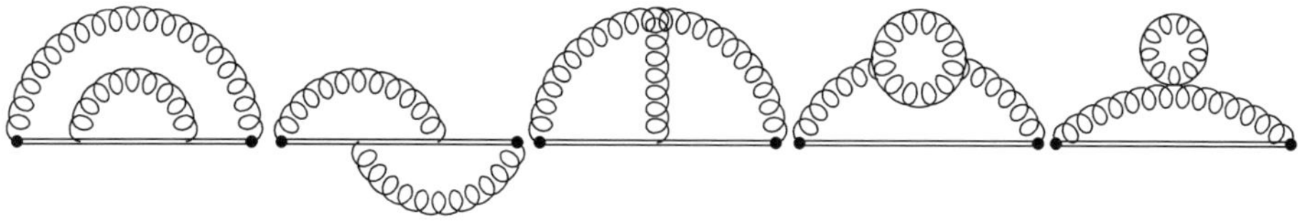

Fig. 10. Diagrams contributing to $\delta\hat{q}$ and $\delta\hat{q}_L$ at NLO. The blobs represent again the field strength insertions. Three- and four-gluon vertices are either bare or resummed HTL vertices.

[u]Caron-Huot obtained this result by applying Eq. (36) to his calculation of $\mathcal{C}(q_\perp)$ at NLO. The two methods are completely equivalent, as Eq. (26) can be formally derived from Eqs. (35) and (36) when $\mathcal{C}(q_\perp)$ is defined from the Wilson loop in Fig. 3, as shown in Ref. 33.

neglected. The leading, $\mathcal{O}((q^+)^0)$ term, once integrated along the contour, will give rise to a linear divergence, which has to match with the one arising from the corresponding mistreated region in the collinear process (see footnote r). Indeed, the calculation[16] confirms that the two divergences cancel.

For what concerns the $\mathcal{O}(1/q^+)$ term, we already encountered such behavior at LO, where it gave rise to the asymptotic mass. In the photon case[14] it was found, rather surprisingly, that the NLO contribution to this term amounted to replacing the quark asymptotic mass m_∞^2 with $m_\infty^2 + \delta m_\infty^2$, where δm_∞^2 is its soft $\mathcal{O}(g)$ correction, in the fermionic equivalent of Eq. (71) and then expanding to linear order in δm_∞^2 (and hence in g). The explicit sum-rule computation[16] of the diagrams in Fig. 10 yields the same (up to the different asymptotic mass), i.e.

$$\frac{m_\infty^2 + \delta m_\infty^2}{q_\perp^2 + m_\infty^2 + \delta m_\infty^2} = \frac{m_\infty^2}{q_\perp^2 + m_\infty^2} + \delta m_\infty^2 \frac{q_\perp^2}{(q_\perp^2 + m_\infty^2)^2} + \mathcal{O}(g^2), \tag{88}$$

where δm_∞^2 is given in Eq. (84). This result can be interpreted physically in the following way: expanding for large (complex) q^+ takes the soft fields to approach their collinear limit, where the only effect of resummation is to introduce an asymptotic mass (at leading order) and the soft correction thereto (at next-to-leading order). For this same reason, HTL vertices become small and not relevant.

Hence we obtain

$$\delta\hat{q}_L = g^2 C_R T \int \frac{d^2 q_\perp}{(2\pi)^2} \frac{q_\perp^2\, \delta m_\infty^2}{(q_\perp^2 + m_\infty^2)^2} = \frac{g^2 C_R T \delta m_\infty^2}{4\pi} \left[\ln\left(\frac{(\mu^{\mathrm{NLO}})^2}{m_\infty^2} \right) - 1 \right], \tag{89}$$

where we have introduced a regulator μ^{NLO}. As we will show, the semi-collinear region will remove the dependence on it, so that it should be taken to obey $gT \ll \mu^{\mathrm{NLO}} \ll \sqrt{g}T$.

5.3. *The semi-collinear region*

As we anticipated before, semi-collinear processes can be seen as $1 \leftrightarrow 2$ splitting processes where the opening angle (and hence the virtuality) are larger. Two examples are drawn in Fig. 11. The scalings of this region are as follows: $K \sim gT$ is soft,

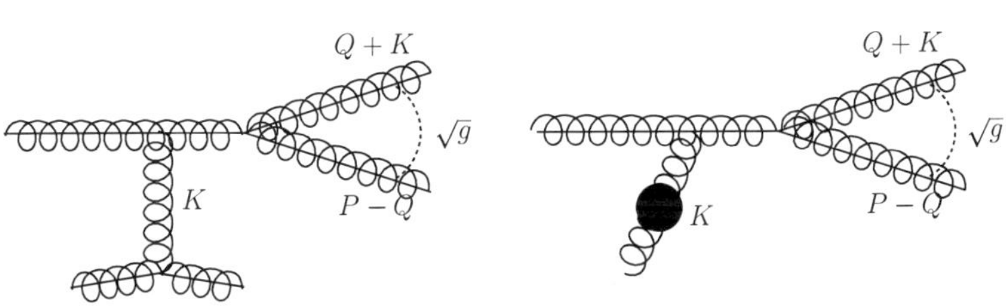

Fig. 11. Diagrams for two typical semi-collinear processes. In the first case the soft gluon is in the spacelike Landau cut, whereas in the second case it is on its timelike plasmon pole, represented by the black blob.

whereas the two final-state particles are collinear, albeit with an increased virtuality and opening angle with respect to the collinear sector. The leading contribution then comes from $q^+ \sim T$, $q^- \sim gT$, $q_\perp^2 \sim gT^2$, $Q^2 \sim gT^2$.

Naive power-counting arguments would suggest that the semi-collinear region should contribute to leading order, as it is the largest slice of phase space where a soft gluon can attach to a $1 \leftrightarrow 2$ process. However, once all diagrams are summed and squared, a cancellation, first noticed in the context of photon radiation,[30] introduces an extra $\mathcal{O}(g)$ suppression. Furthermore, since $K \sim gT$ in all components, the contribution from timelike soft gluons, e.g. plasmons, is now allowed. This is contrasted by the collinear region, where kinematics enforce $k^- \sim \delta E \sim g^2T \ll k^+, k_\perp$, thus restricting soft gluons to the space-like domain only.

The contribution $\delta C_{\text{semi-coll}}$ to the collision operator can be written in the same way as the collinear one, as given by Eq. (42), with the replacement of the collinear rates with semi-collinear ones. In pure glue it reads[16]

$$\delta C_{\text{semi-coll}}[\mu] = \int_{-\infty}^{+\infty} d\omega \left[f_{p+\omega} \frac{d\Gamma(p+\omega,\omega)}{d\omega}\bigg|_{\text{semi-coll}} - f_p \theta(p-2\omega) \frac{d\Gamma(p,\omega)}{d\omega}\bigg|_{\text{semi-coll}} \right], \tag{90}$$

The derivation of the semi-collinear rates then requires the evaluation of processes of the form of Fig. 11, with $p, q^+ \gg q_\perp \gg k_\perp, k^+$. This results in the following factorized form[16]

$$\frac{d\Gamma(p,\omega)}{d\omega}\bigg|_{\text{semi-coll}} = \frac{g^2 C_A}{\pi p} \frac{(1-x+x^2)^3}{x(1-x)} \times (1+n(\omega))(1+n(p-\omega)) \int \frac{d^2q_\perp}{(2\pi)^2} \frac{\hat{q}(\delta E)}{q_\perp^4} \tag{91}$$

where

$$\delta E = \frac{p q_\perp^2}{2\omega(p-\omega)}, \tag{92}$$

and $\hat{q}(\delta E)$, which was first introduced for photon radiation,[14] is a modified version of $\hat{q}$ which keeps track of the changes in the small light-cone components p^- and q^-, induced by the interaction with the soft degrees of freedom.

In the field-theoretical language of Sec. 2, $\hat{q}(\delta E)$ is given by[v]

$$\hat{q}(\delta E) = \frac{g^2 C_R}{d_A} \int_{-\infty}^{\infty} dx^+ e^{ix^+\delta E} \langle v^\mu F_\mu{}^{\nu,a}(x^+) U_A^{ab}(x^+,0) v^\rho F^b_{\rho\nu}(0) \rangle, \tag{93}$$

where we have made use of the simplification discussed in Eq. (85). The dependence on the minus components appears as a simple phase accumulation during the eikonal

[v] Strictly speaking, this is the version that applies to photon radiation and in this case we would need a more complicated three-pole expression, along the lines of the discussion that followed Eq. (43). However, we are evaluating it at the leading-order, one-gluon exchange level, so that the more complicated form is not necessary.

propagation, so that one can also easily see how Eq. (93) goes into Eq. (26) for $\delta E \to 0$. Equation (93) can also be evaluated using Euclidean techniques, yielding[14]

$$\hat{q}(\delta E) = g^2 C_R T \int \frac{d^2k_\perp}{(2\pi)^2} \left[\frac{m_D^2 k_\perp^2}{(k_\perp^2 + \delta E^2)(k_\perp^2 + \delta E^2 + m_D^2)} + \frac{2\delta E^2}{k_\perp^2 + \delta E^2} \right]. \quad (94)$$

As we have anticipated at the beginning of this section, a counterterm from the large-angle scattering region needs to be subtracted, as well as a collinear one (see footnote r). The former arises from taking the $m_D \to 0$ limit of Eq. (94) (there is no resummation taking place on external legs at LO) and the latter by taking the $\delta E \to 0$ limit (changes in the minus component are not accounted for in the collinear limit). Once these subtractions are performed, the semi-collinear rate becomes

$$\left.\frac{d\Gamma(p,\omega)}{d\omega}\right|_{\text{semi-coll}} = \frac{g^4 C_A^2 T}{\pi p}(1+n(\omega))(1+n(p-\omega)) \int \frac{d^2q_\perp}{(2\pi)^2} \int \frac{d^2k_\perp}{(2\pi)^2} \frac{1}{q_\perp^4} \times \frac{(1-x+x^2)^3}{x(1-x)} \left[\frac{m_D^2 k_\perp^2}{(k_\perp^2 + \delta E^2)(k_\perp^2 + \delta E^2 + m_D^2)} - \frac{m_D^2}{k_\perp^2 + m_D^2} \right]. \quad (95)$$

We stress that the semi-collinear collision operator is given by Eq. (90).

The $q_\perp$ integration in Eq. (95) is to be understood as IR-regulated by μ^{NLO}. One can show[16] how the small-ω-and-$q_\perp$ region gives rise to IR logarithms that cancel the μ^{NLO} dependence of diffusion processes. Details of how the transverse integrations can be carried out analytically are also given there. The ω integration remains to be performed numerically.

6. Summary and Conclusions

In this review we have shown how at leading and next-to-leading order the propagation, energy loss, and momentum diffusion of high-energy particles in the QGP can be cast in the form of a Boltzmann equation describing the interactions between the jet particles and the hard and soft constituents of the plasma (Eqs. (1) and (79)). In Sec. 2 we introduced this formalism at leading order, where three processes contribute. These processes include $2 \leftrightarrow 2$ scatterings of the jet with the hard particles, drag and (longitudinal and transverse) momentum diffusion induced by the soft highly occupied gluonic background, and collinear bremsstrahlung (and merging) induced by the transverse momentum diffusion. The $2 \leftrightarrow 2$ scatterings are dealt with in a standard kinetic fashion (Eq. (10)), the collinear splittings and joinings give rise to a somewhat simpler rate term (Eq. (42)), and the interactions with the soft background are described by a Fokker–Planck equation (Eq. (21)) or equivalently by a Langevin equation (Eq. (23)).

We have furthermore shown how the contributions of the soft degrees of freedom to the kinetic picture (the diffusion coefficients in the Fokker–Planck equation and the transverse scattering kernel in the collinear rate) can be cleanly defined as correlators of Wilson lines, or Wilson loops, supported on the lightcone or on a light front (see Eqs. (26), (27) and (39)). These Wilson lines are to be evaluated using the HTL effective theory. We devote Sec. 3 to the illustration of the recent understanding of thermal field theory (including HTLs) on the lightcone. We show how the causal properties of amplitudes at light-like separations bring about a dramatic simplification in their calculation.[12] To put it simply, jet and hard particles, which move at $v \sim c$, probe the medium in a very simple (thermodynamical) way, since the soft background "does not have the time" to respond to the perturbation they cause. At a practical level, we show how the computation of certain lightlike correlators, such as transverse momentum diffusion, "euclideanizes", that is, it reduces to a much simpler calculation in the three-dimensional Euclidean theory EQCD (see Eqs. (64) and (65)). We also show how a different set of lightlike correlators, such as longitudinal diffusion, does not euclideanize. This case also simplifies, however, and the same causal properties on the lightcone make it sensitive only to the thermal mass and soft corrections thereto (see Eqs. (70) and (71)).

In Secs. 4 and 5 we reviewed two major consequences of these lightcone simplifications. First, we showed how operators that euclideanize can be measured on the lattice. Second, we reviewed the consequences on perturbation theory. At leading order the lightcone simplifications result in simple analytical closed forms in place of relatively straightforward numerical integrals. At NLO the impact is much more far-reaching — calculations that would have been extremely intricate, if not impossible, can now be performed with relative ease.

In more detail, in Sec. 4 we reviewed the basics of EQCD, and described the first non-perturbative calculations of $\hat{q}$ and the transverse momentum scattering kernel using this effective theory.[13] We would like to emphasize how these first results open a new avenue of research — all other Euclidean operators can be measured on the lattice in the same way. This creates the tantalizing possibility of a factorized approach to kinetics, where perturbation theory is used at the hard scale to compute the $2 \leftrightarrow 2$ scatterings and the splitting processes, while the 3D lattice is used at the soft and ultrasoft scales to compute the diffusion processes and the transverse scatterings leading to bremsstrahlung.

In Sec. 5 we give an overview of the NLO generalization of the LO Boltzmann equation. We showed how the three processes at LO are corrected at NLO (see Eq. (79)). Besides the corrections to the LO diffusion and collinear processes, a new *semi-collinear* process enters, which appears as a bridge between the others, being collinear but with relaxed constraints, i.e. going beyond strict collinearity. In the evaluation of each of these processes we sketched how the lightcone simplifications described above make the NLO analysis possible. We described the NLO calculations of transverse[12] and longitudinal[16] momentum diffusion (Eq. (87) and Eq. (89)), and introduced two related Euclidean operators: the condensate Z_g responsible for the thermal mass (Eq. (80)), and $\hat{q}(\delta E)$, a modified version of $\hat{q}$ relevant for the semicollinear bremsstrahlung (Eq. (93)). The corrections to the longitudinal diffusion and drag coefficients at NLO are due to corrections to the thermal mass (or Z_g). Finally, we presented complete expressions for the NLO rates in the Boltzmann equation for pure gauge theory.

The extension of the NLO Boltzmann equation to include light quarks requires one extra process at leading- and next-to-leading order, the *conversion process*, which can be seen as a fermionic analogue of diffusion.[16] The interaction of hard or jet particles with soft quarks results in a conversion from a quark to a gluon with approximately the same momentum and vice versa. The computation of these rates involve fermionic correlators on the light-cone. It turns out that the lightcone simplifications of the fermionic HTLs are strikingly similar to the longitudinal diffusion described in Subsec. 3.3, and consequently the conversion rates are sensitive only to the (fermionic) asymptotic mass and its soft corrections. These corrections to the fermion mass are computable with the 3D effective theory, in much the way that the effective theory can be used to compute corrections to Z_g.

One very important question that we leave to future works is the impact of these developments on calculations of jet energy loss and their comparison to experimental data. We remark that the Monte Carlo event generator MARTINI[75] currently implements a kinetic approach that is very close to the leading-order picture sketched in Sec. 2. It is then an ideal candidate for the inclusion of the developments discussed in Secs. 4 and 5. The reorganization in terms of $2 \leftrightarrow 2$ scattering, diffusion and collinear processes, as well as the implementation of the NLO corrections discussed in Sec. 5, is already underway. This can be easily complemented by the inclusion of non-perturbative input. Besides the measured transverse scattering kernel, lattice calculations for $\hat{q}(\delta E)$ and m_∞^2 could easily be included into MARTINI, should they become available in the future. It would also be interesting to see the results of this numerical implementation for the angular structure of jets, for which recent order-of-magnitude estimates from perturbation theory are available.[76]

As a general remark, we find it difficult to tell *a priori* the impact the NLO corrections of Sec. 5 will have. The recent calculation to NLO of the thermal photon rate,[14] (which includes many of the ingredients presented here, such as Euclideanizations, light-cone causality, collinear and semi-collinear processes) showed how the NLO corrections naturally divide into two large, and largely canceling, contributions — see Fig. 12. Clearly the NLO corrections are modest over a significant range of photon momentum. The positive contribution arose from collinear processes enhanced by a NLO increase of the scattering kernel, and a NLO decrease of the thermal mass, while the negative contribution arose from the semi-collinear and NLO soft processes. The large cancellation between these two corrections is largely accidental and is dependent on the details of the medium, such as the number of colors and flavors. Nevertheless, we anticipate a similar cancellation for the current more complicated case of parton energy loss.

Finally, we note that the kinetic approach we have outlined is often used to compute the transport coefficients of QCD plasmas.[22,77] While an extension to NLO of these calculations requires more than what is outlined in Sec. 5, a computation of transport coefficients in thermal QCD beyond leading order will almost certainly

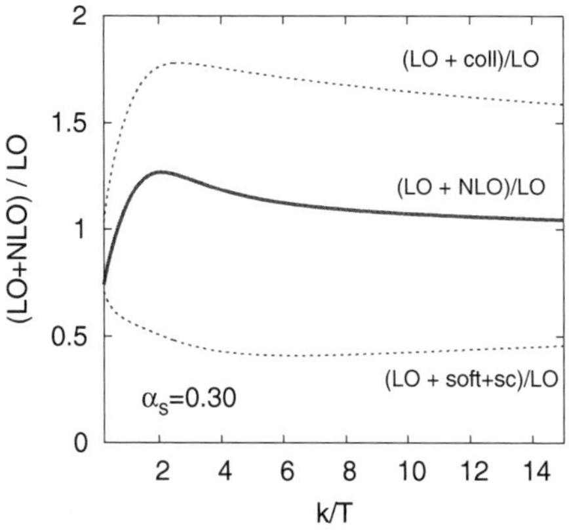

Fig. 12. The differential photon emission rate $d\Gamma_\gamma/dk$ relative to the leading order rate as a function of the photon momentum k/T at a fixed coupling $\alpha_s = 0.3$ in a $N_f = 3$ QGP. The full next-to-leading order rate[14] (LO + NLO) is a sum of the leading order rate (LO), a collinear correction (coll), and a soft+semi-collinear correction (soft + sc). The dashed curve labeled LO + coll shows the ratio of rates when only the collinear correction is included, with the analogous notation for the LO + soft + sc curve. Figure taken from the original reference.[14]

make considerable use of the euclidean light-cone simplifications described in this review.

Acknowledgments

We would like to thank Guy Moore for collaboration in Ref. 16. We also thank Simon Caron-Huot, Aleksi Kurkela, Mikko Laine and Marco Panero for useful conversations. This work was supported in part by the Swiss National Science Foundation (SNF) under grant 200020_155935 and a grant from the U.S. Department of Energy, DE-FG-02-08ER4154.

References

1. G. Roland, K. Safarik and P. Steinberg, *Prog. Part. Nucl. Phys.* **77** (2014) 70–127.
2. D. d'Enterria, Jet quenching, in *Landholt-Boernstein* **23** (2009) 471.
3. A. Majumder and M. Van Leeuwen, *Prog. Part. Nucl. Phys. A* **66** (2011) 41–92.
4. Y. Mehtar-Tani, J. G. Milhano and K. Tywoniuk, *Int. J. Mod. Phys. A* **28** (2013) 1340013.
5. R. Baier, D. Schiff and B. Zakharov, *Ann. Rev. Nucl. Part. Sci.* **50** (2000) 37–69.
6. G. Qin and X.-N. Wang, Jet quenching in heavy-ion collisions, this volume (2015).
7. J.-P. Blaizot and Y. Mehtar-Tani, Color decoherence and jet structure inside the QGP, this volume (2015).
8. E. Braaten and R. D. Pisarski, *Nucl. Phys. B* **337** (1990) 569.
9. E. Braaten and R. D. Pisarski, *Phys. Rev. D* **45** (1992) 1827–1830.
10. J.-P. Blaizot and E. Iancu, *Phys. Rept.* **359** (2002) 355–528.
11. P. B. Arnold, G. D. Moore and L. G. Yaffe, *JHEP* **0301** (2003) 030.
12. S. Caron-Huot, *Phys. Rev. D* **79** (2009) 065039.
13. M. Panero, K. Rummukainen and A. Schäfer, *Phys. Rev. Lett.* **112** (2014) 162001.

14. J. Ghiglieri, J. Hong, A. Kurkela, E. Lu, G. D. Moore and D. Teaney, *JHEP* **1305** (2013) 010.
15. J. Ghiglieri and G. D. Moore, *JHEP* **1412** (2014) 29.
16. J. Ghiglieri, G. D. Moore and D. Teaney, Jet-Medium interactions at NLO in a weakly-coupled quark-gluon plasma, arXiv:1509.07773.
17. O. K. Kalashnikov and V. V. Klimov, *Sov. J. Nucl. Phys.* **31** (1980) 699.
18. H. A. Weldon, *Phys. Rev. D* **26** (1982) 1394.
19. J.-P. Blaizot, A. Ipp, A. Rebhan and U. Reinosa, *Phys. Rev. D* **72** (2005) 125005.
20. S. Caron-Huot, *Phys. Rev. D* **79** (2009) 125002.
21. B. Svetitsky, *Phys. Rev. D* **37** (1988) 2484–2491.
22. P. B. Arnold, G. D. Moore and L. G. Yaffe, *JHEP* **0011** (2000) 001.
23. J. Hong and D. Teaney, *Phys. Rev. C* **82** (2010) 044908.
24. P. B. Arnold, *Phys. Rev. E* **61** (2000) 6091–6098.
25. P. B. Arnold, *Phys. Rev. E* **61** (2000) 6099–6102.
26. P. B. Arnold and W. Xiao, *Phys. Rev. D* **78** (2008) 125008.
27. C. Kittel, *Elementary Statistical Physics*, Dover Books on Physics Series (Dover Publications, 2004). ISBN 9780486435145.
28. P. B. Arnold, *Phys. Rev. D* **79** (2009) 065025.
29. R. Baier, A. H. Mueller, D. Schiff and D. Son, *Phys. Lett. B* **502** (2001) 51–58.
30. P. B. Arnold, G. D. Moore and L. G. Yaffe, *JHEP* **0111** (2001) 057.
31. P. B. Arnold, G. D. Moore and L. G. Yaffe, *JHEP* **0206** (2002) 030.
32. F. D'Eramo, H. Liu and K. Rajagopal, *Phys. Rev. D* **84** (2011) 065015.
33. M. Benzke, N. Brambilla, M. A. Escobedo and A. Vairo, *JHEP* **1302** (2013) 129.
34. S. Jeon and G. D. Moore, *Phys. Rev. C* **71** (2005) 034901.
35. R. Baier, Y. L. Dokshitzer, S. Peigne and D. Schiff, *Phys. Lett. B* **345** (1995) 277–286.
36. R. Baier, Y. L. Dokshitzer, A. H. Mueller, S. Peigne and D. Schiff, *Nucl. Phys. B* **483** (1997) 291–320.
37. B. Zakharov, *JETP Lett.* **63** (1996) 952–957.
38. B. Zakharov, *JETP Lett.* **65** (1997) 615–620.
39. P. B. Arnold, D. T. Son and L. G. Yaffe, *Phys. Rev. D* **59** (1999) 105020.
40. D. F. Litim and C. Manuel, *Phys. Rept.* **364** (2002) 451–539.
41. S. Caron-Huot, *JHEP* **0904** (2009) 004.
42. E. Braaten, R. D. Pisarski and T.-C. Yuan, *Phys. Rev. Lett.* **64** (1990) 2242.
43. S. Caron-Huot and G. D. Moore, *JHEP* **0802** (2008) 081.
44. P. B. Arnold, G. D. Moore and L. G. Yaffe, *JHEP* **0112** (2001) 009.
45. V. Alves, A. K. Das and S. Perez, *Phys. Rev. D* **66** (2002) 125008.
46. H. A. Weldon, *Phys. Rev. D* **67** (2003) 128701.
47. H. A. Weldon, *Phys. Rev. D* **67** (2003) 128701.
48. P. Aurenche, F. Gelis and H. Zaraket, *JHEP* **0205** (2002) 043.
49. E. Braaten and M. H. Thoma, *Phys. Rev. D* **44** (1991) 1298–1310.
50. D. Besak and D. Bödeker, *JCAP* **1203** (2012) 029.
51. S. Peigne and A. Peshier, *Phys. Rev. D* **77** (2008) 014015.
52. M. Laine and A. Rothkopf, *JHEP* **1307** (2013) 082.
53. A. Majumder, *Phys. Rev. C* **87** (2013) 034905.
54. E. Braaten, *Phys. Rev. Lett.* **74** (1995) 2164–2167.
55. E. Braaten and A. Nieto, *Phys. Rev. D* **51** (1995) 6990–7006.
56. E. Braaten and A. Nieto, *Phys. Rev. D* **53** (1996) 3421–3437.
57. K. Kajantie, M. Laine, K. Rummukainen and M. E. Shaposhnikov, *Nucl. Phys. B* **458** (1996) 90–136.
58. K. Kajantie, M. Laine, K. Rummukainen and M. E. Shaposhnikov, *Nucl. Phys. B* **503** (1997) 357–384.

59. A. D. Linde, *Phys. Lett. B* **96** (1980) 289.
60. A. Hietanen, K. Kajantie, M. Laine, K. Rummukainen and Y. Schroder, *Phys. Rev. D* **79** (2009) 045018.
61. M. Laine, Electroweak phase transition beyond the standard model, *Strong and Electroweak Matter 2000 Proceedings* (2000) pp. 58–69.
62. M. Laine, G. Nardini and K. Rummukainen, *JCAP* **1301** (2013) 011.
63. M. Laine and O. Philipsen, *Phys. Lett. B* **459** (1999) 259–264.
64. M. Laine, *Eur. Phys. J. C* **72** (2012) 2233.
65. K. Kajantie, M. Laine, K. Rummukainen and Y. Schröder, *Phys. Rev. D* **67** (2003) 105008.
66. M. Laine and Y. Schroder, *JHEP* **0503** (2005) 067.
67. S. Borsanyi, G. Endrodi, Z. Fodor, S. Katz and K. Szabo, *JHEP* **1207** (2012) 056.
68. M. D'Onofrio, A. Kurkela and G. D. Moore, *JHEP* **1403** (2014) 125.
69. J. Casalderrey-Solana, D. C. Gulhan, J. G. Milhano, D. Pablos and K. Rajagopal, *JHEP* **1410** (2014) 19.
70. B. Brandt, A. Francis, M. Laine and H. Meyer, *JHEP* **1405** (2014) 117.
71. P. Aurenche, F. Gelis, G. Moore and H. Zaraket, *JHEP* **0212** (2002) 006.
72. I. Ghisoiu and M. Laine, *JHEP* **1410** (2014) 83.
73. T. Liou, A. Mueller and B. Wu, *Nucl. Phys. A* **916** (2013) 102–125.
74. J.-P. Blaizot, F. Dominguez, E. Iancu and Y. Mehtar-Tani, *JHEP* **1406** (2014) 075.
75. B. Schenke, C. Gale and S. Jeon, *Phys. Rev. C* **80** (2009) 054913.
76. A. Kurkela and U. A. Wiedemann, *Phys. Lett. B* **740** (2014) 172–178.
77. P. B. Arnold, G. D. Moore and L. G. Yaffe, *JHEP* **0305** (2003) 051.

Jet Quenching in High-Energy Heavy-Ion Collisions

Guang-You Qin* and Xin-Nian Wang*,†

*Institute of Particle Physics and Key Laboratory, of Quark and Lepton Physics (MOE), Central China Normal University, Wuhan, 430079, P. R. China

†Nuclear Science Division, MS 70R0319, Lawrence Berkeley National Laboratory, Berkeley, CA 94720, USA

†xnwang@lbl.gov

Jet quenching in high-energy heavy-ion collisions can be used to probe properties of hot and dense quark–gluon plasma. We provide a brief introduction to the concept and framework for the study of jet quenching. Different approaches and implementation of multiple scattering and parton energy loss are discussed. Recent progresses in the theoretical and phenomenological studies of jet quenching in heavy-ion collisions at RHIC and LHC are reviewed.

1. Introduction

One of the main goals of ultra-relativistic nucleus–nucleus collisions, such as those performed at the Relativistic Heavy-Ion Collider (RHIC) at Brookhaven National Laboratory (BNL) and the Large Hadron Collider (LHC) at European Organization for Nuclear Research (CERN), is to create the hot and dense quark–gluon plasma (QGP) and study its various novel properties. QCD jets produced from early stage collisions of beam quarks and gluons from two nuclei play an essential role in studying transport properties of the QGP produced in these energetic collisions. During their propagation through the hot and dense medium, the interaction between hard jets and the colored medium will lead to parton energy loss (jet quenching).[1–3] There have been many striking experimental signatures of jet energy loss observed at RHIC and the LHC, such as the suppression of inclusive hadron spectra at high transverse momentum (p_T),[4–6] the modification of back-to-back hadron–hadron and direct photon–hadron correlations,[7–10] and the modification of reconstructed jets[11–14] and jet substructure,[15–17] as compared to the expectations from elementary proton–proton collisions.

During the last decades, extensive efforts have been devoted to the study of jet quenching phenomena in ultra-relativistic heavy-ion collisions. The theoretical study of jet quenching in a dynamically evolving QGP in heavy-ion collisions involves a variety of components, such as the initial jet production spectrum, a realistic description of the density profile of the medium, the propagation of hard jets in the medium, and the hadronization (fragmentation) of jet partons. Production rates for hard jets may be calculated within the framework of perturbative QCD due to the hard scales involved in the jet production. The medium density profiles are usually obtained from relativistic hydrodynamic calculations which have been very successful in describing the dynamical evolution of the bulk matter and especially in explaining the observed anisotropic collective flows.[18–22] The central topic of jet quenching in heavy-ion collisions is concentrated on the detailed mechanisms for the interaction of energetic partons with the hot and dense nuclear medium, and the manifestation of medium modification of jets in the final state observables.

2. Jet Quenching: General Framework

Due to large scales involved, the production of jets and the interaction of jets with dense nuclear matter may be studied within the framework of perturbative QCD. In perturbative QCD, processes involving large momentum transfer, such as the production of high transverse momentum hadrons, can be described with controlled uncertainties as the convolution of the incoming parton distribution functions (PDFs), hard partonic scattering process, and the final state fragmentation function (FFs), as illustrated Fig. 1. The cross-section of inclusive hadron production at high p_T can be obtained as follows:

$$d\sigma_{pp\to hX} \approx \sum_{abjd} \int dx_a \int dx_b \int dz_j f_{a/p}(x_a, \mu_f) \otimes f_{b/p}(x_b, \mu_f) \otimes d\sigma_{ab\to jd}(\mu_f, \mu_F, \mu_R) \otimes D_{j\to h}(z_j, \mu_F), \tag{1}$$

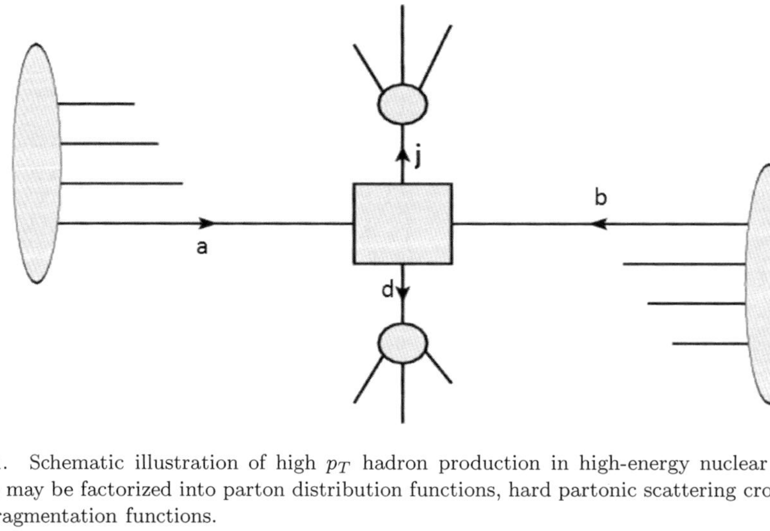

Fig. 1. Schematic illustration of high p_T hadron production in high-energy nuclear collisions which may be factorized into parton distribution functions, hard partonic scattering cross-section and fragmentation functions.

where $x_a = p_a/P_A$, $x_b = p_b/P_B$ are the initial momentum fractions carried by the interacting partons, $z_j = p_h/p_j$ is the momentum fraction carried by the final observed hadron. $f_{a/p}(x_a, \mu_f)$ and $f_{b/p}(x_b, \mu_f)$ are two parton distribution functions, $d\sigma_{ab\to jd}(\mu_f, \mu_F, \mu_R)$ is the differential cross-section for parton scattering process, and $D_{j\to h}(z_j, \mu_F)$ is the fragmentation function for parton j to hadron h. There are three different scales involved in the calculation: μ_f and μ_F are factorization scales and μ_R is the renomalization scale; they are usually taken to be the same $\mu_f = \mu_F = \mu_R$ as a typical hard scale Q involved in the process, such as the hadron p_T. The non-perturbative PDFs and FFs are universal, and their scale dependences obey the Dokshitzer–Gribov–Lipatov–Altarelli–Parisi (DGLAP) evolution equations.[23–25] They are usually determined by the global fit to the experimental data from elementary collisions, such as e^+e^- experiments, deep inelastic scatterings and proton–proton collisions.

The above formula is written at the leading order (LO) perturbative QCD. For LO calculation of single inclusive hadron production in proton–proton collisions, one needs LO $2 \to 2$ hard cross-sections, LO resummed PDFs and FFs, and one-loop expression for the strong coupling constant α_s. Since hard cross-sections at LO only have $2 \to 2$ scattering processes, there is no dependence on the factorization scale to compensate the scale dependence in resummed PDFs and FFs. This situation can be greatly improved when we go beyond LO and perform full next-to-leading order (NLO) calculation.

At NLO, one needs to calculate real $2 \to 3$ diagrams and $2 \to 2$ one-loop virtual diagrams for hard scattering cross-sections (see e.g., Ref. 26). Collinear singularities in both real $2 \to 3$ processes and virtual one-loop $2 \to 2$ processes are absorbed in PDFs and FFs at the factorization scales. Soft infrared singularities in real $2 \to 3$ processes are canceled by virtual one-loop $2 \to 2$ processes. The ultraviolet poles in virtual diagrams are removed by the renormalization of the coupling constant α_s at the renormalization scale. Therefore, for NLO calculation of single inclusive hadron production in proton–proton collisions, one needs, NLO hard cross-sections, NLO resummed PDFs and FFs, and two-loop expression for the strong coupling α_s.

Figure 2 shows the production cross-section for high p_T hadrons in proton–proton collisions at $\sqrt{s_{NN}} = 200\,\text{GeV}$, compared to NLO pertubative QCD calculations.[26] We can see that NLO perturbative QCD can describe the production of single inclusive hadrons at high p_T very well, which indicates that the hard jets in vacuum are under well control on both experimental and theoretical sides. This serves as the baseline for studying jet-medium interaction and the nuclear modification of jets in high-energy nucleus–nucleus collisions.

Often in many phenomenological studies (especially on jet quenching), LO expression is used with a multiplicative K factor introduced to mimic the NLO corrections to hard jet production. But a constant K-factor usually over-predicts the curvature of single inclusive hadron spectra, especially at small p_T. This may be partially corrected by the incorporation of the intrinsic k_T-smearing of partons which originates from initial state radiation. To perform this, one can introduce the

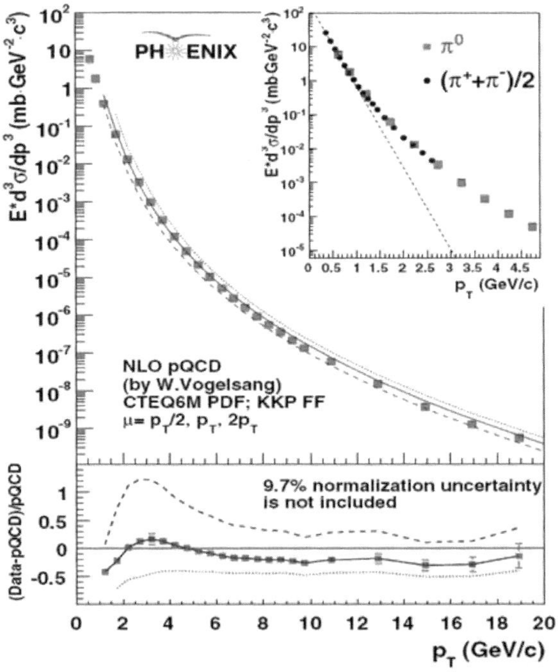

Fig. 2. (Color online) The cross-section for high p_T hadron production in proton–proton collisions at RHIC, compared to NLO perturbative QCD calculations. The figure is taken from Ref. 27.

so-called generalized parton distribution function:

$$\tilde{f}(x, k_\perp, Q^2) = f(x, Q^2) g(k_\perp), \tag{2}$$

where $g(k_T)$ is the intrinsic k_T smearing function, which is often taken to be Gaussian, $g(k_T) = e^{-k_T^2/\langle k_T^2\rangle}/(\pi\langle k_T^2\rangle)$. After taking into account the intrinsic k_T, the single inclusive hadron production can be calculated as:

$$d\sigma_{pp\to hX} \approx \sum_{abjd} \int dx_a \int dx_b \int dz_j \int d^2k_{T,a} \int d^2k_{T,b} \tilde{f}_{a/p}(x_a, k_{T,a}, Q^2)$$
$$\otimes \tilde{f}_{b/p}(x_b, k_{T,b}, Q^2) \otimes d\sigma_{ab\to jd}(Q^2) \otimes D_{h/j}(z_j, Q^2). \tag{3}$$

In the above equation, all the scales have been taken to be the same Q. Note that with the inclusion of intrinsic k_T, there will also be the modification of kinematics in addition to the integrals $\int d^2k_{T,a} \int d^2k_{T,b}$.[28,29]

When studying jet energy loss and jet quenching in ultra-relativistic heavy-ion collisions, two types of nuclear effects need to be taken into account. First, the PDF in nucleus $f_{a/A}$ is different from free proton PDF $f_{a/p}$ used in proton–proton collisions. Such effect is often called cold nuclear matter (CNM) effect. To take this into account, one may define the nuclear modification factor $R_i^A(x, Q^2)$ for the

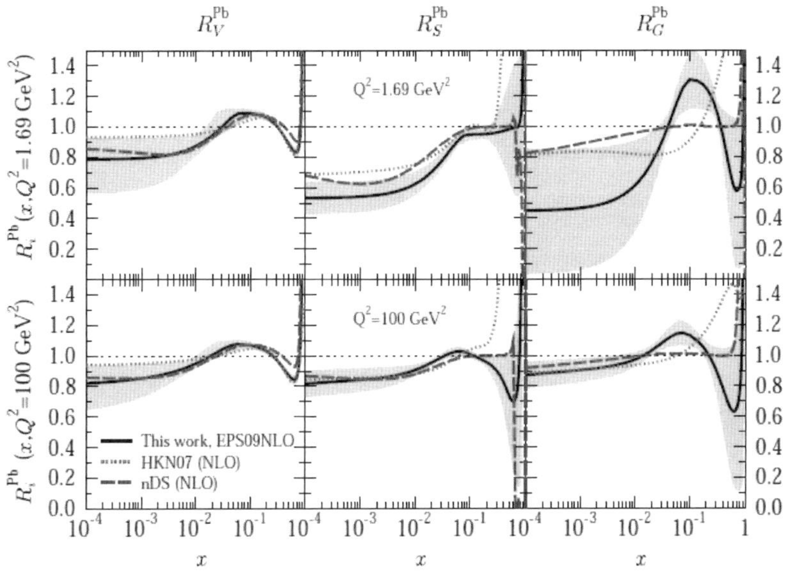

Fig. 3. (Color online) Nuclear modification factor for the valence and sea quark, and gluon distribution in Pb nucleus at $Q^2 = 1.69\,\text{GeV}^2$ and $Q^2 = 100\,\text{GeV}^2$ from EPS09NLO,[30] HKN07,[31] and nDS.[32] The figure is taken from Ref. 30.

parton distribution function:

$$R_a^A(x, Q^2) = \frac{f_{a/A}(x, Q^2)}{f_{a/p}(x, Q^2)}. \tag{4}$$

The PDF nuclear modification factor $R_i^A(x, Q^2)$ is often obtained by the global fit to the data from deep inelastic scattering off a nucleus, proton–nucleus and deuteron–nucleus collisions. Figure 3 shows the nuclear modification factors $R_i^A(x, Q^2)$ for the valence quarks, sea quarks, and gluons in lead (Pb) nucleus at $Q^2 = 1.69\,\text{GeV}^2$ (upper panels) and $Q^2 = 100\,\text{GeV}^2$ (lower panels) from EPS09NLO,[30] HKN07,[31] and nDS[32] parametrizations. One can see that the nuclear modification of parton distribution function generally has four different x regimes: shadowing, anti-shadowing, EMC-effect, and Fermi-motion.

The second effect in high-energy nucleus–nucleus collisions is the production of the hot and dense QGP medium. This is called hot nuclear matter effect and is schematically illustrated in Fig. 4. Partonic jets which are produced from the initial hard scattering processes have to travel through and interact with the hot and dense nuclear matter before fragmenting into final state hadrons which are observed by the detectors. By taking into account both cold and hot nuclear effects, the above factorized formula for single inclusive hadron production may be modified as

$$d\tilde{\sigma}_{AB\to hX} \approx \sum_{abjj'd} f_{a/A}(x_a) \otimes f_{b/B}(x_b) \otimes d\sigma_{ab\to jd} \otimes P_{j\to j'} \otimes D_{h/j'}(z_{j'}), \tag{5}$$

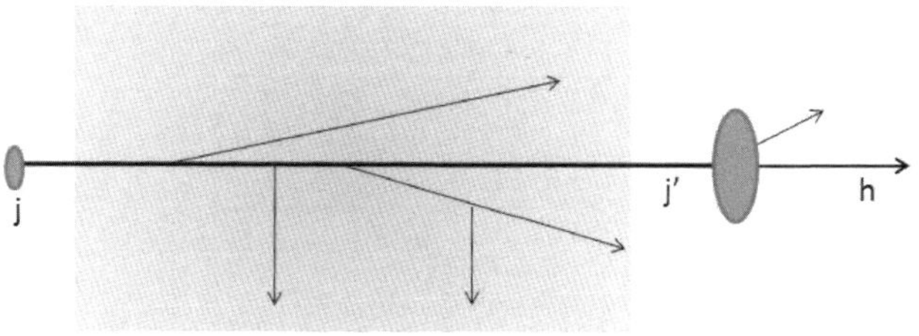

Fig. 4. (Color online) Schematic illustration of jet modification and parton energy loss in QGP medium before the fragmentation into hadrons.

where the additional piece $P_{j\to j'}$ describes the effect of the hard partons j interacting with the colored medium before fragmenting into hadrons. Here for brevity, the scale dependences of various quantities are not explicitly written. In phenomenological study, one often defines the so-called medium-modified FF for convenience, by convoluting the vacuum FF with the parton-medium interaction function,

$$\tilde{D}_{h/j}(z_j) \approx \sum_{j'} P_{j\to j'}(p_{j'}|p_j) \otimes D_{h/j'}(z_{j'}). \tag{6}$$

This is schematically shown in Fig. 4. One may also define the medium-modified parton cross-section by combining the vacuum cross-section with the parton-medium interaction function.

The above factorized formula has been widely used in phenomenological studies of jet quenching in ultra-relativistic heavy-ion collisions. However, currently there is no formal proof of such factorization yet. The assumption of factorization is motivated by the fact that jet energy loss observed in nucleus–nucleus collisions is a dynamical final state effect which occurs after the hard process. It is also consistent with various phenomenological jet quenching studies.

As has been mentioned, the interaction of hard jets and the dense nuclear medium usually leads to jet energy loss. One important observable is the suppression of the single inclusive hadron yield at high p_T in nucleus–nucleus collisions as compared to the expectation from proton–proton collisions. In order to quantify such nuclear modification effect, one may define the nuclear modification factor R_{AA}:

$$R_{AA}(p_T, y, \phi_p) = \frac{1}{N_{\rm coll}} \frac{\dfrac{dN_{AA}}{dp_T dy d\phi_p}}{\dfrac{dN_{AA}}{dp_T dy d\phi_p}}, \tag{7}$$

where $N_{\rm coll}$ is the number of binary collisions for a given collision centrality class. Similar nuclear modification quantities may be defined for the production of single inclusive jets, as well as jets/hadrons triggered by jets/hadrons or direct photons.

Figure 5 shows the nuclear modification factor R_{AA} as a function of p_T for charged hadrons, direct photons and other identified particles. The left shows PHENIX measurement for most central Au+Au collisions at $\sqrt{s_{NN}} = 200\,\text{GeV}$,

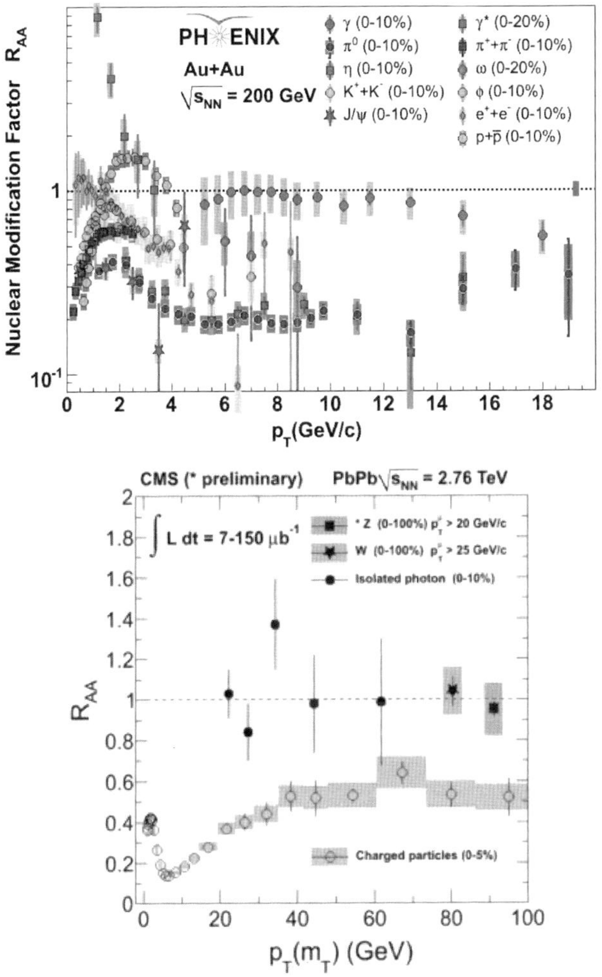

Fig. 5. (Color online) Upper: R_{AA} for identified hadrons, direct photons and non-photonic electrons measured by PHENIX in most central Au+Au collisions at $\sqrt{s_{NN}} = 200\,\text{GeV}$.[33] Lower: R_{AA} for charged hadrons, direct photons, and Z/W bosons measured by CMS in most central Pb+Pb collisions at $\sqrt{s_{NN}} = 2.76\,\text{TeV}$.[34]

and the right shows CMS measurements for most central Pb+Pb collisions at $\sqrt{s_{NN}} = 2.76\,\text{TeV}$. One can see that the yields of high p_T hadrons are strongly suppressed in nucleus–nucleus collisions compared to those in proton–proton collisions, whereas the nuclear modification factor R_{AA} for high p_T photons is consistent with unity. Since photons carry no color charge, they only interact with surrounding matter electromagnetically. Photons, once produced, will fly to the detectors without further rescattering due to the fact that their mean free paths are much larger than the medium size. This indicates that the observed strong suppression for high p_T hadron production is due to the final state effect, i.e., the interaction of partonic jets with the colored medium which usually causes jets to lose a fraction of their energy.

3. Parton Energy Loss Formalisms

Hard partonic jets may lose energy via a combination of both elastic and inelastic collisions with the constituents of the hot and dense nuclear medium. The energy loss of the primary hard parton occurred in $2 \to 2$ elastic collisions is usually called elastic or collisional energy loss. Multiple scatterings with the medium constituents may induce additional radiation which takes away a fraction of energy from the primary parton; this is called radiative energy loss. Figure 6 shows the typical diagrams for the calculation of jet energy loss originating from collisional (left panel) and radiative (right panel) processes.

3.1. *Radiative energy loss*

Radiative energy loss has been regarded as the most important component in studying parton energy loss in nuclear medium and jet quenching in relativistic heavy-ion collisions. Even in absence of a medium, high-p_T partons produced from the initial hard collisions between the incoming partons will undergo vacuum splitting processes and reduce their virtuality (offshellness). The presence of the hot and dense medium will modify the gluon radiation or parton splitting processes as compared to the vacuum case due to the rescatterings of the propagating parton shower with the medium constituents.

One important effect in the study of medium-induced gluon emission or parton splitting processes is the so-called Landau–Pomeranchuk–Migdal (LPM) effect.[35,36] For small angle or collinear radiation, the primary parton and the radiated gluon propagate along similar paths, and a finite time period (the formation time $\tau_f \sim 2\omega/k_\perp^2$, with ω and $k_\perp$ the energy and transverse momentum of the radiation), is required for the radiation process to complete. If the formation time τ_f of the radiation is larger than the mean free path λ of the propagating parton, the multiple scatterings experienced by the propagating parton can no longer be treated as

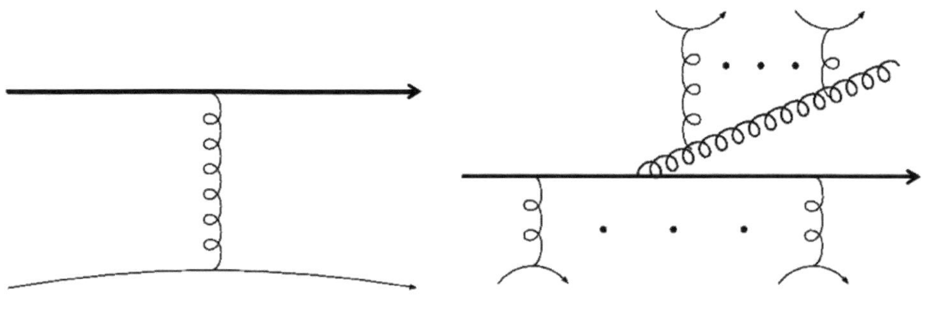

Fig. 6. Typical diagrams for collisional (left) and radiative (right) energy losses of an energetic parton propagating through a hot and dense nuclear medium.

independent. Such quantum interference between successive scatterings caused by the LPM effect will lead to the suppression of the radiation spectrum as compared to the Bethe–Heitler spectrum which assumes incoherent multiple scatterings. For the case of QCD, since the emitted gluons also carry color charge, their rescatterings with soft gluons in the medium will produce more dominant contribution to the modification of the radiation spectrum.

A number of parton energy loss formalisms have been developed to study medium-induced gluon bremsstrahlung and radiative process in dense nuclear medium, namely, Baier–Dokshitzer–Mueller–Peigne–Schiff–Zakharov (BDMPS-Z),[37–39] Gyulassy, Levai and Vitev (GLV),[40–42] Amesto–Salgado–Wiedemann (ASW),[43,44] Arnold–Moore–Yaffe (AMY)[45,46] and higher twist (HT)[47,48] formalisms. In the following, we provide some basic information about different jet quenching formalisms.

3.1.1. *Single gluon emission*

In most jet quenching formalisms, the starting point or the central goal is to calculate the single gluon emission kernel (e.g., the differential radiation spectrum $dN_g/d\omega dk_\perp^2 dt$ or its variants). In Sec. 4, we will provide more details about the single gluon emission kernel formula when presenting the phenomenological applications of jet quenching formalisms and the comparison to experimental data. In this section we only give a brief general description of different formalisms and focus on the similarity and differences among them. In the calculation of single gluon emission kernel, various approaches differ in two main aspects: the assumptions about the medium and the treatments of multiple scatterings per emission. A systematic comparison of different jet energy loss models has been performed in the context of a "brick" of QGP in Ref. 49. More details may be found there and the references therein.

BDMPS-Z and ASW-MS formalisms

In the Baier–Dokshitzer–Mueller–Peigne–Schiff–Zakharov (BDMPS-Z)[37–39] formalism, the calculation of gluon radiation process is formulated in terms of a path-integral in which the scatterings on multiple heavy static colored scattering centers are resummed. The interference between vacuum and medium-induced radiation in such a path-integral formulation is included later by Wiedemann.[43,44] Most analytical and numerical calculations from the BDMPS-Z formalism utilize a saddle point approximation, which assumes that the primary parton interacts with the medium mainly via multiple soft scattering processes. The numerical implementations of the multiple soft scattering version of the BDMPS-Z formalism are based on the work on calculating the quenching weights by Amesto, Salgado and Wiedemann[50] (usually denoted as ASW-MS). In such an implementation, the medium is fully characterized by the jet transport coefficient $\hat{q} = d\langle \Delta p_T^2 \rangle / dL$ which characterizes the mean of the transverse momentum squared exchanged between the propagating parton and the colored medium per unit path length.

DGLV and ASW-SH formalisms

The opacity expansion was developed by Gyulassy, Levai and Vitev (GLV)[40–42] and independently by Wiedemann.[43,44] Such approach is based on a systematic expansion in terms of the number of scatterings experienced by the propagating parton. The interference between the vacuum and medium-induced radiation is included in this formalism. In most phenomenological calculations, only the leading term ($N = 1$) in the expansion is included and usually called single hard scattering limit. The medium effect in this approach is characterized by two parameters: the density ρ of the scattering centers (or the mean free path λ of the propagating parton) and the Deybe screening mass μ_D which is introduced to regulate the infrared behavior of single scattering cross-section. The opacity expansion was first developed using heavy static scattering centers to characterize the medium, and the extension to the medium with dynamical scattering centers was later performed by Djordjevic and Heinz.[51] There are two different model implementations for this approach, namely, DGLV[51] and the single hard scattering limit of ASW formalism (ASW-SH).[50]

AMY formalism

Arnold–Moore–Yaffe (AMY) approach[45,46] is a quantum field theory formulation of parton energy loss in a weakly-coupled medium in thermal equilibrium. The medium that the hard propagating parton interacts is formulated as a thermal equilibrium state in the framework of Hard Thermal Loop (HTL) finite temperature field theory. The properties of the colored medium are characterized fully by its temperature. In phenomenological calculations, the model parameter is usually taken to be the strong coupling constant $\alpha_s = g_s^2/(4\pi)$. The AMY formalism was first developed assuming infinite length for the medium, and was recently improved to include the effect of finite medium length by Caron-Huot and Gale.[52] In this approach, the incoming parton is assumed to be almost on-shell, thus the vacuum radiation of incoming partons is not included, and only the medium-induced effect is taken into account. Since the perturbative description of the thermal QGP medium is applied, the AMY formalism in principle only applies at very high temperature.

HT formalism

Higher-twist parton energy loss formalism was pioneered by Guo and Wang.[47,48] In this approach, the properties of the medium enter into the calculation in terms of higher-twist matrix elements. In phenomenological applications, one usually factorizes out the nuclear PDFs of the incoming stuck parton from the higher-twist matrix elements, and the remainder of the matrix elements describes the interaction between the final state propagating parton and the dense nuclear medium. The formalism was first developed for single scattering and was later extended to include multiple scatterings per emission by Majumder.[53] Similar to the GLV formalism, most numerical implementations of the HT approach still make use of single scattering per emission approximation. However, the characterization of the medium in the HT scheme is a little different from GLV. While GLV takes the full functional form for the single scattering cross-section, HT takes the leading moment of the exchanged transverse momentum distribution. Therefore the medium in the HT formalism is solely characterized by the jet quenching parameter $\hat{q}$. Note that a recent study has shown that the stimulated radiation process may be modified by the longitudinal momentum exchange with the medium constituents (such as the longitudinal drag and diffusion).[54]

Despite many differences among various parton energy loss formalisms as has been mentioned above, there are three main common assumptions that are made in most calculations. The first is the eikonal approximation which assumes both the incoming parton energy E and radiated gluon energy ω are much larger than the transverse momentum $q_\perp$ exchanged with the medium constituents, i.e., $E, \omega \gg q_\perp$, so the partons are traveling along eikonal trajectories. Additionally, soft radiation approximation is often imposed in some approaches, so that the gluon energy ω is assumed to be much smaller the primary parton energy E, i.e., $x = \omega/E \ll 1$ (Note that such approximation is not made in the AMY calculation). The second common approximation is the collinear or small-angle radiation approximation which assumes that the transverse momentum $k_\perp$ of the radiated gluon is much smaller than its energy ω, i.e., $k_\perp \ll \omega$. The third approximation assumes that the momentum transfer between the hard parton and the medium is localized. This means that the mean free path λ of the propagating parton is much larger than the Debye screening length $\lambda_D = 1/\mu_D$. More extensive discussions on the effects of different approximations on the calculation of medium-induced gluon radiation and parton energy loss may be found in Ref. 49.

3.1.2. *Multiple gluon emission*

In general, multiple parton splitting may occur in medium, similar to the case of vacuum parton shower. A full calculation of multiple parton final state would include the interference between different emissions. There has been some effort made along this direction, such as the calculations of the interference between two emitters (the antenna problem),[55] but the generalization of these results to the case of multiple parton emission is still not clear at present. In all available phenomenological studies, a common practice to calculate multiple gluon emission is based on the repeated application of the single gluon emission kernel (as schematically illustrated in Fig. 7). In the following we will present different prescriptions used for the calculation of multiple gluon emission.

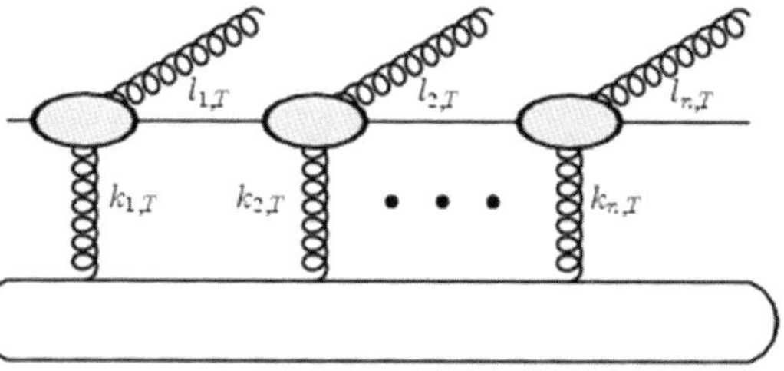

Fig. 7. Schematically illustration of multiple gluon emission reduced to the repeated application of single gluon emission kernel.

Poisson ansatz

One popular method to obtain multiple gluon emission is the use of Poisson convolution which assumes that the number of emitted gluons in a given path follows a Poison distribution. The mean of the Poisson distribution is obtained by the integral of single gluon emission kernel, while the energy distribution of each emitted gluon obeys the single gluon emission spectrum. This method has been widely applied in phenomenological studies using GLV and ASW formalisms.[50,56,57] The central quantity using Poisson ansatz is to calculate the probability distribution $P(\Delta E)$ of parton energy loss, which may be obtained from single gluon emission kernel $dN_g/d\omega$ as follows:

$$P(\Delta E) = \sum_{n=0}^{\infty} \frac{e^{-\langle N_g \rangle}}{n!} \left[\prod_{i=1}^{n} \int d\omega \frac{dN_g(\omega)}{d\omega}\right] \delta\left(\Delta E - \sum_{i=1}^{n} \omega_i\right), \tag{8}$$

where $dN/d\omega$ is the spectrum for single gluon emission and $\langle N_g \rangle = \int d\omega dN_g/d\omega$ is the average number of radiated gluons. In most jet energy loss calculations that use Poisson ansatz, the degrading of the primary parton energy during its propagation is not dynamically updated after each emission, therefore there could be finite probability for the total radiated energy larger than the incoming parent parton energy. In practice such finite probability is treated as the probability for jet to lose its total energy.

Transport models

Transport approach is another popular method in the study of parton evolution in medium. In the AMY formalism, the transport rate equations have been solved to obtain the time evolution of jet energy (momentum) distributions $f(p) = dN(p)/dp$.[58,59] The coupled rate equations can be schematically written as:

$$\frac{df(p,t)}{dt} = \int dk \left[f(p+k,t)\frac{d\Gamma(p+k,k,t)}{dkdt} - f(p,t)\frac{d\Gamma(p,k,t)}{dkdt}\right], \tag{9}$$

where $d\Gamma(p,k,t)/dkdt$ is the differential rate for a parton with momentum p to lose momentum k (the parton species may change and the indices for parton species are not explicitly written in the above generic formula). The above transport equation includes the coupling between quark and gluon distributions of the jets (f_q and f_g), and keep track of the change of parton energy during its propagation. By performing the $\int dk$ integration from $-\infty$ to ∞, the above rate equation also takes into account the absorption of thermal partons from the medium, thus preserve the detailed balance.

The transport approach for jet propagation in medium has also been recently implemented in a Linear Boltzmann Transport (LBT) model[60–62] which not only follows the propagation of jet shower partons but also keeps track of medium recoil partons so that energy and momentum are conversed throughout the transport processes. Within the LBT model, the propagation of jet shower partons and medium excitation is simulated according to a linearized Boltzmann equation

$$p_1 \cdot \partial f_1(p_1) = -\int dp_2 dp_3 dp_4 (f_1 f_2 - f_3 f_4)|M_{12\to 34}|^2 \times (2\pi)^4 \delta^4(p_1 + p_2 - p_3 - p_4), \tag{10}$$

where $dp_i = d^3p_i/[2E_i(2\pi)^3]$, $f_i = 1/(e^{p\cdot u/T} \pm 1)$ $(i = 2, 4)$ are parton phase-space distributions in a thermal medium with local temperature T and fluid velocity $u = (1, \vec{v})/\sqrt{1 - \vec{v}^2}$, and $f_i = (2\pi)^3\delta^3(\vec{p} - \vec{p_i})\delta^3(\vec{x} - \vec{x_i} - \vec{v_i}t)$ $(i = 1, 3)$ are the parton phase-space densities before and after scattering. For elastic scatterings, the amplitudes $|M_{12\to 34}|^2$ are given by the lowest order (LO) pQCD parton scattering processes. Partons are assumed to propagate along classical trajectories between two adjacent scatterings. The probability of scattering in each time step Δt is determined by,

$$P_a = 1 - \exp\left[-\Delta t \frac{p \cdot u}{E} \Gamma_a(p \cdot u, T)\right], \tag{11}$$

where Γ_a is the total scattering rate for parton type a with four-momentum p in the local fluid comoving frame. The backreaction in the Boltzmann transport is accounted for by subtracting initial thermal partons (p_2), denoted as "negative" partons, from the final observables after been transported also according to the Boltzmann equation. These negative partons are considered as part of the recoiled partons that are responsible for jet-induced medium excitations.[60] Induced radiation accompanying each elastic scattering is simulated according to the HT approach and multiple gluon emissions within each time step are included according to the Poisson distribution in Eq. (8).

Modified DGLAP evolution

HT formalism was developed using perturbative QCD power expansion. It solves the following DGLAP-like evolution equations to obtain the medium-modified fragmentation function $\tilde{D}(z, Q^2)$[63–66]:

$$\frac{\partial \tilde{D}(z, Q^2, q^-)}{\partial \ln Q^2} = \frac{\alpha_s}{2\pi} \int \frac{dy}{y} P(y) \int d\zeta^- K(\zeta^-, Q^2, q^-, y) \tilde{D}\left(\frac{z}{y}, Q^2, q^- y\right), \tag{12}$$

where q^- is the light-cone energy of the hard jet, y the fractional energy of the radiation, and ζ^- is the location of the jet. $P(y)$ is the vacuum splitting function and $K(\zeta^-, Q^2, q^-, y)$ is the kernel for the parton-medium scattering which induces additional radiation process (see Sec. 4 for its detailed expression). In DGLAP-like evolution, the coupling between quark and gluon fragmentation function (D_q and D_g) is included, and the change of parton energy during its propagation may be traced as well. The contribution from vacuum radiation is also implicitly included in the above equation.

There are some conceptual difference between the above three methods. AMY rate equations, LBT model and HT DGLAP-like evolution equations have two aspects of improvement as compared to the Poisson convolution used by BDMPS, GLV and ASW formalism, First, the evolution equations used by the AMY, LBT and HT formalisms keep track of all shower partons while the Poisson convolution only trace the lost energy of the leading parton. In addition, AMY, LBT and HT include the change of the emission probability due to the degrading of jet energy during its propagation, but the Poisson convolution does not. Thus the use of Poisson convolution may lead to finite probability to lose more energy than the parent parton (such issue may be fixed by using Monte Carlo method to calculate the probability distribution $P(\Delta E)$, as done in Ref. 66).

In spite of these differences, there is one important point that is not well addressed in all parton energy loss models: the gluon radiation process is not a purely local phenomenon, but has a finite formation time τ_f. Therefore, the evolution of jet energy loss should be accompanied at the same by an evolution in the coordinate space, i.e., as the jet energy degrades along its propagation, the remaining path length through the medium decreases as well. Many jet quenching models have taken into account the evolution of the medium and the corresponding change in the local medium information (such as the temperature, the energy/entropy density and the flow velocities), but the combination of the local information and the true finite size effect has not yet been included at present.

3.2. *Collisional energy loss*

Collisional energy loss of hard jets propagating through nuclear matter was first studied by Bjorken in 1982[1] and then by a few others in recent years.[57,59,67–71] Many calculations of collisional energy loss were performed in the framework of perturbative finite temperature field theory. Collisional energy loss may be treated in the same footing as radiative energy loss in the AMY formalism[59,71] and the LBT model.[60–62] In BDMPS-Z, GLV and ASW approaches, the medium is modeled as a collection of heavy static scattering centers, therefore by construction these formalisms neglect the recoil effects and thus conceptually does not allow for collisional energy loss for the propagating hard partons. In phenomenological application, collisional energy loss in these formalisms is added independently from radiative energy loss. It is noted that the GLV formalism has been extended to the medium with dynamical scattering centers (DGLV), thus both collisional and radiative energy loss may be formulated in the same setup (DGLV).[51,70] In the HT formalism, elastic collisions may be added as additional component characterized by the longitudinal drag and diffusion of the propagating hard partons.[64,72,73]

Compared to the radiative component of jet energy loss, collisions energy loss is usually considered to be small for light flavor (leading) partons, especially when the energy of the jet is sufficiently high.[57,59,71] However in realistic calculation of nuclear modification factor R_{AA} at RHIC and the LHC energies, collisional energy loss may give sizable contribution and cannot be simply neglected.[57,59] For heavy quarks, elastic collisions are usually considered as the dominant energy loss mechanism, especially at low and intermediate energy regimes,[68,75] due to the large finite masses of heavy quarks which tend to reduce the phase space of medium-induced gluon radiation (this is usually called the dead-cone effect[76]). When going to the high-energy regimes where heavy quarks become ultra-relativistic as well, collisional energy loss alone is no longer sufficient for simulating the in-medium evolution of heavy quarks, and radiative contribution becomes more significant.[77–80] This can be seen from Fig. 8 which shows the comparison between radiative and collisional energy loss: the left panel for light partons and the right for bottom quarks. We can see that for bottom quarks, below around 17 GeV collisional energy loss dominates; above this transition energy, the effect from the finite mass becomes small and radiative components starts to dominate.[74] As we will discuss in Sec. 5, collisional energy loss may play an important role in understanding the energy loss and nuclear modification of full jet in dense nuclear medium.[81] It is also an essential ingredient when studying the response of the medium to propagating hard jet[82,83] (see Sec. 6).

4. Recent Developments on Phenomenology and Theory: $\hat{q}$, etc.

The main purpose of jet quenching study is to understand the detailed mechanisms of jet-medium interaction and their manifestation in the final state observables. The ultimate goal is to extract various novel properties of the hot and dense QGP produced in ultra-relativistic heavy-ion collisions. To achieve this goal, one important strategy is to perform systematic phenomenological studies of jet quenching and compare to a wealth of experimental measurements of final state observables, such as single inclusive hadron production at high p_T, the reconstructed jets and their substructures, and hadrons/jets correlated with high-p_T hadrons/jets or direct photons, and so on.

Recently much attention of jet quenching study has been paid to the quantitative extraction of various jet transport coefficients, such as the transverse momentum diffusion rate $\hat{q}$, the elastic energy loss rate $\hat{e} = dE/dt$ and the diffusion of elastic energy loss $\hat{e}_2 = d(\Delta E)^2/dt$, etc.[37,72,73] At leading order, these parameters quantify the transverse and longitudinal momenta transferred between the propagating hard

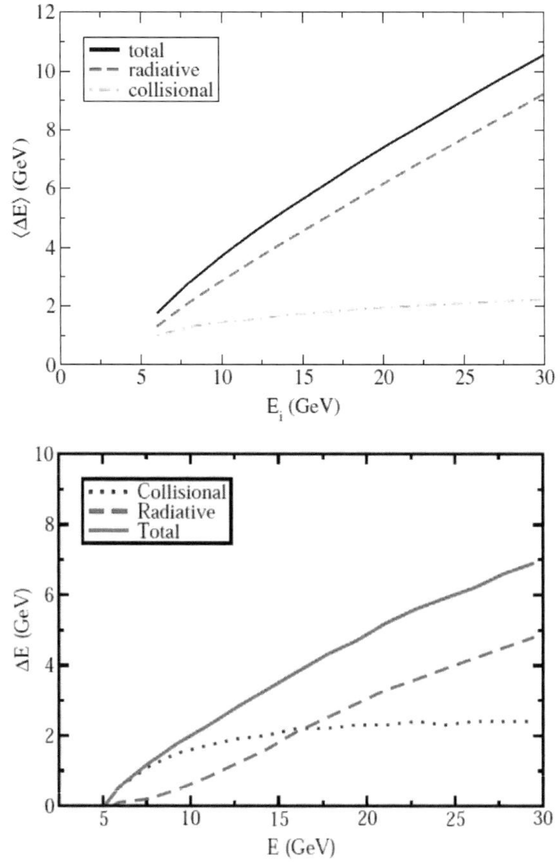

Fig. 8. (Color online) Upper: Average energy loss of a light quark produced at the center of medium passing through the nuclear medium created in most 0–5% central collisions at RHIC.[59] Lower: Average energy loss of a bottom quark passing through the nuclear medium created in most 0–7.5% central collisions at the LHC.[74]

partons and the dense nuclear medium via $2 \to 2$ elastic collisions. Since jet transport coefficients can generally be written as the correlations of gluon fields of the dense nuclear medium, they should be able to provide a lot of information about the internal structures of and the interaction nature of the QGP matter produced in high-energy heavy-ion collisions.

In terms of the three-momentum distribution $f(l_q^-, \mathbf{l}_{q\perp}, L^-)$ of the propagating parton, where l_q^-, $\mathbf{l}_{q\perp}$ and L^- are light-cone energy, the transverse momentum, and the light cone time (length), the effect of multiple elastic scatterings experienced by the hard parton can be described via the following evolution equation,[73,84]

$$\frac{\partial f}{\partial L^-} = \left[D_{L1}\frac{\partial}{\partial l_q^-} + \frac{1}{2}D_{L2}\frac{\partial^2}{\partial^2 l_q^-} + \frac{1}{2}D_{T2}\nabla^2_{l_{q\perp}} + \cdots \right] f(L^-, l_q^-, \mathbf{l}_{q\perp}). \quad (13)$$

The above equation is cut off up to the second order in the gradient expansion of exchanged momentum, and the higher order terms are contained in "⋯". The coefficients D_{L1}, D_{L2} and D_{T2} are defined in the light-cone coordinate; they are related to $\hat{e}$, $\hat{e}_2$ and $\hat{q}$ as follows:

$$\hat{e} = D_{L1}, \quad \hat{e}_2 = \frac{D_{L2}}{\sqrt{2}}, \quad \hat{q} = 2\sqrt{2}D_{T2}. \tag{14}$$

Using the initial condition $f(l_q^-, \mathbf{l}_{q\perp}, L^- = 0) = \delta(l_q^- - q^-)\delta^2(\mathbf{l}_{q\perp})$, the solution for $f(l_q^-, \mathbf{l}_{q\perp}, L^-)$ at a later time L^- can be obtained as,

$$f(l_q^-, \mathbf{l}_{q\perp}, L^-) = \frac{e^{-(l_q^- - q^- + D_{L1}L^-)^2/(2D_{L2}L^-)}}{\sqrt{2\pi D_{L2}L^-}} \frac{e^{-l_{q\perp}^2/(2D_{T2}L^-)}}{2\pi D_{T2}L^-}. \tag{15}$$

One can see that the parton momentum distribution is a Gaussian, due to the fact that only up to the second order terms are included in the evolution equation. From the momentum distribution, one may obtain

$$\langle l_q^- \rangle = q^- - D_{L1}L^-, \quad \langle (l_q^-)^2 \rangle - \langle l_q^- \rangle^2 = D_{L2}L^-, \quad \langle l_{q\perp}^2 \rangle = 2D_{T2}L^-. \tag{16}$$

Therefore, the three terms in Eq. (13) represent the contributions from the longitudinal momentum loss and diffusion, and the diffusion of transverse momentum of the propagating hard parton.

Particular interest has been paid to the jet quenching parameter $\hat{q}$. It not only quantifies the transverse momentum squared exchanged between propagating jets and the nuclear medium, but also plays an important role in describing medium-induced radiative energy loss. The medium modification of radiative process is thought to be dominated by the transverse momentum exchange between the propagating hard jet with the constituents of the dense medium. The importance of jet quenching parameter $\hat{q}$ has also been emphasized earlier in Ref. 85 in which a general relation between the jet parameter $\hat{q}$ and the shear viscosity η is derived for weakly-coupled partonic plasmas such as the QGP at sufficiently high temperature. In a weakly-coupled scenario, the properties of the medium can be described perturbatively on the basis of an appropriate, partonic quasi-particle picture. Since the shear viscosity to entropy ratio η/s saturates in the strong coupling limit as suggested by the Super-Yang–Mills (SYM) calculations, while $\hat{q}/T^3$ continues to increase, it is argued that $\hat{q}/T^3$ may serve as a better and more broadly applicable measure of the coupling strength of a QGP than the transport coefficient η/s. In addition, a general relation between $T^3/\hat{q}$ and η/s is conjectured in Ref. 85 as follows:

$$\frac{T^3}{\hat{q}} = \begin{cases} \approx \dfrac{\eta}{s}, & \text{weakly-coupled}, \\ \ll \dfrac{\eta}{s}, & \text{strongly-coupled}. \end{cases} \tag{17}$$

This relation indicates that the precise determination of both quantities ($\hat{q}/T^3$ and η/s) by systematic phenomenological studies will allow us to quantitatively estimate the interaction strength (nature) of the QGP produced in ultra-relativistic heavy-ion collisions, e.g., when and how a weakly-coupled quark gluon system at sufficiently high temperatures turns into a strongly-coupled fluid at the energies achieved by RHIC and the LHC heavy-ion experiments.

4.1. *Phenomenological studies at RHIC and the LHC*

The study of jet quenching in the dynamically evolving heavy-ion collisions needs a variety of components: the initial hard parton spectrum, a realistic description of the density profiles of the medium, the modeling of jet propagation in dense nuclear medium, and the hadronization (fragmentation) of jets after passing through the medium. In spite of the complexity involved, tremendous phenomenological studies have been performed in recent years, see e.g., Refs. 86–95, in which various jet quenching observables have been investigated utilizing different jet energy loss model calculations.

A significant collaborative effort on the phenomenological study of jet quenching at RHIC and the LHC was recently performed within the framework of JET Collaboration.[96] In this survey, within five different existing approaches to parton propagation and medium-induced energy loss in dense medium, systematic phenomenological studies was carried out on the experimental data on the nuclear modification of single inclusive hadrons at large p_T in relativistic heavy-ion collisions at the RHIC and the LHC . The space–time evolution of the QGP medium used in the study was simulated by the $(2+1)$-dimensional or $(3+1)$-dimensional hydrodynamic models which have been constrained by the experimental data on the bulk hadron spectra and anisotropic collective flow. The goal of this collaborative study is to quantitatively extract the jet quenching parameter $\hat{q}$ and its systematic uncertainties due to model dependence by using the constraint from the experimental data. The five parton energy loss approaches used for such study are: McGill-AMY,[59] Martini-AMY,[97] HT-M,[65] HT-BW,[98] DGLV-CUJET.[99] In the following we provide some details about different model implementations.

DGLV-CUJET model

The DGLV-CUJET code[99] was developed at Columbia University as part of JET Collaboration. It includes both radiative energy loss and elastic energy loss based on DGLV formalism with dynamical scattering centers. The code takes into account the effects of multi-scale running of the QCD coupling $\alpha_s(Q^2)$ in the DGLV opacity expansion series.

At the first order in the opacity expansion, the medium-induced radiative gluon distribution in the running coupling DGLV (rcDGLV) formalism is given by,

$$x\frac{dN_g}{dx}(\mathbf{r},\tau)=\int d\tau\rho_g(\mathbf{r}+\hat{n}(\phi)\tau,\tau)\int\frac{d^2q_\perp}{\pi}\frac{d\sigma}{d^2q_\perp}\int\frac{d^2k_\perp}{\pi}\alpha_s(k_\perp^2/x(1-x))$$
$$\times\frac{12(\mathbf{k}_\perp+\mathbf{q}_\perp)}{(\mathbf{k}_\perp+\mathbf{q}_\perp)^2+\chi(\tau)}\cdot\left(\frac{\mathbf{k}_\perp+\mathbf{q}_\perp}{(\mathbf{k}_\perp+\mathbf{q}_\perp)^2+\chi(\tau)}-\frac{\mathbf{k}_\perp}{\mathbf{k}_\perp^2+\chi(\tau)}\right)$$
$$\times\left(1-\cos\left[\frac{(\mathbf{k}_\perp+\mathbf{q}_\perp)^2+\chi(\tau)}{2x_+E}\right]\right),\tag{18}$$

where x and $\mathbf{k}_\perp$ are the fractional energy and transverse momentum of the radiated gluon with respect to the parent parton, $\mathbf{q}_\perp$ is the transverse momentum exchanged with medium constituents in a single scattering. $\rho_g(\mathbf{r}+\hat{n}\tau,\tau)$ is the local gluon density of the medium, with $\mathbf{r}$ the location of the hard parton, $\hat{n}(\phi)$ the propagation direction in transverse plane, and τ the evolution time. The effective differential quark–gluon cross-section $d\sigma/d^2q_\perp$ is given by

$$\frac{d\sigma}{d^2q_\perp}=\frac{\alpha_s^2(q_\perp^2)}{(q_\perp^2+f_E^2\mu^2(\tau))(q_\perp^2+f_M^2\mu^2(\tau))},\tag{19}$$

where $\mu^2(\tau)=4\pi\alpha_s(4T^2)T^2$ is the local HTL color electric Debye screening mass squared calculated for a pure gluonic plasma. In the above equation, the scale $\chi(\tau)=M^2x_+^2+f_E^2\mu^2(T)(1-x_+)/\sqrt{2}$ is utilized to regulate infrared behaviors and control the dead cone and LPM destructive interference effects due to finite quark mass M and a asymptotic thermal gluon mass $m_g=f_E\mu(T)/\sqrt{2}$. (f_E,f_M) is the HTL deformation parameters which controls the electric and magnetic screening scales related to the HTL. The vacuum running coupling $\alpha_s(Q^2)=\min[\alpha_{\max},2\pi/9\ln(Q^2/\Lambda^2)]$ is used in the calculation. In CUJET-GLV, there are three main model parameters: $(\alpha_{\max},f_E,f_M)$, and the default values for (f_E,f_M) are taken to be $(1,0)$.

In CUJET-GLV code, for each production vertex and propagation direction $(\mathbf{r},\hat{n})$, the Monte Carlo integration method is used to evaluate the gluon emission spectra dN_g/dx and average number of gluons $\langle N_g\rangle$. A Poisson ansatz is then utilized to estimate the effect of multiple gluon emission and to obtain the probability distribution of energy loss $P_{\mathrm{rad}}(\Delta E\,|\,E_0,\mathbf{r},\hat{n})$,

$$P_{\mathrm{rad}}(\Delta E\,|\,E_0,\mathbf{r},\mathbf{n})=\sum_{n=0}^{\infty}\frac{e^{-\langle N_g\rangle}}{n!}\left[\prod_{i=1}^{n}\int dx\frac{dN_g}{dx}\right]\delta\left(\Delta E-\sum_{i=1}^{n}x_iE_0\right).\tag{20}$$

The code also computes the elastic energy loss probability $P_{\mathrm{el}}(\Delta E\,|\,E_0,\mathbf{r},\hat{n})$, which is convoluted with radiative contribution to get total energy loss distribution $P_{\mathrm{tot}}(\Delta E\,|\,E_0,\mathbf{r},\hat{n})$. The final total energy loss distribution $P_{\mathrm{tot}}(\Delta E\,|\,E_0,\mathbf{r},\hat{n})$ is then convoluted with initial parton jet spectra $dN_{pp}/dp_Tdy(p_T,y\,|\,\mathbf{r},\hat{n})$ to get the final modified parton spectra $dN_{AA}/dp_T dy(p_T, y\,|\,\mathbf{r}, \hat{n})$. By averaging over initial jet production configuration (which is determined by Glauber model) and fragmentation (and decay) functions, one gets final spectra and nuclear modification factors for hadrons (or leptons) to be compared with the experimental data.

HT-BW model

In Higher–Twist–Berkeley–Wuhan (HT-BW) model,[98] the medium modified fragmentation function is given by

$$\tilde{D}_{q\to h}(z_h, Q^2) = D_{q\to h}(z_h, Q^2) + \frac{\alpha_s}{2\pi}\int_0^{Q^2}\frac{dl_\perp^2}{l_\perp^2}\int_{z_h}^1\frac{dz}{z} \times \left[\Delta\gamma_{q\to qg}(z, l_\perp^2) D_{q\to h}\left(\frac{z_h}{z}\right) + \Delta\gamma_{q\to gq}(z, l_\perp^2) D_{g\to h}\left(\frac{z_h}{z}\right)\right], \quad (21)$$

where z and $l_\perp$ are the momentum fraction and transverse momentum of the daughter partons with respect to the parent parton. The above medium modified fragmentation functions are very similar to the vacuum radiation corrections which lead to vacuum DGLAP evolution equation, except that now the splitting function $\Delta\gamma_{q\to qg}(z, l_\perp^2)$ and $\Delta\gamma_{q\to gq}(z, l_\perp^2)$ depend on the properties of the medium via jet transport coefficient $\hat{q}$. More specifically, the medium-modified splitting function for a quark jet $\Delta\gamma_{q\to qg}(z, l_\perp^2)$ is given by

$$\Delta\gamma_{q\to qg}(z, l_\perp^2) = C_A \frac{1}{l_\perp^2}\frac{1+z^2}{1-z}\left(1 - \frac{1-z}{2}\right)\int dy^- \hat{q}(y^-) 4\sin^2\left[\frac{l_\perp^2 y}{4Ez(1-z)}\right], \quad (22)$$

where $\hat{q}(y^-)$ is the transport coefficient for a quark jet at a given location y^-. Note that an extra factor of $1-(1-z)/2$ is to include the corrections beyond the helicity amplitude approximation.[100]

In HT-BW calculation, the jet transport coefficient $\hat{q}$ is assumed to be proportional to the local parton density in the QGP phase or the local hadron density in the hadronic phase:

$$\hat{q}(\tau, \mathbf{r}) = \left[\hat{q}_0 \frac{\rho_{\rm QGP}(\tau, \mathbf{r})}{\rho_{\rm QGP}(\tau_0, 0)}(1 - f_h) + \hat{q}_h(\tau, \mathbf{r}) f_h\right] \times \left(\frac{p\cdot u}{p_0}\right), \quad (23)$$

where τ and $\mathbf{r}$ are the time and the location. $\hat{q}_0$ is the jet transport parameter at the center of the bulk medium in QGP phase at initial time τ_0, and ρ_{QGP} is the parton density in an ideal gas at a given temperature, f_h is the fraction of hadronic phase at any given time τ and space location $\mathbf{r}$. The extra factor $(p\cdot u/p_0)$ is to take into account the effect of a dynamically evolving medium on the jet transport coefficient, with u^μ the local four flow velocity of the fluid. Assuming a hadron resonance gas, the jet transport parameter in the hadron phase is approximated as:

$$\hat{q}_h = \frac{\hat{q}_N}{\rho_N}\left[\frac{2}{3}\sum_M \rho_M(T) + \sum_B \rho_B(T)\right], \tag{24}$$

where ρ_M and ρ_B are the meson and baryon density at a given temperature, and $\rho_N = 0.17\,\text{fm}^{-3}$ is the nucleon density in the center of a large nucleus. $\hat{q}_N$ is the jet transport parameter for a quark through a large nucleus and taken to be $\hat{q}_N = 0.02\,\text{GeV}^2/\text{fm}$ from recent studies of deeply inelastic scattering (DIS) compared to HERMES data.[101]

HT-M model

The Higher-Twist-Majumder (HT-M) approach[65] is similar to the HT-BW model, but it includes multiple gluon emission through a set of effective modified QCD evolution equations. The medium-modified fragmentation functions are obtained by solving the DGLAP-like evolution equations with a vacuum splitting kernel plus a medium-induced contribution. This model explicitly assumes the factorization of hard scattering cross-section from the final fragmentation function.

In HT-M model, there are two contributions to the medium-modified fragmentation functions: one from vacuum evolution and the other from medium-modified evolution. For the quark fragmentation function, the vacuum DGLAP evolution equations is given by,

$$\frac{\partial D_{q\to h}(z,Q^2)}{\partial \ln Q^2} = \sum_i \frac{\alpha_s}{2\pi}\int_z^1 \frac{dy}{y} P_{q\to i}(y) D_{i\to h}\left(\frac{z}{y},Q^2\right). \tag{25}$$

The medium modified DGLAP-like evolution equations read as follows:

$$\begin{aligned}\frac{\partial D_{q\to h}(z,Q^2,q^-)|_{\zeta_i^-}^{\zeta_f^-}}{\partial \ln Q^2} &= \sum_i \frac{\alpha_s}{2\pi}\int_z^1 \frac{dy}{y} P_{q\to i}(y) \\ &\times \int_{\zeta_i^-}^{\zeta_f^-} d\zeta K(\zeta^-,y,Q^2,q^-) D_{i\to h}(\frac{z}{y},Q^2,q^- y)|_{\zeta^-}^{\zeta_f^-}.\end{aligned} \tag{26}$$

In the above equations, $P_{q\to i}(y)$ is the regular vacuum splitting function. The medium modification effect is contained in the scattering kernel $K(\zeta^-,y,Q^2,q^-)$, with ζ^- the location of the scattering. ζ_i^-, ζ_f^- are the original and final locations of the hard parton. The contribution to the scattering kernel $K(\zeta^-,y,Q^2,q^-)$ from the leading power correction is given by

$$\begin{aligned}&K(\zeta^-,y,Q^2,q^-)\\ &= \frac{[\hat{q}_A(\zeta^-) - (1-y)\hat{q}_A/2 + (1-y)^2\hat{q}_F]}{Q^2} 4\sin^2\left[\frac{Q^2(\zeta^- - \zeta_i^-)}{4q^- y(1-y)}\right].\end{aligned} \tag{27}$$

The jet transport coefficient $\hat{q}_A$ and $\hat{q}_F$ are for a gluon jet and a quark jet, and they are related to each other as $\hat{q}_F/\hat{q}_A = C_F/C_A$.

In the HT-M model, the jet quenching parameter $\hat{q}$ is assumed to scale with the entropy density of the dense nuclear medium (either QGP or hadronic phase),

$$\frac{\hat{q}(s)}{\hat{q}_0} = \frac{s}{s_0}, \tag{28}$$

where s_0 is the maximum entropy density achieved at an initial time τ_0 at the center of the medium (produced in most central collisions at RHIC).

When solving both vacuum and medium modified evolution equations, the input fragmentation functions are needed. In HT-M model, the inputs are taken as the vacuum fragmentation functions at the input scale $Q_0^2 = p_T/L$, where p_T is the transverse momentum of the parton and the length factor L is the mean escape length of the hard jet obtained via using the single gluon emission kernel. Such input vacuum fragmentation functions are obtained according to the vacuum evolution equations from $Q_0^2 = 1\,\mathrm{GeV}^2$.

McGill-AMY model

In McGill-AMY model,[59] the medium-induced radiation and elastic scattering processes are described by the high temperature thermal field theory. In the model calculation, the first step is to calculate the production of initial stage partons before the formation of thermal QGP. The second step is to calculate the medium effects on the produced partons which propagate through the hot and dense QGP. The model solves the a set of coupled transport rate equations for parton momentum distributions to obtain the medium modified parton momentum distributions. The third step is to convolute the medium modified parton momentum distribution with vacuum fragmentation functions to give final hadron spectra in relativistic heavy-ion collisions.

In McGill-AMY model, the following transport rate equations are solved for the parton momentum distribution $f(p,t) = dN(p,t)/dp$:

$$\frac{df_j(p)}{dt} = \sum_{a,b} \int dk \left[f_a(p+k) \frac{d\Gamma_{a\to j}(p+k,k)}{dkdt} - f_j(p) \frac{d\Gamma_{j\to b}(p,k)}{dkdt} \right], \tag{29}$$

where $d\Gamma_{j\to b}(p,k)/dkdt$ is the transition rate for the partonic process $j \to a$, with p the initial parton energy and k the lost energy in the splitting process. The transition rates are calculated in the full leading order thermal QCD with the inclusion of the HTL effects and LPM interference. In the model calculation, all the splitting processes are included and all the partons in the shower are kept track of until the partons fragment into hadrons outside the thermal medium. The energy gained by the propagating parton due to the absorption of the thermal partons is included in the $k < 0$ integration part. The contribution from $2 \to 2$ elastic collisions with the medium constituents are included in a similar way.

In the model, the properties of the thermal QGP medium enter through the local temperature and flow velocity when calculating the parton transition rates.

The interaction between the hard parton the medium is controlled by the HTL resummed elastic collision rates:

$$\frac{d\Gamma_{\rm el}}{d^2\mathbf{q}_\perp} = \frac{C_s}{(2\pi)^2}\frac{g_s^2 m_D^2 T}{q_\perp^2(q_\perp^2 + m_D^2)}, \tag{30}$$

where $g_s = \sqrt{4\pi\alpha_s}$ is the coupling constant, and $m_D^2 = (N_c/3 + N_f/6)g^2T^2$ is the Debye screening mass squared. The factor C_s is the color factor of the propagating parton. Currently in the calculation, the coupling constant α_s is taken to be constant for all interaction vertices. The scale dependence of the coupling as well as the finite size effect have not yet been included at present.

MARTINI-AMY model

MARTINI-AMY model[97] is a Monte Carlo event generator for simulating jet production and the subsequent evolution in relativistic heavy-ion collisions. The production of initial jet partons is obtained by running PYTHIA-8 code from each nucleon–nucleon collision with the spatial distribution of the jet partons following the Glauber geometry. The propagation of jet partons in the thermal medium is carried out by solving the same coupled rate equations as in McGill-AMY model, but using Monte Carlo techniques. In addition, Martini-AMY code also includes the scale dependence of the strong coupling α_s for the radiation vertices via $\langle k_T^2\rangle = (\hat{q}k)^{1/2}$. The parton transition rates also include the finite-size effects via a parametrization of the rates derived in Ref. 52.

Within the framework of JET Collaboration, the above five jet quenching models are used to calculate the nuclear modification factor R_{AA} for large p_T single inclusive hadrons in most central collisions at RHIC and the LHC. The results are shown in Fig. 9, where the left panels are for RHIC R_{AA} while the right for the LHC R_{AA}. From the top to the bottom are the results from five different models: CUJET-GLV, HT-BW, HT-M, McGill-AMY and MARTINI-AMY. Different curves in each plot are the calculations with different model parameters within each jet quenching approach. For DGLV-CUJET the model parameters are $(\alpha_{\rm max}, f_E, f_M)$, for two HT calculations jet quenching coefficient $\hat{q}_0$ is the model parameter, and for two AMY calculations, the model parameter is strong coupling α_s. For all the calculations, the realistic ideal/viscous hydrodynamics simulations[102–104] are utilzed for the space–time evolution of the bulk nuclear matter produced in RHIC and the LHC heavy-ion collisions. The initial time for relativistic hydrodynamics evolution is set as $\tau_0 = 0.6\,$fm/c with the initial temperatures set as $T_0 = 346$–$373\,$MeV (477–486 MeV) at the center the hot and dense medium produced in most central Au+Au collisions at $\sqrt{s_{NN}} = 200\,$GeV at RHIC (most central Pb+Pb collisions at $\sqrt{s_{NN}} = 2.76\,$TeV at the LHC).

By comparing to the experimental measurements of single inclusive hadron nuclear modification factor R_{AA} at large p_T, one may obtain a tight constraint on the model parameters for each jet energy loss model calculation. In DGLV-CUJET model, the best fits to the experimental data on R_{AA} at RHIC and the LHC give:

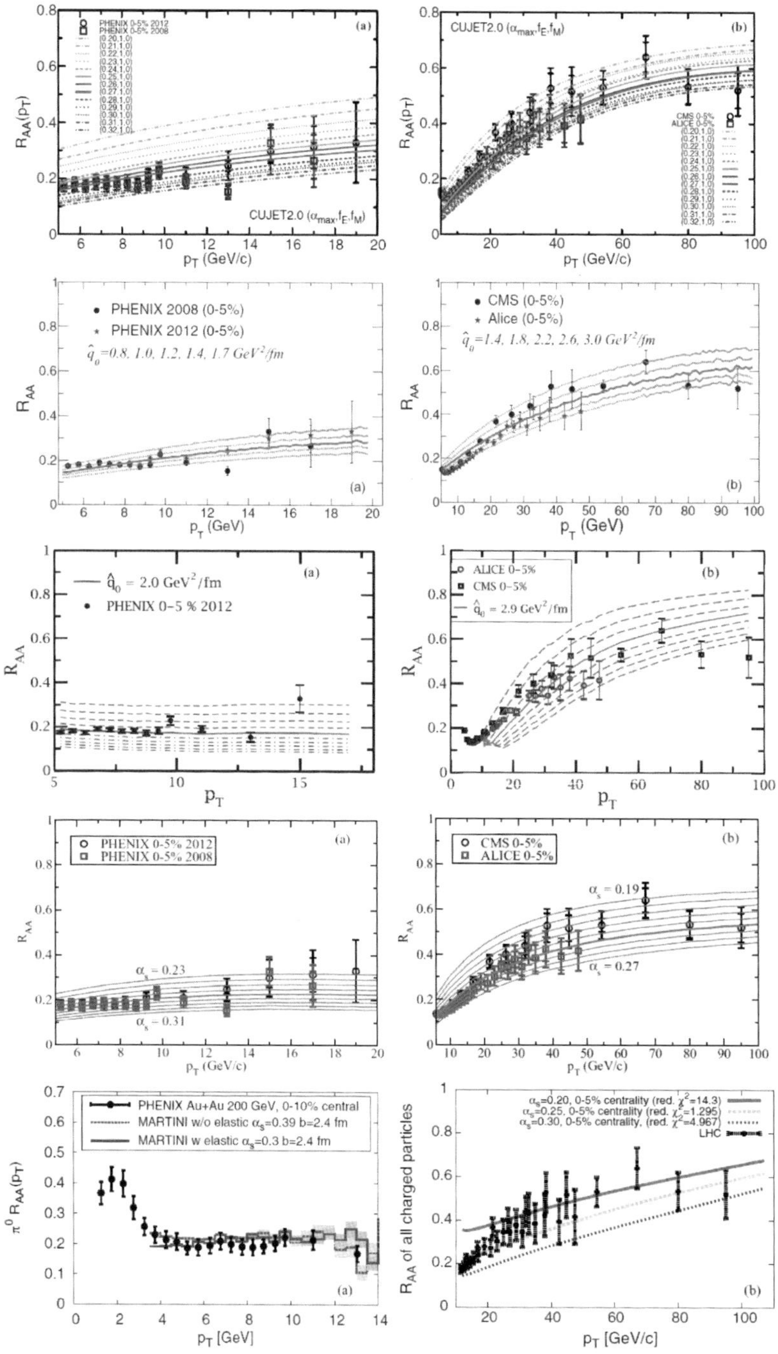

Fig. 9. (Color online) The nuclear modification factor R_{AA} at both RHIC (left panels) and the LHC (right panels) compared with different model calculations. From top to bottom, the models are: CUJET-GLV, HT-BW, HT-M, McGill-AMY and MARTINI-AMY. Figures are taken from Ref. 96 done by JET Collaboration.

$\alpha_{\max} = 0.28$ for $T < 378\,\text{MeV}$ and $\alpha_{\max} = 0.24$ for $378 < T < 486\,\text{MeV}$. For HT-BW model, the best fits produce $\hat{q}_0 = 1.2\,\text{GeV}^2/\text{fm}$ at RHIC, and $\hat{q}_0 = 2.2\,\text{GeV}^2/\text{fm}$ at the LHC for the quark jet transport coefficients. For HT-M model, the gluon jet transport coefficients are obtained from the best fits as: $\hat{q}_0 = 2.0\,\text{GeV}^2/\text{fm}$ at RHIC, and $\hat{q}_0 = 2.9\,\text{GeV}^2/\text{fm}$ at the LHC. For McGill-AMY model, the best fits give $\alpha_s = 0.27$ for RHIC and $\alpha_s = 0.24$ for the LHC. For Martini-AMY model, $\alpha_s = 0.3$ for RHIC and $\alpha_s = 0.24$ for the LHC are obtained.

4.2. *Quantitative extraction of* $\hat{q}$

With the model parameters fixed by the experimental data, one may obtain the effective jet transport coefficient $\hat{q}$ for each model. For CUJET calculation, it is obtained as follows:

$$\hat{q}(E, T, \alpha_{\max}) = \rho_g(T) \int^{q_{\max}} dq_\perp^2 q_\perp^2 \frac{d\sigma}{dq_\perp^2}, \tag{31}$$

where $q_{\max}$ is the ultra-violet cutoff and taken to be $q_{\max} = 6ET$. One can see that the jet transport coefficient depends on jet energy E and the medium temperature T. The variation of the temperature T also affects the electric and magnetic screening mass deformation parameters (f_E, f_M). For McGill-AMY and MARTINI-AMY models, the values of $\hat{q}$ may be obtained as follows:

$$\hat{q} = C_s \alpha_s m_D^2 T \ln\left(\frac{1 + q_{\max}^2}{m_D^2}\right). \tag{32}$$

For HT-BW and HT-M models, the comparisons to experimental data directly give the jet quenching parameter $\hat{q}$.

In Fig. 10, the extracted values for jet quenching parameter $\hat{q}$ scaled by T^3 are shown for a quark jet with an energy $E = 10\,\text{GeV}$ at the center of most central

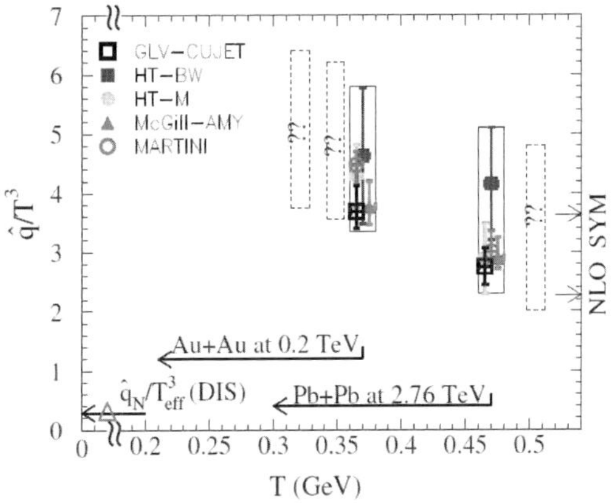

Fig. 10. (Color online) The extracted values of scaled jet transport parameter $\hat{q}/T^3$ by using single inclusive hadron suppression factor R_{AA} at both RHIC and LHC. The shown values are for a quark jet with initial energy of 10 GeV at the center of most central A–A collisions at an initial proper time $\tau_0 = 0.6\,\text{fm/c}$. The figure is taken from Ref. 96 done by JET Collaboration.

A–A collisions at an initial proper time $\tau_0 = 0.6\,\text{fm/c}$ at RHIC and the LHC energies (denoted by two solid boxes). The range of $\hat{q}$ values as constrained by the measured single hadron nuclear modification factors R_{AA} at RHIC and the LHC is obtained as:

$$\frac{\hat{q}}{C_s T^3} = \begin{cases} 3.5 \pm 0.9, & T \approx 370\,\text{MeV (at RHIC)}, \\ 2.8 \pm 1.1, & T \approx 470\,\text{MeV (at the LHC)}. \end{cases} \tag{33}$$

Translating these values for $\hat{q}$, one obtains: $\hat{q} \approx 1.2$ and $1.9\,\text{GeV}^2$ fm for a quark jet with energy $E = 10\,\text{GeV}$ at the highest temperatures reached in the most central Au+Au collisions $\sqrt{s_{NN}} = 200\,\text{GeV}$ at RHIC and Pb+Pb collisions $\sqrt{s_{NN}} = 2.76\,\text{TeV}$ at the LHC.

One can clearly see that the scaled dimensionless quantity $\hat{q}/T^3$ is temperature dependent. Such temperature dependence can be explored further by extending the current study to the future higher energy Pb–Pb collisions at the LHC and lower energy collisions at RHIC. The expected values of $\hat{q}/T^3$ in A–A collisions at 0.063 ATeV, 0.130 ATeV and 5.5 ATeV are shown by dashed boxes. As a comparison, the figure also shows the value of $\hat{q}_N/T^3_{\text{eff}}$ in cold nuclei (indicated by the triangle in the figure) as extracted from jet quenching studies in DIS,[105] where the value of $\hat{q}_N = 0.02\,\text{GeV}^2/\text{fm}$ is taken, and the effective temperature is taken assuming an idea quark gas with 3 quarks within each nucleon in a large nucleus. We can see that the values of $\hat{q}$ in the hot and dense QGP matter are much higher than those in cold nuclei.

The above extracted values of $\hat{q}$ are also compared to the values obtained from a next-to-leading order AdS/CFT calculation[106] (shown by two arrows on the right axis of the figure). By utilizing the AdS/CFT correspondence, one may calculate the jet quenching parameter $\hat{q}$ in a $\mathcal{N} = 4$ supersymmetric Yang–Mills (SYM) plasma at the strong coupling limit. At the leading order (large N_c and large 't Hooft coupling λ), $\hat{q}$ may be obtained as:[107]

$$\hat{q}^{\text{LO}}_{\text{SYM}} = \frac{\pi^{3/2}\Gamma\left(\frac{3}{4}\right)}{\Gamma\left(\frac{5}{4}\right)} \sqrt{\lambda} T^3_{\text{SYM}}, \tag{34}$$

where $\lambda = g^2_{\text{SYM}} N_c$. Including the corrections from finite 't Hooft coupling, the jet quenching parameter at next-to-leading order is obtained as,[106]

$$\hat{q}^{\text{NLO}}_{\text{SYM}} = \hat{q}^{\text{LO}}_{\text{SYM}} \left(1 - \frac{1.97}{\sqrt{\lambda}}\right). \tag{35}$$

The SYM values shown in the figure are obtained for $\alpha_{\text{SYM}} = 0.22$–$0.31$ and have taken into account different numbers of degrees of freedom in SYM and 3 flavor QCD. One can see that the values of $\hat{q}_{\text{SYM}}$ from SYM calculations are within the range of $\hat{q}$ extracted from comparing phenomenological studies to experimental data on R_{AA} at RHIC and the LHC done by JET Collaboration.

In order to fully take into account the systematic uncertainties due to model dependence, more jet quenching calculations (see e.g., Refs. 94, 95, 108–111) should be included in the future for such systematic phenomenological studies as done by the JET Collaboration. More experimental observables, besides the single inclusive hadron R_{AA}, should be incorporated for analysis so that one can obtain tighter constraints on the modeling of jet quenching and on the quantitative extraction of jet transport coefficients.

4.3. $\hat{q}$ *from lattice QCD*

While lattice QCD calcuation cannot provide a full description of the dynamical evolution of heavy-ion collisions and jet transport in the hot and dense nuclear medium, it can provide important guidance for jet quenching studies and many insights for our understanding of jet-medium interaction. For example, jet transport coefficients (such as $\hat{q}$) can in principle be computed with lattice QCD; this will provide very useful reference for phenomenological jet quenching studies.

The first calculation of jet quenching parameter $\hat{q}$ within the framework of finite temperature lattice gauge theory was performed in Ref. 112. The operator expression for $\hat{q}$ is quoted here:

$$\begin{aligned}\hat{q} = \frac{4\pi^2\alpha_s}{N_c}\int \frac{dy^- d^2y_\perp d^2k_\perp}{(2\pi)^3} e^{i\frac{k_\perp^2}{2q^-}y^- - i\mathbf{k}_\perp\cdot\mathbf{y}_\perp} \\ \times \langle \mathrm{Tr}[t^a F_\perp^{a+\mu}(y^-,\mathbf{y}_\perp)U^\dagger(\infty^-,\mathbf{y}_\perp;0^-,\mathbf{y}_\perp)T^\dagger(\infty^-,\infty_\perp;\infty^-,\mathbf{y}_\perp) \\ \times T(\infty^-,\infty_\perp;\infty^-,\mathbf{0}_\perp)U(\infty^-,\mathbf{0}_\perp;0^-,\mathbf{0}_\perp)t^b F_{\perp\mu}^{b+}(0^-,\mathbf{0}_\perp)]\rangle, \qquad (36)\end{aligned}$$

where U denotes a Wilson line along the "−" light-cone direction and T denotes a Wilson line along the transverse "⊥" light-cone direction. Reference 112 utilizes the operator product expansion method, and relates the jet transport parameter $\hat{q}$ as obtained in the physical regime of jet momenta to an infinite series local operators in an unphyical regime of jet momenta via dispersion relations. Then these local operators are computed in quenched $SU(2)$ lattice gauge theory. After extrapolating the calculation in quenched $SU(2)$ to the case of $SU(3)$ with 2 flavors of quarks, $\hat{q} = 1.3$–$3.3\,\mathrm{GeV}^2/\mathrm{fm}$ is obtained for a gluon jet at a temperature $T = 400\,\mathrm{MeV}$. This value is within the range of $\hat{q}$ extracted from phenomenological studies done by JET Collaboration as shown in Subsec. 4.2. In a recent non-perturbative lattice calculation[113] based on a dimensionally reduced effective theory (electrostatic QCD), which only includes the contribution from soft QGP modes, a value of $\hat{q} = 6\,\mathrm{GeV}^2/\mathrm{fm}$ was obtained at RHIC energies. This is about twice the value obtained from NLO perturbative QCD result.[114]

It should be noted that jet quenching parameter $\hat{q}$ only encodes the second moment of the probability distribution $P(k_\perp, L)$ for the transverse momentum transfer,

$$\hat{q} = \frac{1}{L}\int \frac{d^2k_\perp}{(2\pi)^2} k_\perp^2 P(\mathbf{k}_\perp, L), \qquad (37)$$

where L is the length that jet traversed. The probability distribution $P(k_\perp, L)$ satisfies the following evolution equation,

$$\frac{dP(\mathbf{k}_\perp, L)}{dL} = \int \frac{d^2 q_\perp}{(2\pi)^2} C(\mathbf{q}_\perp)[P(\mathbf{k}_\perp - \mathbf{q}_\perp) - P(\mathbf{k}_\perp)], \tag{38}$$

where $C(q_\perp)$ is the transverse collision kernel. A full distribution of the probability distribution $P(k_\perp)$ or the collision kernel $C(k_\perp)$ will provide more detailed information about jet-medium interaction. A first-principle calculation of $C(k_\perp)$ and $P(k_\perp, L)$ was carried out in Ref. 115 and the results are shown in Fig. 11. It is found that the shape of the transverse collision kernel $k_\perp^3 C(k_\perp)$ is consistent with a Gaussian at small transverse momentum $k_\perp$. As $k_\perp$ becomes large, the calculation breaks down as it only includes the soft modes.

The Gaussian form at small $k_\perp$ for the probability distribution $P(k_\perp)$ is expected in both weakly-coupled and strongly-coupled pictures. However at large $k_\perp$, the distribution $P(k_\perp)$ takes very different form in these two scenarios: for weakly coupled quark–gluon plasma, the large $k_\perp$ power tail is proportional to $1/k_\perp^2$, while in the plasma of $\mathcal{N} = 4$ SYM theory in the large-N_c and strong coupling limits, there is no power tail at all.[116] So it would be very interesting and useful if lattice calculation as well as phenomenological jet quenching studies can provide the full distribution for the collision kernel or the probability distribution.

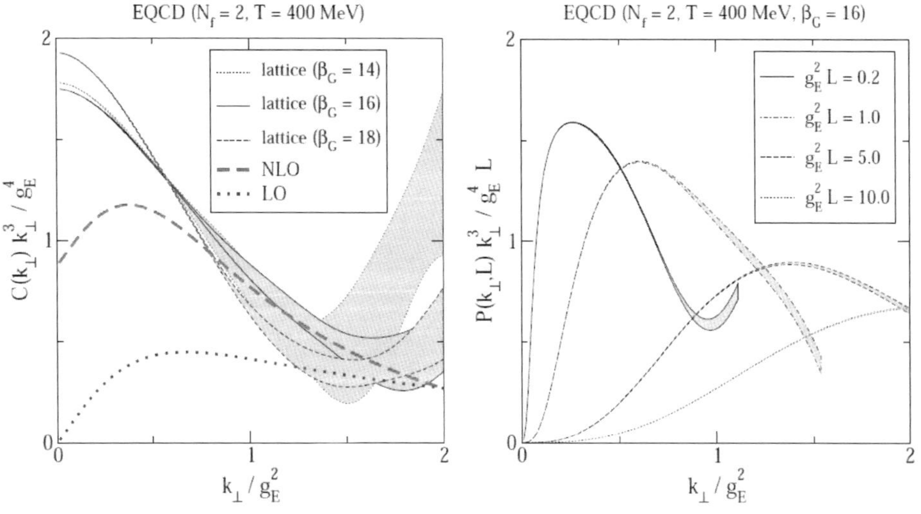

Fig. 11. (Color online) The transverse collision kernel $C(k_\perp)$ (left) using lattice data from Ref. 113 compared with the NLO calculation result from Ref. 114 and the probability distribution $P(k_\perp, L)$ (right). $g_E^2 \sim g^2 T$ is the effective coupling of the dimensionally reduced EQCD effective theory. β_G is a dimensionless number and related to the lattice spacing a as: $\beta_G = 2N_c/(g^2 Ta)$. Figures are taken from Ref. 115.

4.4. *Renormalization of $\hat{q}$ and NLO developments*

Parton energy loss models based on perturbative QCD techniques have been very successful in studying jet quenching phenomena in relativistic heavy-ion collisions at RHIC and the LHC. The comparisons between systematic phenomenological studies using various theoretical models and experimental data have led to a tight determination of jet transport coefficients. However, there still exist strong model dependence due to various approximations applied in the jet quenching calculations. One important reason is that all the phenomenological jet quenching studies have utilized leading-order parton energy loss formalisms. In fact, many model differences have been found to originate from to some specific approximations made in the leading order calculation of the gluon radiation spectrum, such as eikonal, soft and collinear approximations.[49]

To constrain jet quenching models and better our understanding of jet-medium interaction, the effects of different approximations need to be explored in more detail, e.g., by relaxing these approximations when building parton energy loss formalisms. The effect of non-eikonal large angle scattering approximation was recently investigated in Ref. 117 and it is found to be small for jet energies larger than 10–15 GeV. Reference 118 re-study the medium-induced radiation by going beyond eikonal approximation and include the finite energy effect in the new formalism. However, in order to achieve a systematic estimation of the errors made in the leading-order calculation, it is necessary to build a fully next-to-leading order (NLO) framework for studying parton propagation in dense nuclear matter and jet quenching in relativistic heavy-ion collisions.

The first attempt along NLO direction was performed in Refs. 119 and 120, in which the NLO radiative correction to the transverse momentum broadening is computed. Consider a high-energy parton propagating through a dense nuclear medium, it experiences both elastic and inelastic collisions with the medium constituents. Focusing on the transverse momentum broadening for the propagating jet parton, the leading order contribution comes from $2 \to 2$ elastic collisions. At next-to-leading order, one has to consider the contribution from medium-induced radiation, as shown in Fig. 12.

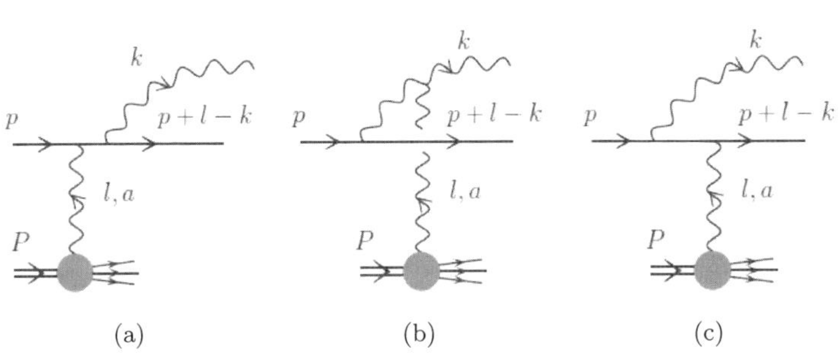

Fig. 12. Induced gluon radiation of a high-energy quark by a single scattering.

The calculation was performed in the framework of BDMPS-Z formalism and it was found that the transverse momentum broadening receives a sizable contribution from radiative correction due to the recoil effect of one-gluon emission in the kinetic region of single scattering. In particular for large media size, the radiative contribution is dominated by the double logarithmic terms. The double logarithmic correction reads as follows:

$$\langle p_\perp^2 \rangle_{\rm rad} = \frac{\alpha_s N_c}{\pi} \hat{q}_0 L \int_{\hat{q}_0 \tau_0}^{\hat{q}_0 L} \frac{dk_\perp^2}{k_\perp^2} \int_{\frac{\hat{q}_0 \tau_0}{k_\perp^2}}^{1} \frac{dx}{x} = \frac{\alpha_s N_c}{2\pi} \hat{q}_0 L \ln^2 \left(\frac{L}{\tau_0} \right), \tag{39}$$

where $\hat{q}_0$ is the transverse momentum broadening from pure elastic collisions, L is the length of the nuclear matter, and τ_0 is some intrinsic size associated with the constituents of the nuclear matter (which was introduced to regulate the logarithmic divergence in the radiative corrections). The jet transport coefficient $\hat{q}$ also receives a double logarithmic correction,

$$\hat{q}_{\rm rad} = \frac{\alpha_s N_c}{\pi} \hat{q}_0 \int_{\tau_0}^{L} \frac{d\tau}{\tau} \int_{\hat{q}_0 \tau}^{\hat{q}_0 L} \frac{dk_\perp^2}{k_\perp^2} = \frac{\alpha_s N_c}{2\pi} \hat{q}_0 \ln^2 \left(\frac{L}{\tau_0} \right). \tag{40}$$

The role of the large double logarithms corrections was further investigated in Refs. 121 and 122. It was argued that such double logarithmic correction may be absorbed by a redefinition/renormalization of the jet quenching parameter $\hat{q}$,

$$\hat{q}(\tau) = \hat{q}_0 \left[1 + \frac{\alpha_s N_c}{2\pi} \ln^2 \left(\frac{\tau}{\tau_0} \right) \right], \tag{41}$$

where τ may be understood as the size of the medium constituents "seen" or probed by the propagating parton. The infrared cut-off τ_0 can be viewed as a factorization scale in this problem. With such redefinition of jet quenching parameter, one may write an evolution equation for $\hat{q}(\tau)$ as a function of the scale τ,

$$\frac{\partial \hat{q}(\tau, Q^2)}{\partial \ln \tau} = \int_{\hat{q}_0 \tau}^{Q^2} \frac{dk_\perp^2}{k_\perp^2} \frac{\alpha_s N_c}{\pi} \hat{q}(\tau, k_\perp^2). \tag{42}$$

It is further argued that the scale evolution of $\hat{q}$ will render additional medium-length dependence for jet quenching parameter, parton transverse momentum broadening as well as parton energy loss as compared to traditional BDMPS jet energy loss formalism, i.e.,

$$\hat{q}(L) \propto L^\gamma, \quad \langle p_\perp^2 \rangle \propto \hat{q}_0 L^{1+\gamma}, \quad \Delta E \propto \hat{q}_0 L^{2+\gamma}, \tag{43}$$

where the extra dimension γ is found to be $\gamma = 2\sqrt{\alpha_s N_c/\pi}$ for large medium size.

The NLO radiative corrections to the transverse momentum broadening and the renormalization of jet quenching parameter $\hat{q}$ have also been studied in Refs. 123 and 124. This work was performed in the framework of hadron production in semi-inclusive deep-inelastic scattering and lepton pair production in p–A collisions (see Fig. 13). The cross-section weighted with $l_{h\perp}^2$ was computed and a factorization formula was obtained after absorbing the collinear divergence into the redefinitions of

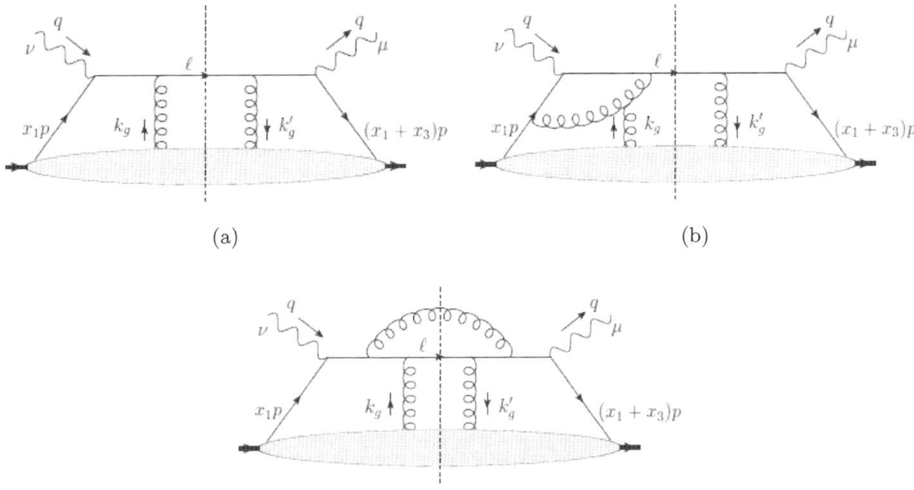

Fig. 13. Some typical diagrams for double scattering contributions to the $k_\perp^2$-weighted differential cross-section from leading-order (a), NLO virtual (b), and NLO real (c) processes.

non-pertubative correlation functions. The factorized form for the NLO correction to $l_{h\perp}^2$-weighted cross-section reads as:

$$\frac{1}{\sigma_0}\frac{d\langle l_{h\perp}^2\sigma\rangle}{dz}\bigg|_{\mathrm{NLO}} = T_{qg}(x,0,0,\mu_f^2)\otimes H_{\mathrm{NLO}}(x,x_B,Q^2,\mu_f^2)\otimes D_{h/q}(z_h,\mu_f^2)+\cdots, \tag{44}$$

where T_{qg} is the twist-4 quark–gluon correlation function, D is the quark-to-hadron fragmentation function, H_{NLO} is the NLO correction to the hard part. The twist-4 quark–gluon correlation function T_{qg} is non-perturbative and can not be directly calculated using perturbative QCD. But its scale dependence may be derived from perturbative QCD, which leads to a new QCD evolution equation (similar to DGLAP equations for PDFs). The scale dependence of twist-4 quark–gluon correlation function reads:

$$\frac{\partial T_{qg}(x_B,0,0,\mu_f^2)}{\partial \ln \mu_f^2} = \frac{\alpha_s}{2\pi}\int_{x_B}^{1}\frac{dx}{x}[\mathcal{P}_{qg\to qg}\otimes T_{qg} + P_{qg}(\hat{x})T_{gg}(x,0,0,\mu_f^2)], \tag{45}$$

where $\mathcal{P}_{qg\to qg}\otimes T_{qg}$ is given by

$$\begin{aligned}\mathcal{P}_{qg\to qg}\otimes T_{qg} =& P_{qq}(\hat{x})T_{qg}(x,0,0) + \frac{C_A}{2}\left(\frac{4}{(1-\hat{x})_+}T_{qg}(x_B,x-x_B,0)\right.\\ &\left. - \frac{1+\hat{x}}{(1-\hat{x})_+}T_{qg}(x,0,x_B-x) + T_{qg}(x_B,x-x_B,x-x_B)\right).\end{aligned} \tag{46}$$

Here $P_{qq}(\hat{x})$ and $P_{qg}(\hat{x})$ are the usual splitting functions in the DGLAP evolution equations, with $\hat{x} = x_B/x$. From the above new QCD evolution equation, the scale dependence of jet quenching parameter $\hat{q}$ may be in principle obtained if one assumes the nucleus is composed of loosely-bounded nucleons. Neglecting the momentum and spatial correlations of two nucleons inside the nucleus, one may relate the twist-4 quark–gluon correlation function T_{qg} to the jet quenching parameter $\hat{q}$ as follows:

$$T_{qg}(x_B, 0, 0, \mu_f^2) \approx \frac{N_c}{4\pi^2\alpha_s} f_{q/A}(x_B, \mu_f^2) \int dy^- \hat{q}(y^-, \mu_f^2). \tag{47}$$

Unfortunately, the above evolution equation is not closed, and a closed evolution equation can only be obtained for certain kinematic limit. In the limit of large x_B ($x_B \to 1$), the formation time of the radiated gluon is much large than the size of the nuclear medium, so gluon radiation is suppressed due to the LPM interference effect. In this case, the splitting kernel $\mathcal{P}_{qg\to qg} \otimes T_{qg}$ reduces to the vacuum one $P_{qq}(\hat{x})T_{qg}(x, 0, 0)$, and one gets scale-independent $\hat{q}$. For intermediate-x_B regime, one may expand the off-diagonal matrix elements T_{qg} around $x = x_B$ and only keep the leading contribution. Then the scale dependence of $\hat{q}$ can be obtained as follows[125]:

$$\frac{\partial \hat{q}(\mu^2)}{\partial \ln \mu^2} = \frac{\alpha_s}{2\pi} C_A \ln\left(\frac{1}{x_B}\right) \hat{q}(\mu^2). \tag{48}$$

One may see that the jet transport coefficient $\hat{q}$ is now not only μ-dependent, but also x_B-dependent. The solution of such equation is

$$\hat{q}(\mu^2) = \hat{q}(\mu_0^2) \exp\left[\frac{\alpha_s}{2\pi} C_A \ln\left(\frac{1}{x_B}\right) \ln\left(\frac{\mu^2}{\mu_0^2}\right)\right]. \tag{49}$$

This μ^2 dependence of $\hat{q}$ essentially is similar to the violation of scaling behavior of normal PDFs. The $x_B = Q^2/(2ME)$-dependence may be translated to the energy dependence of $\hat{q}$ in the target rest frame, which is consistent with the earlier expectations.[126]

5. Full Jet Evolution and Modification

The study of full jets and their nuclear modification can provide many insights into the mechanisms of jet-medium interaction and parton energy loss in ultra-relativistic heavy-ion collisions. The basic idea of reconstructing full jets is to recombine final state hadron fragments and to infer the information about the original hard partons and the medium effects. Since both leading and sub-leading fragments of full jets are included in the reconstructed full jets, they are more differential than single inclusive observables. Therefore they are expected to have the ability to put more stringent constraints on various approximations and assumptions made in jet energy loss calculations, which will be very helpful to eliminate the large systematic uncertainties in the quantitative extraction of the jet transport parameters. One of the biggest challenges in studying reconstructed full jets in ultra-relativistic heavy-ion collision is the contamination from fluctuations of large bulk background, as compared to elementary proton–proton collisions. The development of various experimental techniques in recent years have now enabled us to disentangle the reconstructed full jets from fluctuating background. The availability of many full jet related experimental observables have triggered a lot of theoretical activities as well. The goal of full jet study is to understand how full jets interact with medium constituents and lose energy, and how the medium effects manifest in the final state observables.

5.1. *Full jet studies at RHIC*

Early studies of reconstructed full jets in relativistic heavy-ion collisions were performed in Au–Au and Cu–Cu collisions at RHIC by both STAR and PHENIX collaborations.[127,128] Many interesting results have been obtained about reconstructed full jets and their medium modification, even though with large uncertainties. For example, Fig. 14 shows the measurements of the cross-sections (left) and the nuclear modification factor R_{AA} (right) for inclusive jet production in Au–Au collisions at $\sqrt{s_{NN}} = 200\,\text{GeV}$. The inclusive jet nuclear modification factor R_{AA}^{jet} is defined as follows:

$$R_{AA}^{\text{jet}} = \frac{\dfrac{dN_{AA}^{\text{jet}}}{dE_T dy}}{\dfrac{N_{\text{coll}} dN_{pp}^{\text{jet}}}{dE_T dy}}. \tag{50}$$

The dependence on different full jet reconstruction algorithms (k_T and anti-k_T) has been explored and turned out to be quite strong. Such strong dependence may originate from different responses to the fluctuating background in heavy-ion collisions

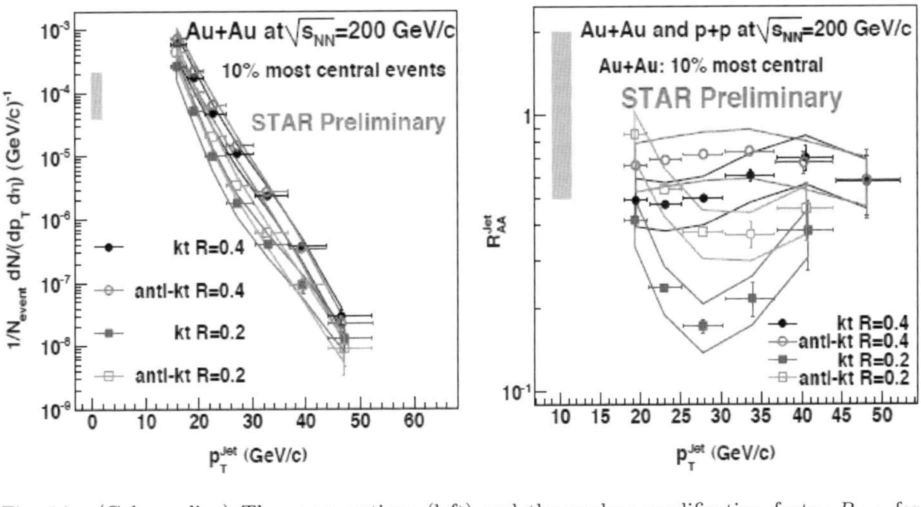

Fig. 14. (Color online) The cross-sections (left) and the nuclear modification factor R_{AA} for inclusive jet production in Au–Au collisions at $\sqrt{s_{NN}} = 200\,\text{GeV}$ measured by STAR Collaboration. Figures are taken from Ref. 127.

when different reconstruction algorithms are utilized. Despite large uncertainties and strong dependence on jet reconstruction algorithms, the results tend to indicate that there are substantial nuclear modification of full jets in heavy-ion collisions as compared to proton–proton collisions. Especially, the strong dependence of jet production and their nuclear modification on the jet resolution parameter R in heavy-ion collisions indicates that full jets are broadened due to their interaction with the hot and dense nuclear medium.

Earlier theoretical studies of full jets and their medium modification also supported the broadening of full jets as observed in heavy-ion collisions at RHIC. In Ref. 129, the medium effect on full jets were studied for Au–Au and Cu–Cu collisions at RHIC energies using the framework of GLV formalism. First, with the assumption of Poisson ansatz for the medium-induced gluon radiation, the differential radiation spectrum $dN_g/d\omega dr$ is utilized to calculate the probability distributions $P_q(\epsilon, E)$ and $P_g(\epsilon, E)$ for high-energy quarks and gluons which lose fraction of energy ϵ. The same differential radiation spectrum is also utilized to calculate the fraction of lost energy that is redistributed inside the jet cone:

$$f_{q,g}^{\rm in} = \frac{\displaystyle\int_0^R dr \int_{p_T^{\rm min}}^{E_T} d\omega \frac{dN_g}{d\omega dr}}{\displaystyle\int_0^{R_\infty} dr \int_{p_T^{\rm min}}^{E_T} d\omega \frac{dN_g}{d\omega dr}}, \tag{51}$$

where $p_T^{\rm min}$ is the momentum cut for the particles i that constitute the jet, and R is the jet cone radius. One expects that as $R \to R_\infty$ and $p_T^{\rm min} \to 0$, the fraction $f_{q,g}^{\rm in} = 1$, i.e., the final state medium effects to inclusive jet cross-section vanishes. The medium-modified jet cross-section per binary nucleon–nucleon scattering can be calculated as follows:

$$\frac{d\sigma^{AA}(R)}{dE_T dy} = \sum_{q,g} \int_0^1 P_{q,g}(\epsilon, E) \frac{1}{[1-(1-f_{q,g}^{\rm in})\epsilon]^2} \frac{d\sigma_{q,g}^{\rm CNM}}{d^2 E_T' dy}, \tag{52}$$

where $(1 - f_{q,g}^{\rm in})\epsilon$ represents the fraction of the lost energy from the parent parton outside the jet cone R, and $E_T' = E_T/[1-(1-f_{q,g}^{\rm in})\epsilon]$ is initial parton energy before interacting with the hot and dense QGP medium. Note that in the above expression, $p_T^{\rm min} = 0$ is used.

The result for inclusive jet $R_{AA}^{\rm jet}$ is presented in Fig. 15 (left panel) for central Au–Au collisions at $\sqrt{s_{NN}} = 200$ GeV at RHIC. The bands represent the uncertainties due to a $\sim 20\%$ change in the rate of parton energy loss relative to the default value. The nuclear modification factor for leading π^0 production is also shown for comparison. One can see clearly the strong dependence of inclusive jet R_{AA} on the jet cone size R: there is more suppression for inclusive jet for smaller jet radius. This indicates full jets is broadened due to additional medium-induced radiation when interacting with the hot and dense nuclear matter created in nucleus–nucleus

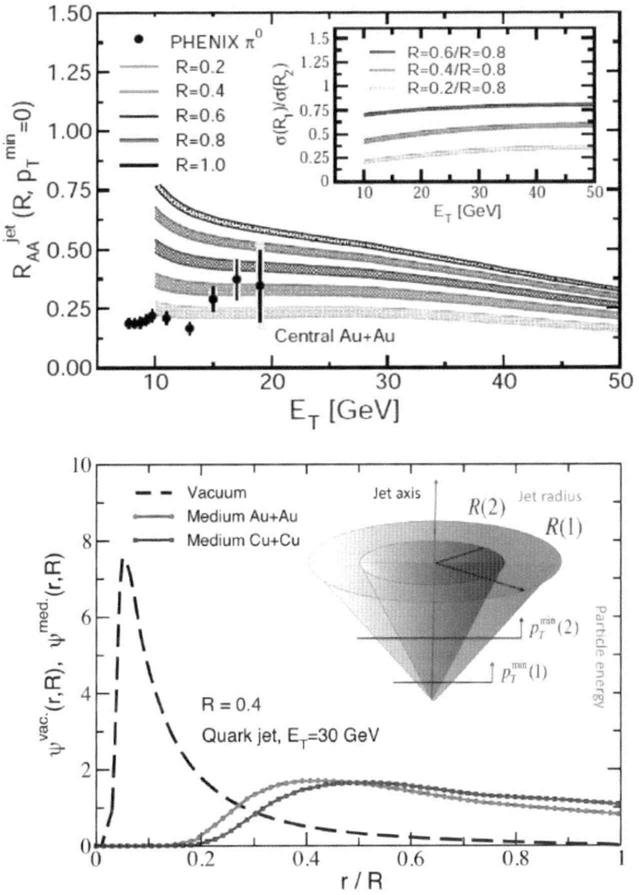

Fig. 15. (Color online) Upper: The nuclear modification factor R_{AA}^{jet} for different cone radii R in $b = 3\,\text{fm}$ Au–Au collisions at RHIC. The inserts show the ratios of jet cross-sections for different R. Lower: The differential jet shape in vacuum compared to the medium-induced jet shape for a quark with $E_T = 30\,\text{GeV}$ in Au+Au and Cu + Cu collisions at $\sqrt{s_{NN}} = 200\,\text{GeV}$. The insert shows a method for studying the substructure of full jets. Figures are taken from Ref. 129.

collisions. The inserted plot shows the ratio of nuclear modification factor for different cone radii in order to reduce the cold nuclear matter (CNM) effect which may contribute about 1/2 of the observed suppression for jet energies $E_T > 30\,\text{GeV}$.

The broadening effect of full jets in heavy-ion collisions may be investigated using jet shape observables. The differential jet shape function is defined as:

$$\psi\left(\frac{r}{R}\right) = \frac{d}{dr}\left[\frac{\sum_i E_{T,i}\theta(r - R_{i,\text{jet}})}{\sum_i E_{T,i}\theta(R - R_{i,\text{jet}})}\right], \tag{53}$$

where the sum over i runs over all the particles inside the full jet. By definition, the differential jet shape function is normalized to unity, $\int_0^R \psi(r,R)dr = 1$. The right plot in Fig. 15 shows the differential jet shape function in vacuum compared to medium-induced jet shape function in Au–Au and Cu–Cu collisions at RHIC. In the plot, the medium-induced jet shape function is directly calculated by using differential gluon bremsstrahlung spectrum with proper normalization: $\psi^{\text{med}} \propto dN_g/d\omega dr$. One can clearly see a characteristic large-angle distribution away from jet axis for medium-induced radiation. It should be noted that the final medium-modified jet shape function has contributions from both vacuum radiation and medium-induced radiation. It may be calculated as follows:

$$\psi^{\text{tot}}\left(\frac{r}{R}\right) = \sum_{q,g}\int_0^1 d\epsilon P_{q,g}(\epsilon, E)\frac{\chi_{q,g}(R, E_T, E'_T)}{[1-(1-f^{\text{in}}_{q,g})]^3} \times [(1-\epsilon)\psi^{\text{vac}}_{q,g}(r/R, E'_T) + \epsilon f^{\text{in}}_{q,g}\psi^{\text{med}}_{q,g}(r/R, E_T)], \tag{54}$$

where $\chi_{q,g} = \frac{d\sigma^{CNM}_{q,g}/d^2E'_T dy}{d\sigma^{AA}/d^2E_T dy}$. The first term in the bracket denotes the vacuum contribution, and the second is the contribution from medium-induced radiation. It is instructive to look at the first moment of the jet shape distribution, or the mean jet width, which may be obtained as follows:

$$\left\langle\frac{r}{R}\right\rangle = \int_0^1 d\left(\frac{r}{R}\right)\left(\frac{r}{R}\right)\psi\left(\frac{r}{R}\right). \tag{55}$$

For jets with transverse energies $E_T = 30\,\text{GeV}$ and cone size $R = 0.4$, one may calculate the broadening effect in terms of the change of mean jet width. For central Au–Au collisions, $\Delta\langle r/R\rangle = \langle r/R\rangle_{\text{tot}} - \langle r/R\rangle_{\text{vac}} = 0.283–0.271 = 0.12$, which is less than 5% increase. This results indicates even with striking suppression pattern for full jets as shown by jet R_{AA}, a very modest growth is seen for the change of mean jet width. More detailed information may be obtained if the full distribution of jet shape function and its medium modification are analyzed (see below).

5.2. *Full jets at the LHC era*

With the launch of the LHC, the center of mass energy of the collisions has been increased by more than a factor of 10 compared to the top collision energies at RHIC. The large kinematics accessed at the LHC have enabled us to investigate the properties of jets and the effects of hot and dense nuclear medium on the propagation of jets with transverse energies over a hundred GeV. One of the most interesting channels for studying the effect of jet-medium interaction is the correlated back-to-back jet pairs. The energy loss of full jets can be studied by measuring the imbalance/asymmetry of the transverse energies between two correlated jets (and the centrality dependence). One may define the energy asymmetry factor A_J for the correlated jet pairs as

$$A_J = \frac{E_{T,1} - E_{T,2}}{E_{T,1} + E_{T,2}}, \tag{56}$$

where $E_{T,i}$ $(i = 1, 2)$ denotes the transverse energy of the leading and sub-leading jets, respectively. The relative azimuthal angle $\Delta\phi = |\phi_1 - \phi_2|$ between two correlated jets can provide information about the degree of deflection that jet experience after passing through the hot and dense QGP medium. The transverse energy asymmetry A_J and the azimuthal angle $\Delta\phi$ of back-to-back jet pairs have been measured by both ATLAS and CMS Collaborations. ATLAS selected events with the leading jets having $E_{T,1} > 120\,\text{GeV/c}$ and the subleading jets having $E_{T,2} > 100\,\text{GeV/c}$, while for CMS $E_{T,1} > 50\,\text{GeV/c}$ and $E_{T,2} > 25\,\text{GeV/c}$. The results from ATLAS measurements are shown in Fig. 16, with the top panels for dijet asymmetry distributions and the bottom panels for azimuthal angle $\Delta\phi$ distribution.

One of the most spectacular results from such measurements is strong modification on dijet transverse energy imbalance distribution in Pb–Pb collisions, as compared to proton–proton collisions, while the distribution of their relative azimuthal angles is largely unchanged. The nuclear modification of dijet asymmetry strongly depends on the collision centrality: the modification is larger in more central collisions. The transverse energy imbalance between full jets and the correlated high p_T direct photons in Pb–Pb collisions have also been measured at the LHC,[13] and very similar results have been obtained. These observations indicated that the subleading jets may experience a significant amount of energy loss after traveling through the hot and dense nuclear matter created in relativistic heavy-ion collisions. These measurements have triggered a lot of theoretical interests on the investigation of parton shower evolution and full jet energy loss in hot QCD matter. Various jet energy loss based models have been utilized to explain the observed transverse energy imbalance between dijets and photon-jet pairs.[61,81,130–138]

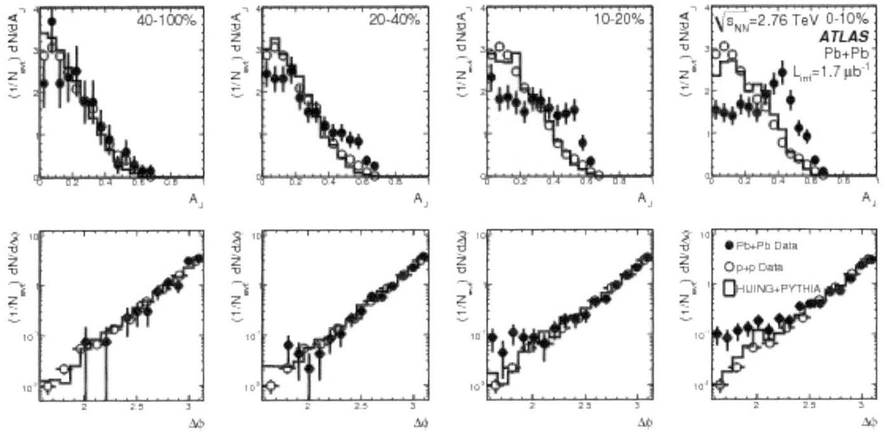

Fig. 16. (Color online) ATLAS 2010 measurements of dijet asymmetry distributions (top panels) and azimuthal angle $\Delta\phi$ distribution (bottom panels) for data (points) and unquenched HIJING + PYTHIA (solid yellow histograms) for different centralities (left to right from peripheral to central). Proton–proton data for 7 TeV is shown as open circles. Figures are taken from Ref. 11.

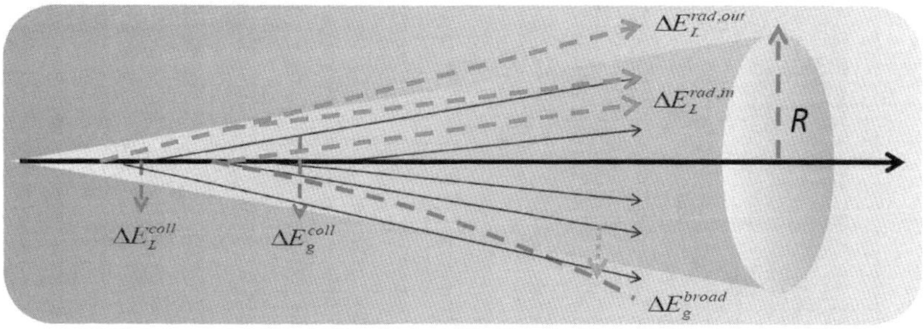

Fig. 17. (Color online) A schematic illustration of the evolution of full jet, and different medium-induced processes that contribute to full jet energy loss and modification in a quark–gluon plasma.

The evolution of full jet in a quark–gluon plasma may be schematically illustrated in Fig. 17, where the thick solid arrow through the center of jet cone represents the leading parton of the jet, and other lines represent the accompanying radiated gluons. One can see that compared to leading hadron observables, a few additional ingredients have to be taken into account when studying the evolution and modification of full jets in hot and dense nuclear medium. One needs to include the interaction of the medium with both the leading parton as well as with the radiated gluons. The radiative gluons may also get deflected by interacting with medium constituents, and some of them may be kicked out of the jet cone and contribute to the energy loss of full jet. Therefore, the total energy loss of energy from the jet cone is the sum of energy loss incurred by the leading parton and the accompanying radiated gluons, as well as the gluons that are scattered out of jet cone.

Based on the above picture, Ref. 81 performed the first quantitative analysis of physical processes that are responsible for full jet energy loss and the observed large di-jet energy asymmetry. The following transport equation is solved for the distribution of radiated gluons:

$$\frac{df_g(\omega, k_\perp, t)}{dt} = -\hat{e}\frac{\partial f_g}{\partial \omega} + \frac{1}{4}\hat{q}\nabla_{k_\perp}^2 f_g + \frac{dN_g^{\rm rad}}{d\omega dk_\perp^2 dt}, \tag{57}$$

where $f_g(\omega, k_\perp, t)$ is the double differential distribution of the accompanying gluons of the full jets. In the above equation, the first and second terms describe the evolution of radiated gluons which may transfer energy into the medium by elastic collisions with medium constituents and accumulate transverse momentum in the process. The sizes of these two effects are controlled by two transport coefficients: elastic energy loss rate $\hat{e}$ and the rate of exchanging transverse momentum squared $\hat{q}$. The last term in the equation represents the additional gluon radiation induced by the interaction of primary hard parton with the medium. These medium-induced radiative gluons, after produced, may also lose energy and accumulate transverse momentum during their propagation.

The above transport equation is combined with the higher-twist formalism which provides the rate for the medium-induced gluon radiation,[48,53]

$$\frac{dN_g^{\rm rad}}{d\omega dk_\perp^2 dt} = \frac{2\alpha_s}{\pi}\frac{xP(x)\hat{q}(t)}{\omega k_\perp^4}\sin^2\left(\frac{t-t_i}{2\tau_f}\right), \tag{58}$$

where $P(x)$ is the vacuum splitting function with $x = \omega/E$ denoting the energy fraction of radiated gluons, and $\tau_f = 2Ex(1-x)/k_\perp^2$ the formation time for the radiated gluons. The transport coefficient $\hat{q}$ depends on the evolution time because of the changing position of the primary parton.

After the above evolution equation for the radiated gluons is solved, one may obtain the information about the full jets after propagating through the medium. From the energy and transverse momentum distribution of the accompanying gluons, the total energy of these gluons contained inside the jet cone can be obtained as follows:

$$E_g(t_f, R) = \int_R d\omega dk_\perp^2 \omega f_g(\omega, k_\perp^2, t_f), \tag{59}$$

where the subscript R means that the integration is taken inside the jet cone, i.e., $k_\perp/\omega < R$. One may calculate the final energy of the leading parton after passing through the medium,

$$E_L(t_f) = E_L(t_i) - \int dt\hat{e}_L(t) - \int dt d\omega dk_\perp^2 \omega \frac{dN_g^{\rm rad}}{d\omega dk_\perp^2 dt}. \tag{60}$$

The total energy inside the jet cone is the sum of the leading parton energy and the energy of the radiated gluon inside the jet cone:

$$E_{\rm jet}(t_f, R) = E_L(t_f) + E_g(t_f, R). \tag{61}$$

Then one may calculate the reduction of the full energy,

$$\Delta E_{\rm jet}(t_f, R) = E_{\rm jet}(t_i, R) - E_{\rm jet}(t_f, R), \tag{62}$$

where $E_{\rm jet}(t_i, R)$ is the initial energy contained within the jet cone. The final expression for the energy loss from the jet cone may be obtained as follows:

$$\begin{aligned}\Delta E_{\rm jet}(t_f, R) = {}& \int dt d\omega dk_\perp^2 \omega \frac{dN_g^{\rm rad}}{d\omega dk_\perp^2 dt} + \int \hat{e}_L(t) dt \\ & - \int_R d\omega dk_\perp^2 \omega [f_g(\omega, k_\perp^2, t_f) - f_g(\omega, k_\perp^2, t_i)],\end{aligned} \tag{63}$$

where the first two terms represent the radiative and collisional energy loss of the leading hard parton. To calculate the loss of the energy from jet cone, one has to subtract the energy carried by the radiated gluons inside the jet cone (which is represented by the last two terms) from the leading parton energy loss. Therefore the energy of the original full jet is now decomposed into several parts:

$$E_{\rm jet} = E_{\rm in} + E_{\rm lost} = E_{\rm in} + E_{\rm out,rad} + E_{\rm out,brd} + E_{\rm coll,th}. \tag{64}$$

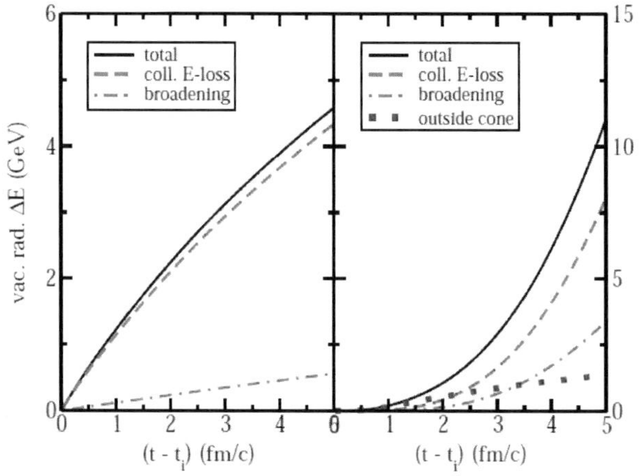

Fig. 18. (Color online) The energy loss experienced by a quark jet with initial energy $E = 100\,\text{GeV}$ traversing a quark–gluon plasma with constant temperature $T = 250\,\text{MeV}$. Contributions from different energy loss mechanisms are compared. The left for the medium effect on vacuum radiation, and the right for the medium effect on medium-induced radiation.[81]

Figure 18 shows the contributions from vacuum radiated gluons (left) and medium-induced radiated gluons (right) to the total energy loss of the jet defined by a cone angle $R = 0.4$. A quark jet is started with initial energy $E = 100\,\text{GeV}$ which radiates gluons in vacuum before $t_i = 1\,\text{fm/c}$, then the medium modification effect is turned on for the jet shower. The vacuum radiation is simulated using PYTHIA, and the effect of the medium on the full jet evolution is calculated via solving the above transport equation. One can see that at early times, the medium-induced energy loss from jet cone is dominated by the medium effect on vacuum radiation, and at later times it is dominated by the medium effect on medium-induced radiation. One interesting observation is that the most significant contribution originates from the collisional energy loss experienced by the radiated gluons $E_{\text{th,coll}}$. The transverse momentum broadening of shower gluons $E_{\text{out,brd}}$ also gives a sizable contribution. The contribution from the radiation directly outside the jet cone $E_{\text{out,rad}}$ is small, since the radiation is mainly dominated by small angle radiation, i.e., the radiated gluons are mostly inside the jet cone.

Figure 19 shows the calculation of dijet energy asymmetry factor A_J in proton–proton and Pb–Pb collisions at $\sqrt{s_{NN}} = 2760\,\text{GeV}$ at the LHC. In proton–proton collisions, A_J distribution for the correlated back-to-back dijets is peaked at $A_J = 0$. After the medium effect is turned on, the dijet energy asymmetry A_J distribution is significantly shifted to the large A_J direction. The asymmetry of dijet energies is more prominent in the most central Pb+Pb collisions (left panel) than less central events (right). The result indicate that the sub-leading jets experience a significant amount of energy loss from the jet cone. The total energy loss is a combination of collisional energy loss experienced by all shower partons, the radiation outside jet cone and the scattering of radiated gluons out of jet cone.

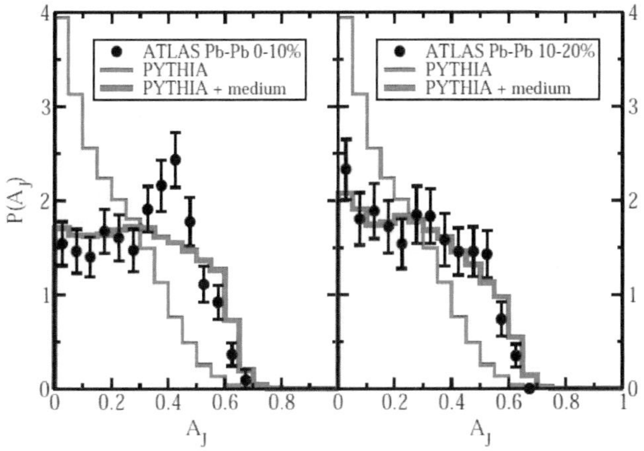

Fig. 19. (Color online) Di-jet energy asymmetry A_J distribution for proton–proton and Pb−Pb collisions at $\sqrt{s_{NN}} = 2760\,\text{GeV}$ at the LHC. The left panel is for 0–10% centrality and the right for 10–20% centrality.[81]

The above finding that the medium absorption or the thermalization of the soft gluon radiation gives the largest contribution to full jet energy loss is not surprising since one expects the soft components of the hard jet or the accompanying gluons at large angles to experience stronger medium modification than the inner hard core of the jet. Such effect is often referred to as jet collimation mechanism, i.e., when jet shower propagates through the hot and dense nuclear medium, soft partons of the jet are easily stripped off by the interaction with the medium constituents.[130]

Similar picture of parton shower evolution and full jet energy loss has also been obtained in other studies.[139–141] The basic idea underlying these studies is that: if the medium color field varies over the jet transverse size, i.e., $\lambda < r_\perp^{\rm jet}$, the interaction between the jet and the medium constituents may destroy the color coherence of the shower partons of the full jet. It is argued that such decoherence effect may greatly increase the phase space for soft and large angle radiation as compared to the traditional BDMPS-Z formalism. To the lowest order approximation, one may consider the successive gluon emissions as independent of each other. This will allows to treat multiple gluon emissions as a probabilistic branching process.

Let us consider the evolution of a gluon shower originating from successive gluon branchings. The differential probability for a gluon with energy ω (or the energy fraction $x = \omega/E$) to split into two gluons is given by

$$\frac{dP_{\rm br}}{dzdt} = \frac{\alpha_s}{2\pi}\frac{P_{g\to g}(z)}{\tau_{\rm br}(z,\omega)}, \tag{65}$$

where $P_{g\to g}(z) = N_c[1-z(1-z)]^2/[z(1-z)]$ is the leading order gluon–gluon splitting function, with z and $1-z$ are energy fractions of two daughter gluons with respect to their parent gluon. The time scale of gluon branching process $\tau_{\rm br}$ is given by

$$\tau_{\rm br}(z,\omega) = \sqrt{\frac{z(1-z)\omega}{[1-z(1-z)]\hat{q}}}, \tag{66}$$

where $\hat{q}$ is the jet quenching parameter. To simplify the following analysis, one may replace the light-cone time t by the dimensionless variable τ, defined as

$$\tau = \frac{\alpha_s N_c}{\pi}\sqrt{\frac{\hat{q}}{E}}t = \frac{\alpha_s N_c}{\pi}\sqrt{\frac{\hat{q}L^2}{E}}\frac{t}{L} = \frac{\alpha_s N_c}{\pi}\sqrt{\frac{2\omega_c}{E}}\frac{t}{L}, \tag{67}$$

where $\omega_c = \hat{q}L^2/2$ is the maximum energy that can be taken away by a single radiated gluon. The branching probability may now be written as

$$\frac{dP_{\rm br}}{dxd\tau} = \frac{\kappa(z)}{2\sqrt{x}} = \frac{1}{2\sqrt{x}}\frac{f(z)}{[z(1-z)]^{3/2}} = \frac{1}{2\sqrt{x}}\frac{[1-z(1-z]^{5/2}}{[z(1-z)]^{3/2}}, \tag{68}$$

where $\kappa(z) = f(z)/[z(1-z)]^{3/2}$ and $f(z) = [1-z(1-z)]^{5/2}$.

To study the multiple gluon branching process, it is convenient to consider the differential gluon spectra $D(x,\tau) = xdN/dx$, with N the number of gluons in the branching cascade. Using the fact that the successive branchings are independent of each other, one may solve the following rate equation for the function $D(x,\tau)$,

$$\frac{\partial D(x,\tau)}{\partial \tau} = \int dz \kappa(z)\left[\sqrt{\frac{z}{x}}D\left(\frac{x}{z},\tau\right) - \frac{z}{\sqrt{x}}D(x,\tau)\right]. \tag{69}$$

The first term in the equation is the gain term which describes the rise in the number of gluons at x due to the emission from gluons at larger values of x. The second term is the loss term which describes the reduction in the number of gluons at x due to the decay into gluons with smaller values of x. Note that the first integral should be taken as $x < z < 1$ since the function $D(x,\tau)$ has support only for $0 \le x \le 1$. Both gain and loss terms have singularities at the ending point $z = 1$, but the singularities exactly cancel each other when combining two terms together.

One may simplify the above evolution equation by changing the kernel to be $\kappa(z) = 1/[z(1-z)]^{3/2}$; such replacement is equivalent to taking $f(z) = 1$. This simplification allows us to solve the equation analytically, but does not affect the singular behavior of the kernel near $z = 0$ and $z = 1$. Furthermore, if one is only interested in the small-x behavior, one may further take $\kappa_0(z) = 1$. For $\tau \ll 1$, one may use iteration method to solve the above equation. Given the initial condition $D^{(0)}(x,\tau = 0) = \delta(x-1)$, which means that a single gluon carries all the energy, the first interaction gives:

$$D^{(1)}(x,\tau) = \frac{\tau}{\sqrt{x}(1-x)^{3/2}} \approx \frac{\tau}{\sqrt{x}}. \tag{70}$$

In the above equation (and equations below in this subsection), the last approximation is taken for $x \ll 1$, i.e., one is interested in the soft components of the shower gluons. Taking $t = L$, the first iteration coincides with the traditional BDMPS-Z

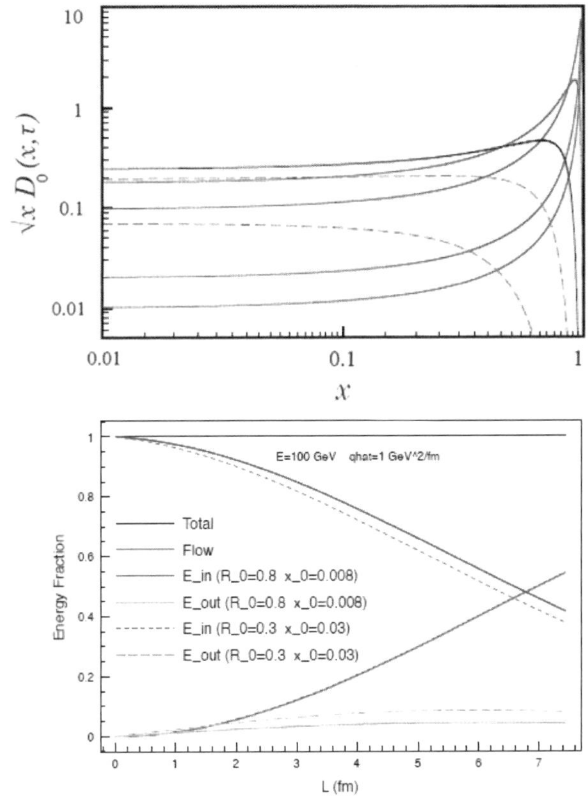

Fig. 20. (Color online) Upper: The τ evolution of $\sqrt{x}D_0 x, \tau$, where $\tau = 0.01, 0.02, 0.1, 0.2, 0.4$ for solid lines from bottom to top and $\tau = 0.6, 0.9$ for dashed lines from top to bottom. Lower: The energy fraction for different components $E_{\rm in}$, $E_{\rm flow}$ and $E_{\rm out}$ as a function of the medium size L for two values of jet size $R_0 = 0.3$ and $R_0 = 0.8$. Figures are taken from Refs. 141 and 142.

spectrum, $\omega dN/d\omega = \alpha_s N_c/\pi\sqrt{2\omega_c/\omega}$. By going beyond the first iteration, one may obtain the complete solution,

$$D(x,\tau) = \frac{\tau}{\sqrt{x}(1-x)^{3/2}} e^{-\pi[(1-x)\tau^2]} \approx \frac{\tau}{\sqrt{x}} e^{-\pi\tau^2}. \tag{71}$$

The above solution is shown in Fig. 20 (left panel), where $\sqrt{x}D(x,\tau)$ is plotted as a function of x for different values of τ: for solid lines $\tau = 0.01, 0.02, 0.1, 0.2, 0.4$ from bottom to top, and for dashed lines $\tau = 0.6, 0.9$ from top to bottom. One may see that for small times, the spectrum at small x increases linearly with τ, besides the exponential factor. This is very different from the DGLAP evolution, in which the spectrum get steeper and steeper at small x with increasing evolution time.

One very interesting and important feature of such branching dynamics is that the total energy contained in the spectrum is not conserved but instead decreases with the evolution time, $\epsilon(\tau) = \int_0^1 dx D(x,\tau) \approx e^{-\pi\tau^2}$. This means that as the energy flows from the higher to lower values of x, there is no energy accumulated at any value at $x > 0$. As the time evolves, the gluon spectrum will eventually disappear into the left moving "shock wave", and the energy will be accumulated into a "condensate" at $x = 0$. Therefore, with increasing evolution time, there could be a substantial fraction of the total jet energy flowing outside the spectrum. The amount of such energy flow is given by

$$\epsilon_{\text{flow}}(\tau) = 1 - \int_0^1 dx D(x,\tau) \approx 1 - e^{-\pi\tau^2}. \tag{72}$$

Besides the energy flow outside the spectrum that contributes to full jet energy loss, there is also contribution from the radiated energy that is emitted directly at large angles. To estimate such contribution, one may introduce a scale x_0 which is related to the radiation angle θ_0 as: $\theta_0 = [(2\hat{q})/(x_0^3 E)]^{1/4}$. The energy fraction transported outside jet cone via the gluon emission at large angles can be calculated as follows:

$$\epsilon_{\text{out}}(\tau) = \int_0^{x_0} D(x,\tau) = \frac{2\tau\sqrt{x_0}}{\sqrt{1-x_0}} e^{-\pi\tau^2} \approx 2\tau\sqrt{x_0} e^{-\pi\tau^2}. \tag{73}$$

One can see that this quantity rapidly decreases with decreasing x_0, therefore the gluon radiation directly outside the jet cone angle seems not be able explain the observed full jet energy loss and dijet energy asymmetry.

Another scale $x_{\text{th}} = T/E$ may be introduced to take into account the fact that when the gluon energies becomes as low as the typical energy scale in the medium, the radiated gluons may "thermalize" and disappear from the jet. Combining all these effects, the total phase space for the branching gluons may be divided into three different parts separated by two scales, x_0 and x_{th}. One may decompose the total energy of the original jet as follows:

$$E_{\text{jet}} = E_{\text{in}} + E_{\text{lost}} = E_{\text{in}}(x > x_0) + E_{\text{out}}(x_{\text{th}} < x < x_0) + E_{\text{flow}}(x < x_{\text{th}}). \tag{74}$$

For $x > x_0$, the branching gluons are inside the jet cone and can be recovered by jet reconstruction. For the radiated gluons with $x_{\text{th}} < x < x_0$, they are radiation directly outside the jet cone. For $x < x_{\text{th}}$, the branching gluons thermalize and flow into the medium (thus disappear from the jet).

Figure 20 (right panel) shows the relative contributions to the total jet energy (loss) from these three different components, E_{in}, E_{flow} and E_{out}, as a function of the medium size L. One can see that with increasing jet cone angle from $R_0 = 0.3$ to $R_0 = 0.8$, the energy contained inside the jet cone increases only slightly. For the medium length $L \geq 4$ fm, more than 20% of the total jet energy is lost from the jet cone. The largest energy loss component is found to be the energy flow out of the gluon spectrum (which is independent of R_0). The radiation directly outside the jet cone gives very small contribution to full jet energy loss. This full jet energy loss picture is very similar to what has been found in Ref. 81.

5.3. *Effects of medium recoils*

During the propagation jet shower partons, the leading parton loses energy and experiences transverse momentum broadening due to multiple scattering. The lost energy and momentum transfer will propagate through the medium via recoiled thermal partons, which lead to jet-induced medium excitations. Some of these recoiled medium partons will be distributed within the jet cone while an increasing fraction will be transported outside the jet cone over the course of the jet transport. Inclusion of the jet-induced medium excitation in the jet reconstruction will affect the final jet energy loss and jet profiles. Such effects of medium recoil can be studied within the LBT model.[60–62] Shown in Fig. 21 is the energy loss of reconstructed jets using the FASTJET algorithm with a cone size $R = 0.3$ as a function of time for an initial gluon with energy $E_0 = 50$ and 100 GeV propagating through a uniform QGP medium with a constant temperature $T = 400\,\text{MeV}$. Only elastic scattering with a constant $\alpha_s = 0.3$ is considered. The open symbols represent jets without "negative" partons due to back-reaction while solid symbols represent jets in which energy of the "negative" partons are subtracted.

Th effect of "negative" partons from back-reaction is negligible during the early stage of the parton propagation. At later times, as the number of "negative" partons grow, they deplete significantly the thermal medium behind leading partons and effectively modify the background underlying the jet. When such modified background is taken into account via the subtraction of "negative" partons from the jet cone, the effective jet energy loss becomes bigger as shown in Fig. 21. The linear

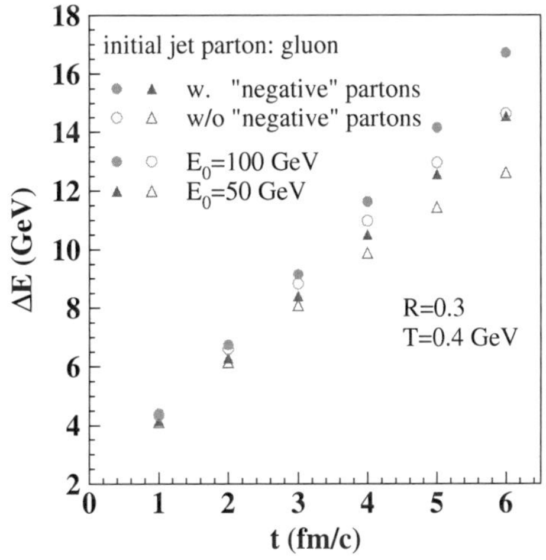

Fig. 21. (Color online) Energy loss of reconstructed jets due to elastic scattering from an initial gluon with energy $E_0 = 50$ and 100 GeV in a uniform QGP medium at a temperature $T = 400$ MeV as a function of propagation time. Jets are reconstructed with all partons (leading and jet-induced medium partons) (solid symbols) or without "negative" partons (open symbols).[62]

distance dependence of the effective jet energy loss is restored only after the energy of "negative" partons is subtracted.

In a realistic case of heavy-ion collisions, one has to include both elastic scattering and induced gluon radiation in an expanding medium as described by relativistic hydrodynamic evolution. In Refs. 60–62, the case of γ-tagged jets in heavy-ion collisions is considered. The initial γ-jet production from HIJING is distributed according to the overlap function of two colliding nuclei with a Wood–Saxon nuclear geometry. The same kinematic cuts for γ-jets are applied as in experiments at LHC. For CMS data,[13] $p_T^\gamma > 60\,\text{GeV}$, $|\eta^\gamma| < 1.44$, $p_T^{\text{jet}} > 30\,\text{GeV}$, $|\eta^{\text{jet}}| < 1.6$, and $\Delta\phi = |\phi^{\text{jet}} - \phi^\gamma| > 7\pi/8$, and for ATLAS data,[143] $60 < p_T^\gamma < 90\,\text{GeV}$, $|\eta^\gamma| < 1.3$, $p_T^{\text{jet}} > 25\,\text{GeV}$, $|\eta^{\text{jet}}| < 2.1$, and $\Delta\phi > 7\pi/8$.

Shown in Fig. 22 (left) is the average total jet energy loss with a cone-size $R = 0.3$ as a function of time in the most 10% central Pb+Pb collisions at $\sqrt{s} = 2.76\,\text{TeV}$ from LBT simulations with $\alpha_s = 0.2$. The space–time evolution of the medium is given by Hirano *et al.*[144] Because of rapid cooling due to $(3+1)$-dimensional expansion, the total jet energy loss flattens out approximately $10\,\text{fm}/c$ after an initial linear rise (solid line). One can see that inclusion of recoiled partons significantly reduces the net jet energy loss as compared to the case when only shower partons (dotted) and radiated gluons (dashed) are included in the jet reconstruction.

Energy loss for a propagating parton is always accompanied by transverse momentum broadening. This should also be reflected in the γ-jet azimuthal correlation. As a consequence of the rapid medium expansion, the γ-jet azimuthal correlation in central Pb+Pb collisions (solid) as shown in Fig. 22 (right) change very little as compared to that in p+p collisions (dotted). This is in agreement with CMS data[13] within errors. There is still, however, significant broadening at large values of azimuthal angle difference $\Delta\phi = \phi - \pi \equiv \phi_{\text{jet}} - \phi_\gamma - \pi$. Future precision data are therefore necessary to measure p_T-broadening of reconstructed jets.

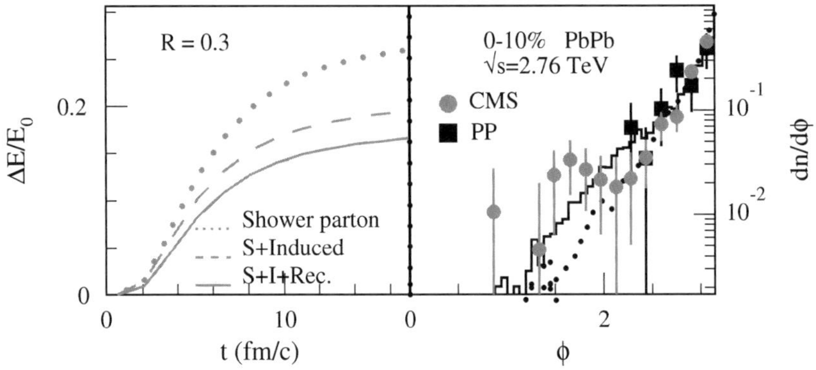

Fig. 22. (Color online) Averaged total energy loss as a function of time (left) and azimuthal jet distribution relative to the γ (right) for γ-tagged jets in central (0–10%) Pb+Pb collisions at $\sqrt{s} = 2.76\,\text{TeV}$.[61]

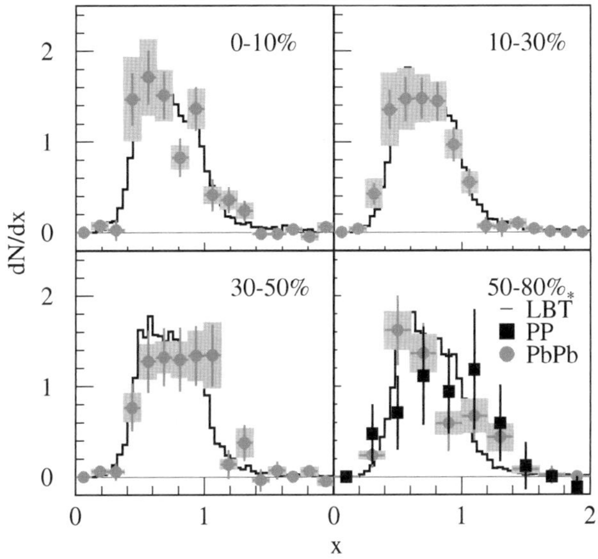

Fig. 23. (Color online) γ-jet asymmetry distribution in $x = p_T^{\rm jet}/p_T^{\gamma}$ in Pb+Pb collisions at $\sqrt{s} = 2.76\,\text{TeV}$ from LBT with $\alpha_s = 0.2$ as compared to CMS data[13] (from Ref. 61).

Shown in Fig. 23 are γ-jet asymmetry distributions dN/dx with $x = p_T^{\rm jet}/p_T^{\gamma}$ from LBT model simulations (histogram) as compared to CMS data[13] (solid circles) in Pb+Pb collisions at $\sqrt{s} = 2.76\,\text{TeV}$ with four different centralities. γ jets are produced with a large momentum asymmetry in $p+p$ and peripheral Pb+Pb collisions due to initial state radiations. In central Pb+Pb collisions, the asymmetry distributions are skewed to smaller values of x due to jet energy loss. The LBT results can fit the experimental data of both $p+p$ and Pb+Pb with different centralities quite well with a fixed value of $\alpha_s = 0.2$.

5.4. *Full jet substructure*

The elastic and inelastic interactions between the hard jet and the dense nuclear medium not only reduce the total energy of the reconstructed jet, but also change the momentum distribution among the constituents inside the jet. In particular, the scatterings experienced by quarks and gluons in the jet shower may result in both energy loss and deflection of jet constituents. The deflection and additional medium-induced radiation may broaden the parton shower and kick the propagating partons out of jet cone. To fully characterize the effects of jet quenching, one needs not only the modification of total full jet energy, but also the detailed information about the substructures of full jets. There are two popular observables for the characterization of jet substructures: jet fragmentation function $D(z)$ and jet shape function $\rho(r)$ (sometimes written as $\psi(r)$). Jet fragmentation function measures the momentum spectrum of hadrons inside the full jet while jet shape function measures the momentum distribution of the jet constituents in the radial direction. Both jet shape function and jet fragmentation function have been measured for Pb–Pb collisions at the LHC by the ATLAS and CMS Collaborations.[15–17]

Jet shape function $\rho(r)$ provides the information about the radial distribution of the momentum carried by the jet constituents (fragments). The differential jet shape function is defined as follows:

$$\rho(r) = \frac{1}{\delta r}\frac{1}{N_{\rm jet}}\sum_{\rm jet}\sum_{h}\frac{p_T^h}{p_T^{\rm jet}}\theta\left[r_h - \left(r - \frac{\delta r}{2}\right)\right]\theta\left[\left(r + \frac{\delta r}{2}\right) - r_h\right], \tag{75}$$

where $r_h = \sqrt{(\eta_h - \eta_{\rm jet})^2 + (\phi_h - \phi_{\rm jet})^2}$, and δr is the bin size. By definition, the differential jet shape function is normalized to unity, $\int \rho(r)dr = 1$. One may define the integrated jet shape function as: $R(r) = \int_0^r \rho(r)dr$. One of the important consequences from jet quenching is the broadening of full jets as compared to vacuum jets, due to a combination of medium-induced radiation and the deflection of shower partons. This effect is usually reflected as an enhancement at large values of r in differential jet shape function $\rho(r)$. Due to the normalization of jet shape function to unity, and an enhancement at large r must be compensated by a depletion at smaller values of r.

Figure 24 shows the nuclear modification of differential jet shape function $\rho(r)$ in Pb–Pb collisions with respect to proton–proton collisions at the LHC measured by CMS Collaborations.[15] A strong centrality dependence is observed for the jet shape functions in Pb–Pb collisions. In most peripheral collisions, the jet shape functions in Pb–Pb collisions are similar to those in proton–proton collisions, which indicates the radial distribution of the particle momentum inside the jets is similar in these

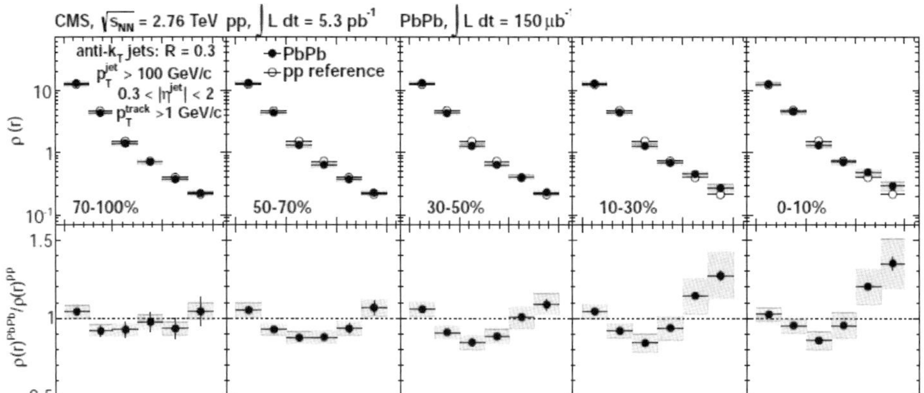

Fig. 24. (Color online) Top: CMS measurements of the differential jet shapes in Pb–Pb collisions (filled circles) as a function of distance from the jet axis for inclusive jets with $p_T^{\rm jet} > 100\,{\rm GeV/c}$ and $0.3 < \eta < 2\,{\rm GeV/c}$ in five Pb–Pb centrality intervals. Charged particles with $p_T^{\rm track} > 1\,{\rm GeV/c}$ are used. Bottom: Jet shape nuclear modification factors, $\rho(r)^{\rm PbPb}/\rho(r)^{pp}$. Figures are taken from Ref. 15.

two collision systems. In more central Pb–Pb collisions (0–10% and 10–30%), one observes an excess of transverse momentum fraction emitted at large radius $r > 0.2$ together with a depletion at intermediate radii, $0.1 < r < 0.2$. This indicates a moderate broadening of the jets after their traversing through the medium. There is very little change at small radii, indicating that the energy distribution in the inner core of the jet is not affected by the jet-medium interaction. The result is consistent with previous CMS measurements in which the lost energy from the jets to the medium is found at large distances from the jet axis outside the jet cone.[12] It is expected from jet quenching picture and consistent with some theoretical calculations.[145,146]

Jet fragmentation function $D(z)$ provides the information about the spectrum of the momentum carried by the jet fragments, where $z = \mathbf{p}_T^h \cdot \mathbf{p}_T^{\text{jet}}/|\mathbf{p}_T^{\text{jet}}|^2$ is the momentum fraction of the jet fragment. Jet fragmentation function is defined as follows:

$$D(z) = \frac{1}{N_{\text{jet}}} \frac{dN_h}{dz}. \tag{76}$$

The interaction between the hard jet and the dense medium usually leads to a soften of jet fragmentation function due to additional medium-induced radiation. This will be reflected in the jet fragmentation function as an enhancement at small values of z. Due to the energy conservation, an enhancement at small z must be accompanied by a depletion at higher values of z.

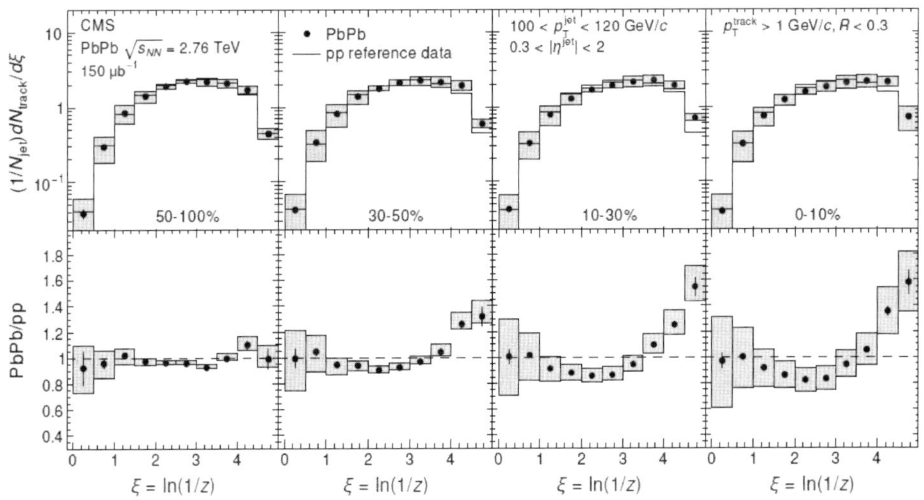

Fig. 25. (Color online) Top: CMS measurements of the fragmentation functions in bins of centrality (increasing from left to right) in Pb–Pb collisions overlaid with proton proton reference data. Jets have $100 < p_T < 120$ GeV/c, and tracks have $p_T > 1$ GeV/c. Bottom: The ratio of each Pb–Pb fragmentation function to its pp reference. Figures are taken from Ref. 16.

The first CMS measurement of jet substructure showed little nuclear modification to the jet fragmentation function (with large uncertainties) in Pb–Pb collisions compared to proton–proton collisions.[15] Later, CMS Collaboration included the charged particles with lower values of p_T in the reconstruction of full jets and a clear nuclear modification of the inclusive jet fragmentation function was observed.[16] Fig. 25 shows the latest CMS measurement of jet fragmentation function (the top panels show jet fragmentation function in Pb–Pb and proton–proton collisions and the lower panels show their ratios, PbPb/pp). One can see that the nuclear modification of the fragmentation function of jets in Pb–Pb collisions grows with increasing collision centrality. The ratio of PbPb/pp is almost flat at unity in the peripheral collisions (50–100% centrality bin), which indicates very little nuclear modification to particle spectrum inside the jet in peripheral Pb–Pb collisions. For the most central 0–10% collisions, a significant excess is observed at high $\xi = \ln(1/z)$ (low z), together with a depletion in the intermediate ξ regime. These results indicate that the spectrum of particles inside the full jet has an enhanced contribution in the soft regime, as compared to that from proton–proton collisions. CMS has also measured jet fragmentation function $D(z)$ for different values of jet transverse momentum $p_T^{\rm jet}$ (up to 300 GeV/c). No significant $p_T^{\rm jet}$ dependence was observed for the modification pattern within statistical and systematic uncertainties.

ATLAS Collaboration has also measured the nuclear modification of jet fragmentation function for Pb–Pb collisions at the LHC. Figure 26 shows the ratios of fragmentation functions, $D(z)$, for central (0–10%) collisions to those in peripheral (60–80%) collisions, for two different values of jet cone size ($R = 0.2$ and $R = 0.3$). Similar to CMS results, an enhanced yield of low z and large z fragments together

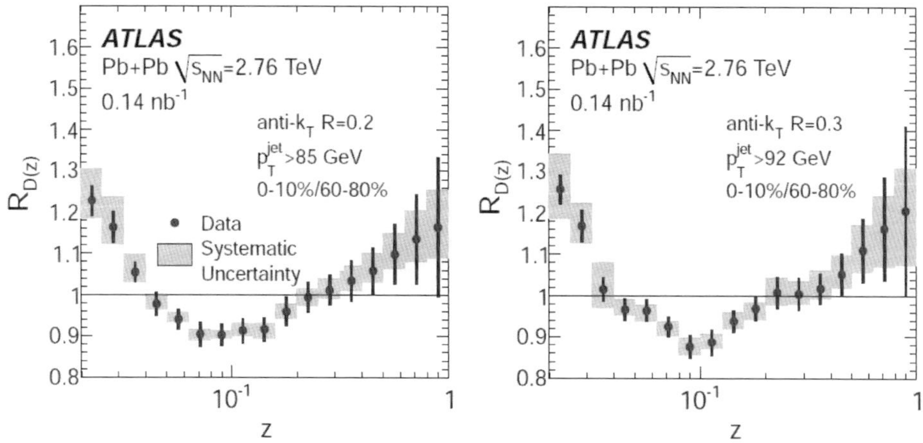

Fig. 26. (Color online) ATLAS measurements of the ratios of fragmentation functions, $D(z)$ for central (0–10%) collisions to those in peripheral (60–80%) collisions for $R = 0.2$ (bottom left) and $R = 0.3$ (bottom right) jets. Figures are taken from Ref. 17.

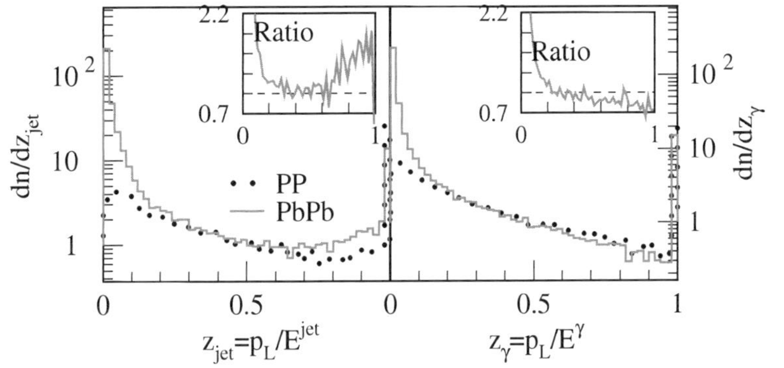

Fig. 27. (Color online) The reconstructed (left) and γ-tagged jet fragmentation functions (right) within a jet cone $R = 0.3$ of γ-tagged jets in central (0%–10%) Pb+Pb collisions at $\sqrt{s} = 2.76$ TeV.[61]

with a suppressed yield of fragments at intermediate z was observed in more central collisions compared to more peripheral collisions.

Jet substructure and their medium modification have also been studied in some details by some parton energy loss models.[61,138,147–149] For example, features of the observed medium modification of jet substructure can be qualitatively reproduced and understood by the LBT model for jet propagation. Shown in Fig. 27 (left) is the jet fragmentation function for partons within the jet cone as a function of momentum fraction $z_{\rm jet}$ that is defined by the reconstructed jet energy $z_{\rm jet} = p_L/E_{\rm jet}$ as in the LHC experimental data on single jets.[17] A large enhancement at small $z_{\rm jet}$ is due to soft partons from the bremsstrahlung and medium recoil within the jet-cone. There is also almost no medium modification at intermediate $z_{\rm jet}$. Similarly to the experimental data, there is an enhancement at large $z_{\rm jet}$. This enhancement is caused by the dominance of leading partons in the reconstructed jet energy whose energy is reduced because a large fraction of nonleading partons is transported outside the jet cone. To illustrate this point, we plot in Fig. 27 (right) γ-tagged jet fragmentation function[150] whose momentum fraction is defined as $z_\gamma = p_L/E_\gamma$ by the γ energy which more or less reflects the initial jet energy before medium modification. As a consequence of parton energy loss, the γ-tagged jet fragmentation function at large z_γ is significantly suppressed. In addition, there is also an enhancement at small z_γ. Therefore, such γ-tagged jet fragmentation function should be more sensitive to parton energy loss and can be used to extract jet transport parameters of the hot medium in heavy-ion collisions.

In general, there are several possible sources that are helpful to explain the observed nuclear modifications and the redistribution of jet energy inside the cone. First, the change in the fraction of jets originating from quarks and gluons may give different energy distribution inside jet cone. The extra radiation associated with inelastic collision process may lead to the broadening of the parton shower, and the softening of the spectrum of the jet fragments. Particles resulting from the medium response to jet transport may also contribute to the energy redistribution in the final reconstructed jets. For example, different ways of treating the recoiled partons may affect the final jet substructure and the energy distribution inside the jet.[61] Fragmentation and recombination mechanisms may lead to different jet fragmentation profiles.[147] In short, the understanding of the interplay between the nuclear modifications in different r of jet shape function and different z of jet fragmentation function can put very tight constraints on modeling jet-medium interaction and medium-induced energy loss. This provides a new window to probe various transport properties of the hot and dense QGP created in relativistic nuclear collisions.

6. Medium Response to Jet Transport

Hard partonic jets lose energy when propagating through QGP medium; some of the lost energy from the hard jets is deposited into the traversed medium. The deposited energy and momentum from the jets may thermalize into the medium, which may contribute to many jet-related observables, such as the angular correlations of hadrons with triggered high-p_T hadrons or direct photons. The space–time evolution and the final anisotropic flow pattern of the bulk matter may be affected by the transport and mementum deposition of the hard jets.[151,152] Particles produced from jet-induced medium excitations may flow into the jet cone and contribute to the energy redistribution in the final reconstructed full jets. Therefore, the study of the medium response to jet-deposited energy and the knowledge of the fate of medium excitations are essential for a complete understanding of jet-medium interaction.

One of the most interesting phenomena in studying the medium response to hard jets is the induced Mach cone structure.[153,154] If observed in heavy-ion experiments, it provides a direct probe to the speed of sound of the hot and dense nuclear medium. But in the dynamically evolving media as produced in relativistic heavy-ion collisions, the Mach cone pattern may be strongly distorted by the large collective flow due to the hydrodynamic expansion of the bulk matter.[155–157] Jet-induced Mach cone structure is also sensitive to many transport properties of the traversed medium, such as the shear viscosity to entropy density ratio.[157,158]

Similar to jet quenching and energy loss studies, the study of the medium response to hard jets may be performed in two different frameworks: full Boltzmann transport approach and jet plus hydrodynamics approach. In the full Boltzmann transport approach such as the AMPT and BAMPS models,[159,160] both the propagation of the jets in the medium and the medium response to jet transport are simulated at the same time in one Monte Carlo transport package. In these transport models, the bulk matter is usually described by a collection of quasi-classical quarks and gluons. The interaction between jets and the medium constituents has the same cross-section as the interaction among medium constituents. In order to

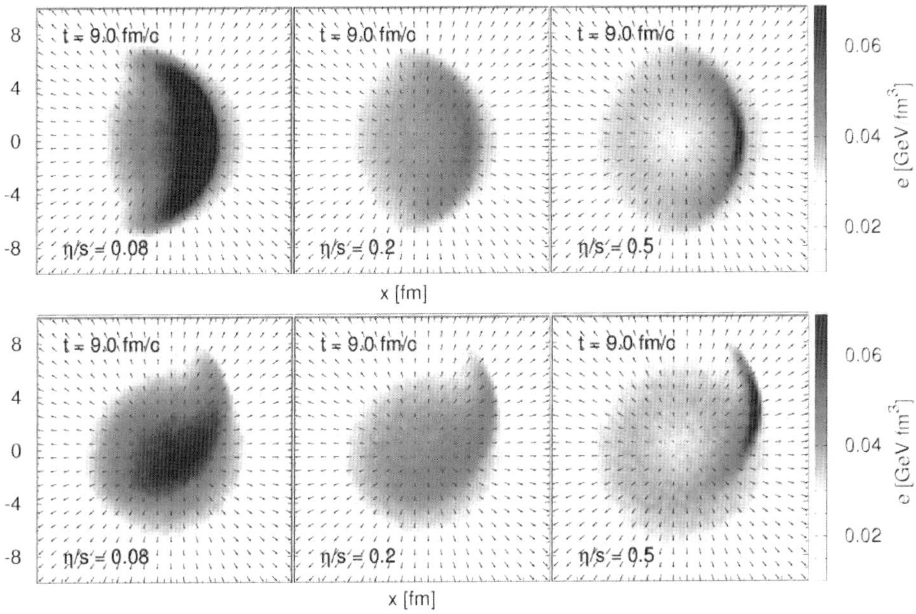

Fig. 28. (Color online) Energy density distribution in the transverse plane around mid-rapidity at $\tau = 0.9$ fm/c, overlaid by the velocity profile indicated by arrows. The results are shown three different values of shear viscosity to entropy density ratio η/s. The parton is initialized at a fixed-angle position of $(r, \phi) = (4\,\text{fm}, \pi)$ (upper panels) and $(r, \phi) = (4\,\text{fm}, 3\pi/4)$ (lower panels) with an initial momentum along the direction of x-axis, $p_x = E = 20\,\text{GeV}$. Figures are taken from Ref. 157.

describe the strong collective flow behavior of the bulk matter as indicated by relativistic hydrodynamics calculation and experimental data, usually a very large partonic cross-section (or large coupling constant) is used in these calculations.

Figure 28 shows a simulation of the medium response to high-energy partons using the BAMPS parton cascade model.[157] In the calculation, the medium is treated as a gas of massless gluons, which follow the classical Boltzmann equation,

$$\left(\frac{\partial}{\partial t} + \frac{\mathbf{p}}{E} \cdot \frac{\partial}{\partial \mathbf{r}}\right) f(\mathbf{x}, \mathbf{p}, t) = C[f]. \tag{77}$$

Particles (both jet and medium partons) collide only via binary collisions with an isotropic cross-section. The effect of shear viscosity is investigated by changing the parton cross-section as follows:

$$\frac{\eta}{s} = \frac{0.4\,e}{(n\sigma)}, \tag{78}$$

where e and n are local energy and particle densities. In the plots, the parton is produced at $(r, \phi) = (4\,\text{fm}, \pi)$ in the upper panels and at $(r, \phi) = (4\,\text{fm}, 3\pi/4)$ in the lower panels, and traveling in the positive x-axis direction. The energy density distribution of the bulk matter in the transverse plane is shown, together with the velocity profiles. Three different values of shear viscosity to entropy density ratio η/s (from left to right) are used. The Mach cone structures are clearly seen for small viscosities, whereas for larger viscosities the characteristic structures are smeared out. When jets propagate in the directions not parallel to the radial flow (lower panels), the Mach cone patterns are strongly distorted by the radial flow.

In jet plus hydrodynamics approach, the space–time evolution of the hot and dense nuclear medium is first simulated by relativistic hydrodynamics. Then one calculates the medium effect on jet transport using the jet-quenching techniques as described in previous sections. The medium response to jet transport is then simulated via solving the hydrodynamic equations with a source term,

$$\partial_\mu T^{\mu\nu}(x) = J^\nu(x). \tag{79}$$

Here the source term $J^\nu(x)$ represents the space–time profiles of the energy and momentum deposited by the propagating hard jets,

$$J^\mu(x) = \left(\frac{dE}{dt}, \frac{d\mathbf{p}_\perp}{dt}, \frac{dp_\parallel}{dt}\right), \tag{80}$$

where $\mathbf{p}_\perp$ ($p_\parallel$) are the momentum components transverse (longitudinal) to the jet propagation direction. Currently, a parallel simulation of both jet evolution and the medium response at the same time is not yet available as it is time-consuming. Most studies only calculate the medium response to jet transport, and simply neglect the effect of the medium response (change) on the subsequent evolution of hard jets.

At a given time and space location, the energy and momentum deposited by the hard jet, the source term $J^\mu(x)$), can be obtained from jet quenching calculations. The energy deposited into the medium is equal to the collisional energy loss experienced by the propagating jet. For a single parton propagating through a QGP medium, the collisional energy loss rate may be calculated within high temperature field theory. Taking the leading logarithmic approximation, the collisional energy loss (energy deposition) rate is given by

$$\left.\frac{dE}{dt}\right|_{\text{dep}} = \left.\frac{dE}{dt}\right|_{\text{coll}} = \frac{1}{4}C_s\alpha_s^2 m_D^2 \left(\frac{4ET}{m_D^2}\right), \tag{81}$$

where $m_D^2 = 4\pi\alpha_s(1+N_f/6)T^2$ is the Debye screening mass squared and C_s is the color factor for the propagating parton. The momentum deposition rate can also be obtained in a similar way.

After the energy deposition profiles $J^\mu(x)$ are obtained from jet quenching calculation, the medium response can be simulated via solving hydrodynamic equation. Figure 29 shows the medium response to the lost energy by a pair of back-to-back partons as simulated via a $(3+1)$-dimensional ideal hydrodynamics simulation.[161] In the calculation, a pair of back-to-back partons are created at ($\tau = 0$, $x_0 = 1.5\,$fm, $y_0 = 0$, $\eta_s = 0$) with the energy $E_0 = 200\,$GeV and travel in the opposite direction along the x axis. The initial location is chosen as $x_0 = 1.5\,$fm to mimic the different

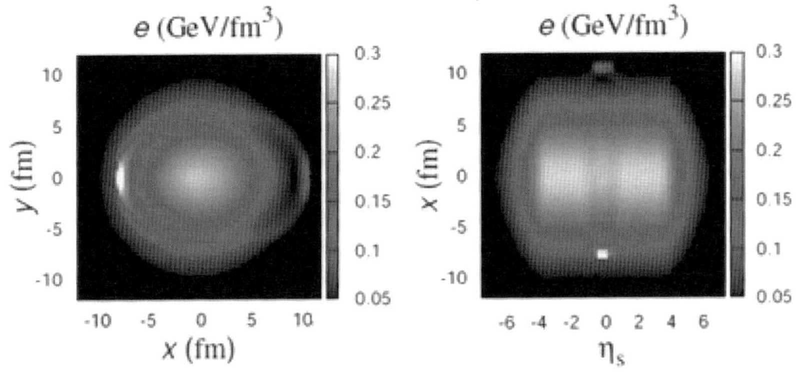

Fig. 29. (Color online) The energy density distribution of the QGP medium at $\tau = 9.6\,\text{fm/c}$ in the transverse plane at $\eta_s = 0$ and the reaction plane at $y = 0$. A pair of partons with the same energy $E_0 = 200\,\text{GeV}$ are produced at ($\tau = 0$, $x = 1.5\,\text{fm}$, $y = 0$, $\eta_s = 0$) and travel in the opposite (positive and negative x-axis) directions. Figures are taken from Ref. 161.

amounts of energy loss between two jets. The interaction between partons and the medium is switched on at $\tau_0 = 0.6\,\text{fm/c}$. The figure shows the energy density profiles of the QGP medium at the time $\tau = 9.6\,\text{fm/c}$ in the transverse (xy) plane at $\eta_s = 0$ (left panel) and the reaction plane ($x\eta$ plane) at $y = 0$ (right panel). One may see clearly the medium response and excitations generated by two energetic partons, as well as the distortion of Mach cone patterns by radial flow in the transverse plane.

Given the hydrodynamic evolution of the medium response, it is very interesting to see how the lost energy and medium excitations redistribute in the final state after hydrodynamic evolution. One may calculate the total p_T-balance of the entire system as follows:

$$\langle p_T^{\parallel} \rangle = -\sum_i \int dp_T d\phi_p p_T \cos(\phi_p - \phi_1) \frac{dN_i}{dp_T d\phi_p}, \tag{82}$$

where the sum is taken over all particles in the entire system. Here the projection "$\parallel$" is taken onto the sub-leading jet axis in the transverse plane. The momentum distribution of particles may be calculated from hydrodynamics using the Cooper–Fry formula.

In fact, this simulation was motivated by the CMS measurements of dijet momentum imbalance, where it is found that the lost energy from the jet are carried by soft particles at large angles.[12] To mimic dijet momentum imbalance A_J, one may change the initial position of the parton pairs x_0 such that two partons lose different amounts of energy loss. The results of the final state momentum distribution are shown in Fig. 30. One can see that the total transverse momentum of the whole system is well balanced (left panel). Inside jet cone (middle panel) the transverse momentum is dominated by hard-p_T particles. Outside jet cone (right panel) particles with $p_T < 2\,\text{GeV}$ dominate, and these low-p_T particles originate

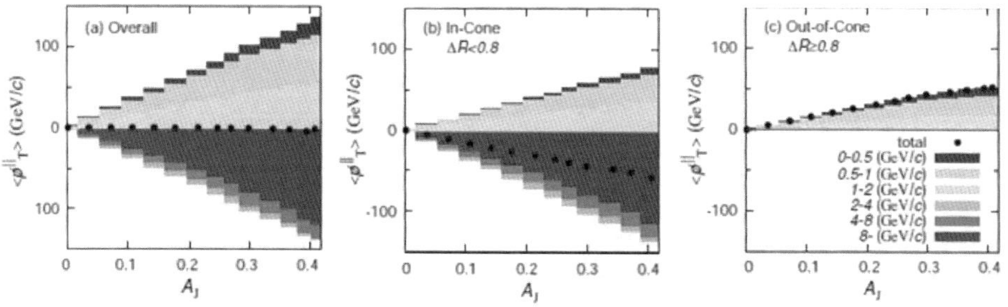

Fig. 30. (Color online) $p_T^{||}$ as a function of the dijet energy asymmetry variable A_J for the whole region (left panel), inside the jet cone $\Delta R < 0.8$ (middle panel), and outside jet cone $\Delta R > 0.8$ (right panel). Six bands represent the contributions in six p_T ranges: 0–0.5, 0.5–1, 1–2, 2–4, 4–8 GeV/c, and $p_T > 8$ GeV/c. The solid circles include the contributions from all p_T (and the propagating partons). Figures are taken from Ref. 161.

from the jet-deposited energy and momentum which are then transported by hydrodynamic fluid. This result is qualitatively consistent with CMS measurements of dijet momentum imbalance and the redistribution of lost energy.[12]

It should be noted that in the above calculations and many other studies of the medium response to jet transport, a simplified energy deposition profile (for a single parton) is taken as the input to simulate the hydrodynamic medium response. In fact, jets are collimated showers of partons which may lose energy by a combination of elastic scatterings with medium constituents and inelastic medium-induced emissions. In fact, both elastic and inelastic collisions play important roles in the study of energy–momentum deposition by the jet shower and the medium response. The radiative emissions can also deposit energy–momentum into the medium via rescattering with the medium constituents, thus the energy–momentum deposition profile for a jet shower is quite different from a single parton. Such effect has been investigated in Refs. 82 and 83 where it is found that the length dependence of energy deposition rate of a jet shower is very different from that of a single parton.

This can be clearly seen in Fig. 31, where the energy (left) and transverse momentum squared (right) deposited into the medium by a single parton which does not radiate are compared with those by a virtual parton which evolves into a parton shower. In the figure, the total energy loss from a single parton is also shown for comparison. To calculate the energy deposition from a parton shower developed from a virtual parton, one may solve the following DGLAP-like evolution equation[82]:

$$\begin{aligned}\frac{\partial \Delta E_j(Q^2, q^-, \zeta_i, \zeta_f)}{\partial \ln Q^2} &= \sum_{k,l} \int \frac{dy}{y} P_{j\to kl}(y) \int_{\zeta_i}^{\zeta_j} d\zeta K_{j\to kl}(Q^2, q^-, y, \zeta) \\ &\quad \times [\Delta E_j(Q^2, q^-, \zeta_i, \zeta) + \Delta E_k(Q^2, q^- y, \zeta, \zeta_f) \\ &\quad + \Delta E_l(Q^2, q^-(1-y)\zeta, \zeta_f)], \end{aligned} \tag{83}$$

where Q^2 and q^- are the virtuality and light-cone energy of the propagating jet. $\Delta E(Q^2, q^-, \zeta_i, \zeta_f)$ is the total energy deposited by a virtual parton traveling from the location ζ_i to the location ζ_f. So the total energy deposition ΔE is now dependent on the virtuality (Q^2) of the jet parton which controls how much parton shower can be developed. The equation describes the change of the total energy deposition due to the change of the virtuality of the parton (which leads to the change of parton shower). Typically the higher virtuality that the initial parton has, the more shower partons can be developed, and the more energy will be deposited into the medium. Similar equation can also been obtained for transverse momentum squared $\Delta p_\perp^2$ deposited into the medium by a virtual parton. The scattering kernel $K_{j\to kl}(Q^2, q^-, y, \zeta)$ in the above equation is the same quantity that controls the energy loss the hard jets as shown in the previous sections. Therefore, both energy loss and deposition profiles can be calculated self-consistently in the same formalism and they are controlled by the same jet transport coefficients. Due to the radiated partons which may serve as additional sources depositing energy and momentum into medium, the length dependence of energy deposition rate is significantly enhanced compared to the case of a single parton.

The medium response to a parton shower with the above energy deposition profiles may be calculated as well via solving hydrodynamic equation with a source. Figure 32 shows the hydrodynamic response to the energy deposited by a single parton (left panel) and by a parton shower (right panel). Note that in this calculation, the medium is assumed to be static and have constant temperature (density),

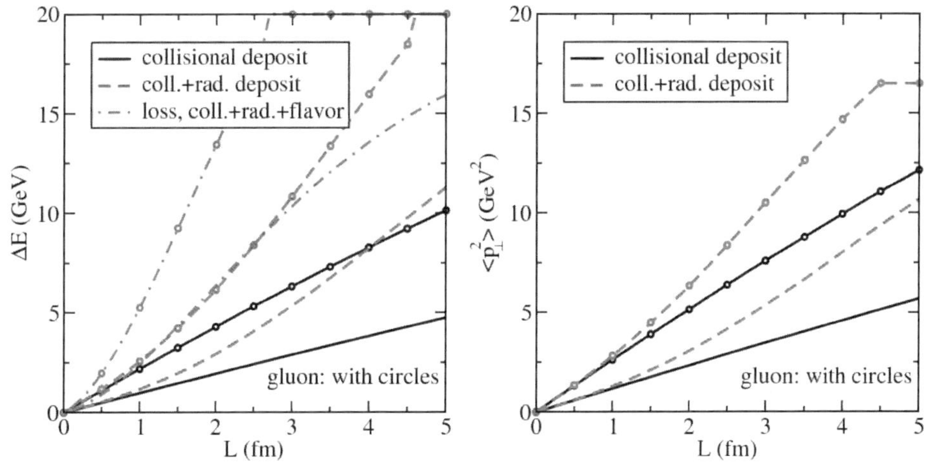

Fig. 31. (Color online) Left panel: Solid lines denote the energy deposited in the medium by a hard parton which does not radiate. Dashed lines denote the energy deposition by a virtual parton evolving into a partonic shower. Dash-dotted lines denote the total energy lost by a hard parton by radiative, elastic together with flavor changing processes. Right panel: Solid lines denote the transverse momentum squared deposited in the medium by a hard parton which does not radiate. Dashed lines denote the transverse momentum squared deposition by a virtual parton evolving into a partonic shower. Figures are taken from Ref. 82.

and the deposited energy–momentum is assumed to equilibrate after a short relaxation time $\sim 1/m_D$. One can clearly see that the Mach cone pattern is seen for both energy deposition cases, but is strongly enhanced for the virtual parton which develops into a parton shower (almost a fact of three).

However, in the above calculation of the energy deposition profiles from a parton shower, it is assumed that the radiative emissions are collinear, and the spatial distribution of the radiation is neglected. The effect of spatial distribution of jet energy and momentum deposition profiles is explored in Ref. 162 where it is found that a wide space distribution, such as the case of large angle radiation, may destroy the nice cone-like structure of medium response, as opposed to the collinear radiation. As an example, Fig. 33 compares the medium response to a parton shower developed from a primary quark with small average gluon emission angle $\theta = 0.1$ (left panel) and with large emission angle $\theta = 0.7$ (right panel). One can see that for nearly collinear emissions, the parton shower can generate a nice Mach cone structure. However, for the larger angle emissions, a well-defined Mach cone is hardly seen; the medium response is more like a superposition of several energy density perturbations. This is due to the fact that each radiatied parton acts as a separate source which deposits energy and momentum into the medium.

Besides the interplay between elastic collisions and inelastic radiations, and the space–time distributions for the radiation and deposition profiles, many other ingredients need to be taken into account when studying the energy–momentum deposition by the hard jets. For example, a cutoff energy is often used in many model calculations to determine which part of radiation phase space is treated as thermalization into the medium.[163,164] Figure 34 (left panel) shows the effect of different choices of such energy cutoff. One can see that the change of such separation scale from T to $4T$ may lead to very different energy loss/deposition profiles for the propagating jet. Another important fact is that the medium probed by the hard jet is dynamically evolving (expanding and cooling) in heavy-ion collisions, so the energy deposition rate is first increasing and then decreasing. Also in realistic event-by-event simulation which includes the fluctuations of the initial states

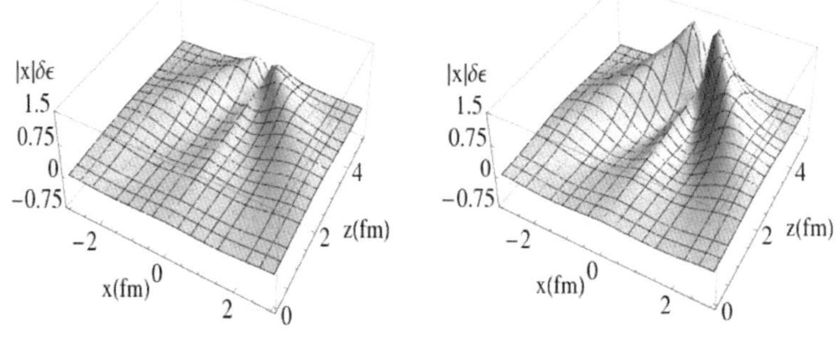

Fig. 32. (Color online) The fluid dynamical response to the energy deposited by a single quark (left) or by a quark-initiated parton shower (right). Figures are taken from Ref. 82.

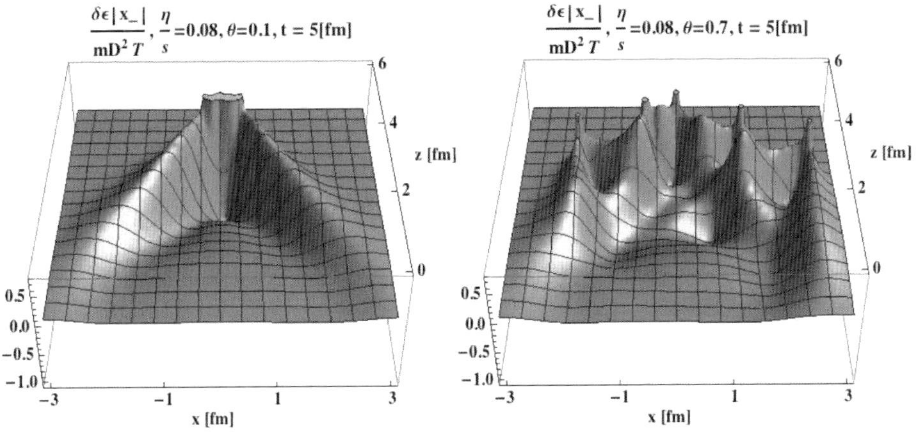

Fig. 33. (Color online) The energy density disturbance generated by a parton shower originating from a primary quark with average gluon emission angle $\theta = 0.1$ (left panel) and $\theta = 0.7$ (right panel). Figures are taken from Ref. 162.

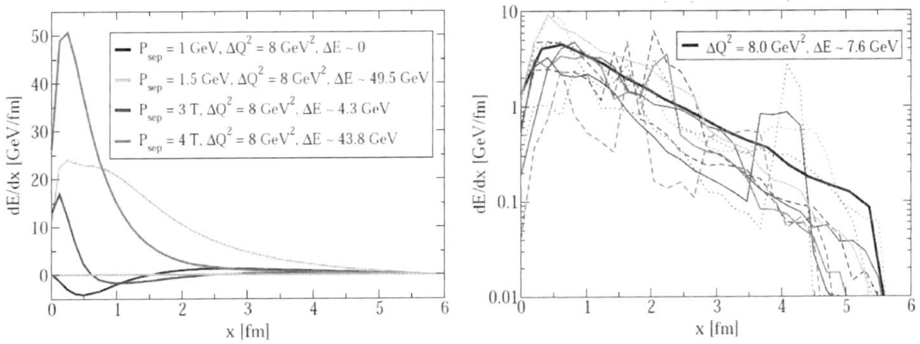

Fig. 34. (Color online) Left: Energy deposition of a parton shower initiated an 120 GeV gluon propagating through an evolving medium. Results for using different separation scales are compared. Right: Energy deposition of a parton shower initiated by an 120 GeV gluon produced propagating through an evolving medium. Results for 10 typical individual events are shown. Figures are taken from Ref. 163.

and jet energy loss, the energy deposition profiles will be very different from one event to another. This is illustrated in Fig. 34 (right panel) which compares the energy deposition profiles for 10 typical individual events which are initiated from a 120 GeV gluon.

7. Summary

In this chapter, we have provided a review on the recent progress in theoretical and phenomenological studies of jet quenching in ultra-relativistic heavy-ion collisions at RHIC and the LHC. After a short introduction, we present the basic framework for studying jet production in elementary proton–proton collisions and the nuclear modification of jets in nucleus–nucleus collisions. Different parton energy loss mechanisms and jet quenching formalisms are discussed in some detail. The recent progress on the phenomenological studies of jet quenching in heavy-ion collisions at RHIC and the LHC is presented. In particular, we have elaborated the effort on the quantitative extraction of jet transport coefficient $\hat{q}$, followed by lattice QCD calculation of $\hat{q}$ and the renormalization of $\hat{q}$. The study of full jets at RHIC and the LHC is then presented, including energy loss of reconstructed jets and the nuclear modification of the jet substructure. In the last section, the progress and some issues about the medium response to jet transport are discussed.

During last years, significant progresses have been achieved in various directions in the studies of jet quenching in heavy-ion collisions, as has been presented in this report. However, there still exist many complications and open questions. For example, in order to narrow down the systematic uncertainties in leading-order jet quenching studies, a fully next-to-leading order framework for studying jet evolution and energy loss needs to be developed. A full understanding of jet substructures and their nuclear modification requires more detailed and systematic investigation. A general framework, with the incorporation of both realistic hydrodynamic models and realistic jet energy loss/deposition calculations, should be developed for simulating the transport of hard jets and the medium response simultaneously at the same time. How exactly the energy and momentum deposited by hard jets thermalize into the medium is not well understood. Future studies along these and other directions will further our understanding of jet-medium interaction, and help to achieve more accurate knowledge about the nature of the hot and dense QCD matter produced in ultra-relativistic nuclear collisions.

Acknowledgments

This work is supported in part by Natural Science Foundation of China Under Grant Nos. 11375072, 11221504 and 11175232, Chinese Ministry of Science and Technology Under Grant No. 2014DFG02050, by the Director, Office of Energy Research, Office of High Energy and Nuclear Physics, Division of Nuclear Physics, of the U.S. Department of Energy Under Contract No. DE-AC02-05CH11231 and within the framework of the JET Collaboration.

References

1. J. D. Bjorken, Energy Loss of Energetic Partons in Quark–Gluon Plasma: Possible Extinction of High $p(t)$ Jets in Hadron–Hadron Collisions, FERMILAB-PUB-82-059-THY.
2. M. Gyulassy and M. Plumer, *Phys. Lett. B* **243** (1990) 432–438.
3. X.-N. Wang and M. Gyulassy, *Phys. Rev. Lett.* **68** (1992) 1480–1483.
4. K. Adcox *et al.*, *Phys. Rev. Lett.* **88** (2002) 022301.
5. C. Adler *et al.*, *Phys. Rev. Lett.* **89** (2002) 202301.
6. K. Aamodt *et al.*, *Phys. Lett. B* **696** (2011) 30–39.
7. C. Adler *et al.*, *Phys. Rev. Lett.* **90** (2003) 082302.

8. J. Adams *et al.*, *Phys. Rev. Lett.* **97** (2006) 162301.
9. A. Adare *et al.*, *Phys. Rev. C* **80** (2009) 024908.
10. B. Abelev *et al.*, *Phys. Rev. C* **82** (2010) 034909.
11. G. Aad *et al.*, *Phys. Rev. Lett.* **105** (2010) 252303.
12. S. Chatrchyan *et al.*, *Phys. Rev. C* **84** (2011) 024906.
13. S. Chatrchyan *et al.*, CMS Collaboration, *Phys. Lett. B* **718** (2013) 773–794.
14. G. Aad *et al.*, ATLAS Collaboration, *Phys. Rev. Lett.* **114** (2015) 7, 072302.
15. S. Chatrchyan *et al.*, *JHEP* **1210** (2012) 87.
16. S. Chatrchyan *et al.*, *Phys. Lett. B* **730** (2014) 243–263.
17. G. Aad *et al.*, ATLAS Collaboration, *Phys. Lett. B* **739** (2014) 320–342.
18. U. Heinz and R. Snellings, *Ann. Rev. Nucl. Part. Sci.* **63** (2013) 123–151.
19. C. Gale, S. Jeon and B. Schenke, *Int. J. Mod. Phys. A* **28** (2013) 1340011.
20. H. Song, *Pramana* **84** (2015) 703–715.
21. P. Romatschke, *Int. J. Mod. Phys. E* **19** (2010) 1–53.
22. P. Huovinen, *Int. J. Mod. Phys. E* **22** (2013) 1330029.
23. Y. L. Dokshitzer, *Sov. Phys. JETP.* **46** (1977) 641–653.
24. V. Gribov and L. Lipatov, *Sov. J. Nucl. Phys.* **15** (1972) 438–450.
25. G. Altarelli and G. Parisi, *Nucl. Phys. B* **126** (1977) 298.
26. B. Jager, A. Schafer, M. Stratmann and W. Vogelsang, *Phys. Rev. D* **67** (2003) 054005.
27. A. Adare *et al.*, *Phys. Rev. D* **76** (2007) 051106.
28. J. Owens, *Rev. Mod. Phys.* **59** (1987) 465.
29. X.-N. Wang, *Phys. Rev. C* **61** (2000) 064910.
30. K. Eskola, H. Paukkunen and C. Salgado, *JHEP* **0904** (2009) 65.
31. M. Hirai, S. Kumano and T.-H. Nagai, *Phys. Rev. C* **76** (2007) 065207.
32. D. de Florian and R. Sassot, *Phys. Rev. D* **69** (2004) 074028.
33. D. Sharma, *J. Phys. G* **38** (2011) 124082.
34. S. Chatrchyan *et al.*, *Eur. Phys. J. C* **72** (2012) 1945.
35. L. Landau and I. Pomeranchuk, *Dokl. Akad. Nauk Ser. Fiz.* **92** (1953) 535–536.
36. A. B. Migdal, *Phys. Rev.* **103** (1956) 1811–1820.
37. R. Baier, Y. L. Dokshitzer, A. H. Mueller, S. Peigne and D. Schiff, *Nucl. Phys. B* **483** (1997) 291–320.
38. R. Baier, Y. L. Dokshitzer, A. H. Mueller, S. Peigne and D. Schiff, *Nucl. Phys. B* **484** (1997) 265–282.
39. B. Zakharov, *JETP Lett.* **63** (1996) 952–957.
40. M. Gyulassy, P. Levai and I. Vitev, *Nucl. Phys. B* **571** (2000) 197–233.
41. M. Gyulassy, P. Levai and I. Vitev, *Phys. Rev. Lett.* **85** (2000) 5535–5538.
42. M. Gyulassy, P. Levai and I. Vitev, *Nucl. Phys. B* **594** (2001) 371–419.
43. U. A. Wiedemann, *Nucl. Phys. B* **588** (2000) 303–344.
44. U. A. Wiedemann, *Nucl. Phys. A* **690** (2001) 731–751.
45. P. B. Arnold, G. D. Moore and L. G. Yaffe, *JHEP* **0111** (2001) 57.
46. P. B. Arnold, G. D. Moore and L. G. Yaffe, *JHEP* **0206** (2002) 30.
47. X.-F. Guo and X.-N. Wang, *Phys. Rev. Lett.* **85** (2000) 3591–3594.
48. X.-N. Wang and X.-F. Guo, *Nucl. Phys. A* **696** (2001) 788–832.
49. N. Armesto *et al.*, *Phys. Rev. C* **86** (2012) 064904.
50. C. A. Salgado and U. A. Wiedemann, *Phys. Rev. D* **68** (2003) 014008.
51. M. Djordjevic and U. W. Heinz, *Phys. Rev. Lett.* **101** (2008) 022302.
52. S. Caron-Huot and C. Gale, *Phys. Rev. C* **82** (2010) 064902.
53. A. Majumder, *Phys. Rev. D* **85** (2012) 014023.
54. G.-Y. Qin and A. Majumder, *Phys. Rev. C* **91** (2015) 044906.

55. N. Armesto, H. Ma, Y. Mehtar-Tani, C. A. Salgado and K. Tywoniuk, *JHEP* **1201** (2012) 109.
56. T. Renk, J. Ruppert, C. Nonaka and S. A. Bass, *Phys. Rev. C* **75** (2007) 031902.
57. S. Wicks, W. Horowitz, M. Djordjevic and M. Gyulassy, *Nucl. Phys. A* **784** (2007) 426–442.
58. S. Jeon and G. D. Moore, *Phys. Rev. C* **71** (2005) 034901.
59. G.-Y. Qin, J. Ruppert, C. Gale, S. Jeon, G. D. Moore and M. G. Mustafa, *Phys. Rev. Lett.* **100** (2008) 072301.
60. H. Li, F. Liu, G.-L. Ma, X.-N. Wang and Y. Zhu, *Phys. Rev. Lett.* **106** (2011) 012301.
61. X.-N. Wang and Y. Zhu, *Phys. Rev. Lett.* **111**(6) (2013) 062301.
62. Y. He, T. Luo, X.-N. Wang and Y. Zhu, *Phys. Rev. C* **91** (2015) 054908.
63. A. Majumder, The in-medium scale evolution in jet modification, arXiv:0901.4516 [nucl-th].
64. G.-Y. Qin and A. Majumder, *Phys. Rev. Lett.* **105** (2010) 262301.
65. A. Majumder and C. Shen, *Phys. Rev. Lett.* **109** (2012) 202301.
66. N.-B. Chang, W.-T. Deng and X.-N. Wang, *Phys. Rev. C* **89**(3) (2014) 034911.
67. E. Braaten and M. H. Thoma, *Phys. Rev. D* **44** (1991) 2625–2630.
68. M. G. Mustafa, *Phys. Rev. C* **72** (2005) 014905.
69. A. Peshier, *Phys. Rev. Lett.* **97** (2006) 212301.
70. M. Djordjevic, *Phys. Rev. C* **74** (2006) 064907.
71. B. Schenke, C. Gale and G.-Y. Qin, *Phys. Rev. C* **79** (2009) 054908.
72. A. Majumder and B. Muller, *Phys. Rev. C* **77** (2008) 054903.
73. G.-Y. Qin and A. Majumder, *Phys. Rev. C* **87**(2) (2013) 024909.
74. S. Cao, G.-Y. Qin and S. A. Bass, *Phys. Rev. C* **88**(4) (2013) 044907.
75. G. D. Moore and D. Teaney, *Phys. Rev. C* **71** (2005) 064904.
76. Y. L. Dokshitzer and D. Kharzeev, *Phys. Lett. B* **519** (2001) 199–206.
77. S. Cao, G.-Y. Qin, S. A. Bass and B. Mller, *Nucl. Phys. A* **904–905** (2013) 653c–656c.
78. J. Uphoff, O. Fochler, Z. Xu and C. Greiner, *Phys. Lett. B* **717** (2012) 430–435.
79. R. Abir, U. Jamil, M. G. Mustafa and D. K. Srivastava, *Phys. Lett. B* **715** (2012) 183–189.
80. M. Nahrgang, J. Aichelin, S. Bass, P. B. Gossiaux and K. Werner, *Phys. Rev. C* **91** (2015) 014904.
81. G.-Y. Qin and B. Muller, *Phys. Rev. Lett.* **106** (2011) 162302.
82. G. Y. Qin, A. Majumder, H. Song and U. Heinz, *Phys. Rev. Lett.* **103** (2009) 152303.
83. R. B. Neufeld and B. Muller, *Phys. Rev. Lett.* **103** (2009) 042301.
84. R. Abir, G. D. Kaur and A. Majumder, *Phys. Rev. D* **90** (2014) 11, 114026.
85. A. Majumder, B. Muller and X.-N. Wang, *Phys. Rev. Lett.* **99** (2007) 192301.
86. S. A. Bass *et al.*, *Phys. Rev. C* **79** (2009) 024901.
87. N. Armesto, M. Cacciari, T. Hirano, J. L. Nagle and C. A. Salgado, *J. Phys. G* **37** (2010) 025104.
88. X.-F. Chen, C. Greiner, E. Wang, X.-N. Wang and Z. Xu, *Phys. Rev. C* **81** (2010) 064908.
89. H. Zhang, J. F. Owens, E. Wang and X.-N. Wang, *Phys. Rev. Lett.* **98** (2007) 212301.
90. T. Renk, *Phys. Rev. C* **78** (2008) 034904.
91. G.-Y. Qin, J. Ruppert, C. Gale, S. Jeon and G. D. Moore, *Phys. Rev. C* **80** (2009) 054909.
92. T. Renk, *Phys. Rev. C* **85** (2012) 044903.
93. B. Betz and M. Gyulassy, *JHEP* **1408** (2014) 090; *JHEP* **1410** (2014) 043.
94. M. Djordjevic, M. Djordjevic and B. Blagojevic, *Phys. Lett. B* **737** (2014) 298–302.

95. Z.-B. Kang, R. Lashof-Regas, G. Ovanesyan, P. Saad and I. Vitev, *Phys. Rev. Lett.* **114** (2015) 9, 092002.
96. K. M. Burke, *et al.*, JET Collaboration, *Phys. Rev. C* **90** (2014) 014909.
97. C. Young, B. Schenke, S. Jeon and C. Gale, *Phys. Rev. C* **86** (2012) 034905.
98. X.-F. Chen, T. Hirano, E. Wang, X.-N. Wang and H. Zhang, *Phys. Rev. C* **84** (2011) 034902.
99. J. Xu, A. Buzzatti and M. Gyulassy, *JHEP* **1408** (2014) 063.
100. B.-W. Zhang and X.-N. Wang, *Nucl. Phys. A* **720** (2003) 429–451.
101. A. Airapetian *et al.*, *Nucl. Phys. B* **780** (2007) 1–27.
102. T. Hirano and K. Tsuda, *Phys. Rev. C* **66** (2002) 054905.
103. H. Song and U. W. Heinz, *Phys. Rev. C* **77** (2008) 064901.
104. Z. Qiu, C. Shen and U. Heinz, *Phys. Lett. B* **707** (2012) 151–155.
105. A. Majumder, E. Wang and X.-N. Wang, *Phys. Rev. Lett.* **99** (2007) 152301.
106. Z.-Q. Zhang, D.-F. Hou and H.-C. Ren, *JHEP* **1301** (2013) 32.
107. H. Liu, K. Rajagopal and U. A. Wiedemann, *Phys. Rev. Lett.* **97** (2006) 182301.
108. A. Ficnar, S. S. Gubser and M. Gyulassy, *Phys. Lett. B* **738** (2014) 464–471.
109. N. Armesto, L. Cunqueiro and C. A. Salgado, *Eur. Phys. J. C* **63** (2009) 679–690.
110. K. Zapp, G. Ingelman, J. Rathsman, J. Stachel and U. A. Wiedemann, *Eur. Phys. J. C* **60** (2009) 617–632.
111. T. Renk, *Int. J. Mod. Phys. E* **20** (2011) 1594–1599.
112. A. Majumder, *Phys. Rev. C* **87** (2013) 034905.
113. M. Panero, K. Rummukainen and A. Scholfer, *Phys. Rev. Lett.* **112** (2014) 162001.
114. S. Caron-Huot, *Phys. Rev. D* **79** (2009) 065039.
115. M. Laine and A. Rothkopf, *PoS* **LATTICE 2013** (2014) 174.
116. F. D'Eramo, M. Lekaveckas, H. Liu and K. Rajagopal, *JHEP* **1305** (2013) 31.
117. R. Abir, *Phys. Rev. D* **87** (2013) 034036.
118. L. Apolinario, N. Armesto, G. Milhano and C. A. Salgado, *JHEP* **1502** (2015) 119.
119. B. Wu, *JHEP* **1110** (2011) 29.
120. T. Liou, A. Mueller and B. Wu, *Nucl. Phys. A* **916** (2013) 102–125.
121. E. Iancu, *JHEP* **1410** (2014) 95.
122. J.-P. Blaizot and Y. Mehtar-Tani, *Nucl. Phys. A* **929** (2014) 202–229.
123. Z.-B. Kang, E. Wang, X.-N. Wang and H. Xing, *Phys. Rev. Lett.* **112** (2014) 102001.
124. Z.-B. Kang, E. Wang, X.-N. Wang and H. Xing, arXiv:1409.1315 [hep-ph].
125. H. Xing, *Talk at Quark Matter*, 2014.
126. J. Casalderrey-Solana and X.-N. Wang, *Phys. Rev. C* **77** (2008) 024902.
127. M. Ploskon, *Nucl. Phys. A* **830** (2009) 255C–258C.
128. Y.-S. Lai, *Nucl. Phys. A* **830** (2009) 251C–254C.
129. I. Vitev and B.-W. Zhang, *Phys. Rev. Lett.* **104** (2010) 132001.
130. J. Casalderrey-Solana, J. G. Milhano and U. A. Wiedemann, *J. Phys. G* **38** (2011) 035006.
131. I. Lokhtin, A. Belyaev and A. Snigirev, *Eur. Phys. J. C* **71** (2011) 1650.
132. C. Young, B. Schenke, S. Jeon and C. Gale, *Phys. Rev. C* **84** (2011) 024907.
133. Y. He, I. Vitev and B.-W. Zhang, *Phys. Lett. B* **713** (2012) 224–232.
134. T. Renk, *Phys. Rev. C* **85** (2012) 064908.
135. J. Casalderrey-Solana, Y. Mehtar-Tani, C. A. Salgado and K. Tywoniuk, *Phys. Lett. B* **725** (2013) 357–360.
136. J.-P. Blaizot, E. Iancu and Y. Mehtar-Tani, *Phys. Rev. Lett.* **111** (2013) 052001.
137. W. Dai, I. Vitev and B.-W. Zhang, *Phys. Rev. Lett.* **110** (2013) 14, 142001.
138. G.-Y. Qin, *Eur. Phys. J. C* **74** (2014) 2959.
139. Y. Mehtar-Tani, C. Salgado and K. Tywoniuk, *Phys. Lett. B* **707** (2012) 156–159.

140. J.-P. Blaizot, F. Dominguez, E. Iancu and Y. Mehtar-Tani, *JHEP* **1406** (2014) 75.
141. Y. Mehtar-Tani and K. Tywoniuk, *Phys. Lett. B* **744** (2015) 284–287.
142. E. Iancu, arXiv:1305.7023 [hep-ph].
143. ATLAS Collaboration, ATLAS-CONF-2012-121.
144. T. Hirano, U. W. Heinz, D. Kharzeev, R. Lacey and Y. Nara, *Phys. Lett. B* **636** (2006) 299–304.
145. I. Vitev, S. Wicks and B.-W. Zhang, *JHEP* **811** (2008) 93.
146. R. Perez-Ramos and T. Renk, *Phys. Rev. D* **90** (2014) 014018.
147. G.-L. Ma, *Phys. Rev. C* **88**(2) (2013) 021902.
148. A. Majumder, *Phys. Rev. C* **88** (2013) 014909.
149. R. Prez-Ramos and T. Renk, arXiv:1411.1983 [hep-ph].
150. X.-N. Wang, Z. Huang and I. Sarcevic, *Phys. Rev. Lett.* **77** (1996) 231–234.
151. R. P. G. Andrade, J. Noronha and G. S. Denicol, *Phys. Rev. C* **90**(2) (2014) 024914.
152. M. Schulc and B. Tomik, *Phys. Rev. C* **90**(6) (2014) 064910.
153. J. Casalderrey-Solana, E. Shuryak and D. Teaney, *J. Phys. Conf. Ser.* **27** (2005) 22–31.
154. J. Ruppert and B. Muller, *Phys. Lett. B* **618** (2005) 123–130.
155. T. Renk and J. Ruppert, *Phys. Rev. C* **73** (2006) 011901.
156. G.-L. Ma and X.-N. Wang, *Phys. Rev. Lett.* **106** (2011) 162301.
157. I. Bouras, B. Betz, Z. Xu and C. Greiner, *Phys. Rev. C* **90** (2014) 024904.
158. R. Neufeld, *Phys. Rev. C* **79** (2009) 054909.
159. Z.-W. Lin, C. M. Ko, B.-A. Li, B. Zhang and S. Pal, *Phys. Rev. C* **72** (2005) 064901.
160. Z. Xu and C. Greiner, *Phys. Rev. C* **71** (2005) 064901.
161. Y. Tachibana and T. Hirano, *Phys. Rev. C* **90** (2014) 021902.
162. R. Neufeld and I. Vitev, *Phys. Rev. C* **86** (2012) 024905.
163. T. Renk, *Phys. Rev. C* **88**(4) (2013) 044905.
164. S. Floerchinger and K. C. Zapp, *Eur. Phys. J. C* **74**(12) (2014) 3189.

Quarkonia in High Energy Nuclear Collisions

Yunpeng Liu
Department of Physics, Tianjin University,
Tianjin 300072, P. R. China

Kai Zhou
Institute for Theoretical Physics,
Johann Wolfgang Goethe-University Frankfurt,
Max-von-Laue-Str.1, 60438 Frankfurt am Main, Germany
zhou@th.physik.uni-frankfun.de

Pengfei Zhuang
Physics Department,
Tsinghua University and Collaborative
Innovation Center of Quantum Matter,
Beijing 100084, P. R. China
Zhuangpf@fsinghua.edu.cn

We first review the cold and hot nuclear matter effects on quarkonium production in high energy collisions, then discuss three kinds of models to describe the quarkonium suppression and regeneration: the sequential dissociation, the statistical production and the transport approach, and finally make comparisons between the models and the experimental data from heavy ion collisions at SPS, RHIC and LHC energies.

1. Introduction

Phase transition is a key question in the study of many body systems in both theory and experiment. For nuclear matter, one of the most important topics is the confinement-deconfinement phase transition where hadrons melt into quarks and gluons at high temperature and/or high baryon density (evidence of the phase transition was found even before the quark model[1]). Although such a phase transition is widely accepted, there are still some open and fundamental questions. On the theoretical side, the confinement is not proved rigorously from the quantum chromodynamics (QCD) itself yet, and most discussions are based on computer simulations or effective models. On the experimental side, free quarks are not observed, and the only experiment we can do on the earth is the relativistic heavy ion collisions, where two heavy nuclei collide with each other at extremely high energy to produce the hot and dense quark–gluon plasma (QGP). However, this new state of matter can not be measured directly, and one needs sensitive probes to show its existence in the early stage of heavy ion collisions.

J/ψ is one of the suggested probes to the QGP in relativistic heavy ion collisions. Unlike light quarks which are largely produced through soft processes in medium, heavy quarks are mainly created through hard processes in the initial stage and can interact with the quark matter strongly and has a chance to survive the hot and/or dense medium. As a result of the color screening effect, J/ψ suppression was suggested by Matsui and Satz in 1986[2] and observed at SPS in 1996,[3] which was regarded as an evidence of the QGP formation.

As a baseline of the quarkonium production in nuclear collisions, the first question to answer is how a quarkonium is produced in nucleon–nucleon collisions. A quarkonium is composed of a pair of heavy quarks, and the production of heavy quarks requires large Q^2, which can be calculated by perturbative QCD. At high energies, the dominant source for the heavy quark production at partonic level is the gluon fusion. Several programs are available like FONLL,[4,5] PYTHIA[6–8] and HIJING.[9,10] While the heavy quark production can well be studied in the pQCD framework, the subsequent formation of a quarkonium bound state as a color singlet hadron is a more complicated nonperturbative process, which is not yet understood clearly from first principle QCD theory. It is worth noting that this soft color neutralization process is much slower compared with the heavy quark hard production, due to the typical time scale hierarchy $\tau_{\text{quarkonium}} \sim 1/\epsilon_{\text{binding}} \gg \tau_{Q\bar{Q}} \sim 1/(2m_Q)$. Therefore, the quantum interference between the two stages are suppressed, which enables us to factorize the heavy quarkonium formation into the heavy quark production and the nonperturbative transition probability. Different phenomenological models and effective field theories are developed to describe this transition, like the color-singlet model (CSM),[11] the color-evaporation model (CEM),[12–15] and the nonrelativistic QCD (NRQCD).[16,17] We refer the readers to Refs. 11, 14, 17–19 for more details.

The quarkonium production in nuclear collisions is strongly affected by the nuclear matter. Before the hard processes for the heavy quark production, the parton distribution in a nucleus is different from the distribution in a nucleon due to the shadowing effect,[20] and the multi scattering of the partons with the surrounding nucleons, called Cronin effect,[21] modifies again the parton momentum distribution. After the $Q\bar{Q}$ pairs are created, they suffer from nuclear absorption[22,23] via the collisions with the spectators. All these nuclear effects happen before the hot medium formation and are called as cold nuclear matter effects. They are widely discussed in many theoretical models[19,24–28] and can be experimentally measured in p–A and d–A collisions[29–32] where a hot medium is not expected at low energies. The produced quarkonia will further be influenced by the hot nuclear matter effects when they pass through the created fireball. On one hand, they are suppressed by inelastic interactions with partons in quark matter and hadrons in hadron matter. On the other hand, quarkonia with low momentum can also be regenerated in the fireball, and the regeneration becomes even dominant when the colliding energy is high enough. To study the properties of quarkonia at finite temperature and density, people take normally lattice QCD and effective field theories. In the spectrum functions from lattice calculation, the peak of J/ψ vanishes at around $1.5T_c$,[33–39] where T_c is the critical temperature of the deconfinement phase transition. For the effective models, we refer the readers to Refs. 40–42.

In this article, we focus on the phenomenological models on nuclear matter effects on quarkonium production and the comparison with the experimental data in high energy nuclear collisions at SPS, RHIC and LHC energies. We first discuss different cold and hot nuclear matter effects, then introduce three kinds of phenomenological models to describe the quarkonium evolution in A+A collisions: the sequential screening, the statistical model, and the transport approach, and finally make comparison between the models and the experimental data. We summarize and give a short outlook in the end.

2. Cold and Hot Nuclear Matter Effects

Although the main goal of the relativistic heavy ion collision is to study the properties of the QGP, it is necessary to discuss the cold nuclear matter effect first, which can hardly be separated in experiments from the hot nuclear matter effects. For quarkonium, there are mainly three cold nuclear matter effects as follows.

2.1. *Nuclear absorption*

The suppression of J/ψ due to hot nuclear matter effects was predicted in theory first.[2,43] Later on the suppression was observed at SPS energy in collisions of small nuclei.[44–47] However, it is found that the suppression can well be explained by the absorption of the nucleons.[48] For uniform nuclei, the length of the path in the two nuclei that the J/ψ passes through after its production is

$$L = \frac{3}{4}(R_1 + R_2), \tag{1}$$

where R_1 and R_2 are the radii of the nuclei. The normal suppression of J/ψ versus the length L is shown in Fig. 1. It can be seen that all the data lies in the same line, and the suppression can be interpreted as the dissociation of the J/ψ by the nucleons successively, so that the survival probability

$$S = e^{-\rho\sigma_{\text{abs}} L}, \tag{2}$$

where ρ is the nuclear density, and σ_{abs} is the nuclear absorption cross section. The value of σ_{abs} is several mb at SPS energy.[43] Such a effect is usually measured in $p + \text{A}$ collisions to determine the σ_{abs} and applied to $\text{A} + \text{A}$ collisions.

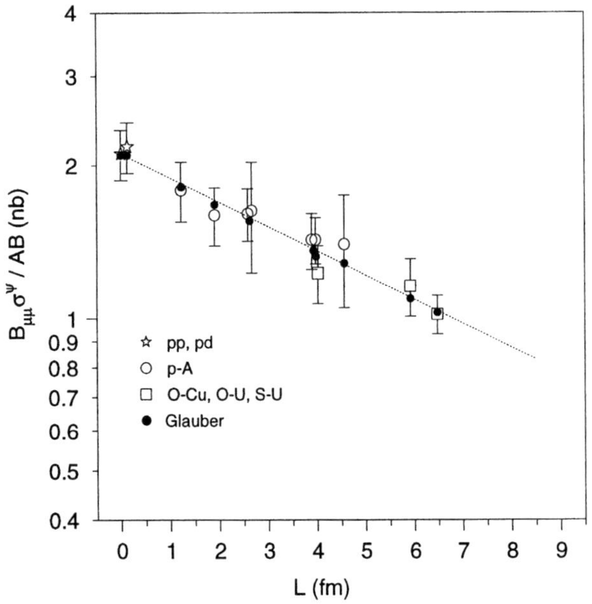

Fig. 1. The normal suppression of J/ψ at SPS energy as a function of the length L. σ^{ψ}, $B_{\mu\mu}$ are the cross section and the branch ratio of J/ψ to dimuon, respectively, and A and B are the mass numbers of the two nuclei. Figure from Ref. 22.

The time for the two nuclei to pass through each other can roughly be estimated as $t_{\text{pass}} \sim 2Rm_N/\sqrt{s_{\text{NN}}}$, where m_N and $\sqrt{s_{\text{NN}}}$ are the mass of a nucleon and the center of mass energy of one nucleon pair. At RHIC energy it is about 0.06 fm/c, which is far smaller than the formation time of J/ψ. At LHC energy, this time is even smaller by one order of magnitude. Therefore, considering a non-zero quarkonium formation time $\tau_{\text{form}} \sim 1/\epsilon_{\text{binding}}$, the σ_{abs} is not simply the cross section between a fully developed J/ψ and a nucleon in vacuum. Instead it reflects the interaction between the nucleons and some preliminary state of J/ψ before it is formed. The nuclear absorption effect is less and less important with increasing collision energy as shown in Fig. 2 for the fitted $\sigma_{\text{abs}}(\sqrt{s_{\text{NN}}})$.[24] It is recently found that the normal suppression of ψ' is much stronger than that of J/ψ at RHIC energy,[49] and a quantitative explanation is still lacking.

2.2. *Nuclear shadowing*

As mentioned in the introduction, the heavy quark production is a perturbative process and is dominated by the gluon fusion at high energy. Thus the yield of the heavy quarkonia is closely related to the parton distribution function of gluons in the initial colliding nuclei. Considering the subprocess

$$g_1 + g_2 \to J/\psi + X, \tag{3}$$

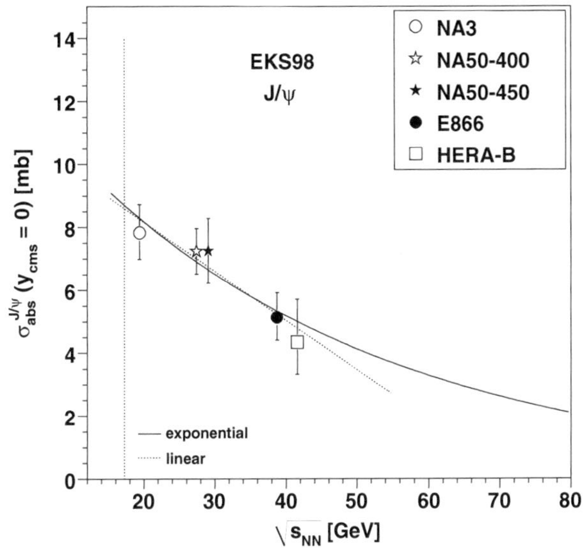

Fig. 2. The collision energy dependence of J/Ψ nuclear absorption cross section $\sigma_{\rm abs}^{J/\psi}$. Figure from Ref. 24.

where g_1 and g_2 are two gluons in the nucleons, then the on shell condition of particle X reads

$$m_X^2 = x_1 x_2 s_{\rm NN} - (x_1 + x_2)\sqrt{s_{\rm NN}} m_T \cosh y \\ + (x_1 - x_2)\sqrt{s_{\rm NN}} m_T \sinh y + m_\psi^2 \tag{4}$$

where it is assumed that both g_1, and g_2 are massless so that the longitudinal momentum and the energy of the subprocess are $(x_1 - x_2)\sqrt{s_{\rm NN}}/2$ and $(x_1 + x_2)\sqrt{s_{\rm NN}}/2$, respectively, with x_1, x_2, $\sqrt{s_{\rm NN}}$ being the momentum fraction of g_1 and g_2 and the center of mass energy of the nucleons, and m_ψ, m_T, y are the mass, transverse mass and rapidity of the J/ψ, respectively. If the particle X (usually being gluon) is also massless ($m_X = 0$), then the above equation gives the relation between x_1 and x_2 as[50]

$$x_2 = \frac{x_1 m_T \sqrt{s_{\rm NN}} e^{-y} - m_\psi^2}{\sqrt{s_{\rm NN}}(x_1\sqrt{s_{\rm NN}} - m_T e^y)}. \tag{5}$$

Furthermore, if the four-momentum of X can be omitted, corresponding to the leading order of the Color Evaporation Model (CEM), both x_1 and x_2 can be determined by the momentum information of the final J/ψ,

$$x_{1,2} = \frac{m_T}{\sqrt{s_{\rm NN}}} e^{\pm y}. \tag{6}$$

Therefore, the transverse momentum and rapidity distribution of J/ψ is determined by not only the cross section of the gluon fusion subprocess, but also the parton distribution functions for gluon in the initial particles. The distribution function f_i^A of parton i in a nucleus differs from a simple superposition of the distribution function f_i in a free nucleon. Such nuclear shadowing is usually described by the modification factor $R_i^A(x,\mu_F) = f_i^A(x,\mu_F)/(Af_i(x,\mu_F))$, where μ_F is the factorization scale for the hard process. Higher energy collisions concern smaller x physics, which usually locate at stronger shadowing region. Therefore, parton shadowing effect plays an important role in high energy collisions.

There are different parameterizations of the change of parton distribution functions: EKS98,[51] EPS08,[52] EPS09,[53] nDS,[54] and so forth. Such a shadowing effect can be measured in p+A collisions, especially in the rapidity dependence of the nuclear modification factor. For more details we refer the readers to Refs. 28, 50–55.

2.3. *Cronin effect*

The momentum distribution of quarkonia in p+A collisions is different from that in $p+p$ collisions. That is the Cronin effect, which was named after Cronin and firstly observed for light hadrons.[21] Due to the Cronin effect, part of the low momentum quarkonia are shifted to the intermediate momentum, so that the yield of the low momentum quarkonia is suppressed, and that of the intermediate ones are enhanced in nuclear collisions.

The Cronin effect of J/ψ can be interpreted as a gluon kicking by the surrounded nucleons in transverse direction before the fusion of the two gluons into a J/ψ. The transverse momentum broadening is proportional to the length l that the gluons pass through in the nuclei, and the transverse momentum distribution in A+A collisions reads

$$f_{AA}(\mathbf{p}_T) = \int d\mathbf{k} f_k(\mathbf{k}) f_{pp}(\mathbf{p}_T - \mathbf{k}), \tag{7}$$

where $\mathbf{k}$ is the transverse momentum increase of the gluons due to the scattering, which follows a Gaussian distribution since it is due to multi-scattering. Taking average over the transverse momentum square on both sides of above equation, one find that

$$\langle p_T^2 \rangle_{AA} = \langle p_T^2 \rangle_{pp} + \langle k^2 \rangle, \tag{8}$$

where the average are for the distributions of f_{AA}, f_{pp} and f_k, respectively, and it is assumed that at least one of f_{pp} and f_k is isotropic, which is usually true. The momentum increase can be taken as Gaussian since it is accumulated from multi-scattering, and it is proportional to l,

$$\langle k^2 \rangle = a_{gN} l, \tag{9}$$

where the parameter a_{gN} is about $0.1\,\mathrm{GeV}^2/\mathrm{fm}$ at RHIC.[56] If the transverse momentum of J/ψ is not too large so that f_{pp} can be described by a Gaussian distribution, f_{AA} is simply

$$f_{AA}(\mathbf{p}_T) = \frac{\langle p_T^2 \rangle_{pp}}{\langle p_T^2 \rangle_{AA}} f_{pp}\left(\sqrt{\frac{\langle p_T^2 \rangle_{pp}}{\langle p_T^2 \rangle_{AA}}}\mathbf{p}_T\right). \tag{10}$$

2.4. *Color screening*

The screening effect is well known in the electric plasma, and it also happens in the QGP. One significant difference is that the temperature T to form QGP is far larger than the masses of its components, while the electric plasma is typically non-relativistic. As a result, the temperature dependence of the screening effect differs qualitatively. It is well known that the screening mass μ of a Coulomb potential can be expressed as

$$\mu = \sqrt{\frac{4\pi\alpha \sum_i Q_i^2 \rho_i}{T}}, \tag{11}$$

where α, Q_i, ρ_i and T are the fine structure constant, the charge number of particle i, the density of particle i, and the temperature of the plasma, respectively. In the non-relativistic case, the density of charged particles is fixed, and the only factor to compete with the screening is the kinetic energy of the charged particles. Therefore the higher the temperature is, the weaker the screening is, and $\mu \propto T^{-1/2}$. In the contrary, the density of the partons increases with temperature in QGP. If the mass of the partons is omitted $\rho_i \propto T^3$, and the screening mass is $\mu \propto T$, so that the higher the temperature is, the stronger the screening effect is. When the temperature is strong enough, the heavy quark pair breaks up and the quarkonium is dissociated. More details will be discussed in Subsec. 3.1.

2.5. *Gluon dissociation*

The photon disintegration of QED bound state like hydrogen atoms is well studied. Similarly, a quarkonium can be dissociated by a gluon. The cross section was calculated first by Peskin with operator production expansion.[57] The result is introduced in Subsec. 3.3. There are three main differences between the hydrogen and the quarkonium dissociation: (1) the coupling constant is small for QED and large for QCD, which makes it difficult to calculate the cross section in high accuracy; (2) the final state interaction is comparable with the attraction in the bound state for the QED, while it is one order of magnitude smaller in quarkonium, because the heavy quarks are in color singlet state in the quarkonium and in color octet state after the quarkonium is dissociated; (3) the three gluon interaction has equal contribution as the quark–gluon interaction at leading order, which is absent in QED. Many efforts were made to understand the quarkonium suppression better after Peskin. The cross section for excited states were calculated later.[58,59] The recoil effect was considered in Ref. 59. The same cross section was reproduced by a pQCD calculation[60] with a quarkonium to heavy quark pair vertex which is from Bathe–Saltpeter equation. By this method, the high-order contribution was also calculated.[61] The correction due to final state interaction was also discussed.[62] The cross section beyond the dipole approximation was calculated in Ref. 63.

2.6. *Quasi-free scattering*

The gluon dissociation $g + (Q\bar{Q}) \to Q + \bar{Q}$ is the leading order process, which is important at low temperature when the binding energy is large. When the temperature is high, the quasi-free scattering is believed to be important[56,64] since the quarks are loosely bounded. In high temperature limit, the heavy quarks can no longer form a quarkonium as in the color screening scenario, which was also implied by the lattice result. A T matrix method was developed to calculate the suppression based on heavy quark potential and the scattering theory, which can adopt the lattice potential directly.[65]

3. Phenomenology

In order to link the properties of quarkonia and the experimental results, plenty of phenomenological research has been done. In the following we will introduce three popular models for quarkonia production and/or suppression in the hot medium.

3.1. *Sequential dissociation model*

One of the earliest models for heavy quarkonium suppression is the sequential dissociation model,[2,66] which assumes that the quarkonia are dissociated sequentially when the temperature of the quark matter increases, since those quarkonia with small binding energy are more fragile. Taking advantage of the heavy mass of the constituent quarks, the interaction between the heavy quarks can be described by a potential. In vacuum, the Cornell potential works well

$$V(r) = -\frac{\alpha}{r} + \sigma r, \tag{12}$$

which contains a Coulomb part at small radius r and a linear part at large r. The former is the result of one gluon exchange potential which comes from the leading order perturbation, and the latter gives the confinement which is non-perturbative. The Schrödinger equation of charm quarks reads

$$\left[-\frac{\nabla^2}{2\mu} + V(r)\right]\psi = E\psi, \tag{13}$$

where μ is the reduced mass of the two heavy quarks. Solving the above equation, it was found in Ref. 67 that the mass spectra of quarkonia can well be reproduced with parameters $m_c = 1.25\,\mathrm{GeV}$, $m_b = 4.65\,\mathrm{GeV}$, and $\alpha = \pi/12$ and $\sigma = 0.2\,\mathrm{GeV}^2$ in the Cornell potential. See Table 1.

At finite temperature, the free energy approximately takes the form[68]

$$\begin{aligned} F(r) = {} & -\frac{\alpha}{r}e^{-\mu r} - \frac{\sigma}{2^{3/4}\Gamma(3/4)}\left(\frac{r}{\mu}\right)^{1/2} K_{1/4}(\mu^2 r^2) - \mu \\ & + \frac{\sigma}{2^{3/2}\mu}\frac{\Gamma(1/4)}{\Gamma(3/4)}, \end{aligned} \tag{14}$$

Table 1. Mass M and binding energy ϵ of heavy quarkonia in vacuum calculated from Cornell potential. Table from Ref. 67.

States	J/ψ	χ_c	ψ'	$\Upsilon(1S)$	$\Upsilon(1P)$	$\Upsilon(2S)$	$\Upsilon(2P)$	$\Upsilon(3S)$
M/GeV	3.10	3.53	3.68	9.46	9.99	10.02	10.26	10.36
ϵ/GeV	0.64	0.20	0.05	1.10	0.67	0.54	0.31	0.20
error/GeV	0.02	−0.03	0.03	0.06	−0.06	−0.06	−0.08	−0.07

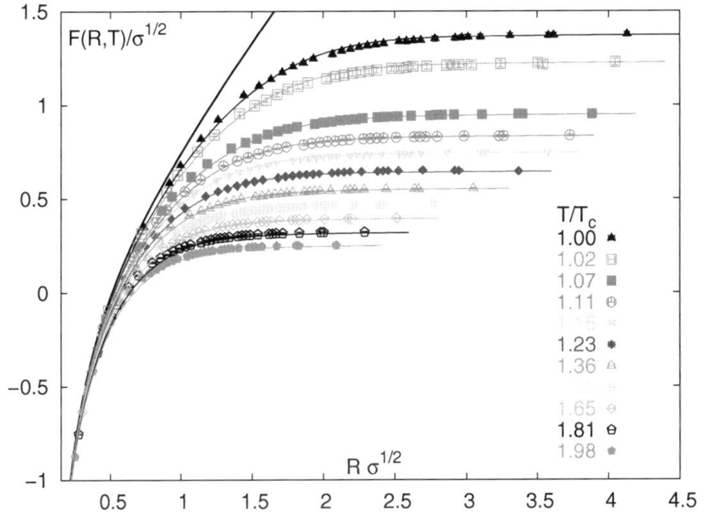

Fig. 3. (Color online) The fit potential of the free energy from Lattice QCD. Figure from Ref. 67.

where all the temperature dependence lies in $\mu = \mu(T)$. It is found that this formula can fit the free energy obtained from the LQCD simulation well,[67] as shown in Fig. 3. We further find that $\mu(T)$ above T_c can be parameterized as

$$\frac{\mu(\bar{T})}{\sqrt{\sigma}} = k\bar{T} + a\sigma_t\sqrt{\frac{\pi}{2}}\left[\mathrm{erf}\left(\frac{b}{\sqrt{2}\sigma_t}\right) - \mathrm{erf}\left(\frac{b-\bar{T}}{\sqrt{2}\sigma_t}\right)\right],$$

where $\bar{T} \equiv T/T_c$, and the dimensionless parameters are $k = 0.587$, $a = 2.150$, $b = 1.054$ and $\sigma_t = 0.07379$.

One can also derive the internal energy

$$U = F - T\frac{\partial F}{\partial T} \tag{15}$$

from the above free energy. The potential is in principle between the free energy limit and the internal energy limit. In the free energy limit, the heavy quark pair can exchange heat with the medium immediately, and in the internal limit the they can not. Taking the potential V in Eq. (13) either as the free energy F or as the potential U will give an extreme limit for the properties of quarkonia. The dissociation temperatures of heavy quarkonia and those of B_c mesons are listed as in Tables 2 and 3, respectively. It can be seen that The dissociation temperature T_d

Table 2. The ratio of the dissociation temperature T_d to the critical temperature T_c of different charmonium and bottomonium states in the internal energy limit $V = U$ and the free energy limit $V = F$. Part of data are from Ref. 67.

States	J/ψ	χ_c	ψ'	$\Upsilon(1S)$	$\Upsilon(1P)$	$\Upsilon(2S)$	$\Upsilon(2P)$	$\Upsilon(3S)$
$V = U$	2.26	1.16	1.11	7.0	3.15	2.98	2.05	1.98
$V = F$	1.16	≤ 1	≤ 1	2.90	1.12	≤ 1	≤ 1	≤ 1

Table 3. The ratio of the dissociation temperature T_d to the critical temperature T_c of different B_c states in the internal energy limit $V = U$ and the free energy limit $V = F$. Part of data are from Ref. 69.

States		$B_c(1S)$	$B_c(1P)$	$B_c(2S)$	$B_c(1D)$
M/GeV		6.36	6.72	6.90	6.98
T_d/T_c	$V = U$	3.27	1.59	1.41	1.17
	$V = F$	1.51	≤ 1	≤ 1	≤ 1

differs significantly between $V = U$ and $V = F$. In the $V = U$ limit, the T_d of J/ψ is above the temperature at SPS, and near to the highest temperature at RHIC. In the $V = F$ limit, it is little above the critical temperature and most of the J/ψs are supposed to dissociate at RHIC energy. As to Υ, no suppression of the 1S state is expected even in the $V = F$ limit. The results for B_c mesons lies between J/ψ and Υ.

Lattice QCD simulation also gives some results on the dissociation temperature of quarkonia from the spectrum function. For J/ψ it is around $1.5T_c$, above which the peak of J/ψ in the spectrum function disappears.[33]

One should keep in mind that the above results are based on the assumption of static heavy quarks. Since the quarkonia produced in high energy nuclear collisions have a typical transverse momentum of several GeV, one should consider the charmonium velocity dependence of the color screening. From the lattice calculation of the charmonium spectral function, the J/ψ suppression is stronger at high p_T,[70] while some phenomenological study including the relaxation time of medium for the color screening indicates an increasing dissociation temperature with J/ψ velocity relative to the medium, which means a less quarkonium suppression at high p_T.[71] The experimental results can be better described by the latter. The velocity dependence of the ratio T_d/T_c predicted by Ref. 71 is shown in Fig. 4.

In the above, the medium is assumed as isotropic, and it can not be true at the early stage of the heavy ion collision. It is found that the screening effect is weaker in an anisotropic medium,[72] which make it easier for the quarkonia to survive the QGP. Recently, the relativistic modification is also discussed in the potential model by solving the Schrödinger equation.[73]

The sequential dissociation model was supported by the experiments for Υ suppression at RHIC[74,75] and the double ratio at LHC.[76]

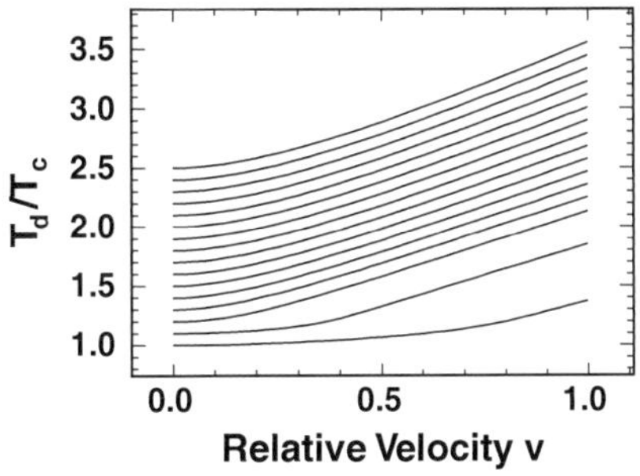

Fig. 4. The velocity dependence of the ratio of dissociation temperature to critical temperature T_d/T_c of heavy quarkonia in QGP.

3.2. *Statistical model*

Soon after the sequential dissociation model, a totally different model was suggested that the yield of J/ψ is determined by the statistics.[77–81] The statistical model was used to describe the light hadrons and made a great success.[82] Then it was applied to charmonia with the consideration of charm quark conservation, which gives the basic equation of the statistical model for heavy flavors[77]

$$N_{Q\bar{Q}} = \frac{1}{2}N_{\mathrm{o}} + N_{\mathrm{h}}, \tag{16}$$

where $N_{Q\bar{Q}}$, N_{o} and N_{h} are the number of initially produced heavy quark pairs, final open heavy flavors and final hidden heavy flavors, respectively. This equation not only require net charmness conservation, but also require conservation of charm and anticharm quarks separately. N_{o} and N_{h} are determined by the statistics.

In the grand canonical ensemble, the numbers on the right-hand side are

$$N_{\mathrm{o}}^{\mathrm{gr}} = g\rho_{\mathrm{o}}V, \tag{17}$$

$$N_{\mathrm{h}}^{\mathrm{gr}} = g^2\rho_{\mathrm{h}}V, \tag{18}$$

where g is the fugacity, and $\rho = \sum_i \rho_i^{th}$ is the total density as a summation of the thermal density over all the open or hidden particles under consideration.

Since the fireball in heavy ion collisions is isolated, a canonical ensemble describes better the system. The difference is negligible when the number of heavy quarks $N_{Q\bar{Q}}$ is large, but it becomes important when $N_{Q\bar{Q}}$ is comparable with or even smaller than unit. In the canonical ensemble the numbers of the heavy flavor is expressed as[81]

$$N_{\mathrm{o}}^{\mathrm{ca}} = g\rho_{\mathrm{o}}V\frac{I_1(g\rho_{\mathrm{o}}V)}{I_0(g\rho_{\mathrm{o}}V)}, \tag{19}$$

$$N_{\mathrm{h}}^{\mathrm{ca}} = g^2\rho_{\mathrm{h}}V, \tag{20}$$

where I_1 and I_0 are modified Bessel functions of the first kind. $I_1(x)/I_0(x)$ approaches to 1 when x approaches to positive infinity, and it approaches to $x/2$ when x approaches to zero. Therefore when the number of charm quarks is large, it approaches the grand canonical ensemble limit. On the other hand, if the number of charm quarks is small then a much larger fugacity g is required to balance the two sides of Eq. (16), which leads to an enhancement of the quarkonium production. Such a canonical enhancement was studied for strangeness before,[83] and it can also be obtained by solving the main equation for number of charm quark distribution, and taking the long time limit to reach the thermal equilibrium, which proved Eq. (19) elegantly with the help of the generating function.[84]

Now let us calculate the yield of hidden heavy flavors explicitly under some approximations. We note that the yield of quarkonia is far smaller than that of the open heavy hadrons, therefore, to determine the fugacity we can safely neglect the N_h term in Eq. (16). As a result we obtain $g = 2N_{Q\bar{Q}}/(\rho_0 V)$ and

$$N_h = \frac{4\rho_h N_{Q\bar{Q}}^2}{\rho_0^2 V} \tag{21}$$

in the large $N_{Q\bar{Q}}$ limit. If $N_{Q\bar{Q}}$ is small instead, we obtain $g^2 = 4N_{Q\bar{Q}}/(\rho_0^2 V^2)$ and

$$N_h = \frac{4\rho_h N_{Q\bar{Q}}}{\rho_0^2 V}. \tag{22}$$

To include both limits, the yield reads

$$N_h = \frac{4\rho_h N_{Q\bar{Q}}^2}{\rho_0^2 V}\left(1 + \frac{1}{N_{Q\bar{Q}}}\right). \tag{23}$$

On the one hand, the basic assumption of the statistical model for heavy flavor is the heavy quark pair conservation, as indicated by Eq. (16). On the other hand, the statistics counts the contribution of arbitrary number of heavy quark pairs. A consequent question comes up: how different number of quark pairs contribute to a single collision. The event by event fluctuation answers this question as suggested by Ref. 85. Considering one event, the yield of heavy quarkonia N_h is proportional to the collision rate between Q and $\bar{Q}$, and thus for a given volume

$$N_h \propto N_Q N_{\bar{Q}} = N_{Q\bar{Q}}^2, \tag{24}$$

where $N_{Q\bar{Q}}$ is a random variable which fluctuates from event to event. When we average events we have $\langle N_h \rangle \propto \langle N_{Q\bar{Q}}^2 \rangle$. Since $N_{Q\bar{Q}}$ comes from plenty of independent nucleon–nucleon collisions, it follows Poisson distribution, and for Poisson distribution we have

$$\langle N_{Q\bar{Q}}^2 \rangle = \langle N_{Q\bar{Q}} \rangle^2 \left(1 + \frac{1}{\langle N_{Q\bar{Q}} \rangle}\right). \tag{25}$$

It means that the correction at small $N_{Q\bar{Q}}$ due to event by event fluctuation is the factor $1 + 1/\langle N_{Q\bar{Q}} \rangle$,[69] which is consistent with Eq. (23) where the average brackets are not written explicitly.

Another question one has to answer is why Eq. (23) is not extensive with the fluctuation considered. Actually the correction term $1/N_{Q\bar{Q}}$ comes from the correlation of the existence of Q and $\bar{Q}$, that is whenever there is a heavy quark, there is also a heavy antiquark in the volume which increases the yield of heavy quarkonia. However, if the quark and antiquark are too far away, then they will not see each other and it no longer helps to produce the quarkonia. Therefore only when they lie in some correlation volume,[83,86] this effect can contribute. Such a correlation volume was taken as the volume of the fireball within unit rapidity bin in most statistical models[80] and was more carefully considered in a transport model.[86] Recently, a detailed study further considered both the correlation in the coordinate space and that in the momentum space to determine the correlation (or effective) volume.[87] It is found that in the classical limit, the correlation volume increases linearly with time at the short time limit, and then it increases with time more slowly, and in the long time limit the correlation volume increases with $t^{3/2}$. Here the "short" time and the "long" time are compared with the average time between collisions of the heavy quark with the particles in the medium. It also implies that the formation time and size of the initial wave package of the heavy quarks can be important in this problem.

The coalescence model made great success for light hadrons[88–92] and were also used to describe the production of heavy quarkonia, whose results is similar to that of the statistical model. The basic idea is to project the quark states to the hadron states with the help of their Wigner functions. For heavy quarkonium production in the coalescence model, one may refer to Refs. 93 and 94. In a recent study, the triply charmed baryon Ω_{ccc} is even predicted in the heavy ion collisions at LHC.[95]

3.3. *Transport model*

Different from the sequential dissociation model and the statistical model, the transport model[64,69,86,96–99] contains the information of time evolution of the quarkonia. As an approximation, the quarkonia can be described by the distribution function $f(\mathbf{x}, \mathbf{p}, t)$ in phase space, which respects the classical Boltzmann equation

$$\delta_t f + \mathbf{v} \cdot \nabla f - (\nabla U) \cdot \nabla_p f = -\alpha f + \beta + C, \tag{26}$$

where $\mathbf{v}$, U, α, β, C are the velocity of the quarkonium, the potential that the quarkonia feels in the medium, the dissociation rate of the quarkonium, the regeneration rate of the quarkonium and the elastic collision term, respectively. Almost all the transport models are solving some simplified version of such a transport equation.

The effect of mean field U was discussed in Ref. 100, and the elastic collision is regarded as small in most of discussions, which is implied by the small v_2 of J/ψ at RHIC.[101] Without the potential term and the elastic collisions term, the above equation can be written in a explicitly boost invariant form as

$$\cosh(y - \eta)\delta_\tau f + \mathbf{v}_T \cdot \nabla_T f + \frac{1}{\tau} \sinh(y - \eta)\delta_\eta f = -\alpha_T f + \beta_T, \tag{27}$$

where $y = (1/2)\ln[(E + p_z)/(E - p_z)]$, $\eta = (1/2)\ln[(t + z)/(t - z)]$, $\mathbf{v}_T = \mathbf{p}_T/E_T$ and $\tau = \sqrt{t^2 - z^2}$ are the momentum rapidity, spacetime rapidity, transverse velocity and the longitudinal proper time of the quarkonium, respectively, and $E_T = \sqrt{M^2 + \mathbf{p}_T^2}$ is the transverse energy. $\mathbf{v}_T = \mathbf{p}_T/E_T$, $\alpha_T = \alpha E/E_T$, and $\beta_T = \beta E/E_T$ are all boost invariant as long as the medium is boost invariant. The solution of the above equation is

$$
\begin{aligned}
f(\mathbf{p}_T, y, \mathbf{x}_T, \eta, \tau) & \\
= f(\mathbf{p}_T, y, \mathbf{X}_T(\tau_0), H(\tau_0), \tau_0) & e^{-\int_{\tau_0}^{\tau} d\tau' \alpha(\mathbf{p}_T, y, \mathbf{X}_T(\tau'), H(\tau'), \tau')/\Delta(\tau')} \\
+ \int_{\tau_0}^{\tau} d\tau' \beta(\mathbf{p}_T, y, & \mathbf{X}_T(\tau'), H(\tau'), \tau')/\Delta(\tau') \\
\times e^{-\int_{\tau'}^{\tau} d\tau'' \alpha(\mathbf{p}_T, y, \mathbf{X}_T(\tau''), H(\tau''), \tau'')/\Delta(\tau'')} &
\end{aligned}
\tag{28}
$$

with

$$
\begin{aligned}
\mathbf{X}_T(\tau') &= \mathbf{x}_T - \mathbf{v}_T[\tau \cosh(y-\eta) - \tau'\Delta(\tau')], \\
H(\tau') &= y - \text{arcsinh}[(\tau/\tau')\sinh(y-\eta)], \\
\Delta(\tau') &= \sqrt{1 + (\tau/\tau')^2 \sinh^2(y-\eta)},
\end{aligned}
\tag{29}
$$

and $\mathbf{x}_T = (x, y)$. Here we have only changed the variables and the physical meaning of f is the same as that in Eq. (26). The two terms of the solution are for the initially produced quarkonia and the regenerated quarkonia, respectively. Both of them suffer anomalous suppression that is represented by the exponential terms.

Taking the approximation $\eta \approx y$, the transport equation can be further simplified to be

$$\delta_\tau f + \mathbf{v}_T \cdot \nabla_T f = -\alpha_T f + \beta_T, \tag{30}$$

and the simplified solution is

$$
\begin{aligned}
f(\mathbf{p}_T, y, \mathbf{x}_T, \eta, \tau) & \\
= f(\mathbf{p}_T, y, \mathbf{X}_T(\tau_0), \eta, \tau_0) & e^{-\int_{\tau_0}^{\tau} d\tau' \alpha(\mathbf{p}_T, y, \mathbf{X}_T(\tau'), \eta, \tau')} \\
+ \int_{\tau_0}^{\tau} d\tau' \beta(\mathbf{p}_T, y, & \mathbf{X}_T(\tau'), \eta, \tau') e^{-\int_{\tau'}^{\tau} d\tau'' \alpha(\mathbf{p}_T, y, \mathbf{X}_T(\tau''), \eta, \tau'')}
\end{aligned}
\tag{31}
$$

with

$$\mathbf{X}_T(\tau') = \mathbf{x}_T - \mathbf{v}_T(\tau - \tau'). \tag{32}$$

If a local equilibrium is reached, the left-hand side of Eq. (30) vanishes, and the corresponding distribution function f_0 satisfies

$$0 = -\alpha_T f_0 + \beta_T. \tag{33}$$

Thus the transport equation (30) can be rewritten as

$$\delta_\tau f + \mathbf{v}_T \cdot \nabla_T f = -\alpha_T (f - f_0). \tag{34}$$

If all the other particles that destroy or produce quarkonia are in thermal equilibrium, then f_0 is the thermal distribution, otherwise f_0 can in general deviate from the thermal distribution. If the coordinate dependence of α_T and β_T is neglected, integrating Eq. (30) over the coordinate space, one obtains

$$\delta_\tau f_p = -\alpha_{Tp} f_p + \beta_{Tp}, \tag{35}$$

which describes the time evolution of the momentum spectrum f_p of quarkonia. If the momentum dependence of α and β is further neglected, namely consider a momentum averaged dissociation and regeneration rates Γ and G, the rate equation is generally written as

$$\frac{dN_\Psi}{d\tau} = -\Gamma N_\psi + G. \tag{36}$$

The second term on the left-hand side of Eq. (30) is kinetic, which leads to the leakage effect[102] so that high p_T quarkonia are more likely to escape the fireball and suffer less suppression. As a result, the transverse momentum increases after the transport.

The two terms on the right-hand side contains the interaction between the quarkonium and the medium. If the temperature is relatively low and quarkonia are tightly bound states, the gluon dissociation process $g + (Q\bar{Q}) \rightarrow Q + \bar{Q}$ is important. Peskin calculated the cross section for the $1S$ state by operator product expansion,[57] and it reads

$$\sigma_{1S} = A_0 \frac{(\omega/\epsilon_{1\mathrm{S}} - 1)^{3/2}}{(\omega/\epsilon_{1\mathrm{S}})^5} \tag{37}$$

with

$$A_0 = \frac{2^{11}\pi}{3N_c^2\sqrt{m_Q^3\epsilon_{1S}}}, \tag{38}$$

where ω, ϵ_{1S}, N_c and m_Q are the energy of gluon in the quarkonia rest frame, the binding energy of the quarkonia, the number of color and the mass of a heavy quark, respectively. For such a process the dissociation rate is

$$\alpha_T = \int \frac{d\mathbf{k}}{(2\pi)^3 E_T} \sigma_{1S} F_{g(Q\bar{Q})} f_g(k), \tag{39}$$

where $F_{g(Q\bar{Q})} = \sqrt{(p \cdot k)^2 - m^2_{(Q\bar{Q})} m_g^2}$ and f_g are the flux factor and the distribution function of gluons, respectively, with k being the momentum of gluons. The cross section of Υ is shown in Fig. 5 as an example, where the cross section for the excited states are from Refs. 58 and 59. It can be seen that the binding energy for the $\Upsilon(1S)$ state is large while that for the excited states are smaller. Meanwhile, the amplitude of the cross section is also much larger for the excited states. Therefore the excited states are more likely to be dissociated in a medium with the temperature to be several hundred MeV. This is qualitatively consistent with the color screening scenario.

However, the gluon dissociation process is not the same as the color screening. To see this more clearly one can consider the classical electric analog, where the

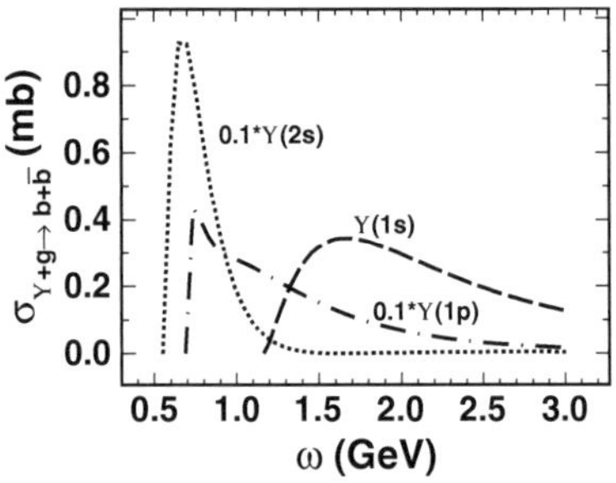

Fig. 5. The cross section for $g + \Upsilon \to b + \bar{b}$.[57–59]

number of charged particles are fixed. In the low temperature limit, the charged particle are closely around the test charge, and the screening is strong. However, in this case the photons are rare and their energies are small which can hardly break a bound state. In the high temperature limit, the screening radius becomes large due to the stronger thermal motion of the particles in the medium, while both the number of photons increases and the average energy of each photon increases, which make it easier to break a bound state. The only reason that the gluon dissociation behaves similar to that of the color screening is that the number of charged particles in heavy ion collisions increases with temperature since the temperature is comparable with the mass of the blocks of the medium. Therefore, it is necessary

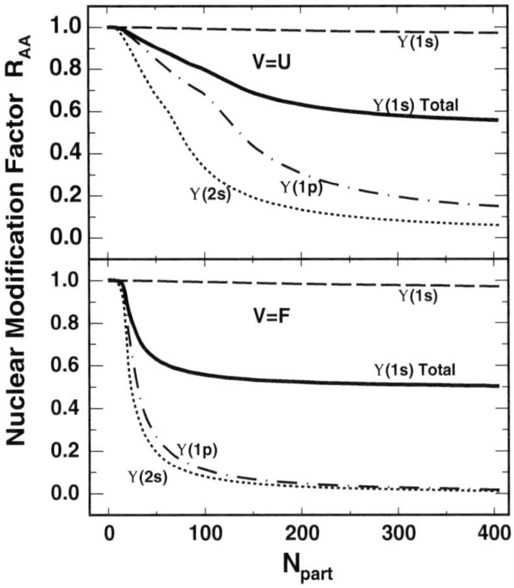

Fig. 6. The nuclear modification factor of $\Upsilon(1S)$ as a function of the number of participant $N_{\rm part}$ predicted by a transport model at RHIC energy. The upper and lower panels are from $V = U$ and $V = F$, respectively. Figure from Ref. 103.

to include the color screening effect besides the scattering process. In some models, this is done by introducing the dissociation temperature T_d of the quarkonia above which a quarkonium dissociates in the medium immediately, and the dissociation temperature can be determined from the potential model as in Subsec. 3.1. The nuclear modification factor for Υ at RHIC is shown in Fig. 6.

The regeneration rate β_T is related to the dissociation rate with the detail balance principle,[98] and thus it depends on the distribution of charm quarks. The motion of charm quarks was widely studied.[104–108] The simplest case is that the momentum distribution of the heavy quarks is thermal. However this may not be true, especially for bottom. An ansatz was suggested that the non-thermal distribution in momentum space will suppress the regeneration process that is related to the relaxation time,[56,109] which was verified by a cascade model[110] later. Recently the modification of initially produced quarkonia due to the change of the distribution function of the heavy quarks from interaction with the medium was

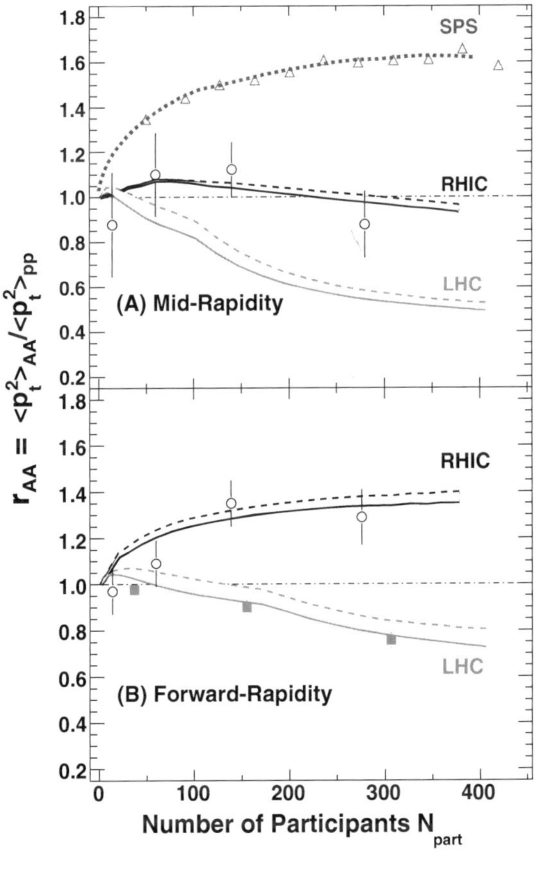

Fig. 7. (Color online) The ratio of $\langle p_T^2 \rangle$ in A–A collisions to that in pp collisions as a function of centrality at different colliders. This figure is from Ref. 116. Data from Refs. 117–120.

discussed in Ref. 111. The influence of the formation time[112] and the event-by-event fluctuation[113] were discussed by the same authors. As in the statistical model, the canonical enhancement should also be considered in the regeneration. The results from a cascade simulation of the heavy quarks[87] shows that the correlation of the heavy quarks in the phase space from the primary hard collisions can also be important, especially when the heavy quarks are rare. Thermal production of the charm quarks were also discussed[114,115] which depends on the initial temperature of the fireball much.

One advantage of the transport model is to study the momentum of quarkonia. It is found that although the nuclear modification factor R_{AA} is similar among SPS energy, RHIC energy and LHC energy at forward rapidity, the ratio

$$r_{AA} \equiv \frac{\langle p_T^2 \rangle_{AA}}{\langle p_T^2 \rangle_{pp}} \tag{40}$$

is sensitive to the origin of J/ψs and thus differs much among SPS, RHIC and LHC.[116] At SPS energy, the initially produced J/ψs are dominant, and therefore the r_{AA} increases with the number of participant $N_{\rm part}$ due to the hot nuclear matter effects and the leakage effect.[102] At RHIC energy, the regenerated J/ψs will play a role in the most central collisions with relatively small transverse momentum due to the thermalization of charm quarks.[98] As a result, the transverse momentum increases with $N_{\rm part}$ first and then decreases with $N_{\rm part}$ when $N_{\rm part}$ is large.[121] At LHC energy, the regenerated J/ψs are much more important, and the contribution is larger and larger for more central collisions, which leads to the continuous decrease of r_{AA}. For the same reason, the r_{AA} is larger at forward rapidity due to less regeneration. This effect was discussed in Ref. 116 and shown in Fig. 7.

4. Comparison with the Experimental Results

In this part we review some results of above models on heavy quarkonium production and suppression compared with the experiments in relativistic heavy ion collisions. Although the J/ψ is mostly discussed, there are a family of quarkonia that we are interested in. Both masses and quantum numbers of them are shown in Table 4. It can be seen that the mass of them are above 3 GeV for charmonia and above 9 GeV for bottomonia, which are high above the temperature of the fireball produced in heavy ion collisions. Only the quarkonia under threshold are listed and the quarkonia of η_c and η_b are not included in this table because they have large width even in vacuum. For example, the width of η_c in vacuum is about 32 MeV that is more than three hundred times larger compared with the width 93 keV of J/ψ. The small widths of J/ψ and $\Upsilon(1S)$ are guaranteed by the J^{PC}, and they can only decay through three gluons in QCD and thus suppressed by the OZI rule,[124] and the excited states can decay directly and/or indirectly to the ground state. Since the binding energy of J/ψ and $\Upsilon(1S)$ is far above the critical temperature T_c for the deconfinement phase transition, they are expected to survive the

Table 4. Mass M, binding energy ϵ, Spin J, parity P, and charge conjugate C of the quarkonium families. The data come from Particle Data Group.[122]

Constituent quarks	Quarkonia	M/GeV	ϵ/GeV	J^{PC}
$(c\bar{c})$	$J/\psi(1S)$	3.097	0.642	1^{--}
	$\chi_{c0}(1P)$	3.415	0.324	0^{++}
	$\chi_{c1}(1P)$	3.511	0.228	1^{++}
	$\chi_{c2}(1P)$	3.556	0.183	2^{++}
	$\psi(2S)$	3.686	0.053	0^{--}
$(b\bar{b})$	$\Upsilon(1S)$	9.460	1.099	1^{--}
	$\chi_{b0}(1P)$	9.859	0.700	0^{++}
	$\chi_{b1}(1P)$	9.893	0.666	1^{++}
	$\chi_{b2}(1P)$	9.912	0.647	2^{++}
	$\Upsilon(2S)$	10.023	0.536	1^{--}
	$\Upsilon(1D)$	10.164	0.395	2^{--}
	$\chi_{b0}(2P)$	10.232	0.327	0^{++}
	$\chi_{b1}(2P)$	10.255	0.304	1^{++}
	$\chi_{b2}(2P)$	10.269	0.290	2^{++}
	$\chi_{b2}(3S)$	10.355	0.204	2^{++}
	$\chi_b(3P)$	10.534	0.025	$?^{?+}$
$(c\bar{b})^\dagger$	B_c^+	6.276	0.873	1^{--}

Note: †Although B_c meson is not a quarkonium as it contains quarks of different flavors, we also list it here because it is similar to heavy quarkonia in the sense that both quarks are heavy in B_c. For discussions on B_c meson, one may refer to Refs. 69, 85 and 123.

QGP as a probe, accompanied with the suppression especially for the excited states whose binding energies are relatively small and thus they are more fragile in the hot medium.

4.1. *Observables*

A quarkonium is a neutral particle, and it can not be observed directly. The decay channel of dileptons are mostly used to measure the quarkonia. The fraction for different channels are shown in Table 5. By measuring the momenta of the lepton pairs, it is possible to reconstruct the quarkonia according to the invariant mass and measure their momenta. Meanwhile the spin of a quarkonium can also be

Table 5. Decay channels and the branch ratios of J/ψ and $\Upsilon(1S)$.

Decay channel	Branch ratio
$J/\psi \to e^+ + e^-$	5.97%
$J/\psi \to \mu^+ + \mu^-$	5.96%
$\Upsilon \to e^+ + e^-$	2.38%
$\Upsilon \to \mu^+ + \mu^-$	2.48%

measured, and this has been done in N–N collisions.[125–127] Although the spin of the quarkonium may help in heavy ion collisions,[128] there is no public data available at present yet. Therefore the current data are all based on either the total yield or the momentum distribution of quarkonia.

Following the original idea by Satz, the most measured observable is the nuclear modification factor

$$R_{AA} \equiv \frac{N_{AB}}{T_{AB}\sigma_{pp}} \tag{41}$$

as a function of centrality, which is usually characterized by the number of participant $N_{\rm part}$ or the number of binary collisions $N_{\rm coll}$. In the above, N_{AB} is the yield of heavy quarkonia in A–A collisions and T_{AB} is the overlap function of the two colliding nuclei, and σ_{pp} is the quarkonium production cross section in pp collisions. This observable is less than unit in nucleus–nucleus collisions from SPS energy to LHC energy up to now.

To look into the details, the momentum dependence is also measured. In the longitudinal direction, the momentum is usually converted to rapidity

$$y \equiv \frac{1}{2}\ln\frac{E+p_z}{E-p_z}, \tag{42}$$

since on one hand it is additive under successive boost along z direction,[a] and on the other hand it is closely related to pseudo-rapidity which is simply a function of the angle between the outgoing particles and the beam and thus related to the position of detectors. For example, the detectors at Phenix only measure the J/ψs at mid rapidity ($|y| < 0.9$), and those at forward rapidity ($1.2 < y < 2.2$).[129] In the transverse direction $R_{AA}(\mathbf{p}_T)$ can be measured, where $\mathbf{p}_T \equiv (p_x, p_y)$ is the transverse momentum of a quarkonium. At RHIC energy the transverse momentum can be measured up to about 10 GeV.[130]

The angular distribution along the transverse direction is described as the elliptic flow

$$v_2 = \left\langle \frac{p_x^2 - p_y^2}{p_x^2 + p_y^2} \right\rangle = \langle \cos 2\phi \rangle, \tag{43}$$

which is the second Fourier coefficients. For light hadrons the elliptic flow has well been measured,[131–134] and discussed a lot, which comes from the non-isotropic properties of the medium and is not expected in N–N collisions.

Besides, we are also interested in the beam energy dependence, which put quarkonia into the fireball with different initial temperatures and densities. In the following we will review some important results from three typical beam energy scale, $\sqrt{s} = 17.3$ A GeV at SPS, $\sqrt{s} = 200$ A GeV at RHIC and $\sqrt{s} = 2.76$ A GeV at LHC, respectively.

[a]In the point of view of group theory, a boost can be expressed as e^{yK}, where K is the boost generator of the Lorentz group, and the rapidity y is the boost parameter, and thus it is additive.

4.2. *SPS*

At SPS energy, J/ψ suppression was observed in experiments[22,46,47] after the theoretical prediction.[2] As shown in Fig. 1, the normal suppression can well be explained by the absorption of J/ψ by the nucleons at SPS. A similar suppression was also observed in d+A or p+A collisions at RHIC and LHC,[49,135] although the formation time of J/ψ is larger than the time for a J/ψ to pass through the primary nuclear matter. The (effective) cross section σ_{abs} is about several mb.[56,86,136,137] It is not well understand why the normal suppression for ψ' is much stronger than that of J/ψ,[49] but some model predicts that the high energy density could lead to even stronger suppression of ψ' in the most central p+Pb collisions at LHC energy due to the hot nuclear matter effects if there is any.[138]

The anomalous suppression was also observed.[3,139] A transport model calculation[109] is shown in Fig. 8 compared with the experimental data.[139] The obvious deviation from the normal suppression implies that J/ψ suffers some other suppression other than the cold nuclear matter effect. It was widely recognized as a smoking gun of the formation of QGP. Note that the anomalous suppression was observed only in central Pb+Pb collisions, but not in smaller system or the peripheral collisions. It can be estimated that the temperature of the fireball is not far above the critical temperature T_c of deconfinement phase transition. It is natural to expect stronger suppression at RHIC and LHC if the only hot nuclear matter effect is color screening. However this is not the fact as it was found later.

Accompanied with the suppression of J/ψ its transverse momentum spectrum also changes.[30] The average transverse momentum square $\langle p_T^2 \rangle$ was found to be larger in central Pb+Pb collisions.[30] Similar effect is also observed in p+A collisions and also for light hadrons at intermediate p_T, known as Cronin effect.[21] The hot nuclear matter effect will also affect the transverse momentum spectrum, since high

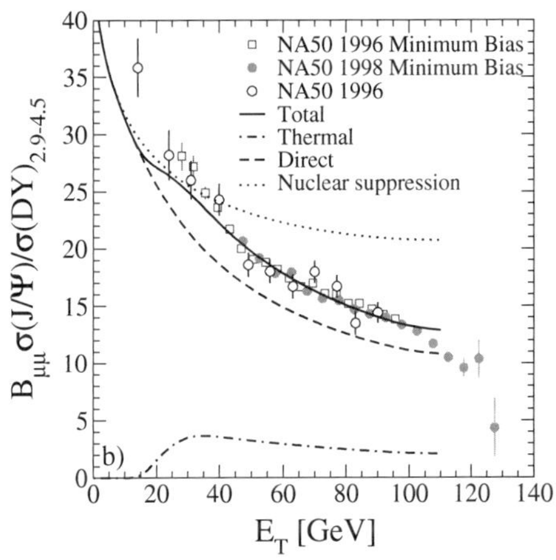

Fig. 8. (Color online) The anomalous suppression of J/ψ. The data is from NA50,[139] and both the calculation and the figure are from Ref. 109.

p_T J/ψ may left the fireball earlier and suffer less suppression, which is called the leakage effect.[102] The behavior of the transverse momentum dependence of the nuclear modification factor is consistent with the above two mechanisms that R_{AA} is larger at high p_T and smaller at low p_T.

ψ' is also measured,[30,140] which makes it possible to compare with the statistical model directly. It can be seen from Fig. 9[141] that the anomalous suppression of ψ' is stronger than that of J/ψ, and the statistical model describes well the ratio of them. It is necessary to keep in mind that excited states are more fragile than the ground state, and when we consider the suppression of inclusive J/ψs as measured in most cases, the suppression of the excited states has significant contribution besides the ground state itself. About 30% of the inclusive J/ψs are from the decay from χ_c, and 10% from ψ'[142,b] whose binding energies are smaller as shown in Table 4. The transverse momentum of ψ' is higher than that of J/ψ at top SPS energy,[30] which implies that a low p_T ψ' is more likely to be destroyed by the hot medium.

4.3. *RHIC*

The centrality dependence of the nuclear modification factor R_{AA} of J/ψ at middle rapidity is shown in Fig. 10. The high transverse momentum calculations ($p_T >$ 5 GeV) are predictions. In the full p_T calculations, the suppression of J/ψ is strong even in the peripheral collisions, and both transport models can describe the data well. For the high p_T results, both transport models predicted a smaller suppression, which also agrees with the data qualitatively. This implies that high p_T J/ψs suffers

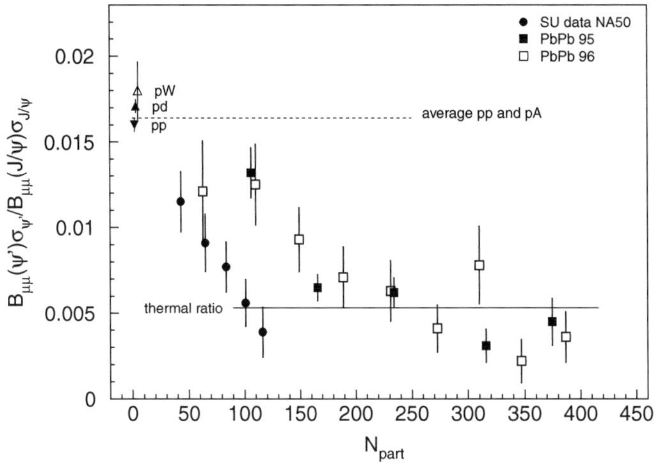

Fig. 9. The anomalous suppression of J/ψ and ψ' at SPS. The "expected" in the figure is the yield with normal suppression included. The data are from NA50,[22,136] and both the calculation and the figure are from Ref. 77.

[b]This ratio is also true for higher energies as implied by the CEM model and the experiments, see Refs. 143 and 144 for example.

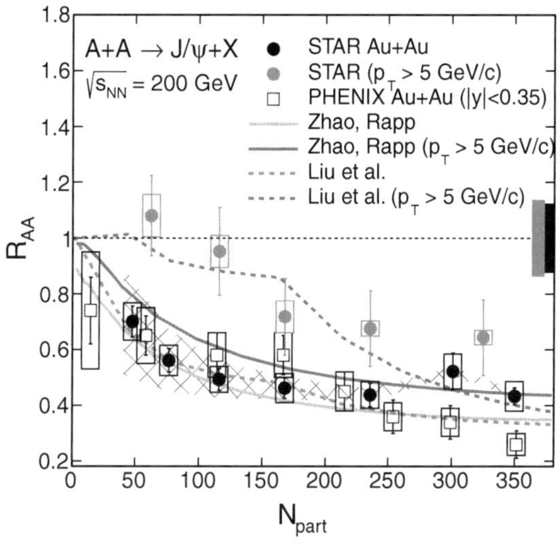

Fig. 10. (Color online) Nuclear modification factor R_{AA} of J/ψ as a function of the number of participant $N_{\rm part}$ at middle rapidity at RHIC. The calculations are from transport models.[121,145] Data from Refs. 130 and 146, and figure from Ref. 147.

less suppression. The centrality dependence of J/ψ suppression in Cu + Cu collisions was also measured, and the $N_{\rm part}$ dependence of R_{AA} is similar to that in Au + Au collisions.[148]

At RHIC energy, quarkonia are measured at both middle rapidity and forward rapidity. Fig. 11 shows the rapidity dependence of R_{AA} both at middle rapidity and at forward rapidity. Different from a naive screening scenario, the suppression at forward rapidity is stronger than that at middle rapidity in both centrality bins.

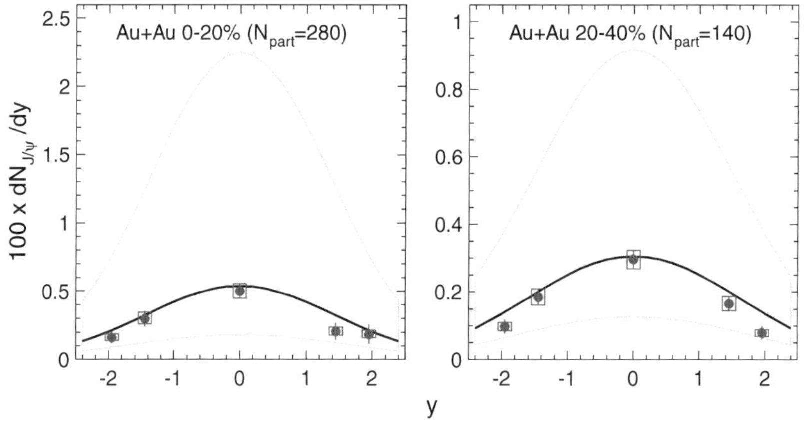

Fig. 11. (Color online) The rapidity dependence of R_{AA} at RHIC from the statistical model[79] compared with experimental data[119] at centralities 0–20% (left) and 20–40% (right). The bands show the uncertainty due to the input cross section of the model. Figure from Ref. 79.

In heavy ion collisions, more energy is deposited at middle rapidity. Therefore the temperature at forward rapidity is not expected to be higher than that at middle rapidity. The results here implies the regeneration of J/ψ as we have discussed in the previous section, since there are more charm quarks at middle rapidity, which leads to larger contribution to the thermal production. When the same idea is applied in the transport model, the rapidity dependence can also be explained qualitatively.[99]

In order to learn more information on the dynamics of J/ψ suppression and regeneration, the transverse momentum dependence of R_{AA} were calculated and measured. It is shown in Fig. 12. Compared with the experimental data are the results from two transport models.[121,145] Both models can explain the data at present. It can be seen that the suppression of J/ψ is stronger at low p_T than that at high p_T as at SPS energy. The PHENIX data also show a valley structure in most centrality bins. This non-monotonic structure can be an evidence of the regeneration,[121] but it needs to be further confirmed since the current data are still with relatively large uncertainty. At $p_T = 3\,\text{GeV}$, there is a large difference between the data from PHENIX and that from STAR in the most central and most peripheral collisions. The nuclear modification factor of J/ψs with high p_T was well measured by STAR,[130] and it confirms that a J/ψ with high p_T suffers less suppression, which is consistent of the leading order gluon dissociation cross section[57,149] and the idea of relaxation of the color screening.[71]

The elliptic flow v_2 is a measure of the collective motion. The v_2 of J/ψ was also measured recently by STAR as shown by Fig. 13. It can be seen from this

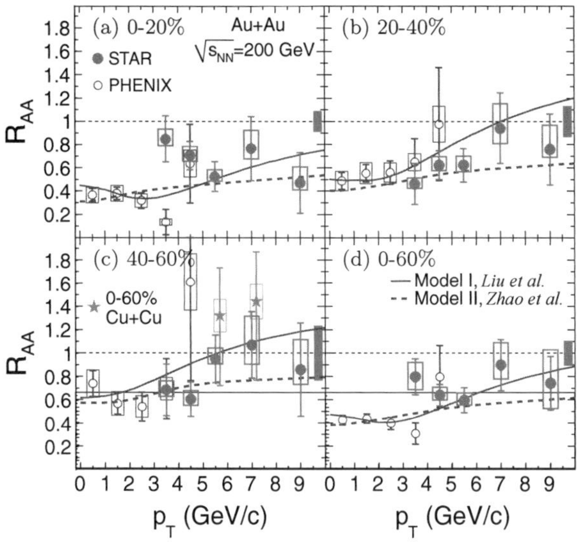

Fig. 12. (Color online) The nuclear modification factor R_{AA} of J/ψ as a function of the transverse momentum p_T at RHIC energy. The data[130] are measured by STAR and PHENIX. The two curves are due to Refs. 121 and 145. Figure from Ref. 147.

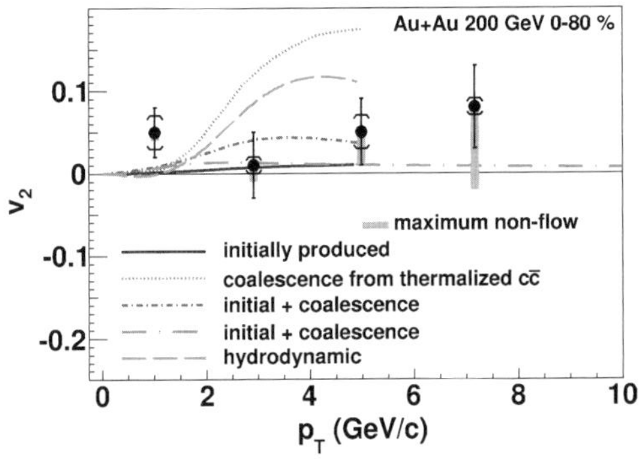

Fig. 13. (Color online) The elliptic flow v_2 of J/ψ as a function of the transverse momentum p_T at RHIC. Figure from STAR in Ref. 101.

figure that the v_2 of J/ψ is far smaller than that of light hadrons like pion, and it is also smaller than what the coalescence model or hydrodynamics predicted, which implies that the J/ψ is far from thermalization at RHIC energy, and the initially produced J/ψ plays an important role in observables. The curves are from models correspond to Refs. 98, 94, 150–152, respectively. We note here that the elliptic flow is measured at peripheral collisions. Although the regeneration is small in this centrality, it can still play an important role in central collisions at RHIC.

Υ was also measured by STAR,[74,75,153] and it is found that the suppression of Υ is consistent with the simple assumption that all the excited Υ states melts up and all the ground state survives as suggested by the sequential dissociation model.

4.4. *LHC*

In peripheral collisions the suppression at LHC is similar to that at middle rapidity at RHIC, while in the most central collisions the R_{AA} is larger than that at RHIC, which is different from the color screening scenario. Similar to RHIC energy, the R_{AA} at middle rapidity at LHC is also larger than that at forward rapidity.[154] The R_{AA} in the most central collisions is about 0.8[154] at middle rapidity, while the R_{AA} for high p_T J/ψs is around 0.2 in the most central collisions.[155] All the observations above implies the important contribution of the regeneration of J/ψ at LHC energy.

The transverse momentum dependence of J/ψ also shows that the R_{AA} is large at low p_T and relatively small at high p_T,[154] which is totally different from that at SPS and RHIC energy. The elliptic flow v_2 of J/ψ at LHC energy[156] is shown in Fig. 14. It can be seen that there is a peek around $p_T = 3\,\text{GeV}$, which was never observed at RHIC energy. Since the elliptic flow is a measure of the collective motion, such a non-zero v_2 implies either the J/ψ get thermalized in the fireball or the regenerated charmonia contribute a lot to the final detected J/ψs. The rapidity dependence of R_{AA} also shows that it decreases with rapidity even at the forward

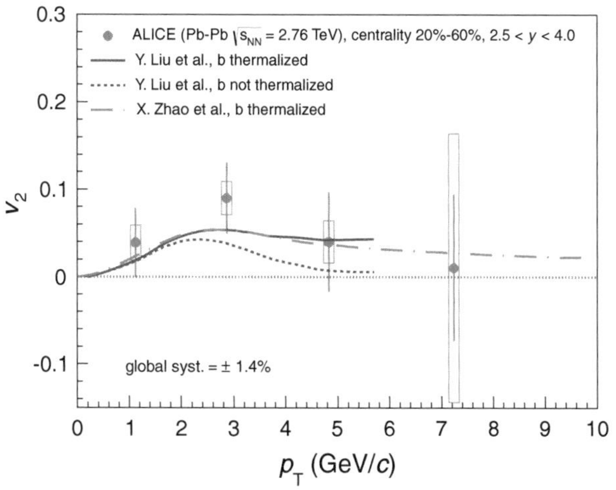

Fig. 14. (Color online) The elliptic flow v_2 of J/ψ at LHC energy as a function of the transverse momentum p_T compared with calculation from transport models.[151,160] Figure from Ref. 156.

bins.[154] All these momentum measures are consistent with the regeneration of J/ψ at LHC.

The excited states were also well measured. The double ratio of Υ shows a significant stronger suppression for excited Υs relative to $\Upsilon(1S)$,[76] which is an analog of the charmonia at SPS. The double ratio of charmonia is measured by ALICE[157] and CMS.[158] According to the CMS data, the double ratio at high p_T increases with $N_{\rm part}$ monotonically, and it is even far larger than unit for 3GeV $< p_T <$ 30GeV and $1.6 < y < 2.4$. Most of the data can be explained by the transport model with B meson decay[159] except those above unit.

5. Outlook

In this paper, we review the progress of heavy quarkonia in relativistic heavy ion collisions. The study of heavy quarkonia has made great progress during the last several decades, and the topic is becoming broader and broader. The study of J/ψ has been extended to its excited states, the Υ families and even B_c mesons. The production of heavy quarks is related to the perturbative QCD, and the formation of quarkonia is related to the hadronization. The suppression and regeneration of quarkonia are related to the interaction inside the hot quark matter, and the motion of heavy quarks and the suppression of heavy quarkonia are similar to jet quenching. The observable of quarkonia is actually dileptons and the production of regenerated quarkonia is also similar to that of the dileptons.

Many observables have been measured, such as the nuclear modification factor $R_{AA}(p_T)$, $R_{AA}(N_{\rm part})$, $R_{AA}(y)$, $R_{AA}(L)$, the double ratio, the transverse momentum square $\langle p_T^2 \rangle$, the elliptic flow $v_2(p_T)$. The beam energy ranges from 17.3 GeV at SPS to 2.76 TeV at LHC. It can be expected that the energy scan at RHIC and the new experiment of FAIR will generate more results about heavy quarkonia.

On the theoretical side, most of the data can be explained qualitatively. Besides the color screening, many new concepts are developed in the theoretical calculations, like the statistical production of quarkonia, the canonical ensemble effect, the regeneration, the velocity dependence of the dissociation temperature, and the correlation between the heavy quarks.

There are still a lot of open questions to be answered in both experiment and theory. For experiments, the direct evidence of the regeneration, that is the enhancement of heavy quarkonia in A–A collisions, may be found for B_c meson[69,123] and Ω_{ccc}[95] at LHC energy. The spin degree of freedom of quarkonia is not measured in A–A collisions, which may contain more information about quarkonium formation. The precision for open heavy flavor and high p_T quarkonia is to be improved. For theorist, the mechanism of heavy quarkonia production is still to be determined. The strong suppression of excited charmonia in p–A and high multiplicity pp collisions at LHC is still to be understood. Of course, a more important question to answer is what properties of the medium we can learn from the heavy quarkonia.

Acknowledgments

The work is supported by the NSFC and MOST grant Nos. 11335005, 2013CB922000 and 2014CB845400.

References

1. R. Hagedorn, *Nuovo Cim. Suppl.* **3** (1965) 147–186.
2. T. Matsui and H. Satz, *Phys. Lett. B* **178** (1986) 416.
3. M. Gonin *et al.*, *Nucl. Phys. A* **610** (1996) 404c–417c.
4. M. Cacciari, M. Greco and P. Nason, *JHEP* **9805** (1998) 007.
5. M. Cacciari, S. Frixione and P. Nason, *JHEP* **0103** (2001) 006.
6. T. Sjostrand, *Comput. Phys. Commun.* **82** (1994) 74–90.
7. T. Sjostrand, S. Mrenna and P. Z. Skands, *JHEP* **0605** (2006) 026.
8. T. Sjostrand, S. Mrenna and P. Z. Skands, *Comput. Phys. Commun.* **178** (2008) 852–867.
9. X.-N. Wang and M. Gyulassy, *Phys. Rev. D* **44** (1991) 3501–3516.
10. X.-N. Wang and M. Gyulassy, *Phys. Rev. D* **45** (1992) 844–856.
11. C.-H. Chang, *Nucl. Phys. B* **172** (1980) 425–434.
12. H. Fritzsch, *Phys. Lett. B* **67** (1977) 217.
13. F. Halzen, *Phys. Lett. B* **69** (1977) 105.
14. J. Amundson, O. J. Eboli, E. Gregores and F. Halzen, *Phys. Lett. B* **372** (1996) 127–132.
15. J. Amundson, O. J. Eboli, E. Gregores and F. Halzen, *Phys. Lett. B* **390** (1997) 323–328.
16. W. Caswell and G. Lepage, *Phys. Lett. B* **167** (1986) 437.
17. G. T. Bodwin, E. Braaten and G. P. Lepage, *Phys. Rev. D* **51** (1995) 1125–1171.
18. J. C. Collins, D. E. Soper and G. F. Sterman, *Adv. Ser. Direct. High Energy Phys.* **5** (1988) 1–91.

19. J. Lansberg, *Int. J. Mod. Phys. A* **21** (2006) 3857–3916.
20. A. H. Mueller and J.-W. Qiu, *Nucl. Phys. B* **268** (1986) 427.
21. J. W. Cronin *et al.*, *Phys. Rev. D* **11** (1975) 3105.
22. M. C. Abreu *et al.*, *Phys. Lett. B* **466** (1999) 408–414.
23. C. Gerschel and J. Hufner, *Phys. Lett. B* **207** (1988) 253–256.
24. C. Lourenco, R. Vogt and H. K. Woehri, *JHEP* **0902** (2009) 014.
25. R. Vogt, *Phys. Rev. C* **81** (2010) 044903.
26. R. Vogt, S. J. Brodsky and P. Hoyer, *Nucl. Phys. B* **360** (1991) 67–96.
27. R. Vogt, *Phys. Rept.* **310** (1999) 197–260.
28. R. Vogt, *Phys. Rev. C* **71** (2005) 054902.
29. M. Leitch *et al.*, *Phys. Rev. Lett.* **84** (2000) 3256–3260.
30. M. Abreu *et al.*, *Phys. Lett. B* **499** (2001) 85–96.
31. A. Adare *et al.*, *Phys. Rev. C* **91** (2015) 014907.
32. R. Aaij *et al.*, *JHEP* **1402** (2014) 072.
33. M. Asakawa and T. Hatsuda, *Phys. Rev. Lett.* **92** (2004) 012001.
34. M. Asakawa, T. Hatsuda and Y. Nakahara, *Prog. Part. Nucl. Phys.* **46** (2001) 459–508.
35. T. Umeda, R. Katayama, O. Miyamura and H. Matsufuru, *Int. J. Mod. Phys. A* **16** (2001) 2215.
36. O. Kaczmarek, F. Karsch, P. Petreczky and F. Zantow, *Phys. Lett. B* **543** (2002) 41–47.
37. R. Morrin, A. O Cais, M. Oktay, M. Peardon, J. Skullerud *et al.*, *PoS* **LAT2005** (2006) 176.
38. O. Kaczmarek and F. Zantow, *Phys. Rev. D* **71** (2005) 114510.
39. H. Ding *et al.*, *Phys. Rev. D* **86** (2012) 014509.
40. A. Bazavov, P. Petreczky and A. Velytsky, Quarkonium at finite temperature, in *Quark–Gluon Plasma 4*, eds. R. C. Hwa and X.-N. Wang (World Scientific, 2010), pp. 61–110.
41. N. Brambilla, A. Pineda, J. Soto and A. Vairo, *Rev. Mod. Phys.* **77** (2005) 1423.
42. N. Brambilla, A. Pineda, J. Soto and A. Vairo, *Nucl. Phys. B* **566** (2000) 275.
43. C. Gerschel and J. Hufner, *Z. Phys. C* **56** (1992) 171–174.
44. C. Baglin *et al.*, *Phys. Lett. B* **220** (1989) 471–478.
45. C. Baglin *et al.*, *Phys. Lett. B* **270** (1991) 105–110.
46. M. Abreu *et al.*, *Z. Phys. C* **38** (1988) 117.
47. M. Abreu *et al.*, *Phys. Lett. B* **449** (1999) 128–136.
48. D. Kharzeev and H. Satz, *Phys. Lett. B* **356** (1995) 365–372.
49. A. Adare *et al.*, *Phys. Rev. Lett.* **111**(20) (2013) 202301.
50. E. Ferreiro, F. Fleuret, J. Lansberg and A. Rakotozafindrabe, *Phys. Rev. C* **81** (2010) 064911.
51. K. Eskola, V. Kolhinen and C. Salgado, *Eur. Phys. J. C* **9** (1999) 61–68.
52. K. J. Eskola, H. Paukkunen and C. A. Salgado, *JHEP* **0807** (2008) 102.
53. K. Eskola, H. Paukkunen and C. Salgado, *JHEP* **0904** (2009) 065.
54. D. de Florian and R. Sassot, *Phys. Rev. D* **69** (2004) 074028.
55. E. Ferreiro, F. Fleuret, J. Lansberg and A. Rakotozafindrabe, *Phys. Lett. B* **680** (2009) 50–55.
56. X. Zhao and R. Rapp, *Phys. Lett. B* **664** (2008) 253–257.
57. M. E. Peskin, *Nucl. Phys. B* **156** (1979) 365.
58. X.-N. Wang and F. Yuan, *Phys. Lett. B* **540** (2002) 62–67.
59. A. Polleri, T. Renk, R. Schneider and W. Weise, *Phys. Rev. C* **70** (2004) 044906.
60. Y.-S. Oh, S. Kim and S. H. Lee, *Phys. Rev. C* **65** (2002) 067901.

61. Y. Park, K.-I. Kim, T. Song, S. H. Lee and C.-Y. Wong, *Phys. Rev. C* **76** (2007) 044907.
62. F. Brezinski and G. Wolschin, *Phys. Lett. B* **707** (2012) 534–538.
63. Y. Liu, C. M. Ko and T. Song, *Phys. Rev. C* **88**(6) (2013) 064902.
64. L. Grandchamp and R. Rapp, *Phys. Lett. B* **523** (2001) 60–66.
65. M. Mannarelli and R. Rapp, *Phys. Rev. C* **72** (2005) 064905.
66. F. Karsch, M. Mehr and H. Satz, *Z. Phys. C* **37** (1988) 617.
67. H. Satz, *J. Phys. G* **32** (2006) R25.
68. V. V. Dixit, *Mod. Phys. Lett. A* **5** (1990) 227.
69. Y. Liu, C. Greiner and A. Kostyuk, *Phys. Rev. C* **87** (2013) 014910.
70. H.-T. Ding, *Nucl. Phys. A* **904–905** (2013) 619c–622c.
71. Y. Liu, N. Xu and P. Zhuang, *Phys. Lett. B* **724** (2013) 73–76.
72. A. Dumitru, Y. Guo, A. Mocsy and M. Strickland, *Phys. Rev. D* **79** (2009) 054019.
73. X. Guo, S. Shi and P. Zhuang, *Phys. Lett. B* **718** (2012) 143–146.
74. R. Reed, *J. Phys. Conf. Ser.* **270** (2011) 012026.
75. L. Adamczyk *et al.*, *Phys. Lett. B* **735** (2014) 127.
76. S. Chatrchyan *et al.*, *Phys. Rev. Lett.* **107** (2011) 052302.
77. P. Braun-Munzinger and J. Stachel, *Phys. Lett. B* **490** (2000) 196–202.
78. A. Andronic, P. Braun-Munzinger and J. Stachel, *Phys. Lett. B* **673** (2009) 142–145.
79. A. Andronic, P. Braun-Munzinger, K. Redlich and J. Stachel, *Nucl. Phys. A* **789** (2007) 334–356.
80. A. Andronic, P. Braun-Munzinger, K. Redlich and J. Stachel, *Phys. Lett. B* **571** (2003) 36–44.
81. M. I. Gorenstein, A. Kostyuk, H. Stoecker and W. Greiner, *Phys. Lett. B* **509** (2001) 277–282.
82. A. Andronic, P. Braun-Munzinger and J. Stachel, *Acta Phys. Polon. B* **40** (2009) 1005–1012.
83. S. Hamieh, K. Redlich and A. Tounsi, *Phys. Lett. B* **486** (2000) 61–66.
84. C. Ko *et al.*, *Phys. Rev. Lett.* **86** (2001) 5438–5441.
85. A. Kostyuk, Double, triple and hidden charm production in the statistical coalescence model, arXiv:nucl-th/0502005.
86. L. Grandchamp, R. Rapp and G. E. Brown, *Phys. Rev. Lett.* **92** (2004) 212301.
87. Y. Liu, C. M. Ko and F. Li, Heavy quark correlations and the effective volume for quarkonia production, arXiv:1406.6648.
88. V. Greco, C. Ko and P. Levai, *Phys. Rev. Lett.* **90** (2003) 202302.
89. V. Greco, C. M. Ko and P. Levai, *Phys. Rev. C* **68** (2003) 034904.
90. R. C. Hwa and C. Yang, *Phys. Rev. C* **67** (2003) 034902.
91. R. Fries, B. Muller, C. Nonaka and S. Bass, *Phys. Rev. Lett.* **90** (2003) 202303.
92. D. Molnar and S. A. Voloshin, *Phys. Rev. Lett.* **91** (2003) 092301.
93. B. Zhang, *Phys. Lett. B* **647** (2007) 249–252.
94. V. Greco, C. M. Ko and R. Rapp, *Phys. Lett. B* **595** (2004) 202–208.
95. H. He, Y. Liu and P. Zhuang, *Phys. Lett. B* **746** (2015) 59–63.
96. R. L. Thews, M. Schroedter and J. Rafelski, *Phys. Rev. C* **63** (2001) 054905.
97. L. Grandchamp, S. Lumpkins, D. Sun, H. van Hees and R. Rapp, *Phys. Rev. C* **73** (2006) 064906.
98. L. Yan, P. Zhuang and N. Xu, *Phys. Rev. Lett.* **97** (2006) 232301.
99. Y. Liu, Z. Qu, N. Xu and P. Zhuang, *J. Phys. G* **37** (2010) 075110.
100. B. Chen, K. Zhou and P. Zhuang, *Phys. Rev. C* **86** (2012) 034906.
101. L. Adamczyk *et al.*, *Phys. Rev. Lett.* **111**(5) (2013) 052301.
102. P.-F. Zhuang and X. L. Zhu, *Phys. Rev. C* **67** (2003) 067901.

103. Y. Liu, B. Chen, N. Xu and P. Zhuang, *Phys. Lett. B* **697** (2011) 32–36.
104. K. Huggins and R. Rapp, *Nucl. Phys. A* **896** (2012) 24–45.
105. J. Uphoff, O. Fochler, Z. Xu and C. Greiner, *Phys. Rev. C* **84** (2011) 024908.
106. S. K. Das, F. Scardina, S. Plumari and V. Greco, *Phys. Rev. C* **90**(4) (2014) 044901.
107. S. Cao, G.-Y. Qin and S. A. Bass, *Phys. Rev. C* **88**(4) (2013) 044907.
108. X. Zhu, N. Xu and P. Zhuang, *Phys. Rev. Lett.* **100** (2008) 152301.
109. L. Grandchamp and R. Rapp, *Nucl. Phys. A* **709** (2002) 415–439.
110. T. Song, K. C. Han and C. M. Ko, *Phys. Rev. C* **85** (2012) 054905.
111. T. Song, *Phys. Rev. C* **89** (2014) 044903.
112. T. Song, C. M. Ko and S. H. Lee, *Phys. Rev. C* **87** (2013) 034910.
113. T. Song, K. C. Han and C. M. Ko, The effect of initial fluctuations on bottomonia suppression in relativistic heavy-ion collisions, arXiv:1112.0613.
114. B.-W. Zhang, C.-M. Ko and W. Liu, *Phys. Rev. C* **77** (2008) 024901.
115. J. Uphoff, K. Zhou, O. Fochler, Z. Xu and C. Greiner, *PoS* **BORMIO2011** (2011) 032.
116. K. Zhou, N. Xu, Z. Xu and P. Zhuang, *Phys. Rev. C* **89**(5) (2014) 054911.
117. N. Topilskaya *et al.*, *Nucl. Phys. A* **715** (2003) 675–678.
118. B. Alessandro *et al.*, *Eur. Phys. J. C* **39** (2005) 335–345.
119. A. Adare *et al.*, *Phys. Rev. Lett.* **98** (2007) 232301.
120. E. Scomparin, Presentation at QM2012 (2012).
121. Y.-P. Liu, Z. Qu, N. Xu, and P.-F. Zhuang, *Phys. Lett. B* **678** (2009) 72–76.
122. K. Olive *et al.*, *Chin. Phys. C* **38** (2014) 090001.
123. M. Schroedter, R. L. Thews and J. Rafelski, *Phys. Rev. C* **62** (2000) 024905.
124. D. Griffiths, *Introduction to Elementary Particles* (Wiley-VCH, Weinheim Germany, 2008).
125. A. Adare *et al.*, *Phys. Rev. D* **82** (2010) 012001.
126. E. T. Atomssa, *Eur. Phys. J. C* **61** (2009) 683–686.
127. J. Lansberg, *Phys. Lett. B* **695** (2011) 149–156.
128. Y. Liu, C. Greiner and C. M. Ko, Spin asymmetry of J/psi in peripheral Pb + Pb collisions at LHC, arXiv:1403.4317.
129. K. Adcox *et al.*, *Nucl. Phys. A* **757** (2005) 184–283.
130. L. Adamczyk *et al.*, *Phys. Lett. B* **722** (2013) 55–62.
131. J.-Y. Ollitrault, *Phys. Rev. D* **46** (1992) 229–245.
132. A. Adare *et al.*, *Phys. Rev. Lett.* **98** (2007) 162301.
133. S. S. Adler *et al.*, *Phys. Rev. Lett.* **91** (2003) 182301.
134. J. Adams *et al.*, *Phys. Rev. Lett.* **95** (2005) 122301.
135. M. Marchisone, *PoS* **EPS-HEP2013** (2013) 185.
136. B. Alessandro *et al.*, *Eur. Phys. J. C* **48** (2006) 329.
137. R. Arnaldi *et al.*, *Phys. Lett. B* **706** (2012) 263–267.
138. Y. Liu, C. M. Ko and T. Song, *Phys. Lett. B* **728** (2014) 437–442.
139. M. Abreu *et al.*, *Nucl. Phys. A* **698** (2002) 127–134.
140. M. Abreu *et al.*, *Phys. Lett. B* **423** (1998) 207–212.
141. B. Alessandro *et al.*, *Eur. Phys. J. C* **49** (2007) 559–567.
142. M. Abreu *et al.*, *Phys. Lett. B* **444** (1998) 516–522.
143. A. Adare *et al.*, *Phys. Rev. Lett.* **98** (2007) 232002.
144. I. Abt *et al.*, *Phys. Lett. B* **561** (2003) 61–72.
145. X. Zhao and R. Rapp, *Phys. Rev. C* **82** (2010) 064905.
146. L. Adamczyk *et al.*, *Phys. Rev. C* **90**(2) (2014) 024906.
147. B. Trzeciak, J/ψ measurements in the STAR experiment, arXiv:1412.7345.
148. A. Adare *et al.* *Phys. Rev. Lett.* **101** (2008) 122301.

149. T. Song, Y. Park, S. H. Lee and C.-Y. Wong, *Phys. Lett. B* **659** (2008) 621–627.
150. X. Zhao and R. Rapp, Charmonium production at high $p(t)$ at RHIC, arXiv:0806.1239.
151. Y. Liu, N. Xu and P. Zhuang, *Nucl. Phys. A* **834** (2010) 317C–319C.
152. U. W. Heinz and C. Shen, Private communication (2011).
153. H. Liu, *Nucl. Phys. A* **830** (2009) 235c–238c.
154. B. B. Abelev *et al.*, *Phys. Lett. B* **734** (2014) 314–327.
155. S. Chatrchyan *et al.*, *JHEP* **1205** (2012) 063.
156. E. Abbas *et al.*, *Phys. Rev. Lett.* **111** (2013) 162301.
157. E. Scomparin, *Nucl. Phys. A* **904–905** (2013) 202c–209c.
158. D. H. Moon, *Nucl. Phys. A* **910–911** (2013) 215–218.
159. B. Chen, Y. Liu, K. Zhou and P. Zhuang, *Phys. Lett. B* **726** (2013) 725–728.
160. X. Zhao, A. Emerick and R. Rapp, *Nucl. Phys. A* **904–905** (2013) 611c–614c.